Nanomaterials for Air and Water Purification

Nanomaterials for Air and Water Purification

Edited by Deepu A. Gopakumar,
Jyotishkumar Parameswaranpillai,
Jinu Jacob George, and Midhun Dominic C.D.

WILEY-VCH

The Editors

Dr. Deepu A. Gopakumar
Cochin University of Science and Technology
Department of Polymer Science and Rubber Technology
Kochi 682022, Kerala
India

Dr. Jyotishkumar Parameswaranpillai
Alliance University
Bengaluru 562106, Karnataka
India

Dr. Jinu Jacob George
Cochin University of Science and Technology
Department of Polymer Science and Rubber Technology
Kochi 682022, Kerala
India

Dr. Midhun Dominic C.D.
Sacred Heart College
Department of Applied Chemistry
Thevara 682013, Kochi, Kerala
India

Cover Image: © SkillUp/Shutterstock

Library of Congress Card No.: applied for

British Library Cataloguing-in-Publication Data
A catalogue record for this book is available from the British Library.

Bibliographic information published by the Deutsche Nationalbibliothek
The Deutsche Nationalbibliothek lists this publication in the Deutsche Nationalbibliografie; detailed bibliographic data are available on the Internet at <http://dnb.d-nb.de>.

Print ISBN: 978-3-527-35052-0
ePDF ISBN: 978-3-527-83803-5
ePub ISBN: 978-3-527-83804-2
oBook ISBN: 978-3-527-83805-9

Typesetting Straive, Chennai, India
Printing and Binding CPI Group (UK) Ltd, Croydon, CR0 4YY

C9783527350520_010224

Contents

Preface

Air and water are essential elements for the occurrence and continuity of life, including humans, plants, and animals. With the increasing needs of modern society and the advancement of technical knowledge from the beginnings to the present day, air and water are increasingly demanded. Water is one of the planet's most important natural resources and is required for all life forms to live. Drinking, farming, and other industrial processes make the most use of water. Currently, industrial sectors use a significant amount of water for their activities and discharge that water as wastewater into the ground or waterways. This leads to a shortage of water and environmental contamination. Wastewater contains many contaminants that are harmful to living things. The contaminated water must therefore be treated before it may be used for other uses. With the growing demands of modern society, water and air are compromised, as the concentration of contaminants increases. Due to their high surface area and chemical reactivity, nanomaterials have gained considerable attention in the fabrication of various functional constructs for air and water purification. Recent breakthroughs in nanotechnology and nanomaterials have opened up new possibilities for the development of processes and devices for air and water purification and lowering associated health risks. Several techniques are now being used to cleanse contaminated waters. Due to their numerous benefits, including their straightforward synthesis, unique chemical and physical properties, distinctive photocatalytic activity, advantageous antimicrobial properties, and high efficiency, use of nanoparticles/composites has attracted lot of attention as an effective purification technology among other purification methods. All nanoparticles, composites, and cutting-edge methods used in water and air purification are thoroughly covered in the book *Nanomaterials for Air and Water purification*, and new insights are provided. Academicians and researchers from prestigious institutions and universities in India and overseas have each contributed a chapter to this textbook with their research or review work.

The sections of this book focus on various aspects dealing with nanomaterials for air and water purification. This book comprises of nanocomposite for effective air and water filtration systems, water and air filter membranes, photocatalytic materials and technologies for air and water purification, carbon-based nanomaterials for air purification, etc. This book is mainly focused on the nanotechnology approach, green chemistry, and sustainable approaches of advanced nanomaterials in the area of air and water purification. This book also focuses comprehensively on nanomaterials and the methodologies already developed for air and water purification. Academicians, researchers, textile and environmental engineers, the pollution control board, policymakers, stakeholders, and governmental and non-profit groups will benefit from this book. This will be an invaluable resource for environmental engineers and materials scientists who wish to study how nanotechnology is used to enhance air quality. This book critically examines and accesses the recent progress in the development of nanomaterials for air and water purification and the present techniques of nanomaterials for air and water purification, together with their negative impacts. This book will be a valuable reference source for university professors, researchers, undergraduate and postgraduate students, and scientists primarily working in the field of materials science, applied chemistry, applied physics, and nanotechnology.

04-12-2023
Kochi, India

Dr. Deepu A. Gopakumar
Cochin University of Science and Technology
India

Dr. Jyotishkumar Parameswaranpillai
Alliance University
India

Dr. Jinu Jacob George
Cochin University of Science and Technology
India

Dr. Midhun Dominic C.D.
Sacred Heart College
India

1

New-Age Nano Adsorbents for Water Purification

Subhadeep Biswas, Abhishek Johri, and Anjali Pal

Civil Engineering Department, Indian Institute of Technology Kharagpur, Kharagpur 721302, West Bengal, India

1.1 Introduction

Rapid urbanization and huge industrialization often lead to the contamination of different water bodies. Irresponsible discharge of dyes, heavy metals, and emerging contaminants into the water often causes enormous pollution. A huge and uncontrolled release of unwanted materials into the water bodies deteriorates the water quality and poses a potential threat to aquatic life. The presence of colored substances, even in a trace quantity, in water is objectionable from an aesthetic point of view. Apart from that, many dye molecules show carcinogenicity and adverse toxic effects. In addition to this, illegal disposal of heavy metal ions present in industrial effluents has worsened the situation. In recent times, many harmful compounds, such as surfactants, phenol derivatives, and pharmaceutical waste, have also been detected in surface and subsurface water. These compounds are termed as emerging pollutants.

Considering all these facts, it is obvious that water purification is one of the most urgent needs to maintain the overall ecological balance. Innovative research activities are continuously carried out all over the globe on various wastewater treatment processes like adsorption, advanced oxidation process, electrocoagulation, electrooxidation, and membrane separation. Adsorption is one of the simplest and most efficient water and wastewater treatment techniques. However, the development of environmentally congenial adsorbents with excellent removal capacity and reusability is still a challenge. For a long time, activated carbon materials have been widely used for adsorption purposes. However, regeneration incapability and high cost associated with activated carbon materials are some of the major drawbacks in their use. Pristine materials like alumina, silica, and bentonite have also been explored in the recent past for adsorption purpose. Ma et al. [1] utilized bentonite-based adsorbents for removal of anionic pollutants from water bodies. Zhu et al. [2] in their recent review article highlighted the superiority of montmorillonite-based adsorbents for pollutant removal purposes. In this regard,

Nanomaterials for Air and Water Purification, First Edition.
Edited by Deepu A. Gopakumar, Jyotishkumar Parameswaranpillai, Jinu Jacob George, and Midhun Dominic C.D.

surface modifications, like acid modification, alkali modification, and surfactant modification, were also tried out for increasing adsorption capacity of different materials. Surfactant-modified alumina and silica have been found to be successful in removal of phenol [3], dyes [4, 5], and heavy metals [6].

In this field, nanomaterials have opened a new dimension for water treatment purpose. Nanoparticles are materials whose dimensions are less than 100 nm. In recent times, there has been a huge interest in the application of nanomaterials as adsorbents due to their excellent surface properties. Recently discovered materials like graphene and MXene have enhanced the scope of nanomaterials by forming compatible composites with them. Neha et al. [7] in their recent review article described the potential of different nanoadsorbents for the eradication of several toxic pharmaceutical compounds from water bodies. Kyzas and Matis documented in their review article the potential of various types of nanoadsorbents for wastewater treatment [8]. They highlighted the fact that the nanoadsorbents show unique physical and chemical characteristics, and the atoms that have high adsorption capacity lie on the surface of the nanomaterials. The current chapter describes the potential of different categories of nanoadsorbents such as metal–organic framework (MOF)-based nanoadsorbents, MXene-based nanoadsorbent, ZVI-based nanoadsorbent, and biochar-based nanoadsorbents for water purification purposes. Different mechanisms for pollutant removal by these nanoadsorbents have been elaborated. The next section deals with the characterization techniques commonly practiced in order to get insight into the adsorption process. Lastly, current challenges are often faced in this domain and the scope for further research has been presented before conclusion.

1.2 Different New-Age Nanoadsorbents

1.2.1 Metal–Organic Framework (MOF)-Based Nanoadsorbents

MOF-based compounds can be described as three-dimensional organic–inorganic complexes having porous structures, high specific surface area, and abundant sites exposed for adsorption [9]. In recent years, MOF-based nanocomposites have been explored by engineers and scientists worldwide for adsorption of different types of pollutants. A novel surfactant-functionalized MOF@MOF nanoadsorbent was reported by Li et al. [10] for Cr(VI) removal purposes. Thus, synthesized nanocomposite showed promising adsorption capacity and also reusability due to the presence of tunable pores along with abundant active sites on the surface. It showed good performance throughout a large pH range of 1–11, and the highest removal occurred at pH 2, with the maximum adsorption capacity being 932.1 $mg\,g^{-1}$. He et al. [11] carried out thermal decomposition of Ce-MOF micro and nanocomposites under N_2 atmosphere at a relatively lower temperature (400–500 °C). The composites thus produced after thermal treatment showed high phosphate removal efficiency. In comparison to ceria, the removal capacity was 2–4 times higher, and the maximum uptake was obtained at 189.4 $mg\,g^{-1}$. Yan et al. [12] reported the preparation and application of various polyoxometalates-MOF nanocomposite for the removal of cationic dye methylene blue (MB). The removal occurred at 98% within five minutes for an initial MB concentration of $100\,mg\,l^{-1}$, and the maximum adsorption capacity obtained was $371\,mg\,g^{-1}$.

Soltani et al. [13] synthesized layered double hydroxide (LDH)-MOF nanocomposite without involvement of any toxic solvent and utilized the same for the remediation of heavy metals such as Cd(II) and Pb(II) from water. The LDH/MOF nanocomposite thus prepared showed promising performance toward heavy metal remediation, and the experimental data fitted well with the Langmuir isotherm model and pseudo-first-order kinetic model. The calculated maximum adsorption capacity of the nanocomposite was found to be 415.3 and $301.4\,mg\,g^{-1}$ toward the removal of Cd(II) and Pb(II), respectively. Yuan et al. [14] in their recent work designed a UiO-66-F-based nanocomposite for trapping carbamazepine from aqueous media. Experimental data fit well with the Langmuir isotherm model and pseudo-second-order kinetic model.

Nguyen et al. [15] reported the feasibility of application of zirconium-organic framework-based nanocomposites for the efficient adsorption of model cationic and anionic dyes MB and methyl orange (MO). The MOF-based nanoadsorbent showed excellent reusability for up to five cycles without any significant loss in uptake capacity.

Zr-based MOF nanocomposites have been explored by different research groups for wastewater treatment purpose. Huang et al. [16] used magnetic Zr-based nanocomposite $Fe_3O_4@SiO_2@UiO\text{-}66$ and its amino derivatives for the abatement of heavy metal ions and dyes from water bodies. It showed a high removal capacity for Pb^{2+} ($102\,mg\,g^{-1}$), MB ($128\,mg\,g^{-1}$), and MO ($219\,mg\,g^{-1}$). The nanoadsorbent has been reported to possess excellent selectivity toward anionic and cationic dyes by adjusting the solution pH. Moreover, it can completely remove both MO and MB at a neutral pH (~7). With respect to regeneration, it can be reused six times without losing efficiency. Some other MOF-based nanoadsorbents include Zr-cluster-based MOF for tetracycline removal [17], cobalt-MOF with morpholine for dye removal [18], and LDH/MOF nanocomposite for Orange II and Cr(VI) removal [19].

1.2.2 Graphene, Graphene Oxide (GO), and Reduced Graphene Oxide (rGO)-Based Nanoadsorbents

Carbon-based materials are often explored for adsorption purpose. In this regard, graphene is the latest member of the family of carbon allotropes. It is nearly one atom thick, has an excessively large surface-to- volume ratio, and is considered to be one of the most attractive new-age adsorbents. Besides graphene, graphene oxide (GO), reduced graphene oxide (rGO), and their composites also emerged as viable options of adsorbents. In one of the recent review articles, Kim et al. [20] investigated the potential of graphene-based nanocomposites for wastewater remediation. Possessing variable structure, chemical strength, and being light in density, the graphene family offers a numerous variety of nanocomposite materials for adsorption purpose. Shoushtarian et al. [21] explored the GO-based nanoadsorbent for efficient removal of Basic Red 46, a cationic dye. Under the optimized condition, the maximum removal capacity reached up to 360 $mg\,g^{-1}$. The prepared GO was reused for four more cycles. However, the adsorption efficiency dropped with consecutive cycles. During four cycles of regeneration, the adsorption capacity dropped from 360 to 209 $mg\,g^{-1}$. Calimli et al. [22] reported the application of reduced GO-supported nano-Ni composite for the removal of MB. The maximum removal capacity was obtained at 946.12 $mg\,g^{-1}$.

ZnO nanoparticles coated with rGO were synthesized by Wang et al. [23] by means of self-assembly and in situ photoreduction technique. Thus synthesized rGO-based nanocomposite was utilized for the removal of rhodamine B (RhB) dye. rGO-supported ZnO nanocomposite showed excellent adsorption performance toward RhB removal. After adsorption, it was further subjected to dye photodegradation. In comparison to bare ZnO nanoparticles, rGO-supported nanoparticles showed much higher efficiency. Lujaniene et al. [24] explored the Prussian blue GO nanocomposite for the remediation of heavy metals and radionuclides from wastewater.

Verma and Dutta [25] synthesized nitrogen-enriched rGO-based nanocomposite for the removal of uranyl ions from water. A detailed experimental investigation showed that the adsorption process was kinetically and thermodynamically feasible. Removal occurred by chemisorption phenomenon, and the maximum uptake capacity reached up to 337.93 $mg\,g^{-1}$. High removal has been attributed to the large specific surface area of the nanocomposite (57.086 $m^2\,g^{-1}$). El-Maghrabi et al. [26] reported the preparation and application of magnetic graphene-based nanocomposite for uranium removal. Excellent removal uptake of 455 $mg\,g^{-1}$ was exhibited by this nanocomposite within 60-minute contact time. Experimental data showed good correlation with pseudo-second-order kinetic model. Moreover, a high degree of sustainability has been shown by the nanocomposites (>5 times profiteering).

1.2.3 ZVI-Based Nanoadsorbents

nZVI is one of the most effective new-age adsorbents possessing several distinguished properties that immensely help in wastewater treatment. However, despite possessing several advantageous characteristics, it often gets agglomerated. Hence, various modifications have been reported in the literature to overcome this problem. Eljamal et al. [27] used four polymers, namely polyacrylamide, carboxymethyl cellulose, polyethylene sorbitan monolaurate, and polyvinylpyrrolidone, to increase the reactivity of nZVI and also to prevent their aggregation. Thus, different nanocomposites were finally deployed for nitrate and phosphorus removal from water. It was found that the combination of polyacrylamide and nZVI was the best for the adsorption of phosphorus.

Zhu et al. [28] applied activated carbon-loaded nZVI for the adsorptive removal of As(III) and As(V) from water. Activated carbon was loaded with ferrous sulfate, followed by reduction with $NaBH_4$. Arsenic removal was fast in the first 12 hours. However, equilibrium was reached at nearly 72 hours. It was seen during the interference study that the presence of different cations enhanced the removal of arsenate, but the presence of Fe^{2+} diminished the uptake of arsenite. In order to prevent the loss of reactivity of nZVI, Bui et al. [29] synthesized core/shell type structure of nZVI/manganese oxide nanocomposite and applied the same for arsenic remediation purpose. It was quite fascinating to observe that the nanocomposite thus produced was superior in adsorption capacity in comparison to the bare nZVI by two to three times. Jiang et al. [30] deployed biochar-supported nZVI for the removal of glyphosphate from water medium. Langmuir isotherm model fitted well with the experimental values, and the kinetic data fitted with the pseudo-second-order kinetic model. Hence, it was concluded that the adsorptive removal process involved both chemical and surface phenomena. Maximum adsorptive uptake capacity was noticed to be 80 $mg\,g^{-1}$, which is nearly twice that of biochar only. Thus loaded, nZVI improved adsorption capacity significantly.

For As(V) removal from wastewater in single-component system as well as in presence of Se(VI), Suazo-Hernandez et al. [31] deployed nZVI and nZVI- functionalized zeolite-based nanocomposites. Studies revealed that the zeolite-based nZVI composite was a much better option for arsenic remediation due to the presence of more specific sites on its surface.

Kadu and Chikate [32] prepared different clay-based nZVI nanocomposites for Cr(VI) removal purpose. The pore diffusion process has been found to be the governing mechanism for chromium abatement. Besides that, the presence of nZVI further facilitated the reduction of adsorbed hexavalent chromium.

1.2.4 Polysaccharide-Based Nanoadsorbents

Polysaccharides are biopolymers where complex carbohydrate molecules, i.e. long chain of monosaccharides, are linked by glycosidic bonds. They are biodegradable, nontoxic, biocompatible, low-cost, renewable, and sustainable, and they contain different functional groups. These polysaccharides have huge potential to be used as adsorbents in the field of wastewater treatment. These materials are popular as adsorbents due to their selectivity for pollutants and the capability of forming chelating compounds for trapping the pollutants using the $-OH$ and $-NH_2$ groups present in their structure. In recent times, nanobiopolymers or nanocomposites with biopolymers have emerged as one of the most interesting materials for wastewater treatment. Different types of biopolymer-based nanocomposites that have been explored for wastewater treatment purpose have been described in the following section.

1.2.4.1 Chitin and Chitosan-Based Adsorbents

Among the existing polysaccharides, chitin is the second most abundant polysaccharide after cellulose. It is an animal polysaccharide. It is obtained from crustacean (shells) as a byproduct. It is a homopolymer of β-(1-4) linked N-acetyl-D-glucosamine, and found in three forms: chitin-α, β, and γ. Chitosan (poliglusam) is obtained by deacetylation of chitin. It is made up of monomers of N-acetyl-D-glucosamine and D-acetyl-glucosamine linked by glycosidic β-(1-4) linkages. It is a cationic biopolymer, semicrystalline in nature and exists in solid state. It is insoluble in water and soluble in dilute acid. Solubility of chitosan depends on many factors, such as degree of acetylation, concentration of acid, distribution of acetyl group on chitosan, molecular weight, and pH of solution.

Karimi et al. [33] deployed the chitosan/Al_2O_3/Fe_3O_4 nanocomposite for the removal of heavy metal ions such as Cd(II), Cu(II), Zn(II) from wastewater. In the case of Cd(II), the removal efficiency was highest, followed by Cu(II) and Zn(II). Hassan et al. [34] reported the application of chitosan/silica/ZnO nanocomposite for the removal of MB dye. Maximum removal capacity reached up to 293.3 $mg\,g^{-1}$ and the experimental data followed a pseudo-second-order kinetic model. Khan et al. [35] reported the application of chitosan/Co-silica nanocomposites for the removal of dyes such as MO, acridine orange, indigo carmine, and congo red (CR).

1.2.4.2 Cellulose-Based Adsorbents

Cellulose is the most abundant naturally occurring biodegradable polysaccharide. It is a homopolymer made up of β-D-glucopyranose units linked by β-1,4 glycosidic linkages. It is semicrystalline in nature, showing both a crystalline and an amorphous nature depending upon the source of its extraction. Although it is nontoxic, renewable, biocompatible, sustainable, and cost-effective, it also has some demerits like poor hydrophilicity and low physical and chemical stability. These demerits can be overcome by reducing the size of polymers up to the nanometer. Reduction of polymer size results in availability of larger surface area, high strength, increased hydroxyl functional groups, and enhanced stability and stiffness of the polymer. Nanocellulose can be categorized into three groups on the basis of its extraction: as nanofibrillated cellulose (NFC), cellulose nanocrystal (CNC), and bacterial cellulose.

In order to remove Cr(VI) from aqueous medium, Dong et al. [36] explored CNC-based adsorbents. Polydopamine nanoparticles of size less than 30 nm were loaded on cellulose surface. Maximum uptake capacity toward Cr(VI) reached up to 205 $mg\,g^{-1}$, which is much higher in comparison to the cellulosic materials. Shoukat et al. [37] synthesized nanocomposites made up of bacterial cellulose and TiO_2. Thus prepared cellulose-based nanocomposite was explored for Pb(II) removal from wastewater. With initial concentration of lead at 100 $mg\,l^{-1}$, the removal efficiency reached 90% within 120 minutes at pH 7. The nanoadsorbent showed good reusability for three cycles.

1.2.4.3 Starch-Based Adsorbents

Starch is a homopolysaccharide made up of glucose monomers linked through glycosidic linkages. It is found in plants, staple crops, crop seeds, and food items such as potatoes, rice, wheat, corn, and maize. It forms straight-chain polymer known as amylose as well as branched-chain polymer called as amylopectin. It is nontoxic, renewable, easily affordable, and biodegradable in nature. Different physical, chemical, genetic, and enzymatic methods are used for modification of starch. Chemical methods used for modification of starch are acid hydrolysis, cross linking, acetylation and esterification, oxidation, grafting, etc. Lawchoochaisakul et al. [38] reported the synthesis and application of cationic starch intercalated

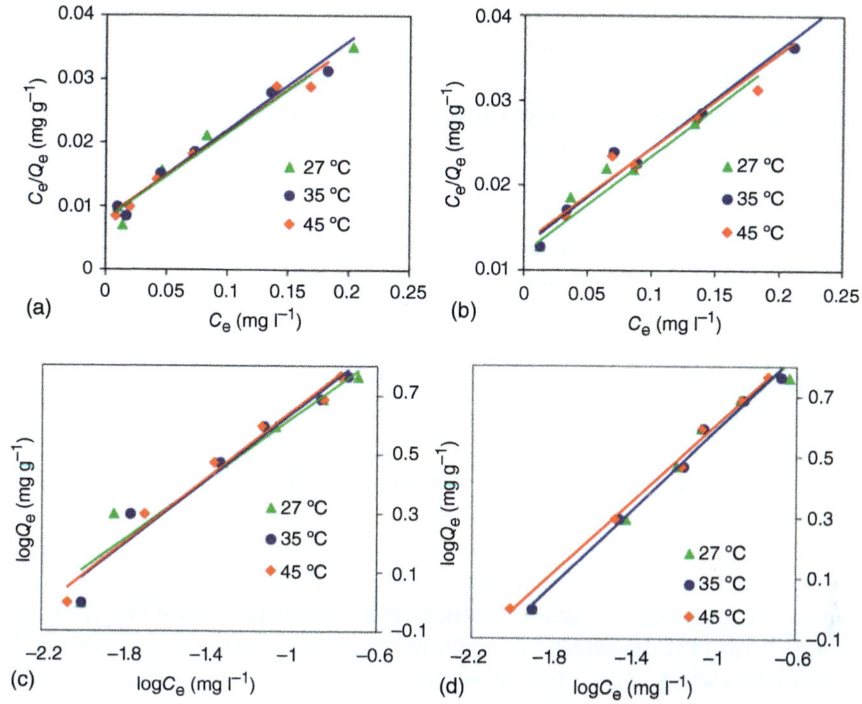

Figure 1.1 Langmuir {(a) and (b)} and Freundlich {(c) and (d)} plots for As(III) uptake by C1 and C2, respectively. Source: Siddiqui et al. [39]/with permission of Elsevier.

montmorillonite nanocomposite for the treatment of dye wastewater. The presence of cationic starch in the nanocomposite enhanced the interlayer spacing, leading to an improvement in adsorption capacity. Siddiqui et al. [39] developed starch-functionalized maghemite nanocomposites for abatement of As(III) from dilute solutions in order to achieve potable water. Starch functionalized nanocomposite was designated as C1 and nonfunctionalized material was denoted as C2. Incorporation of starch in the maghemite enhanced adsorption capacity. Adsorption data fitted well with the Freundlich isotherm model as well as with the pseudo-second-order and film diffusion model. A graphical presentation of experimental data has been shown in Figure 1.1.

1.2.4.4 Pectin-Based Adsorbents

Pectin is a plant-based linear polysaccharide found in cell walls of plants. It comprises monomers of galacturonic acid linked by α-(1-4) glycosidic linkage. The composition of pectin depends on the method of extraction from its sources. It is an anionic polymer containing hydroxyl and carboxyl groups present in its structure, which can be used as adsorption sites for removal of heavy metals and dyes. It shows gel-forming ability.

Babaladimath and Badalamoole [40] incorporated Ag nanoparticles inside the pectin hydrogel matrix to prepare a novel adsorbent for MB dye. Thus prepared pectin-based nanocomposite showed much swelling due to the presence of Ag nanoparticles. After three cycles of usage, it showed promising performance, reaching 80% of its initial efficiency. Kodoth and Badalamoole [41] tested Ag nanoparticles embedded in pectin biopolymer nanocomposite for the adsorption of dyes and metal ions. The maximum adsorption capacity obtained was 1950 $mg\,g^{-1}$ for crystal violet (CV) dye, 111 $mg\,g^{-1}$ for Cu(II), and 130 $mg\,g^{-1}$ for Pb(II).

1.2.4.5 Alginate-Based Adsorbents

Alginate is a natural copolymer made up of two epimer uronic acids, i.e. 1,4-(β-D) mannuronic acid (M) and 1,4- (α-L) guluronic acid (G). It is obtained from alginic acid, which is extracted from brown algae. It is biocompatible, nontoxic, highly water permeable, and used as a stabilizing agent for nanoparticles.

Alginate-based nanocomposites (alginate-clay, alginate-phosphate, and alginate-activated charcoal) were synthesized by Aziz et al. [42] for the removal of Cd(II) from wastewater. The highest adsorption efficiency was obtained by alginate-activated charcoal nanocomposites. For an initial concentration of Cd(II) equal to 1000 $mg\,l^{-1}$, the maximum adsorption capacity reached up to 137 $mg\,g^{-1}$. Moreover, due to its excellent stability and high removal capacity, it has been recommended for usage in continuous-flow real wastewater treatment systems. Majhi and Patra [43] described the applicability of nanocomposites synthesized from polyaniline and sodium alginate toward organic dye removal. Thus prepared alginate-based nanoadsorbents showed promising efficiency toward both cationic and anionic dyes. Mozaffari et al. [44], in their recent work, incorporated Cu(II) in an alginate-based hydrogel polymer matrix to prepare a novel nanocomposite useful for dye removal from wastewater. Malachite green (MG) and CV were tried out as the model dyes. Experimental data showed good fitting to the Freundlich isotherm model as well as to pseudo-second-order kinetic model. Moreover, one of the fascinating observations regarding this nanoadsorbent is that it showed good performance even after eight cycles.

1.2.5 MXene-Based Nanoadsorbents

MXenes are undoubtedly one of the most wonderful materials of recent times, possessing several promising qualities required for water treatment purpose. They are 2D carbides of transition metals with unique physicochemical properties, high hydrophilicity, and oxidation resistance. Moreover, cations can be intercalated through its layers. Jeon et al. [45] in their recent review highlighted the importance of MXene-based nanomaterials as one of the promising new-age adsorbents for pollution abatement purpose. Kim et al. [46] explored the adsorption capacity of $Ti_3C_2T_x$MXene for the removal of pharmaceutical compounds like amitriptyline, verapamil, carbamazepine, 17 α-ethinyl estradiol, ibuprofen, and diclofenac. It was observed that sonicated MXene nanocomposite exhibited better performance in comparison to the pristine MXene. Fard et al. [47] used 2D $Ti_3C_2T_x$ nanosheets for the adsorptive removal of barium from water bodies. Under optimized conditions, removal efficiency was achieved up to 90% within first ten minutes.

Sodium-intercalated $Ti_3C_2T_x$-based nanoadsorbent was explored by Ghani et al. [48] for the adsorptive removal of ciprofloxacin from wastewater. Sodium intercalated nanoadsorbent showed better removal efficiency in comparison to the pristine Mxene.

Jun et al. [49] explored the potency of the MXene-based adsorbent for elimination of radioactive Cs^+ from low-level radioactive wastewater. It showed an excellent adsorption capacity of 148 mg g^{-1} within one hour. Moreover, the MXene-based adsorbent could be reused for at least four cycles, which undoubtedly makes it suitable for practical purpose.

Cai et al. [50] explored the potency of phytic acid-based MXene-based nanocomposites for the dye removal from sewage. MB and RhB were taken as the model dyes for the adsorption purpose. The nanocomposites showed excellent adsorptive removal capacity, and even after 12 cycles, MB removal efficiency remained at 85%. Shahzad et al. [51] deployed MoS_2-MXene nanocomposite for the adsorption of mercury. The maximum adsorption capacity reached 7.21 mmol g^{-1}.

1.2.6 Biochar-Based Nanoadsorbents

In the current decade, special attention has been paid to biochar-based composite materials for wastewater treatment purposes. Biochar refers to the class of organic substances that are produced by heating organic waste mass in presence of low oxygen or totally in absence of oxygen, which is termed as pyrolysis. The main purpose of producing biochar is to reduce the volume of waste and also produce energy. As they are resistant to degradation under ambient conditions, have large surface area, and possess negative surface charge, they are one of the most interesting classes of adsorbents for cationic pollutant removal from water bodies. Sizmur et al. [52] and Tang et al. [53] reviewed the application of biochar-based composites for wastewater treatment.

Liu et al. [54] reported the utility of biochar-supported carbon nanotubes and GO for the adsorption of Pb(II) and Cd(II) from water media. The adsorption capacity of the nanocomposite was higher than that of the bare biochar.

By means of ball milling, MgO/biochar nanocomposites were synthesized by Zheng et al. [55] and have been applied for the adsorptive removal of cationic dye and anionic phosphate from water medium. Li et al. [56] used agroforestry waste materials for the huge production of nano- and micro-sized biochar composites for several applications. Thus synthesized nanobiochar has been applied for Pb^{2+} removal with varying initial concentrations (20–180 mg l^{-1}) at pH 7. The maximum adsorption capacities for upper and lower layer biochar have been reported as 55.56 and 41.67 mg g^{-1}, respectively. Shaikh et al. [57] prepared biochar-based nanocomposite consisting of doped AgNP and utilized it for the removal of toxic dyes from water bodies. Silver-doped biochar-based nanocomposite showed promising performance toward the removal of dyes like CR and RhB (>90%). Dye adsorption was found to be a spontaneous and exothermic process, and the experimental data fitted well with the Freundlich isotherm and pseudo-second-order model. Biochar-based nanocomposite was prepared by the same research group [58] from waste tea leaves and explored for toxic dye removal from wastewater.

1.2.7 Hybrid Nanoadsorbents

In many cases, MOF, GO, Mxenes, biochars, and polysaccharides are often blended to produce nanocomposites of desired properties. Faradonbeh et al. [59] explored MOF-graphene hybrid nanocomposite for the purpose of Ni(II) removal from water media. For 300 ppm nickel solution, maximum removal efficiency reached up to 96% and the experimental data fitted well with the pseudo-second-order kinetic model. Among different hybrids prepared, maximum removal was observed for the nanohybrid synthesized by incorporating 30 wt% graphene into Cu-organic framework, and the highest capacity reached 1152 $mg\,g^{-1}$.

Hou et al. [60] utilized MOF (MIL-101 (Cr)) loaded nZVI composite for the removal of tetracycline antibiotic from water bodies. Maximum adsorption capacity of the nanocomposite was reported as 625 $mg\,g^{-1}$ and the experimental data fitted well with the pseudo-second-order kinetic model. Ventura et al. [61] synthesized novel hybrid nanoadsorbent composed of MOFs, Fe_3O_4, and GO. Hybrid nanoadsorbent thus produced was further applied for the removal of MB dye, and also tested for the adsorption of various metal ions such as Na^+, Ca^{2+}, Mg^{2+}, SO_4^{2-}, and SiO_3^{2-} from a model brackish water. In comparison to the parent MOF material, the adsorption capacity toward MB increased by an amount of 30.52 and 13.75 $mg\,g^{-1}$ for the Co and Ni nanocomposites.

Mittal et al. [62] reported the application of hybrid nanocomposite made up of GO and xanthan gum toward CV removal. Thus prepared nanocomposite showed excellent performance toward CV removal with a maximum adsorption capacity of 1566.97 $mg\,g^{-1}$. Experimental data showed good correlation with Langmuir isotherm model and pseudo-second-order kinetic model.

To remove heavy metal ions such as Cu(II), Pb(II), and Cr(VI) from the water medium, Najafabadi et al. [63] applied chitosan/GO nanocomposite. It was revealed that the double exponential kinetic model and Redlich–Petersen isotherm best fit the experimental data. For Pb(II), Cu(II), and Cr(VI) the maximum adsorption capacity obtained was 461.3, 423.8, and 310.4 $mg\,g^{-1}$, respectively. Moreover, the nanocomposites could be reused without any loss of efficiency.

1.2.8 Other Nanoadsorbents

Apart from the above-mentioned nanocomposite materials, there are also other nanomaterials reported in the literature for pollutant removal. Abraham et al. [64] used the nanosize thorium(IV) metavandate for the removal of Ni(II) from water media. Kaolin and kaolin/ZnO nanocomposites have been explored by Mustapha et al. [65] for tannery wastewater treatment purpose. Nanocomposites were prepared by sol–gel method followed by wet impregnation technique. In comparison to kaolin, kaolin/ZnO nanocomposites showed better efficiency toward wastewater treatment. Under the same experimental conditions, Cr(VI) removal

Table 1.1 List of some nanoadsorbents with their adsorption capacities.

Adsorbent	Pollutant	Adsorption Capacity (mg g^{-1})	Adsorption Isotherm Model	Reference
CTAB-surface-functionalized magnetic MOF@MOF	Cr (VI)	932.1	—	[10]
Ce – MOF- 500(S)	Phosphate	189.4	—	[11]
LDH/MOF nanocomposite	Cd (II)	415.3	Langmuir	[13]
	Pb (II)	301.4		
Zr–MFC–N	Pb (II)	102	Langmuir	[16]
	MB	128		
Zr–MFC–O	MO	219		
GO nano adsorbent	BR 46	370.4	Langmuir	[21]
Ni–rGO	MB	946.12	—	[22]
Amine- reduced graphene oxide nanocomposite (CARGO)	Uranyl ions	337.93	Langmuir	[25]
Magnetic ferberite graphene nanocomposite	U (VI)	455	Langmuir	[26]
nZVI/activated carbon	As (III)	18.2	Langmuir, Freundlich	[28]
	As (V)	12		
nZVI/Mn oxide	As (III)	29.4	Langmuir	[29]
	As (V)	35.7		
Biochar-nZVI	Glyphosate	80	Langmuir	[30]
Magnetic chitosan Al_2O_3/Fe_3O_4 nanocomposite	Cu (II)	35.97	Langmuir	[33]
	Cd (II)	85.39		
	Zn (II)	32.89		
Chitosan silica zinc oxide nanocomposite	MB	293.3	Langmuir	[34]

(Continued)

Table 1.1 (Continued)

Adsorbent	Pollutant	Adsorption Capacity (mg g^{-1})	Adsorption Isotherm Model	Reference
Polydopamine- coated cellulose nanocrystals	Cr (VI)	205	Langmuir	[36]
Cationic starch- intercalated montmorillonite	BB66	49.7	Langmuir	[38]
	BY1	49.9		
Functionalized maghemite	As (III)	8.6	Freundlich	[39]
Pectin-based nanocomposite (Pec-g-PAMPS)	MB	62	Freundlich	[40]
Pec-g-PAMPS-SN		89		
Pec-g-poly (AMPS-*co*-Aam)/Ag	CV	1950	Langmuir	[41]
	Cu (II)	111		
	Pb (II)	130		
Sodium alginate-based nanocomposite (SA-C Beads)	Cd (II)	112	Langmuir	[42]
SA-P Beads		105		
SA-Ch beads		137		
Polyaniline sodium nanocomposite	MB	555.5	Langmuir	[43]
	RhB	434.78		
	O-II	476.19		
	MO	416.66		
Nanocomposite hydrogel	MG	—	Freundlich	[44]
	CV			
Mxene	Ba	9.3	Freundlich	[47]
Mxene ($Ti_3C_2T_x$)	Cs	148	Freundlich	[49]
PA-Mxene-12	MB	42	Langmuir	[50]
	RhB	22		

Biochar-supported carbon nanotubes and GO nanocomposites	Pb (II)	38.6	Dubinnin Radushke-vich	[54]
	Cd (II)	10.8		
Biochar-based silver nanocomposite	RhB	23.39	Freundlich	[57]
	CR	23.55		
MOF–5Cu–Graphene hybrid 30%	Ni (II)	1152	Freundlich	[59]
nZVI/MIL–101(Cr)	TC	625	Langmuir	[60]
$Ni_3(BTC)_2$–MOF	MB	67	Langmuir	[61]
$Co_3(BTC)_2$–MOF		54		
XG–HNC	CV	1566.97	Langmuir	[62]
Chitosan/graphene oxide nanofibrous adsorbent	Pb (II)	461.3	Redlich -Peterson	[63]
	Cu (II)	423.8		
	Cr (VI)	310.4		
Core shell bimagnetic NPs $CoFe_2O_4$ @ γ Fe_2O_3	P	71.2	Freundlich	[66]
Magnetite – maghemite nanoparticles	As (III)	3.69	Freundlich	[68]
	As (V)	3.71		

by the nanocomposite was nearly 100% whereas it was around 78% with kaolin. The increase in treatment efficiency was due to the higher surface area of the nanocomposite.

Guerra et al. [66] utilized core@shell bimagnetic nanoparticles ($CoFe_2O_4$@γ-Fe_2O_3) for adsorptive removal of phosphorus from water. Two different-sized nanocomposites were prepared. Maximum adsorption capacity for the smaller-sized nanocomposite reached up to 71.2 $mg\,g^{-1}$ while it was 46.9 $mg\,g^{-1}$ for the bigger-sized nanocomposite at pH = 2.

Alias et al. [67] immobilized commercial TiO_2 and hydrothermally synthesized nano-flower like TiO_2 on basil seed support for dye adsorption purpose. Hydrothermally prepared nanoTiO_2 composites exhibited better performance in comparison to the commercial ones. Moreover, after one cycle of use, when the nanocomposite was further subjected to reuse, its efficiency got enhanced significantly.

Chowdhury and Yanful [68] developed a useful and practical method for arsenic and chromium removal from water by using a mixture of maghemite and magnetite nanoparticles. In pH-controlled environment, almost all the arsenic and chromium species in the water were adsorbed. Maximum adsorption capacity was reported as 3.69 $mg\,g^{-1}$ for As(III), 3.71 $mg\,g^{-1}$ for As(V), and 2.4 $mg\,g^{-1}$ for Cr(VI) at pH 2, while their initial concentration in the solution was 1.5 $mg\,l^{-1}$ for As(III) and As(V) and 1 $mg\,l^{-1}$ for Cr(VI).

A list of recently reported nano adsorbents has been compiled in Table 1.1, along with their maximum adsorption capacities toward certain pollutants.

1.3 Different Mechanisms

Different interesting mechanisms have been put forward by different research groups regarding adsorption by nanocomposites. Adsorption of pollutants by the carbon material occurs through electrostatic and non-electrostatic interactions. Fard et al. [47] explained the mechanism of Ba^{2+} removal by MXene-based nanoadsorbent. The removal occurred through physisorption as well as chemisorption. After the etching process, the surface of the MXene got covered with OH, F, and O groups. These groups enhanced the removal efficiency of barium. Furthermore, from D-R model it was clear that physisorption as well as chemisorption are involved in the removal process. Mechanisms can be well understood from Figure 1.2. Barium removal by MXene-based nanocomposite was proposed by the following two equations:

$$Ba^{2+} + 2OH^- \rightarrow Ba(OH)_2$$

$$Ba^{2+} + 2F^- \rightarrow BaF_2$$

Shaikh et al. [57] explained the CR uptake as surface complexation accompanied by electrostatic attraction. On the other hand, removal of RhB occurred only through surface complexation. The mechanism has been elucidated in Figure 1.3.

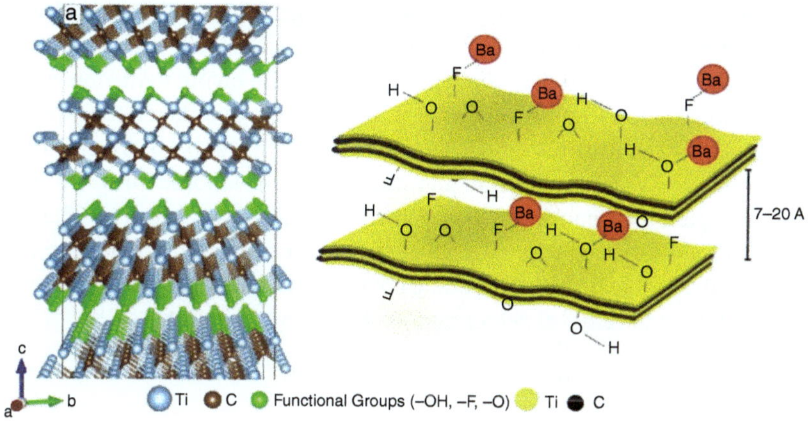

Figure 1.2 Adsorption mechanism of barium ion on surface of MXene layers. Source: Fard et al. [47]/with permission of Elsevier.

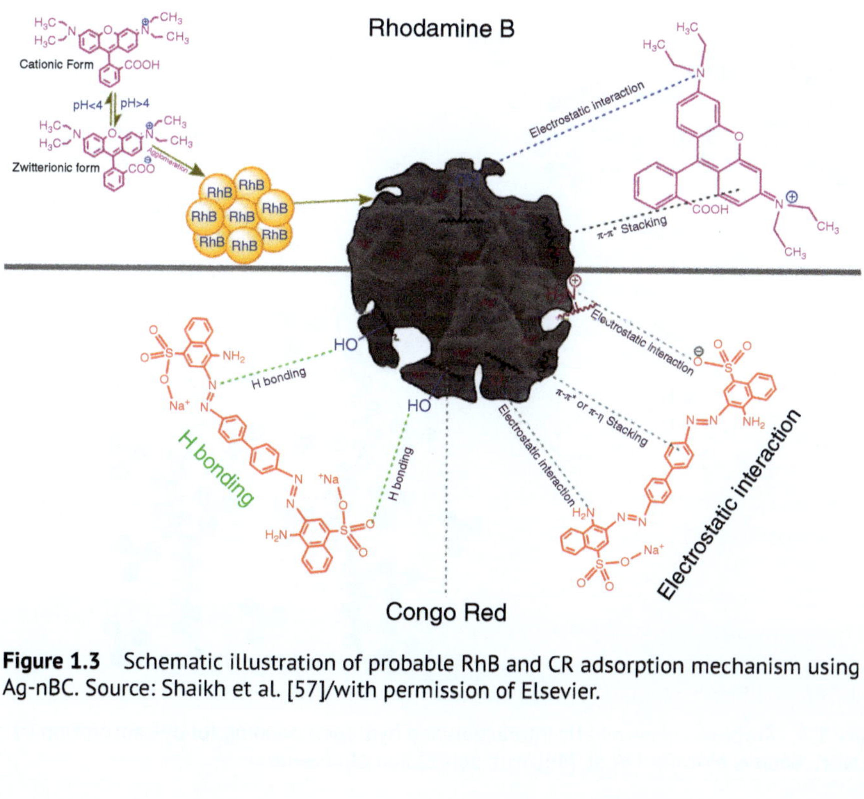

Figure 1.3 Schematic illustration of probable RhB and CR adsorption mechanism using Ag-nBC. Source: Shaikh et al. [57]/with permission of Elsevier.

Mozaffari et al. [44] provided a detailed mechanism behind the dye uptake by their newly synthesized tetraamine Cu(II) nanoparticles embedded in a hydrogel matrix adsorbent termed as novel nanocomposite hydrogel (NCH). The nanoadsorbent worked well for CV and MG dye. Enhanced dye removal capacity was attributed due to the increase in surface area of the hydrogel composite owing to the incorporation of the nanoparticles. Considering pK_a values of all the dyes taken for experimental purpose (CV, MG, RhB, CR, and MO), it was found that at pH 6, CV and MG existed as cationic form, RhB as neutral, and MO and CR as anionic form. On the other hand, hydrogel-based nanocomposite matrix existed as anion at pH 6. Therefore, cationic dyes got removed in greater percentages in comparison to the anionic dyes. For anionic dyes, only hydrogen bonding between the adsorbate and adsorbent was responsible for small removal percentage as shown in Figure 1.4.

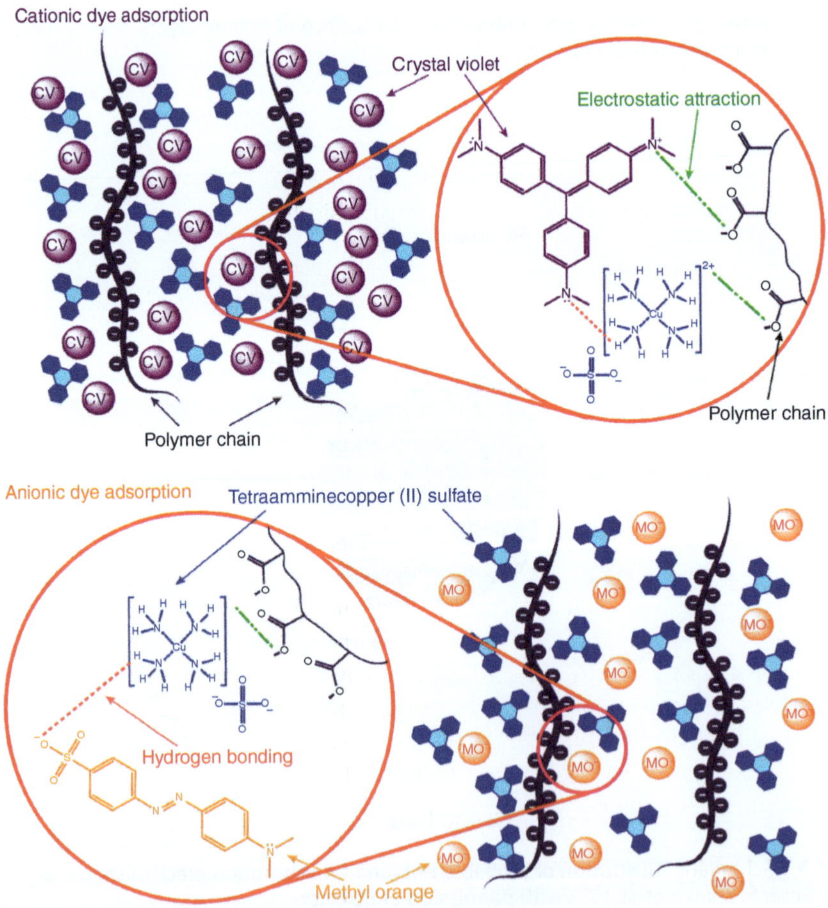

Figure 1.4 Proposed electrostatic interaction and hydrogen bonding for dye adsorption by the NCH. Source: Mozaffari et al. [44]/with permission of Elsevier.

1.4 Different Characterization Techniques

1.4.1 FTIR Analysis

FTIR analyses are often done in order to get an idea of the functional groups involved in the adsorption phenomenon. Shaikh et al. [58] explained the dye removal by nanosilver modified biochar composites by FTIR analysis. From the FTIR analysis, it was clear that –OH, C≡N, and S=O groups were involved in the adsorption phenomenon. Furthermore, it was also confirmed that the C=C group on the surface of the nanocomposite contains a π-bond, which creates a π–π stacking with the delocalized π-aromatic bonds of RhB and CR molecules resulting in enhanced adsorption of these two dyes.

Das et al. [69] conducted FTIR analysis of GO/chitosan-PVA hydrogel nanocomposites. Strong peak at ~3270 cm^{-1} appeared due to the –OH functional group and $-NH_2$ group of chitosan. Peaks at ~2939 and 1377 cm^{-1} have been assigned corresponding to the C–H stretching vibration. Due to the reaction that occurred between hydroxyl groups of PVA and the NH_2 group of chitosan, C=N has been formed, and a peak has been obtained at ~1647 cm^{-1}.

1.4.2 Electron Microscopic Analysis

In order to investigate the interesting morphological features of the nanoadsorbents, scanning electron microscopic (SEM) and transmission electron microscopic (TEM) analysis is often done. Jiang et al. [30] carried out the SEM and TEM analysis of biochar, bare nZVI and biochar-supported nZVI. It was observed that the surface of the original biochar was smooth consisting of lamellar multilayer structure of size around 0.5 μm. Bare nZVI was spherical in appearance with a size of 20–100 nm. After loading on the biochar surface, nZVI particles were uniformly dispersed and showed less sign of reunion. Moreover, the outer layer of the biochar-nZVI composite looked brown, possibly due to the oxidation. It prevented the interior nZVI from being oxidized and thus retained its reactivity.

From thc SEM analysis of tea leaf-supported Ag nanoparticles [58], it was visualized that the surface of the nanocomposite exhibited porous, irregular surface with the presence of numerous cavities on it. It has been proposed that volatilization of lignin and holocellulose contributed to the pore formation, which ultimately aided in enhanced adsorption capacity of the nanoadsorbent.

In the work of Mozaffari et al. [44], TEM analysis showed that the size of the nanoparticles is within the range of 5–30 nm. From the TEM images of different weight percentages (0.5% and 0.7%) of chitosan/GO nanocomposite [63], it was clear that in 0.7% composite, GO sheets were embedded in the nanofibers, while in 0.5% composite, no such evidence was there.

1.4.3 X-Ray Diffraction (XRD) Analysis

Powder X-ray diffraction (XRD) is an analytical technique commonly practiced in order to determine the phase of a crystalline material and also to get an idea about the oxidation state of the metal ions involved in the adsorption process. From the XRD spectrum of Co-MOF, it was found that the nanocomposites were crystalline in nature possessing lattice structure and unvarying morphology [18]. In the XRD spectrum of pristine GO [21], the diffraction peaks were obtained at $2\theta = 11.2°$ and 42.6° corresponding to (001) and (100) reflections, respectively.

In the work of Verma and Dutta [25], the XRD analysis confirmed the reduction of GO by L-cysteine. First, formation of GO from graphite was clear from the diffraction peak at $2\theta = 9.58°$ (corresponding to 002 plane of graphite), with an interplanar distance of 9.22 Å, which was larger than that of the graphite (3.34 Å). An increase in the d-spacing occurred due to the introduction of oxygen-containing moieties and water molecules between the graphene layers. Further introduction of L-cysteine amide shifted the diffraction peaks of the GO (of [002] plane) toward higher 2θ value. A decrease in d spacing indicated reduction of GO.

1.4.4 UV-Diffuse Reflectance Spectra (DRS) Analysis

Diffuse reflectance spectroscopy is an advanced technique often explored for the determination of bandgap energy of nanoparticles. Abraham et al. [64] conducted DRS analysis of thorium metavandate nanoparticles. From the Tauc plot, the bandgap energy of the nanoparticles was found to be 2.36 eV. Mukherjee et al. [70] carried out the DRS study of Cu^{2+} engrafted $MgAl_2O_4$ nanoparticles ($Mg_{1-x}Cu_xAl_2O_4$, $x = 0$, 0.1, 0.3, and 0.5). These nanocomposites are designated as MCA0, MCA1, MCA3, and MCA5. Strong peaks were noticed at wavelengths of 275, 280, and 236 nm, respectively. Moreover, a blue shift was observed in the spectra of the nanocomposites as the percentage of Cu^{2+} increased due to the surface defects.

Arshadi [71] carried out the UV–Vis DRS analysis of Mn nanoparticles supported on mixed metal oxides (SA@MnNP) applied for Hg(II) removal purpose. Strong absorptions have been noticed at 253, 269, and 315 nm because of the occurrence of the $\pi \rightarrow \pi^*$, $n \rightarrow \pi^*$ transitions of the ligands. On the other hand, many low-intensity bands have been found in the visible region (500–700 nm), which may be due to $d \rightarrow d$ transitions for Mn(II)-Salen ligand interaction. DRS plot of the nanocomposite both before and after Hg(II) removal has been shown in Figure 1.5.

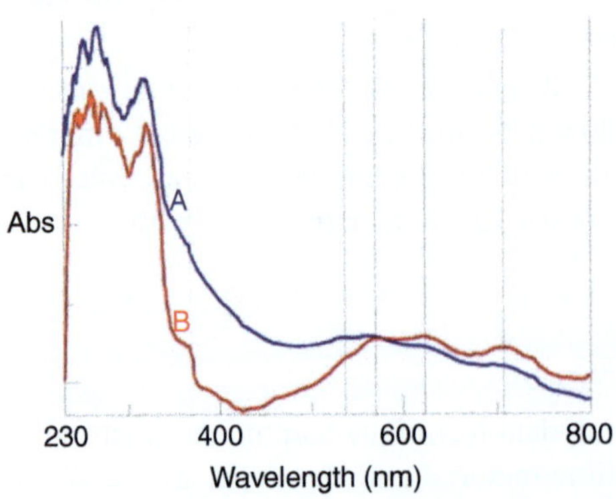

Figure 1.5 UV–Vis diffuse reflectance spectra (DRS) of the Si/Al-PAEA = SA@MnNP before (A) and after adsorption of Hg(II) ions (B) from aqueous solution. Source: Arshadi [71]/with permission of Elsevier.

1.4.5 Brunauer–Emmett–Teller (BET) Analysis

Brunauer–Emmett–Teller (BET) analysis is often conducted to find out the specific surface area of nanocomposites. Shoushtarian et al. [21] found the specific surface area of pristine GO to be 465 $m^2 g^{-1}$.

Ahmad et al. [18] conducted the BET analysis of the MOF used for dye removal. N_2 adsorption–desorption isotherm indicated that it was a type IV isotherm and the BET surface area of the MOF was obtained as 153.45 $m^2 g^{-1}$. Verma and Dutta [25] carried out the BET analysis of the nanocomposite made up of rGO and L-cysteine amide (CARGO). Depending upon the ratio of the rGO and L-cysteine, the nanoadsorbents were named as CARGO-1, CARGO-2, and CARGO-3 with the increase of L-cysteine. Among the three, CARGO-1 possessed the highest specific area of 57.088 $m^2 g^{-1}$ followed by CARGO-2 and CARGO-3.

1.5 Current Challenges and Future Perspectives

Nanomaterials-based adsorbents are undoubtedly one of the most promising new-age materials for water purification. However, their applications often suffer from various drawbacks. Lata and Samadder described various challenges and drawbacks related to application of nanoadsorbents for arsenic remediation [72]. One of the most challenging aspects of using nanoadsorbents is the difficulty in their separation. Second, most of the nanoadsorbents are prepared via suspension and obtained as fine powders. However, these fine powders are not suitable as column materials as they possess low hydraulic conductivity.

Gomez-Pastora et al. [73], in their recent review article on the implementation of magnetic nanoparticles as adsorbents, described the pros and cons of applying nanoparticles as adsorbents. It was documented that the successful application of nanoadsorbent lay behind the regeneration capacity and also on the waste management technique of the spent solution used for regeneration. Moreover, it was also highlighted that the dose required by these nanoadsorbents was quite low in comparison to the other traditional adsorbents, and contact time was also relatively less. So, the apparent cost might look lesser for nanoadsorbent. But in some scenarios, the cost associated with the low-cost adsorbent might be lower than the nanoparticles. In one such report regarding the usage of nZVI, it was reported that the cost of commercially available nZVI oscillated from 225 to 2255 \$ kg^{-1} [68]. It was much higher compared to that of the agro-waste-based adsorbents. Therefore, cost estimation for these nanoparticles should be optimized depending on the scenario.

Gusain et al. [74] in one of their recent review articles described the potency of carbon-based new-age nanoadsorbents for water purification. Apart from discussing several merits possessed by these nanocomposites, the authors highlighted several issues that need to be addressed seriously. Many carbon-based nanomaterials in their pristine form often fail to serve the desired purpose. Prone to segregation, fouling effects and poor adsorption capacities are often encountered with those materials. However, these problems can be handled by means of synthesizing macroscopic structure containing carbon nanomaterials as the building blocks. But these macrostructure or superstructures often suffer from poor mechanical strength. Hence, reinforcement with nanofibers, crosslinking with other reagents, or embedding in the polymer matrix are often carried out.

In most of the studies, nanomaterials are applied for small-scale study. However, large-scale studies in real wastewater scenario are extremely necessary from practical point of view. In addition to this, environmental impact assessment of application of these nanoadsorbents should be thoroughly done. Understanding the fate and transport of these nanoparticles are essential, as after adsorption the pollutant-loaded nanoparticles may affect the ecological system in a negative manner.

Hence, this chapter discusses the positive aspects of different kinds nanoadsorbents that are often used for wastewater treatment. Different modifications and several mechanisms are discussed. Finally, some recommendations are highlighted for carrying out future research in this field.

Acknowledgment

The authors are grateful to IIT Kharagpur and Ministry of Education, Government of India, for providing financial support.

Conflicts of Interest

The authors declare no conflict of interest.

References

1 Ma, J., Qi, J., Yao, C. et al. (2012). A novel bentonite-based adsorbent for anionic pollutant removal from water. *Chem. Eng. J.* 200–202: 97–103.

2 Zhu, R., Chen, Q., Zhou, Q. et al. (2016). Adsorbents based on montmorillonite for contaminant removal from water: a review. *Appl. Clay Sci.* 123: 239–258.

3 Adak, A., Pal, A., and Bandyopadhyay, M. (2006). Removal of phenol from water environment by surfactant-modified alumina through adsolubilization. *Colloids Surf., A* 277: 63–68.

4 Adak, A., Bandyopadhyay, M., and Pal, A. (2005). Removal of crystal violet dye from wastewater by surfactant-modified alumina. *Sep. Purif. Technol.* 44: 139–144.

5 Koner, S., Pal, A., and Adak, A. (2010). Cationic surfactant adsorption on silica gel and its application for wastewater treatment. *Desalin. Water Treat.* 22: 1–8.

6 Khobragade, M.U. and Pal, A. (2014). Investigation on the adsorption of Mn(II) on surfactant-modified alumina: batch and column studies. *J. Environ. Chem. Eng.* 2: 2295–2305.

7 Neha, R., Adithya, S., Jayaraman, R.S. et al. (2021). Nanoadsorbents an effective adsorbent for removal of toxic pharmaceutical compounds from aqueous environment: a critical review on emerging trends. *Chemosphere* 272: 129852.

8 Kyzas, G.Z. and Matis, K.A. (2015). Nanoadsorbents for pollutant removal: a review. *J. Mol. Liq.* 203: 159–168.

9 Joseph, L., Jun, B., Jang, M. et al. (2019). Removal of contaminants of emerging concern by metal-organic framework nanoadsorbents: a review. *Chem. Eng. J.* 369: 928–946.

10 Li, L., Xu, Y., Zhong, D., and Zhong, N. (2020). CTAB-surface-functionalized magnetic MOF@MOF composite adsorbent for Cr(VI) efficient removal from aqueous solution. *Colloids Surf. A.* 586: 124255.

11 He, J., Xu, Y., Wang, W. et al. (2020). Ce(III) nanocomposites by partial thermal decomposition of Ce-MOF for effective phosphate adsorption in a wide pH range. *Chem. Eng. J.* 379: 122431.

12 Yan, A., Yao, S., Li, Y. et al. (2014). Incorporating polyoxometalates into a porous MOF greatly improves its selective adsorption of cationic dyes. *Chem. Eur. J.* 20: 6927–6933.

13 Soltani, R., Pelalak, R., Pishnamazi, M. et al. (2021). A novel and facile green synthesis method to prepare LDH/MOF nanocomposite for removal of Cd(II) and Pb(II). *Sci. Rep.* 11: 1609.

14 Yuan, N., Gong, X., and Han, B. (2021). Hydrophobic fluorous metal-organic framework nanoadsorbent for removal of hazardous wastes from water. *ACS Appl. Nano Mater.* 4: 1576–1585.

15 Nguyen, L.H.T., Nguyen, H.T.T., Le, B.Q.G. et al. (2021). Microwave-assisted solvothermal synthesis of defective zirconium-organic framework as a recyclable nano-adsorbent with superior adsorption capacity for efficient removal of toxic organic dyes. *Colloid Interf. Sci. Commun.* 46: 100511.

16 Huang, L., He, M., Chen, B., and Hu, B. (2018). Magnetic Zr-MOFs nanocomposites for rapid removal of heavy metal ions and dyes from water. *Chemosphere* 199: 435–444.

17 Fu, M., Deng, X., Wang, S. et al. (2022). Scalable robust nano-porous Zr-based MOF adsorbent with high-capacity for sustainable water purification. *Sep. Purif. Technol.* 288: 120620.

18 Ahmad, K., Shah, H., Parveen, S. et al. (2021). Metal Organic Framework (KIUB-MOF-1) as efficient adsorbent for cationic and anionic dyes from brackish water. *J. Mol. Struct.* 1242: 130898.

19 Soltani, R., Marjani, A., and Shirazian, S. (2020). A hierarchial LDH/MOF nanocomposite: single, simultaneous and consecutive adsorption of a reactive dye and Cr(VI). *Dalton Trans.* 49: 5323–5335.

20 Kim, S., Park, C.M., Jang, M. et al. (2018). Aqueous removal of inorganic and organic contaminants by graphene-based nanoadsorbents: a review. *Chemosphere* 212: 1104–1124.

21 Shoushtarian, F., Moghaddam, M.R.A., and Kowsari, E. (2020). Efficient regeneration/reuse of graphene oxide as a nanoadsorbent for removing basic Red 46 from aqueous solutions. *J. Mol. Liq.* 312: 113386.

22 Calimli, M.H., Nas, M.S., Burhan, H. et al. (2020). Preparation, characterization and adsorption kinetics of methylene blue dye in reduced-graphene oxide supported nanoadsorbents. *J. Mol. Liq.* 309: 113171.

23 Wang, J., Tsuzuki, T., Tang, B. et al. (2012). Reduced graphene oxide/Zno composite: reusable adsorbent for pollutant management. *ACS Appl. Mater. Interfaces* 4: 3084–3090.

24 Lujaniene, G., Semcuk, S., Lecinskyte, A. et al. (2017). Magnetic graphene oxide based nano-composites for removal of radionuclides and metals from contaminated solutions. *J. Environ. Radioact.* 166: 166–174.

25 Verma, S. and Dutta, R.K. (2017). Development of cysteine amide reduced graphene oxide (CARGO) nano-adsorbents for enhanced uranyl ions removal from aqueous medium. *J. Environ. Chem. Eng.* 5: 4547–4558.

26 El-Maghrabi, H.H., Abdelmaged, S.M., Nada, A.A. et al. (2017). Magnetic graphene based nanocomposite for uranium scavenging. *J. Hazard. Mater.* 322: 370–379.

27 Eljamal, R., Eljamal, O., Maamoun, I. et al. (2020). Enhancing the characteristic and reactivity of nZVI: polymers effect and mechanisms. *J. Mol. Liq.* 315: 113714.

28 Zhu, H., Jia, Y., Wu, X., and Wang, H. (2009). Removal of arsenic from water by supported nano zero-valent iron on activated carbon. *J. Hazard. Mater.* 172: 1591–1596.

29 Bui, T.H., Kim, C., Hong, S.P., and Yoon, J. (2017). Effective adsorbent for arsenic removal: core/shell structural nano zero-valent iron/manganese oxide. *Environ. Sci. Pollut. Res.* 24: 24235–24242.

30 Jiang, X., Ouyang, Z., Zhang, Z. et al. (2018). Mechanism of glyphosphate removal by biochar supported nano-zero-valent iron in aqueous solutions. *Colloids Surf., A* 547: 64–72.

31 Suazo-Hernandez, J., Sepulveda, P., Manquian-Cerda, K. et al. (2019). Synthesis and characterization of zeolite-based composites functionalized with nanoscale zero-valent iron for removing arsenic in the presence of selenium from water. *J. Hazard. Mater.* 373: 810–819.

32 Kadu, B.S. and Chikate, R.C. (2013). nZVI based nanocomposites: role of noble metal and clay support on chemisorptive removal of Cr(VI). *J. Environ. Chem. Eng.* 1: 320–327.

33 Karimi, F., Ayati, A., Tanhaei, B. et al. (2022). Removal of metal ions using a new magnetic chitosan nano bio-adsorbent; a powerful approach in water treatment. *Environ. Res.* 203: 111753.

34 Hassan, H., Salama, A., El-ziaty, A.K., and El-Sakhawy, M. (2019). New chitosan/silica/zinc oxide nanocomposite as adsorbent for dye removal. *Int. J. Biol. Macromol.* 131: 520–526.

35 Khan, S.A., Khan, S.B., Kamal, T. et al. (2016). Antibacterial nanocomposites based on chitosan/Co-MCM as a selective and efficient adsorbent for organic dyes. *Int. J. Biol. Macromol.* 91: 744–751.

36 Dong, L., Deng, R., Xiao, H. et al. (2019). Hierarchial polydopamine coated cellulose nanocrystal microstructures as efficient nanoadsorbents for removal of Cr(VI) ions. *Cellulose* 26: 6401–6414.

37 Shoukat, A., Wahid, F., Khan, T. et al. (2019). Titanium oxide-bacterial cellulose bioadsorbent for the removal of lead ions from aqueous solution. *Int. J. Biol. Macromol.* 129: 965–971.

38 Lawchoochaisakul, S., Monvisade, P., and Siriphannon, P. (2021). Cationic starch intercalated montmorillonite nanocomposites as natural based adsorbent for dye removal. *Carbohydr. Polym.* 253: 117230.

39 Siddiqui, S.I., Singh, P.N., Tara, N. et al. (2020). Arsenic removal from water by starch functionalized maghemite nano-adsorbents: thermodynamics and kinetics investigations. *Colloid Interf. Sci.* 36: 100263.

40 Babaladimath, G. and Badalamoole, V. (2019). Silver nanoparticles embedded pectin-based hydrogel: a novel adsorbent material for separation of cationic dyes. *Polym. Bull.* 76: 4215–4236.

41 Kodoth, A.K. and Badalamoole, V. (2020). Silver nanoparticle-embedded pectin-based hydrogel for adsorptive removal of dyes and metal ions. *Polym. Bull.* 77: 541–564.

42 Aziz, F., Achaby, M.E., Lissaneddine, A. et al. (2020). Composites with alginate beads: a novel design of nano-adsorbents impregnation for large scale continuous flow wastewater treatment pilots. *Saudi. J. Biol. Sci.* 27: 2499–2508.

43 Majhi, D. and Patra, B.N. (2020). Polyaniline and sodium alginate nanocomposite; a pH-responsive adsorbent for the removal of organic dyes from water. *RSC Adv.* 10: 43904–43914.

44 Mozaffari, T., Vanashi, A.K., and Ghasemzadeh, H. (2021). Nanocomposite hydrogel based on sodium alginate, poly (acrylic acid), and tetraamminecopper(II) sulfate as an efficient dye adsorbent. *Carbohydr. Polym.* 267: 118182.

45 Jeon, M., Jun, B., Kim, S. et al. (2020). A review on MXenes-nanomaterials as adsorbents in aqueous solution. *Chemosphere* 261: 127781.

46 Kim, S., Gholamirad, F., Yu, M. et al. (2021). Enhanced adsorption performance for selected pharmaceutical compounds by sonicated $Ti_3C_2T_x$MXene. *Chem. Eng. J.* 406: 126789.

47 Fard, A.K., Mckay, G., Chamoun, R. et al. (2017). Barium removal from synthetic natural and produced water using MXene as two dimensional (2-D) nanosheet adsorbent. *Chem. Eng. J.* 317: 331–342.

48 Ghani, A.A., Shahzad, A., Moztahida, M. et al. (2021). Adsorption and electrochemical regeneration of intercalated $Ti_3C_2T_x$ MXene for the removal of ciprofloxacin from wastewater. *Chem. Eng. J.* 421: 127780.

49 Jun, B., Jang, M., Park, C.M. et al. (2020). Selective adsorption of Cs^+ by MXene ($Ti_3C_2T_x$) from model low-level radioactive wastewater. *Nucl. Eng. Technol.* 52: 1201–1207.

50 Cai, C., Wang, R., Liu, S. et al. (2020). Synthesis of self-assembled phytic acid-MXene nanocomposites via a facile hydrothermal approach with elevated dye adsorption capacities. *Colloids Surf., A* 589: 124468.

51 Shahzad, A., Jang, J., Lim, S., and Lee, D.S. (2020). Unique selectivity and rapid uptake of molybdenum-disulphide-functionalized MXene nanocomposite for mercury adsorption. *Environ. Res.* 182: 109005.

52 Sizmur, T., Fresno, T., Akgul, G. et al. (2017). Biosorption modification to enhance sorption of inorganics from water. *Bioresour. Technol.* 246: 33–47.

53 Tan, X., Liu, Y., Gu, Y. et al. (2016). Biochar-based nano-composites for the decontamination of wastewater: a review. *Biresour. Technol.* 212: 318–333.

54 Liu, T., Gao, B., Fang, J. et al. (2016). Biochar-supported carbon nanotube and graphene oxide nanocomposites for Pb(II) and Cd(II) removal. *RSC Adv.* 6: 24314–24319.

55 Zheng, Y., Wan, Y., Chen, J. et al. (2020). MgO modified biochar produced through ball milling: a dual-functional adsorbent for removal of different contaminants. *Chemosphere* 243: 125344.

56 Li, L., Zhang, K., Chen, L. et al. (2017). Mass preparation of micro/nano-powders of biochar with water-dispersibility and their potential application. *New J. Chem.* 41: 9649–9657.

57 Shaikh, W.A., Islam, R.U., and Chakraborty, S. (2021). Stable silver nanoparticle doped mesoporous biochar-based nanocomposite for efficient removal of toxic dyes. *J. Environ. Chem. Eng.* 9: 104982.

58 Shaikh, W.A., Kumar, A., Chakraborty, S. et al. (2022). Biochar-based nanocomposite from waste tea leaf for toxic dye removal: from facile fabrication to functional fitness. *Chemosphere* 291: 132788.

59 Faradonbeh, M.R., Dadkhah, A.A., Rashidi, A. et al. (2018). Newly MOF-graphene hybrid nanoadsorbent for removal of Ni(II) from aqueous phase. *J. Inorg. Organomet. Polym. Mater.* 28: 829–836.

60 Hou, X., Shi, J., Wang, N. et al. (2020). Removal of antibiotic tetracycline by metal-organic framework MIL-101 (Cr) loaded nano zero-valent iron. *J. Mol. Liq.* 313: 113512.

61 Ventura, K., Arrieta, R.A., Marcos-Hernandez, M. et al. (2020). Superparamagnetic MOF@GO Ni and Co based hybrid nanocomposites as efficient water adsorbents. *Sci. Total Environ.* 738: 139213.

62 Mittal, H., Alili, A.A., Morajkar, P.P., and Alhassan, S.M. (2021). Graphene oxide crosslinked hydrogel nanocomposites of xanthan gum for the adsorption of crystal violet dye. *J. Mol. Liq.* 323: 115034.

63 Najafabadi, H.H., Irani, M., Rad, L.R. et al. (2015). Removal of Cu^{2+}, Pb^{2+} and Cr^{6+} from aqueous solutions using a chitosan/graphene oxide composite nanofibrous adsorbent. *RSC Adv.* 5: 16532–16539.

64 Abraham, S.D., Bennie, R.B., Joel, C. et al. (2020). Synthesis and microstructural analysis of nanosized $Th(VO_3)_4$ for the uptake of Ni(II) ions from aqueous solution. *Sep. Sci. Technol.* 55: 247–256.

65 Mustapha, S., Tijani, J.O., Ndamitso, M.M. et al. (2020). The role of kaolin and kaolin/ZnO nanoadsorbent in adsorption studies for tannery wastewater treatment. *Sci. Rep.* 10: 13068.

66 Guerra, A.A.A.M., Campos, A.F.C., Lima, R.M. et al. (2020). Efficient uptake of phosphorus from water by core@shell bimagnetic nanoadsorbents. *J. Environ. Chem. Eng.* 8: 103888.

67 Alias, S.S., Harun, Z., Azhar, F.H. et al. (2020). Comparison between commercial and synthesised nano flower-like rutile TiO_2 immobilised on green super adsorbent towards dye wastewater treatment. *J. Cleaner Prod.* 251: 119448.
68 Chowdhury, S.R. and Yanful, E.K. (2010). Arsenic and chromium removal by mixed magnetite-maghemite nanoparticles and the effect of phosphate on removal. *J. Environ. Manage.* 91: 2238–2247.
69 Das, L., Das, P., Bhowal, A., and Bhattacharjee, C. (2020). Synthesis of hybrid hydrogel nano-polymer composite using graphene oxide, chitosan and PVA and its application in wastewater treatment. *Environ. Technol. Innov.* 2020: 100664.
70 Mukherjee, A., Adak, M.K., Dhak, P., and Dhak, D. (2020). A simple chemical method for the synthesis of Cu^{2+} engrafted $MgAl_2O_4$ nanoparticles: efficient fluoride adsorbents, photocatalyst and latent finger print detection. *J. Environ. Sci.* 88: 301–315.
71 Arshadi, M. (2015). Manganese chloride nanoparticles: a practical adsorbent for the sequestration of Hg(II) ions from aqueous solution. *Chem. Eng. J.* 259: 170–182.
72 Lata, S. and Samadder, S.R. (2016). Removal of arsenic from water using nano adsorbents and challenges: a review. *J. Environ. Manage.* 166: 387–406.
73 Gomez-Pastora, J., Bringas, E., and Ortiz, I. (2014). Recent progress and future challenges on the use of high performance magnetic nano-adsorbents in environmental applications. *Chem. Eng. J.* 256: 187–204.
74 Gusain, R., Kumar, N., and Ray, S.S. (2020). Recent advances in carbon nanomaterial-based adsorbents for water purification. *Coord. Chem. Rev.* 45: 213111.

2

Nanomaterials: Prospects for Water Purification and Relevance of Biosensing in Drinking Water

Seemesh Bhaskar[1,2,3] and Sai Sathish Ramamurthy[4]

[1] *Carl R. Woese Institute for Genomic Biology, University of Illinois at Urbana-Champaign, Urbana, IL 61801, USA*
[2] *Nick Holonyak Jr. Micro and Nanotechnology Laboratory (HMNTL), University of Illinois at Urbana-Champaign, Urbana, IL 61801, USA*
[3] *Department of Electrical and Computer Engineering, University of Illinois at Urbana-Champaign, Urbana, IL 61801, USA*
[4] *STAR Laboratory, Department of Chemistry, Sri Sathya Sai Institute of Higher Learning, Prasanthi Nilayam, Puttaparthi, Anantapur 515134, Andhra Pradesh, India*

2.1 Introduction: Water Pollution and Relevance of Biosensing

The crucial need for water for human existence as well as its prominence for other undertakings such as domestic purposes, agriculture, fishing, navigation, hydropower generation, flood prevention, recreation, tourism and conservation have exponentially increased the demand for water with the ever growing population [1-5]. As rightly said by the poet W.H. Auden "Thousands have lived without love, not one without water," water has become part and parcel of human existence, just like air sustains breath. Unfortunately, complications with regard to basic management of water resources have been a long-standing challenge, which are debatable in terms of individual impact as well as the difficulties and gaps in governmental policies, plans, and societal practices [1]. For many years' environmental pollution (air, water, and soil) has been the topic of utmost concern, specifically water pollution. Being a major global problem, water pollution is not only raising apprehension within the countries but also among transboundary water resources, where the water resources are shared transnationally, thereby multiplying the existing issues within the nations [1-3]. Industrialization and technological development sometimes seem to overlook the long-term burden and hazards caused by water pollution. Consequently, it is apt to mention that aspects of water pollution have grown highly sophisticated and complex, which demands immediate attention and recovery. Essentially, the point to be noted is that uncontrolled water pollution is an increasing threat to a healthy environment, including terrestrial and aquatic ecosystems [4-7].

Nanomaterials for Air and Water Purification, First Edition.
Edited by Deepu A. Gopakumar, Jyotishkumar Parameswaranpillai, Jinu Jacob George, and Midhun Dominic C.D.

By and large the water pollution is defined as, "Any direct or indirect alteration of the physical, thermal, chemical, biological, radioactive properties of any part of the environment by, discharge, emission or deposit of wastes so as to affect any beneficial use adversely or to cause a condition, which is hazardous to public health, safety or welfare of animals, birds, wildlife, aquatic life or to plants of every description" (Environment Act of 1980) [3]. It is instructive to briefly mention the causes and effects of water pollution before deliberating upon the need for sensing ions and molecules in water bodies. In general, the causes of water pollution are categorized as point source or nonpoint source [8]. While the former is associated with the pollution occurring from single location such as discharge pipes or channels attached to a factory/industry, the latter is by definition associated to the pollutants that enter the environment from many different places at once. Examples of nonpoint source pollution include waste runoff from construction sites and land disturbances to name a few. Major causes of water pollution include industrialization, urbanization, unsustainable land, water exploitation, and population growth. Consequently, oil pollution, oil spills, intensive farming with hazardous pesticides, herbicides, and fungicides, fertilizers, nuclear/radioactive waste, household waste, industrial waste water, food contaminants, soap and detergents, pharmaceutical by-products, marine dumping, and sewers with heavy metals (mercury, lead, and zinc) have all different magnitude of contribution toward water pollution in developed as well as developing countries [1-8]. The effects of water pollution have been disastrous including the contamination of food and drinking water sources, water bodies turn inhospitable for aquatic life, development of water-borne diseases to name a few. Figure 2.1 provides a broad overview of the amount of available water and the different sources and effects of water pollution. Consequently, it is

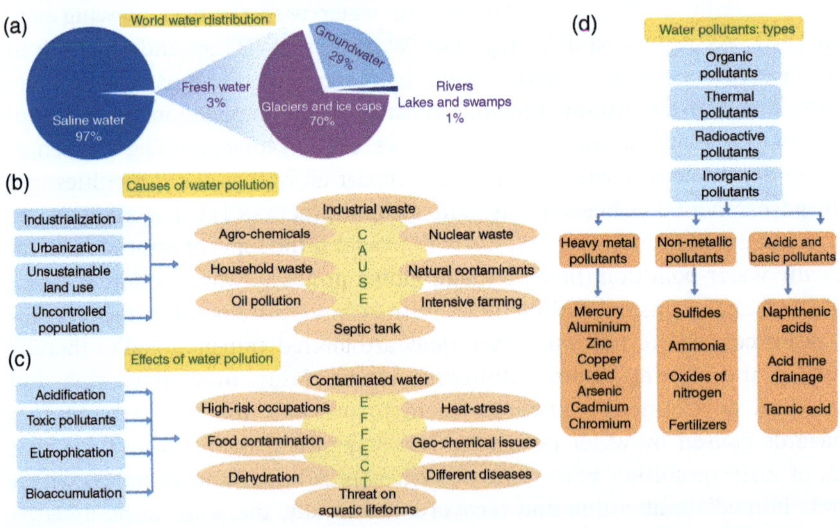

Figure 2.1 Schematic representation of (a) earth's total water distribution emphasizing the low availability of fresh water resources, (b) causes of water pollution, (c) effects of water pollution, and (d) types of water pollution.

important to answer the question as to which stage of water pollution demands immediate action [9]. In order to address this, it is essential for humans to gather information about the purity levels of various water bodies. Therefore, the extent of water contamination needs to be evaluated using different techniques. In this perspective, different methodologies have been developed for evaluating the purity of water at the early stages of pollution. Techniques that can detect slight variations in the concentration of pollutants in water are being developed in this direction, where nanoscience and nanotechnology have played a significant role [10].

2.2 Utility of Nanomaterials for Biosensing

Early detection of water pollutants is need of the hour for early monitoring, regulation, and treatment. Conventionally, there are two techniques using which the quality of the water can be measured: (i) chemical and (ii) biological indicators [1-3]. The former includes measuring the concentrations of different chemicals in the water samples collected from the bodies of interest. If the concentration of the analyte (molecule or ion) is beyond the recommended limit, then the water is said to be polluted. The latter method includes examination of fish, insects, and other invertebrates that live in such water bodies. Depending on which life forms can be supported by the water bodies and also by closely examining such lifeforms, the quality of the water can be determined.

Quantitative as well as qualitative research in understanding the water pollution is highly important for maintaining public health and societal welfare. Recently simple, selective, and real-time analysis of a variety of organic and inorganic environmental pollutants using cost-effective techniques are on the rise [11-13]. Biosensors based on nanomaterials offer advantages in terms of sensitivity, selectivity, and portability. Different techniques have been developed where monitoring environmental pollutants in air and water is achieved in an expedited way with the use of a variety of nanomaterials and nanohybrids [12-14]. Rapid, reliable, and early detection of such pollutants require a simple yet elegant mechanism of interaction of the pollutants with the nanomaterial directly or indirectly. Nanomaterials such as biocompatible supramolecular structures such as cellulose nanofibrils (CNFs), cellulose nanocrystals (CNCs) as well as functionalized metal nanoparticles, semiconductor quantum dots, carbon dots, nanotubes, and a variety of nanocrystals are being utilized in this regard [10-14].

So far, we have deliberated on the relevance of biosensing in drinking water samples, in general. As detailed in the abstract, the first half of the chapter presents details on the bourgeoning field of photo-plasmonics, which assists in biosensing of environmentally, and biologically relevant ions and molecules, particularly in drinking water. The terminology "photo-plasmonics" is broadly adopted to discuss the phenomena that involve the interaction of photons with the plasmons, which are charge density oscillations on metal-dielectric interface [15-18]. Additionally, the expansion of the term "photo" in "photo-plasmonics" can also be applied to the photonic crystal substrates that are widely used to guide the photons with the help of surface trapped states and internal optical modes (IOMs) [17]. By and large, it can be considered that the overall concept of photo-plasmonics is attributed to the study of light matter interactions and related phenomena between light/photons and photonic crystals as well as plasmons. Recently numerous methodologies are being reported where the photonic crystal platforms and the metal/plasmonic platforms are being utilized for biosensing of ions and molecules in different systems [18-22]. On account of the ability of these systems to respond to slight changes in the local refractive index of the medium, the sensitivity achieved is significantly larger than the other techniques such as absorbance and fluorescence.

2.3 Utility of Surface Plasmon-Coupled and Photonic Crystal-Coupled Emission Platforms

Plasmonics techniques have been immensely synergized with conventional spectroscopic techniques and developed into a well-known "plasmon-enhanced spectroscopy" field [23-25]. In the year 2004, Lakowicz and coworkers developed a versatile sensing technology termed surface plasmon-coupled emission (SPCE) [23, 26, 27]. This platform yields >50 % exceedingly directional and polarized emission as compared to <1 % collection efficiency in conventional luminescence techniques. The SPCE technology presents high background suppression and spectral resolution and are used for several biosensing applications as well as to study processes such as real-time polymerization process, kinetics of adsorption and excitation-emission synchronization [28-34]. These scientific explorations have presented and explained novel insightful biophysicochemical concepts comprehensively.

However, surface plasmon resonance (SPR), SPCE, metal-dielectric-metal (MDM) [35], and Tamm state-coupled emission (TSCE) technology [36] present a major drawback of unavoidable parasitic Ohmic losses due to metal-dependent subject platform [37]. In order to address the limitation of loss in SPCE platform, in the year 2020, STAR laboratory developed photonic crystal-coupled emission (PCCE) platform using one-dimensional photonic crystal (1DPhC) as an essential lossless analogue to SPCE substrate [15-18]. This is made of alternating layers of high refractive index (HRI) and LRI nanolayers fabricated such that it sustains a photonic band gap, where a set of wavelengths are not allowed at normal excitation [17]. The fabrication details of 1DPhC along with cross-sectional imaging and experimental reflectance data are presented in our recent reports [15-18].

In conventional fluorescence spectrophotometer, the isotropic emission obtained from the sample placed in a cuvette is captured at 90° as shown in Figure 2.2a. In the case of the emitter dipoles placed at the glass-water interface (Figure 2.2b) the emission in the relatively HRI region (glass, $n_g = 1.52$) develops into a partially directional emission [26], on account of the effect of critical angle (θ_C). Although this off-normal (nonisotropic) emission is partially directional it is important to note that it is not polarized emission. While these are the preliminary observations with regard to the

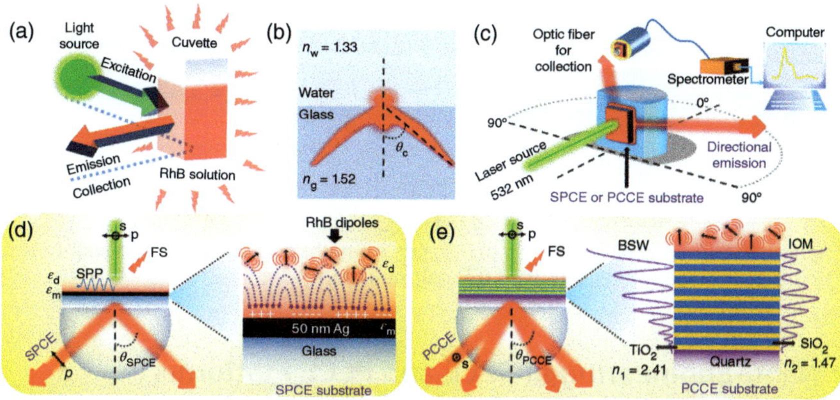

Figure 2.2 Conceptual schematic of (a) fluorescence emission (of RhB) recorded in conventional fluorescence spectrophotometer, (b) the angular dispersion of fluorescence emission as observed in water-glass interface. The angle shown is critical angle (θ_C) of emission. (c) Optical setup used for SPCE and PCCE experimental work with reverse Kretschmann (RK) configuration. Conceptual schematic of coupling of emission with (d) propagating surface plasmon polaritons (SPPs) of metal thin film (with θ_{SPCE} as the angle of emission) and (e) surface-trapped modes of 1DPhC (with θ_{PCCE} as angle of emission). The magnified images on the right side of (d) and (e) display the coupling at the nanointerface. Source: Reproduced with permission from Bhaskar et al. [18], Copyright 2021, De Gruyter, Nanophotonics/CC BY 4.0/Public domain.

emission in cuvette and glass-water interface, the emission is sharply channelized and polarized in the case of SPCE and PCCE substrates (Figure 2.2c–e) [15-18]. In all the sensing applications presented in this chapter, reverse Kretschmann (RK) optical configuration is utilized where the laser light illuminates the engineered SPCE or PCCE substrates directly from the free space (FS) side (and not from the curved surface of prism side) (Figure 2.2c). Appropriate filters and polarizers are used to collect the emission. Both the FS and coupled emission (from SPCE and PCCE platforms) were recorded using USB4000 fiber optic spectrometer connected to Ocean Optics SpectraSuite software. Poly-vinyl alcohol (PVA) has been used as the dielectric nanolayer for doping additional nanomaterial across all the experiments, in order to maintain generality.

Furthermore, the SPCE and PCCE substrates include 50 nm Ag thin film and 1DPhC, respectively (Figure 2.2d,e). The interfacial coupling phenomena of the radiating/emitter dipoles with the SPCE and PCCE substrates is also presented on the right side of Figure 2.2d,e. While the surface plasmons couple with radiating dipoles in the case of SPCE platform (θ_{SPCE}), the Bloch surface waves (BSWs) and IOMs of PCCE substrate couple with the radiating dipoles in the case of PCCE platform [15-18]. Different functional nanomaterials are doped in PVA matrix in different nanointerfaces such as spacer, cavity and extended (ext.) cavity nanointerface, to develop nanoengineered SPCE and PCCE substrates.

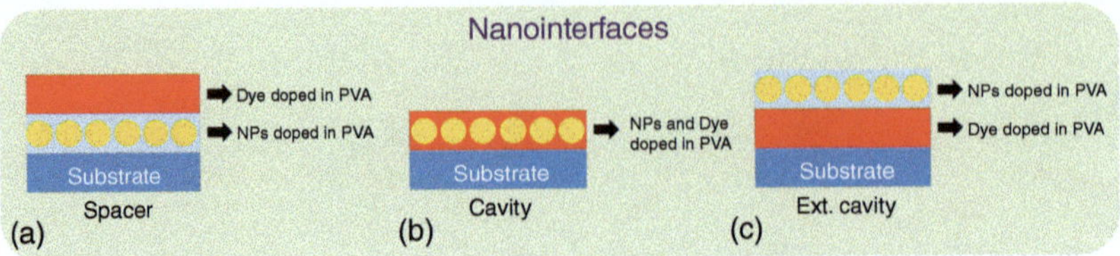

Figure 2.3 Schematic representation of (a) spacer, (b) cavity and (c) ext. cavity nanointerfaces. Source: Reproduced with permission from Bhaskar et al. [18], Copyright 2021, De Gruyter, Nanophotonics/CC BY 4.0/Public domain.

2.4 Nano-Engineering in SPCE and PCCE Platforms

The conceptual schematic of these nanointerfaces are presented in Figure 2.3a–c, and the details are elaborated in our earlier reports [38-48]. In brief, spacer nanointerface is fabricated by first spin coating the NPs doped PVA (30 nm), followed by the radiating dipoles doped PVA (30 nm). Here the nanomaterials act as functional spacer between the radiating dipoles and the metal thin film. The cavity nanointerface is fabricated by spin coating the mixture of dye and NPs as a single layer, as a result of which the dipoles sandwiched in the infinitesimal nanogaps between the substrate and the NPs experience high field intensity. The ext. cavity nanointerface is a reverse format of spacer, where the radiating dipoles doped in PVA are spin coated first, followed by the NPs doped in PVA. Depending on the intrinsic property of the nanomaterial, desired nanointerface needs to be adopted to obtain optimum plasmonic coupling. By and large we have adopted these nanointerfaces for detection of ions and molecules in drinking water samples, as described in Sections 2.5 and 2.6.

2.5 Case Study: Tannic Acid Sensing in Drinking Water Using SPCE Platform

The HRI Nd_2O_3 nanorods (NRs) yielded high (118-fold) SPCE enhancements on account of its loss less nature in SPCE platform [49]. This high SPCE enhancement was utilized for detection of environmentally hazardous organic pollutant, tannic acid (TA) in drinking water samples. On account of the strong affinity TA toward Nd_2O_3, surface adsorption, and breakdown of Nd_2O_3 to smaller NPs occurs and a typical trend in the SPCE enhancements were observed as presented in Figure 2.4a. Upon addition of TA, increased dispersion of Nd_2O_3 NRs assisted significant coupling and increased the SPCE enhancements. Insignificant concentrations of TA from 1 to 100 fM did not alter the SPCE enhancements (as this concentration did not affect Nd_2O_3 NRs), hence, limited TA detection to 10 pM [49]. Furthermore, SPCE enhancements decreased with very high concentrations of TA on account of increased agglomeration with PVA thereby resulting in higher FS emission. In order to utilize economically viable and portable detectors in place of expensive spectrometer detectors, the experiments were performed using the cost-effective smartphone

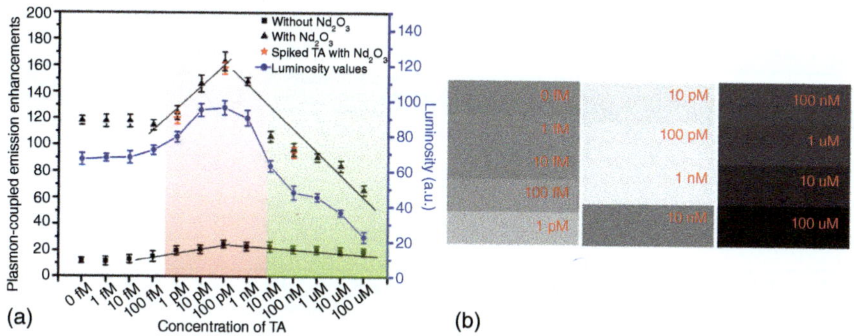

Figure 2.4 (a) Left *y*-axis: modulation in plasmon-coupled emission enhancements with increasing TA concentration (red stars indicate spiked samples). Right *y*-axis: luminosity plot presenting alteration in the emission intensity from Rh6G with increase in TA concentration. (b) Gray scale shade card for different concentrations of TA. Source: Reproduced with permission from Bhaskar and Ramamurthy [49], Copyright 2019, American Chemical Society.

camera [49]. The directional emission was collected using the smartphone and processed using ColorGrab app. The luminosity values (right *y*-axis) obtained therefrom exhibited excellent correlation with the SPCE enhancements (left *y*-axis) and the corresponding gray scale shade cards are presented in Figure 2.4b.

2.6 Case Studies of Sensing Ions such as Cu^{2+}, Zn^{2+}, Hg^{2+}, Al^{3+}, and I^- in Drinking Water

The need for the detection of Cu^{2+}, Zn^{2+}, Hg^{2+}, Al^{3+}, and I^- ions in drinking water samples is elaborately presented in the earlier works [17, 50-52]. More than 600-fold emission enhancements obtained using AgPt hybrid soret colloids (AgPtSC) nano-assemblies was utilized for mobile phone-based picomolar detection of copper ions in drinking water [50]. In a 5.5 pH Britton–Robinson (BR) buffer (Composition: 0.04 M phosphoric acid, 0.04 M boric acid, 0.04 M acetic acid, and 0.2 M sodium hydroxide) alizarin red S (ARS) (fluorescent molecule) shows maximum fluorescence emission. The same experimental conditions were adopted to develop the sensor where a simple complexation of ARS with Cu^{2+} ions were utilized for it's sensing. ARS fluorescent molecule absorbs at ~400 nm and emits at ~560 nm. With the use of 532 nm laser excitation no visible emission was observed from 0.5 μM ARS in SPCE platform without any NPs, since the absorption coefficient is negligible at this excitation wavelength. Upon addition of 1 mM of Cu^{2+} ion concentration to ARS (without AgPtSC) ~40-fold SPCE enhancements is observed on account of the shift absorbance of ARS toward ~510 nm due to complexation with Cu^{2+} ions. Furthermore, upon addition of Cu^{2+} ions to the above mixture 610-fold SPCE enhancements were observed with of AgPtSC in spacer nanointerface. As the concentration of the Cu^{2+} ions were reduced the enhancements also decreased as shown in Figure 2.5a, with a linear range from1 mM to 1 pM. Decreasing the

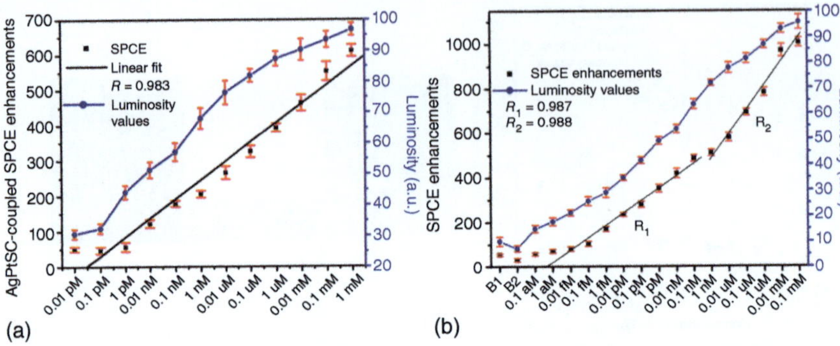

Figure 2.5 (a) AgPtSC-coupled emission enhancements (left *y*-axis) and luminosity values (right *y*-axis) obtained for different concentrations of Cu^{2+} ion under study. (b) Overlap of SPCE enhancements (left *y*-axis) obtained from an Ocean Optics detector and the luminosity values (right *y*-axis) extracted from the mobile phone-based detector for the detection of Zn^{2+} ions using AgAu nanohybrids. Source: (a) Reproduced with permission from Bhaskar et al. [50], Copyright 2021, Elsevier. (b) Reproduced with permission from Rai et al. [51], Copyright 2021, American Chemical Society.

concentration of Cu^{2+} ions lower than this limit did not fit into the linear range and displayed ~35-fold emission (due to the presence of AgPtSC).

Furthermore, as AgAu nanohybrids yielded the highest SPCE enhancements in ext. cavity nanointerface, this interface was utilized for detection of Zn^{2+} ions [51]. As mentioned earlier, fluorescence observed was negligible with the study of only ARS under 532 nm excitation. Upon addition of AgAu nanohybrids, 54-fold SPCE enhancement was observed from ARS on account of the hotspots (blank B1). Addition of 0.1 mM Zn^{2+} ion (without any NPs) yielded ~30-fold SPCE enhancements (blank B2) (Figure 2.5b). This is on account of the shift in the absorbance maximum of ARS from 400 to 510 nm (closer to the laser excitation). With the mixture of 0.1 mM Zn^{2+} ions and AgAu nanohybrids, ~1000-fold SPCE enhancements were observed [51]. Upon decreasing the concentration of Zn^{2+} ions, the SPCE enhancements also decreased, displaying two linear ranges from 1 aM to 0.1 nM ($R = 0.987$) and from 0.1 nM to 0.1 mM ($R = 0.988$). Additionally, an excellent correlation between the SPCE enhancements and the luminosity values were observed. It is important to note that in the case of Zn^{2+} ion sensing, better sensitivity is observed (as compared to that observed in Cu^{2+} ion sensing), as high 1000-fold enhancements are observed in this case [51]. The details of percentage recovery, spiking studies, as well as interference studies are detailed in the earlier works [50, 51].

Furthermore, in another work [52] the high >900-fold SPCE enhancements was utilized for detection of Hg^{2+} ions at attomolar concentrations by using the interaction of sharp edged AgNPs with mercury ions (Figure 2.6a). While the AgNPs synthesized via frugal approach [52] was mixed with the Hg^{2+} ions, a drastic change in the UV–Vis absorbance was observed as presented in Figure 2.6b. These samples were further studied in SPCE platform along with the nano-engineered surface, thereby assisting in attomolar sensitivity of Hg^{2+} ions, the details of which are captured in

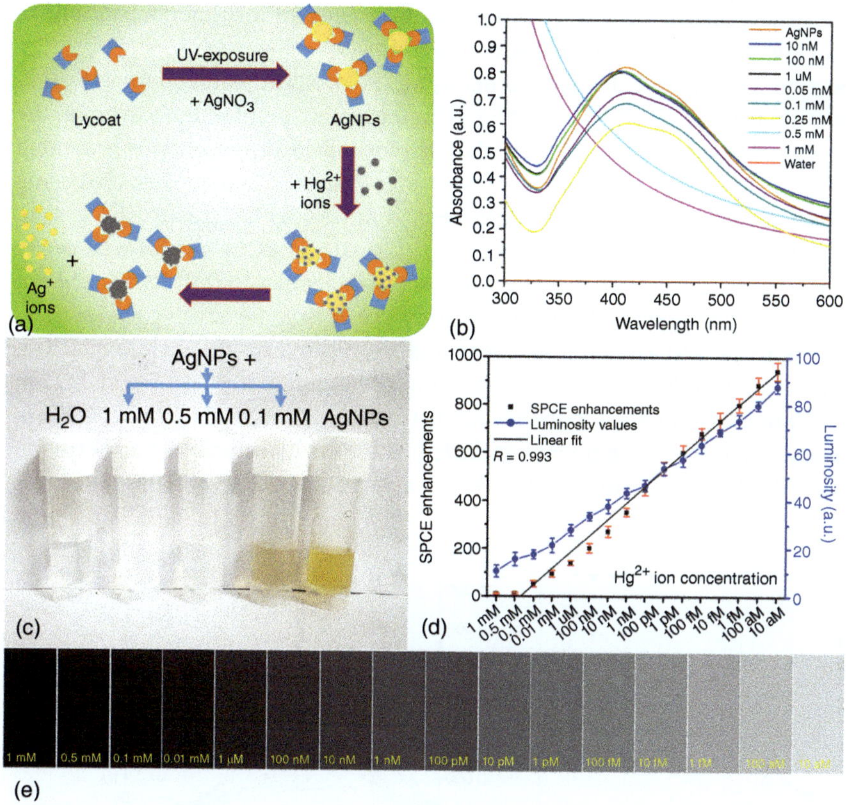

Figure 2.6 (a) Schematic illustration of the effect of addition of Hg^{2+} ions to AgNPs capped with LYCOAT. (b) UV–Vis absorbance pattern obtained upon the addition of different concentrations of Hg^{2+} to AgNPs. (c) Visual change in the color of the solution with and without Hg^{2+} ions. (d) SPCE enhancements and luminosity values obtained using different concentrations of Hg^{2+} ions in the SPCE platform. (e) Shade cards corresponding to the luminosity values at different Hg^{2+} ion concentrations. Source: From Rathnakumar et al. [52], ©2021/Reproduced with permission from American Chemical Society.

Figure 2.6. Upon addition of Hg^{2+} ions to AgNP solutions, the color of the solution changes (Figure 2.6c) due to the disintegration of stable AgNPs into Ag^+ ions (this process is presented schematically in Figure 2.6a). The absence of the localized surface plasmon resonance (LSPR) mode of AgNPs upon Hg^{2+} ions addition aids in confirming this understanding. In order to provide a clear comparison, the SPCE enhancements (gathered by spectrophotometer) and the luminosity values (gathered by smartphone-based analysis) are plotted in left and right *y*-axis (Figure 2.6d), presenting an excellent correlation between them. The shade cards corresponding to that of the respective luminosity values are shown in Figure 2.6e. We observe that the SPCE enhancements decreased upon increasing concentration of Hg^{2+} ions, on account of the surface etching and replacement of Ag^0 with Hg^0. Consequently, the ability of the sharp-edged nanoarchitectures to generate the plasmonic hotspots is reduced, thereby resulting in reduced enhancements. As the proposed sensor did not yield observable changes in the SPCE enhancements for very low concentrations (1 aM) of Hg^{2+} ions, the limit of detection is 10 aM.

In Section 2.6, we have deliberated upon the use of SPCE platform for detection of ions such as Cu^{2+}, Zn^{2+}, and Hg^{2+} in drinking water samples. Furthermore, the higher performance of lossless PCCE platform for sensing application is presented in Figure 2.7. Al^{3+} ions (Figure 2.7a,b) and I^- ions (Figure 2.7a,b) are detected using different nanosystems in PCCE platform. Environmentally hazardous Al^{3+} ions in drinking water samples were monitored using AgNPs in PCCE platform in cavity nanointerface. The ability of the ARS fluorescent molecules to shift its absorbance from 430 to 504 nm upon addition of Al^{3+} ions was utilized for sensing the latter.

Since the 1 mM ARS (without Al^{3+} ions and AgNPs) did not yield observable fluorescence signal, AgNPs were incorporated in cavity nanointerface. Upon 532 nm laser excitation 17-fold SPCE enhancements were observed and is termed as blank in Figure 2.7a, as there are no NPs included in the cavity interface. Furthermore, when 1 mM Al^{3+} ions were added to ARS solution, fold SPCE enhancement was

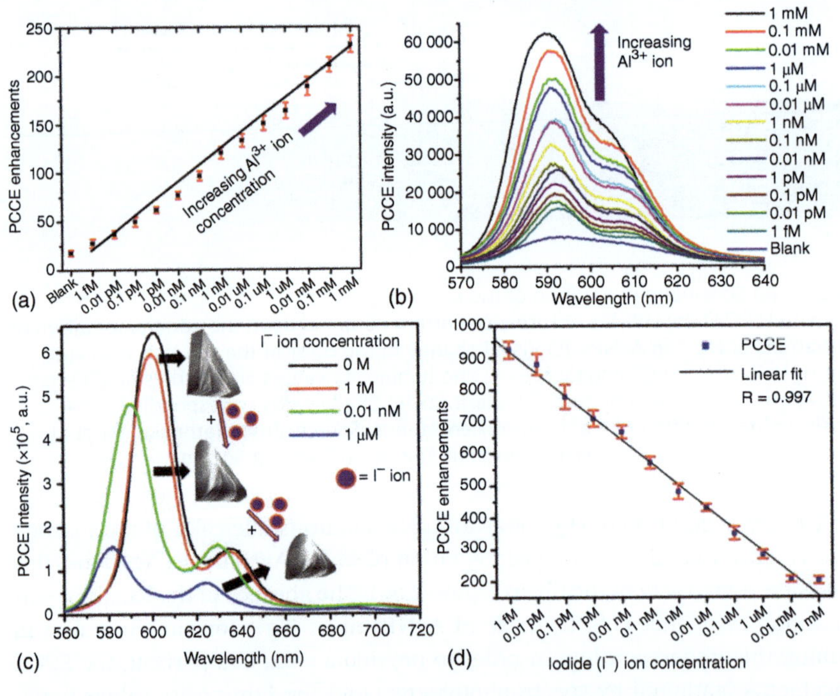

Figure 2.7 (a) PCCE enhancements were obtained with different concentrations of Al^{3+} ions from 1 mM to 1 fM with 10 times dilution at each step, with the use of 1 mM ARS in a 60 nm PVA matrix. Blank indicates 17-fold enhancements obtained with ARS and AgNPs. (b) PCCE spectra obtained using different concentrations of Al^{3+} ions from 1 mM to 1 fM with ten times dilution for each spectrum. (c) Representative PCCE intensity plot displaying a correlation between the effect of iodide concentration, the shape of AgNPr, and the emission profile. (d) Variation in the emission intensity with the increasing concentration of iodide ions. Source: Reprinted with permission from Reference [15, 17], Copyright 2020, American Chemical Society.

observed on account of the shift in the absorption peak toward 500 (which is nearer to the laser excitation source). Amplified 230-fold PCCE enhancements obtained upon addition of AgNPs to the above solution was utilized for sensing Al^{3+} ions, by further by decreasing its concentration from 1 mM to 1 fM with 10 times dilution. From Figure 2.7a, we clearly observe a linear sensing range from 1 mM to 1 fM (for triplicate measurements). The change in the emission spectra observed with changing concentrations of Al^{3+} ions is shown in Figure 2.7b. The results pertaining to spiking studies and recovery analysis in drinking water samples are elaborately presented in an earlier work [17].

Nanomaterials are extensively utilized for various sensing applications [53-56]. Interestingly, sharp-edged nanomaterials are recently being increasingly explored for sensing of pollutants in water bodies [57-60]. Silver nanoprisms (AgNPrs) yielded 947-fold PCCE enhancements on account of sharp-edged surfaces presenting intense hotspots [15]. The close interaction of the iodide (I^-) ions with the sharp-edged AgNPrs was utilized for sensing of I^- ions. The PCCE intensity profile as shown in Figure 2.7c presents a correlation between the effect of iodide concentration and the shape of AgNPr. As seen in Figure 2.7c, a linear relationship is observed from 0.01 mM to 1 fM with an R value of 0.997. We also see that extremely low concentrations of I^- ions (1 fM concentration) did not show any noteworthy change in the PCCE enhancements, thereby displaying the inability to detect the same [15]. It was observed that with the addition of concentrations more than 0.01 mM the PCCE enhancements remained as a constant of ~200-fold (as it may be observed with pristine AgNPs). At this juncture, it is important to mention that the recommendations for iodine intake as given by World Health Organization varies from 90 to 200 $\mu g\,ml^{-1}$ (or from 0.1 to 1.96 μM) [15]. Hence, we conclude that the developed PCCE sensor demonstrates a safe zone (from 0.1 to 2 μM) as well as a harmful zone (2 μM). These two ranges can be utilized for monitoring patients with hypo- and hyperthyroidism, respectively. In this context, the developed PCCE platform with augmented sensitivity can be utilized for on-site and real-time monitoring of thyroid conditions in a rapid, sensitive, reliable, and reproducible manner.

2.7 Water Purification via Defluoridation: Introduction

In the first half of the chapter, we have deliberated upon the need for sensing biologically and environmentally relevant ions and molecules. In this section, we focus on different methodologies that may be adopted for removal of a particular inorganic pollutant, considering the case of fluoride ion. Access to pure drinking water has remained the elixir of human existence. Over one billion people globally suffer from lack of access to safe portable water. The consumption of asymptotic fluoride in drinking water has resulted in major public health problems in the most populous developing countries such as India and China [61]. Since fluoride is of geogenic origin, the scenario in Indian subcontinent has remained deleterious on account of fluoride menace, where the quality of groundwater that sustains 80% (in rural areas) and 50% (in urban areas) of domestic needs are contaminated [62]. As per the recommendation standards of the WHO, the suitable fluoride concentration in drinking water is in the range of 1.0–1.5 $mg\,l^{-1}$ [63]. Interestingly, the altered fluoride concentrations have both tonic and toxic effects, subject to the concentrations that are consumed. On the one hand, this inorganic anion acts as an essential material for healthy teeth, preventing dental caries, and is hence recommended in drinking water at desirable concentrations [64]. Moreover, calcification of the dental enamel in newborns is supported by fluoride, at a concentration of 1.0 $mg\,l^{-1}$ [64, 65]. On the other hand, superfluous fluoride concentration has been reported to present predominantly three types of fluorosis: (i) dental, (ii) skeletal, and (iii) soft tissue fluorosis, thereby leading to precarious biological, long-lasting, and lethal effects [66-69].

In addition to the above-mentioned geogenic leaching of fluoride into groundwater, there has been a catastrophic upsurge in the fluoride ion concentration in groundwater and several other water bodies on account of anthropogenic causes. Coal power plants, fertilizer manufacturing, uranium enrichment, semiconductor industry, glass and ceramic production, as well as electroplating, are the principal reasons for generation of fluoride-rich discharges [66-69]. In this context, effective techniques for detection and quantification of fluoride ions in environmental and industrial wastewaters is need of the hour. While several inhabitants across the Indian subcontinent suffer from this most widespread geochemical disease, the states of Andhra Pradesh, Gujarat, and Rajasthan are severely affected [61]. As the fluoride concentration in these regions ranges from 0.5 to 48 $mg\,l^{-1}$, there is an urgent need to address jeopardy [66, 70]. Furthermore, while certain countries advise municipal water supplies to add fluoride (as their natural concentrations are lower than recommended), certain other countries have levels in natural water bodies more than 1.5 $mg\,l^{-1}$, thereby globally demanding an efficient and suitable (novel, simple, and economical) methodology for removal of excess fluoride concentration.

Currently, there are majorly three methodologies that are being adopted for the defluoridation of drinking water, namely (i) precipitation, (ii) membrane, and (iii) adsorption processes [65, 66, 71]. All the three techniques possess certain advantages and disadvantages. The first technique involves the use of salts (calcium and aluminum) for the formation of precipitate or coprecipitate of fluoride. This technique presents a major limitation with regard to the high chances of contamination of the drinking water on account of undesirable chemicals and thereby resulting in difficulties in the waste removal. The second technique involves methodologies including nanofiltration, reverse osmosis, Donnan dialysis, and electrodialysis [72-74]. These techniques are generally not economical and involves high maintenance budget. In this background, increasing attention is being devoted to the third methodology wherein adsorption on alumina, zeolites, alum, lime, spent catalyst, clay minerals, coconut shell carbon (CSC), charcoal, and ash are utilized and the water passage through a suitable contact bed aids in substantial removal of fluoride through a simple and inexpensive route [65-68, 75-77]. Recently, biomass of biopolymer chitosan, *Eichhornia crassipes*, aligned carbon nanotubes (CNTs), amorphous alumina supported on CNTs, or a hybrid process combining Donnan dialysis and adsorption has been exponentially explored, and the focus has been toward the development of low-cost adsorbents [65-68]. In this chapter, we specifically discuss some of the key highlights of the study, wherein low-cost coconut shell and fiber carbon have been explored for fluoride removal.

Furthermore, research is focused on identification of different novel low-cost adsorbents that exhibit high adsorption efficiency on account of metal ion impregnation. In this direction, lanthanum-modified chitosan and metal-ion-loaded zeolites have been studied in the past [78]. Moreover, in conventional colorimetric based analysis of fluoride, zirconium and its metal complexes are widely used on account of the extraordinary affinity of zirconium toward fluoride ion [79]. This effect is an offshoot of high charge density of fluoride ion that in turn assists the formation of stable fluoro complexes with zirconium(IV), which is a tetravalent metal ion. Activated carbons are effective for fluoride adsorption and removal. Based on earlier works, while these are impregnated with metal ions, the efficiency of fluoride removal increases by three to five times that of plain activated carbon [65, 80]. Coconut shell (*Cocos nucifera* L.), is widely available in rural India from the agronomic sector [81]. Consequently, the expenditure involved is nil and hence is considered as a low-cost carbon preparation technique. Activated carbon impregnated with zirconium ion has been earlier reported for treatment of phosphate, arsenic, selenium, mercury, and chromium from water samples [65]. This chapter broadly covers the effect of different metal ions ($ZrOCl_2$, CaO, Alum, $CaCl_2$, and Borax) and sources (groundnut shell, coconut shell, and coconut fiber) of activated charcoal for defluoridation [65-68]. Of the various metal ions that have been explored, zirconium ion impregnated coconut shell carbon (ZICSC) and zirconium ion impregnated coconut fiber carbon (ZICFC) have proved to be the most effective defluoridating agent for treating effluents as compared the groundnut shell charcoals [67-69]. In the case of ZICFC, maximum defluoridation was attained at an original pH value of 4.0 with a quick 93% adsorption being achieved within 10 minutes. In the case of ZICSC, for the entire pH range of 2–9 the adsorption was above 90% with the 91% of the adsorption being achieved within 10 minutes. Both these results were obtained for the initial fluoride concentration of 10 $mg\,l^{-1}$ [66-68]. In summary, the scope and objective of the current chapter are to investigate and summarize the fluoride adsorption capacities of ZICSC and ZICFC by pursuing batch equilibrium and kinetic experiments, wherein the effects of various parameters such as pH, agitation time, and adsorbent dosage have been elaborated. We also comment on the futuristic scope of research opportunities in this direction.

2.8 Development of Material for Adsorption Studies and Characterization

The activated carbon was prepared using the standard procedure from the coconut shell as well as the coconut fiber. Both the materials were completely carbonized by subjecting the respective crushed 100 g of samples to low temperature muffle furnace at 573–673 K, for three hours. This carbonized material was cooled, powdered, and transferred to a beaker of 2 l capacity. Slowly 200 ml of concentrated sulfuric acid was added to the beaker with continuously stirring to make sure of thorough mixing. Furthermore, the beaker was kept in a hot air oven (373 K for five hours). The activated carbon was then cooled and left overnight and washed free of acid and dried at 110 °C for two hours, then sieved using 40 mesh. Then the activated carbon was immersed in 1% Na_2CO_3 solution and washed free of alkali, providing the preferred adsorbent for impregnation.

In line with the earlier literature [82-84] zirconium ion impregnation was carried out, where 5% $ZrOCl_2$ solution (pH 1.6) was stirred with activated carbon (solution/solid ratio 1.5 : 2) at room temperature (298 K) for five days. In order to confirm the effluent free from zirconium the impregnated carbon was then filtered, rinsed, dried in an oven at 333 K and subsequently used for defluoridation studies. In order to study the likelihood of zirconium ion leaching into the treated water sample, quality assurance and quality control checks were performed on ZICSC and ZICFC. The energy dispersive X-ray (EDX) spectra was obtained before and after the use of these samples use in the batch equilibrium and kinetic experiments. This quantitative analysis confirmed that there was absolutely no variation in the zirconium ion concentration, thereby confirming that that no leaching of zirconium has occurred. Furthermore, it is important to note that zirconium is normally risk-free as an element or in compounds and the oral toxicity is low: OSHA standards for pulmonary exposure specify a TLV of 5 mg zirconium per m^3 [65-68]. In accordance with the solubility product of zirconyl hydroxide ($K_{sp} = 6.3 \times 10^{-49}$), its solubility is 26.7×10^{-15} g l^{-1} [65]. Consequently, amount of zirconium in our effluents is well below the safe limit. The fluoride adsorption capability of ZICSC and ZICFC was then examined by studying the batch equilibrium and kinetic experiments. All the reagents used for the study were of analytical grade. A stock solution of 1000 mg l^{-1} fluoride ion was prepared and various other fluoride standard test solutions were obtained by successive dilution of the stock. All the experiments were carried out at room temperature. Orion potentiometer, Model SA 720, with a fluoride ion selective electrode (FISE) was used to measure fluoride ion concentration in water samples using potentiometric method. pH measurements and sample agitation were performed using pH meter (Systronics) and Remi shaking machine, respectively. Leica S-440I microscope fitted with an EDX spectrometer and Link ISIS detector was utilized for obtaining scanning electron microscopy (SEM) images and EDX spectrum. Quantachrome Autosorb-1 BET surface analyzer was utilized for pore size and surface area evaluation.

The physiochemical properties on zirconium ion impregnate samples are presented in Figure 2.8. In EDX (Figure 2.9), the samples were impacted with

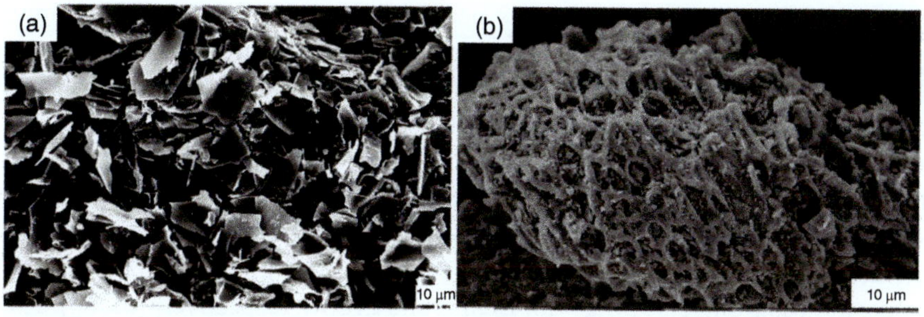

Figure 2.8 SEM image of (a) ZICFC, (b) ZICSC. Source: Sathish et al. [65, 66]/Taylor & Francis.

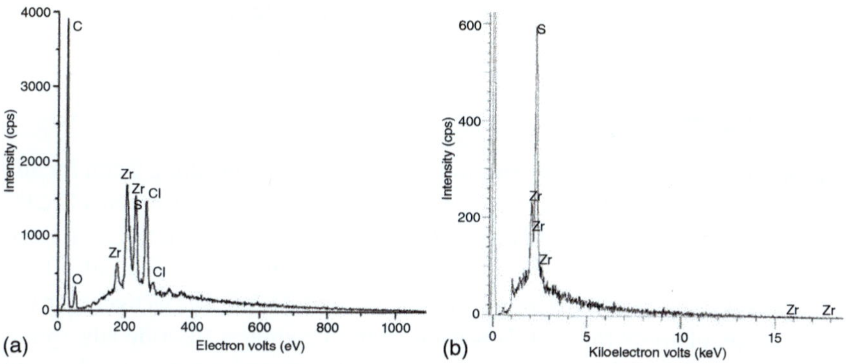

Figure 2.9 Energy dispersive X-ray (EDX) spectrum of (a) ZICFC and (b) ZICSC. Source: Sathish et al. [65, 66]/Taylor & Francis.

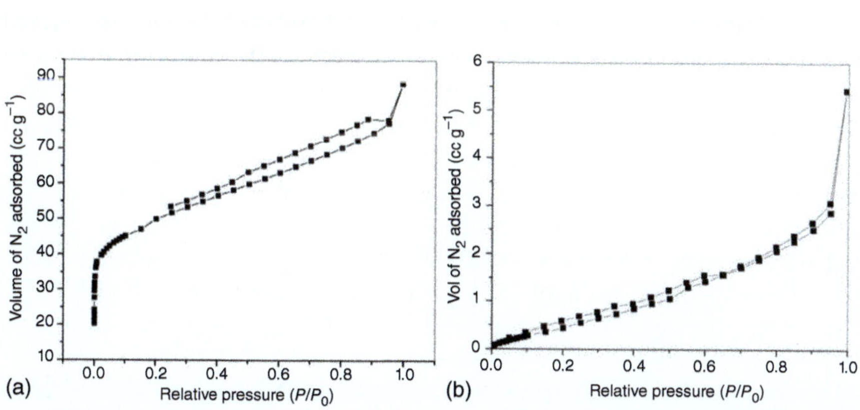

Figure 2.10 BET surface area analysis of (a) ZICFC and (b) ZICSC. Source: Sathish et al. [65, 66]/Taylor & Francis.

an electron beam of specific energy that knocked out the inner shell electrons. Consequently, the outer shell electrons occupied the inner gap emitting X-rays that were characteristic of zirconium, confirming its presence. As the samples used for zirconium ion impregnation was previously sulphonated, intense signals corresponding sulfur was also observed [65, 66].

For BET measurements, the sample taken in a glass holder was degassed at around 423 K to desorb any of the adsorbed gases on the sample. Relative pressures from 0 to 1 and at a 77.3 K adsorption temperature was used to measure nitrogen adsorption isotherms. The surface area was calculated from the BET plots (Figure 2.10) and the values are 163.2 and 2.82 $m^2\,g^{-1}$ for ZICFC and ZICSC, respectively [65, 66]. The total pore volume of ZICFC and ZICSC varies in the range of 5–10 A° and 30– 300 A°, respectively, thereby contributing to the variable surface area.

2.9 Batch Equilibrium and Kinetic Experiments to Investigate the Fluoride Adsorption Capacity

2.9.1 Effect of pH

It is well-known that the pH of the solution is an important controlling factor for any adsorption process. The influence of pH on the sorption rate for adsorbents ZICFC and adsorbent ZICSC are shown in Figure 2.11a,b, respectively. It was found that the q_e, which is the amount of fluoride ions adsorbed per unit weight of the adsorbent ($mg\,g^{-1}$), was above 1.6 and 0.9 $mg\,g^{-1}$ for ZICFC and ZICSC, respectively for the entire pH range of study [65, 66]. However, as the maximum fluoride adsorption capacity, 3 mg F/0.5 g of zirconium ion impregnated activated carbon, was obtained at pH 4.0 [65, 66], this pH was adopted for further studies. In the case of acidic pH, the hydronium ion concentration was optimal to ensure positively charged adsorbent surface and hence was available for fluoride ions. Interestingly, a strong decline in the degree of fluoride adsorption with rise in pH (10–12) of the solution was observed in the case of ZICFC as well as ZICSC. This is on accout of the competition of active

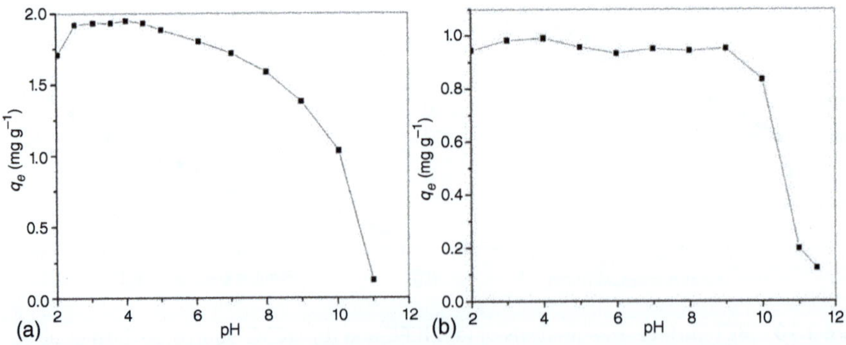

Figure 2.11 Effect of pH on fluoride adsorption onto (a) ZICFC and (b) ZICSC at room temperature. Source: Sathish et al. [65, 66]/Taylor & Francis.

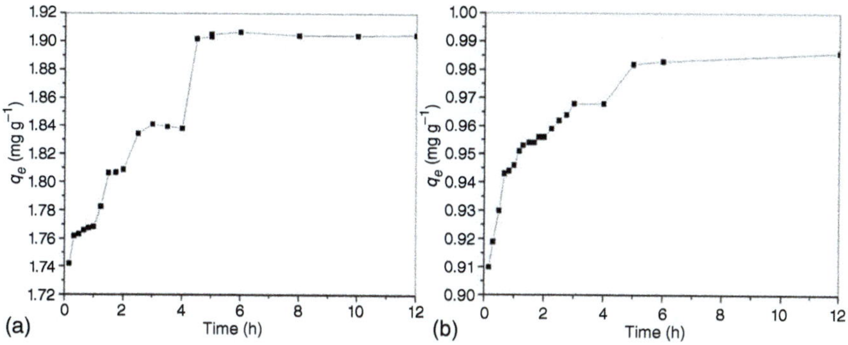

Figure 2.12 Effect of agitation time on fluoride adsorption onto (a) ZICFC and (b) ZICSC at room temperature and pH 4. Source: Sathish et al. [65, 66]/Taylor & Francis.

sites on ZICFC and ZICSC surface between OH^- and F^- [65, 66]. It is important to note that for a wide pH window, i.e. from strongly acidic to weakly basic conditions the samples under study displayed excellent performance. Hence, both ZICFC and ZICSC are promising adsorbent material for defluoridation where no pH control required in real time application to water treatment [65, 66].

2.9.2 Effect of Agitation Time

The effect of agitation time on the adsorption of fluoride ions on the adsorbent for ZICFC and ZICSC are shown in Figure 2.12a,b, respectively. A great extent of fluoride ion affinity and adsorption was observed with q_e of 1.77 and 0.94 mg g^{-1} (attained in one hour) for ZICFC and ZICSC, respectively [65, 66]. The extent of removal of fluoride ions from water increases with an increase in agitation time. However, this trend was observed till a particular limit where the available adsorption sites on the adsorbent are utilized. After this particular time of six hours the increase in agitation time does not bring about increased fluoride uptake. Hence, an initial rapid adsorption successively resulted in equilibrium and saturation, on account of deficiency of accessible active sites on the adsorbent surface at six hours, which was considered as the equilibrium time [65, 66].

2.9.3 Effect of Adsorbent Dose

The adsorption of fluoride as a function of ZICSC dosage for ZICFC and ZICSC is shown in Figure 2.13a,b, respectively. The results presented in Figure 2.13 specify that as the dosage concentration is increased, the efficiency of fluoride removal increases on account of the increase in surface area or adsorption sites. However, this increase was observed until an optimum dosage was reached (20 and 10 g l^{-1} for ZICFC and ZICSC, respectively) [65, 66]. After this point, the fluoride uptake has no noteworthy change with ZICFC or ZICSC dosage, and the adsorption density decreases. With these studies, the optimum conditions required for effective defluoridation of fluoride-rich water were obtained as 10 and 20 g l^{-1} of ZICFC and ZICSC adsorbent dosages with a pH of four and six hours of agitation [65, 66].

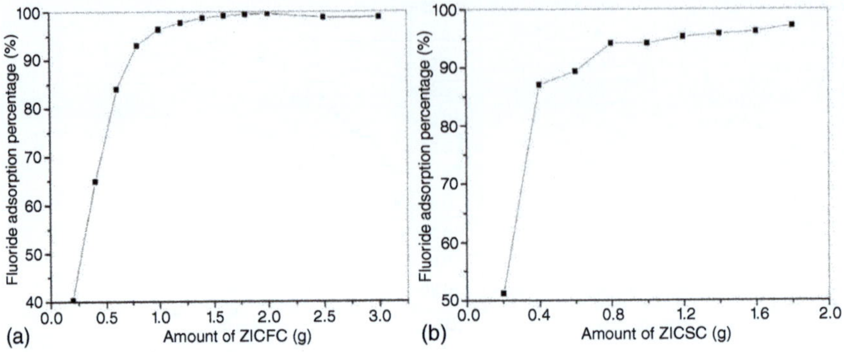

Figure 2.13 Effect of adsorbent dose on fluoride adsorption onto (a) ZICFC and (b) ZICSC at room temperature and pH 4, with an agitation time of six hours. Source: Sathish et al. [65, 66]/Taylor & Francis.

2.10 Adsorption Isotherms

Langmuir, Freundlich, Redlich–Peterson, and Temkin sorption isotherm models were used to test the experimental data for fluoride adsorption on ZICFC and ZICSC, and an elaborate discussion is presented in earlier works. In this chapter, we present the highlights of these results, providing a comprehensive comparative analysis. The parameter values of the Langmuir and Freundlich isotherms obtained with the use of ZICFC and ZICSC are presented in Tables 2.1 and 2.2, respectively. It is observed that the correlation coefficient R^2 value for the Langmuir model is near unity for fluoride uptake by ZICFC as well as ZICSC. Hence, it can be concluded that defluoridation process follows the Langmuir isotherm with a monolayer coverage and constant adsorption energy [65, 85] (Figure 2.14).

Table 2.1 Langmuir and Freundlich isotherm parameters for Fluoride uptake by ZICFC.

pH	Langmuir parameters			Freundlich parameters		
	q_m (mg g^{-1})	b (1 mg^{-1})	R^2	k_f (mg g^{-1})	n	R^2
2.0	16.026	0.003	0.999	0.068	1.132	0.999
4.0	40.016	0.001	0.999	0.061	1.073	0.997
6.0	20.109	0.003	0.999	0.069	1.124	0.997
8.0	8.844	0.005	0.997	0.077	1.231	0.994
10.0	8.565	0.003	0.964	0.057	1.270	0.949

Source: Reproduced with permission from Sathish et al. [66], © 2008, Taylor & Francis.

Table 2.2 Langmuir and Freundlich isotherm parameters for Fluoride uptake by ZICSC.

pH	Langmuir parameters			Freundlich parameters		
	q_m (mg g^{-1})	b (1 mg^{-1})	R^2	k_f (mg g^{-1})	n	R^2
2.0	6.77	0.02	0.98	0.20	0.69	0.93
4.0	6.41	0.02	0.98	0.22	0.67	0.94
6.0	7.51	0.02	0.98	0.18	0.71	0.94
8.0	7.51	0.01	0.97	0.15	0.75	0.94
10.0	6.64	0.01	0.97	0.17	0.70	0.92

Source: Reproduced with permission from Sathish et al. [65], © 2007, Taylor & Francis.

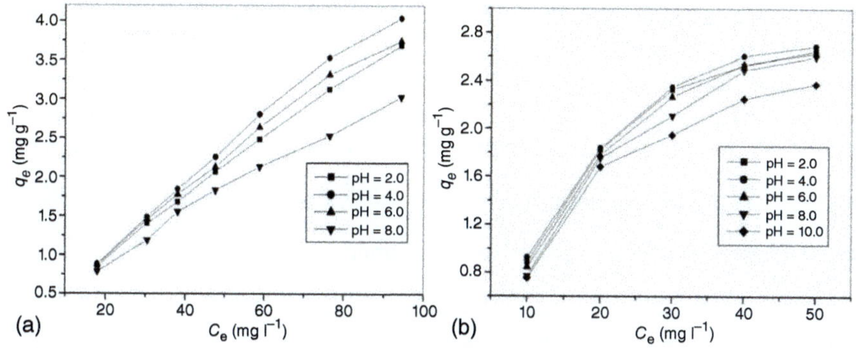

Figure 2.14 Adsorption isotherms of (a) ZICFC and (b) ZICSC as a function of pH values at room temperature. Source: Sathish et al. [65, 66]/Taylor & Francis.

2.11 Adsorption Kinetics

The kinetics of sorption, describing the solute uptake rate, govern the residence time of the sorption reaction. Hence, kinetics is an important feature defining the efficiency of sorption. In order to understand the behavior of ZICFC and ZICSC, we have performed the kinetics of fluoride removal. This is studied by testing the kinetics of fluoride adsorption on ZICFC and ZICSC using the first order rate equation, pseudo-first-order rate equation, and pseudo-second-order rate equation, and the details are presented in the earlier publications. In summary, it was observed that the kinetics of fluoride adsorption on both adsorbents were best described by a pseudo-second-order model [86, 87].

In this regard, the adsorbent (ZICFC or ZICSC)-fluoride ion reaction may be represented as:

$$ZrO(OH)Cl(\text{adsorbed on CFC or CSC}) + 2F^- \Leftrightarrow ZrOF_2 \tag{2.1}$$

For a pseudo-second-order reaction, an integrated rate law may be derived centered on the fluoride adsorption equilibrium as follows:

$$t/q_t = (1/k_2 q_e^2) + (1/q_e)t$$

where k_2 is the rate constant for sorption (g mg^{-1} min); h is the initial sorption rate and is defined as $k_2 q_e^2$. The applicability of the pseudo-second-order kinetic model was confirmed by the linear graphs obtained by plotting (t/q_t) vs. t at various fluoride concentrations, which are elaborately captured in the earlier works [65, 66]. The k_2 and q_e values are obtained as the intercept and slope, respectively. Further, the k_2 and h values of the pseudo-second-order reaction are presented in Tables 2.3 and 2.4 for ZICFC and ZICSC, respectively, with correlation coefficient R^2 values for this model being near unity. Therefore, we conclude that the defluoridation process using ZICFC and ZICSC follows the pseudo-second-order as compared to pseudo-first-order and first order kinetic models. The large values of h indicate that the initial fluoride adsorption rate was exceptionally fast, in line with our understanding from experimental observations. The process of fluoride uptake by ZICFC (93%) and ZICSC (91%) was rapid within 10 minutes of contact at the optimum pH of 4.0 [65, 66].

Table 2.3 Pseudo-second-order parameters for fluoride adsorption by ZICFC.

Fluoride (mg l^{-1})	k_2 (g mg^{-1} h^{-1})	h (mg g^{-1})	R^2
20	383.75	361.01	0.99
30	365.14	757.58	1.00
40	176.16	666.67	1.00
50	201.80	1195.55	1.00
60	50.68	401.61	0.99

Source: Reproduced with permission from Sathish et al. [66], © 2008, Taylor & Francis.

Table 2.4 Pseudo-second-order kinetic model parameters for fluoride adsorption by ZICSC.

Fluoride (mg l^{-1})	k_2 (g mg^{-1} h^{-1})	h (mg g^{-1})	R^2
10	214.676	186.553	0.999
20	17.176	62.385	0.994
30	12.119	73.585	0.994
40	17.082	124.692	0.998
50	19.036	145.984	0.999

Source: Reproduced with permission from Sathish et al. [65], © 2007, Taylor & Francis.

From the extensive experimentation and analysis (kinetic modeling and adsorption studies) presented above, it may be concluded that the fluoride exclusion process using both ZICFC and ZICSC follows Langmuir adsorption isotherms and pseudo-second-order kinetics [65, 66]. Moreover, based on earlier works, it can be mentioned that the defluoridation process follows a synergy of chemisorption and physisorption processes via monolayer formation in agreement with the Langmuir isotherm process [65, 66]. As both CSC and coconut fiber carbon (CFC) present numerous Lewis acid sites, the fluoride ions, being the Lewis base, coordinate markedly with the zirconium species that are previously adsorbed on CSC. In the chemisorption mechanism involved, exchange of the hydroxide and chloride of the ZrO(OH)Cl species adsorbed on CSC and CFC occurs. This leads to the formation of $ZrOF_2$ (as presented in Eq. (2.1)) and is hence a monolayer surface adsorption phenomenon. It must be noted that the zirconium ions that are adsorbed on the CSC and CFC are also monolayers. In this context, after the chemisorption of monolayer of fluoride ions on the CSC and CFC, all the active sites are completely occupied, leaving no sites for further adsorption of fluoride ions [65, 66]. On account of the Van der Waals interaction between the fluoride ion and activated carbon, it is expected that the physisorption process may be followed by the chemisorption process. This physisorption process is propelled by the intraparticle diffusion of the hydrated fluoride ion (3.52 A° in size) into the pores of ZICFC (which varies from 5 to 10 A°) and ZICSC (which varies from 30 to 300 A°) [65, 66]. These results confirm our earlier understanding that although the fluoride adsorption by ZICFC and ZICSC was fast with q_e at 1.74 and 0.91 mg g^{-1} respectively, within 10 minutes of the contact, subsequently with the increase in time an equilibrium and saturation was observed only at six hours [65, 66]. This indicates that the diffusion process is be present in the rate limiting step of the process under study. Furthermore, the electrostatic forces of repulsion between the existing physisorbed fluoride ions and subsequently adsorpted ions once again ensures the formation of monolayer.

The materials under study, such as ZICSC and ZICFC, were regenerated by eluting a 0.02 M sodium hydroxide solution. This assists in quantitatively desorbing all fluoride without damaging the original zirconium species adsorbed on CSC and CFC. It is interesting to note that the impregnated charcoal showed comparable results after four regenerations, in accordance with the first set of defluoridation experiments. The developed economical adsorbents can be discarded safely and effortlessly, as they can be used as filler material in landfills and also in the fabrication of bricks [65, 66].

2.12 Concluding Remarks

In summary, this chapter provides a comprehensive analysis of the need for sensing ions and molecules in drinking water as well as discusses certain strategies to remove hazardous ions such as fluoride ions. To begin with various causes and effects of water pollution are discussed along with different types involved. SPCE and PCCE substrates are introduced as effective substrates for detection of ions and molecules in drinking water samples. Different nanointerfaces are discussed, providing insights into the plasmonic coupling phenomenon. Case studies are discussed with regard to hazardous TA, representing an example of a molecule. Case studies with regard to ions are performed using Cu^{2+}, Zn^{2+}, Hg^{2+}, Al^{3+}, and I^- in drinking water using SPCE and PCCE platforms. The experiments demonstrated excellent sensitivity with reliable and reproducible results. The radiative decay life of the fluorescent molecules was modulated by incorporating different nanomaterials into different nanointerfaces. The developed nano-engineered SPCE and PCCE substrates with sharp-edged nanointerfaces offer newer opportunities for the design of biosensors for environmental water monitoring. The proposed methodologies offer several advantages, including low cost, user-eco-friendly, convenient, and fast identification of bioanalytes of interest.

In the second part of the chapter, we focused on the ability of ZICSC and ZICFC materials to serve as effective adsorbents for removal of fluoride. Extensive analysis with regard to batch equilibrium and kinetic experiments is presented, along with the study of various guiding parameters such as pH, agitation time, and adsorbent dosage. The excellent defluoridation properties of ZICFC and ZICFC are attributed to the chemisorption process occurring in combination with physisorption and intraparticle diffusion. Furthermore, the processes followed Langmuir adsorption isotherm and the pseudo-second-order chemical reaction kinetics. It was observed that ZICSC sample could be utilized for preconcentration of phosphate, arsenic, selenium, mercury, and chromium(VI) and also for fluoride rich water samples. In summary, it was observed that the utility of ZICFC exhibited comparatively higher performance as compared to earlier works with alumina, chitin, and phosphorus-free carbon [65, 66] as well as superior performance as compared to ZICSC [66]. The experimental data presented in this chapter with effective comparative analysis is believed to be useful for development of an appropriate technology in the planning and execution of treatment plants for water bodies rich in fluoride content. We firmly believe that the concepts developed and the results discussed in this chapter would be of immediate relevance for researchers developing biosensing frameworks for detection of hazardous ions and molecules in drinking water as well as for water purification, with an emphasis on relevance in developing countries.

Consent for Publication

Authors provide consent for the publication.

Acknowledgment

The authors would like to acknowledge support from Tata Education and Development Trust (TEDT/MUM/HEA/SSSIHL/2017–2018/0069-RM-db), Prasanthi Trust, Inc., U.S. (22-06-2018), DST-Technology Development Program (IDP/MED/

19/2016), Life Sciences Research Board (LSRB), and DST-Inspire Fellowship (IF180392), Government of India. SB thanks Prof. Brian T. Cunningham, Department of Electrical and Computer Engineering, USA, Illinois Cancer Center, USA, Woese Institute for Genomic Biology, USA, Holonyak Micro and Nanotechnology Laboratory, USA, University of Illinois at Urbana-Champaign, USA for research support and guidance. S.B. is supported by a postdoctoral fellowship from the Woese Institute for Genomic Biology, USA. We especially acknowledge SSSIHL-CRIF for extending the usage of the required instrumentation facility. Guidance from Bhagawan Sri Sathya Sai Baba is gratefully acknowledged.

Conflict of Interest

Authors have no conflict of interest

References

1 Guo, R. (2021). *Cross-Border Resource Management*. Elsevier.
2 Levin, S.A. (2013). *Encyclopedia of Biodiversity*. Elsevier Inc.
3 Sivaperuman, C., Velmurugan, A., Singh, A.K., and Jaisankar, I. (ed.) (2018). *Biodiversity and Climate Change Adaptation in Tropical Islands*. Academic Press.
4 Goel, P.K. (2006). *Water Pollution: Causes, Effects and Control*. New Age International.
5 Abel, P.D. (2002). *Water Pollution Biology*. CRC Press.
6 Wang, Z., Walker, G.W., Muir, D.C., and Nagatani-Yoshida, K. (2020). Toward a global understanding of chemical pollution: a first comprehensive analysis of national and regional chemical inventories. *Environ. Sci. Technol.* 54 (5): 2575–2584.
7 Patel, M., Kumar, R., Kishor, K. et al. (2019). Pharmaceuticals of emerging concern in aquatic systems: chemistry, occurrence, effects, and removal methods. *Chem. Rev.* 119 (6): 3510–3673.
8 Peirce, J.J., Vesilind, P.A., and Weiner, R. (1998). *Environmental Pollution and Control*. Butterworth-Heinemann.
9 Richardson, S.D. (2009). Water analysis: emerging contaminants and current issues. *Anal. Chem.* 81 (12): 4645–4677.
10 Xue, X.Y., Cheng, R., Shi, L. et al. (2017). Nanomaterials for water pollution monitoring and remediation. *Environ. Chem. Lett.* 15 (1): 23–27.
11 Soni, R., Soni, M., and Shukla, D.P. (2020). Emerging techniques and materials for water pollutants detection. In: *Sensors in Water Pollutants Monitoring: Role of Material* (ed. D. Pooja, P. Kumar, P. Singh, and S. Patil), 277–297. Singapore: Springer.
12 Teodoro, K.B., Sanfelice, R.C., Migliorini, F.L. et al. (2021). A review on the role and performance of cellulose nanomaterials in sensors. *ACS Sens.* 6 (7): 2473–2496.

13 Walekar, L., Dutta, T., Kumar, P. et al. (2017). Functionalized fluorescent nanomaterials for sensing pollutants in the environment: a critical review. *TrAC Trends Anal. Chem.* 97: 458–467.

14 Aragay, G., Pino, F., and Merkoçi, A. (2012). Nanomaterials for sensing and destroying pesticides. *Chem. Rev.* 112 (10): 5317–5338.

15 Bhaskar, S., Singh, A.K., Das, P. et al. (2020). Superior resonant nanocavities engineering on the photonic crystal-coupled emission platform for the detection of femtomolar iodide and zeptomolar cortisol. *ACS Appl. Mater. Interfaces* 12 (30): 34323–34336.

16 Bhaskar, S., Das, P., Srinivasan, V. et al. (2022). Plasmonic-silver sorets and dielectric-Nd_2O_3 nanorods for ultrasensitive photonic crystal-coupled emission. *Mater. Res. Bull.* 145: 111558.

17 Bhaskar, S., Das, P., Srinivasan, V. et al. (2020). Bloch surface waves and internal optical modes-driven photonic crystal-coupled emission platform for femtomolar detection of aluminum ions. *J. Phys. Chem. C* 124 (13): 7341–7352.

18 Bhaskar, S., Das, P., Moronshing, M. et al. (2021). Photoplasmonic assembly of dielectric-metal, Nd_2O_3-gold soret nanointerfaces for dequenching the luminophore emission. *Nanophotonics* 10 (13): 3417–3431.

19 Nair, R.V. and Vijaya, R. (2010). Photonic crystal sensors: an overview. *Progr. Quantum Electron.* 34 (3): 89–134.

20 Dodero, A., Lova, P., Vicini, S. et al. (2020). Sodium alginate cross-linkable planar 1d photonic crystals as a promising tool for Pb^{2+} detection in water. *Chemosensors* 8 (2): 37.

21 Mou, F.A., Rahman, M.M., Islam, M.R., and Bhuiyan, M.I.H. (2020). Development of a photonic crystal fiber for THz wave guidance and environmental pollutants detection. *Sens. Bio-Sens. Res.* 29: 100346.

22 Asher, S.A., Sharma, A.C., Goponenko, A.V., and Ward, M.M. (2003). Photonic crystal aqueous metal cation sensing materials. *Anal. Chem.* 75 (7): 1676–1683.

23 Lakowicz, J.R., Ray, K., Chowdhury, M. et al. (2008). Plasmon-controlled fluorescence: a new paradigm in fluorescence spectroscopy. *Analyst* 133 (10): 1308–1346.

24 Geddes, C. D. (Ed.). (2017). *Surface Plasmon Enhanced, Coupled and Controlled Fluorescence.* Wiley.

25 Li, J.F., Li, C.Y., and Aroca, R.F. (2017). Plasmon-enhanced fluorescence spectroscopy. *Chem. Soc. Rev.* 46 (13): 3962–3979.

26 Dutta Choudhury, S., Badugu, R., and Lakowicz, J.R. (2015). Directing fluorescence with plasmonic and photonic structures. *Acc. Chem. Res.* 48 (8): 2171–2180.

27 Gryczynski, I., Malicka, J., Gryczynski, Z., and Lakowicz, J.R. (2004). Radiative decay engineering 4. Experimental studies of surface plasmon-coupled directional emission. *Anal. Biochem.* 324 (2): 170–182.

28 Weng, Y.H., Xu, L.T., Chen, M. et al. (2019). In situ monitoring of fluorescent polymer brushes by angle-scanning based surface plasmon coupled emission. *ACS Macro Lett.* 8 (2): 223–227.

29 Badugu, R., Szmacinski, H., Ray, K. et al. (2015). Metal–dielectric waveguides for high-efficiency coupled emission. *ACS Photon.* 2 (7): 810–815.

30 Zhao, Y., Liu, Y.H., Cao, S.H. et al. (2020). Excitation–emission synchronization-mediated directional fluorescence: insight into plasmon-coupled emission at vibrational resolution. *J. Phys. Chem. Lett.* 11 (7): 2701–2707.

31 Rangełowa-Jankowska, S., Jankowski, D., Bogdanowicz, R. et al. (2012). Surface plasmon-coupled emission of rhodamine 110 aggregates in a silica nanolayer. *J. Phys. Chem. Lett.* 3 (23): 3626–3631.

32 Matveeva, E., Gryczynski, Z., Gryczynski, I. et al. (2004). Myoglobin immunoassay utilizing directional surface plasmon-coupled emission. *Anal. Chem.* 76 (21): 6287–6292.

33 Pan, X.H., Cao, S.H., Chen, M. et al. (2020). In situ and sensitive monitoring of configuration-switching involved dynamic adsorption by surface plasmon-coupled directional enhanced Raman scattering. *Phys. Chem. Chem. Phys.* 22 (22): 12624–12629.

34 Koo, W.H., Jeong, S.M., Nishimura, S. et al. (2011). Polarization conversion in surface-plasmon-coupled emission from organic light-emitting diodes using spontaneously formed buckles. *Adv. Mater.* 23 (8): 1003–1007.

35 Dutta Choudhury, S., Badugu, R., Ray, K., and Lakowicz, J.R. (2013). Steering fluorescence emission with metal-dielectric-metal structures of Au, Ag, and Al. |*J. Phys. Chem. C* 117 (30): 15798–15807.

36 Badugu, R., Descrovi, E., and Lakowicz, J.R. (2014). Radiative decay engineering 7: Tamm state-coupled emission using a hybrid plasmonic–photonic structure. *Anal. Biochem.* 445: 1–13.

37 Jahani, S. and Jacob, Z. (2016). All-dielectric metamaterials. *Nat. Nanotechnol.* 11 (1): 23–36.

38 Bhaskar, S., Moronshing, M., Srinivasan, V. et al. (2020). Silver soret nanoparticles for femtomolar sensing of glutathione in a surface plasmon-coupled emission platform. *ACS Appl. Nano Mater.* 3 (5): 4329–4341.

39 Bhaskar, S., Kowshik, N.C.S., Chandran, S.P., and Ramamurthy, S.S. (2020). Femtomolar detection of spermidine using au decorated SiO_2 nanohybrid on plasmon-coupled extended cavity nanointerface: a smartphone-based fluorescence dequenching approach. *Langmuir* 36 (11): 2865–2876.

40 Bhaskar, S., Patra, R., Kowshik, N.C.S. et al. (2020). Nanostructure effect on quenching and dequenching of quantum emitters on surface plasmon-coupled interface: a comparative analysis using gold nanospheres and nanostars. *Physica E* 124: 114276.

41 Bhaskar, S., Visweswar Kambhampati, N.S., Ganesh, K.M. et al. (2021). Metal-free, graphene oxide-based tunable soliton and plasmon engineering for biosensing applications. *ACS Appl. Mater. Interfaces* 13 (14): 17046–17061.

42 Rai, A., Bhaskar, S., and Ramamurthy, S.S. (2021). Plasmon-coupled directional emission from soluplus-mediated AgAu nanoparticles for attomolar sensing using a smartphone. *ACS Appl. Nano Mater.* 4 (6): 5940–5953.

43 Rai, A., Bhaskar, S., Ganesh, K.M., and Ramamurthy, S.S. (2021). Engineering of coherent plasmon resonances from silver soret colloids, graphene oxide and Nd_2O_3 nanohybrid architectures studied in mobile phone-based surface plasmon-coupled emission platform. *Materials Lett.* 304: 130632.

44 Bhaskar, S. and Ramamurthy, S.S. (2021). Synergistic coupling of titanium carbonitride nanocubes and graphene oxide for 800-fold fluorescence enhancements on smartphone based surface plasmon-coupled emission platform. *Materials Lett.* 298: 130008.

45 Rai, A., Bhaskar, S., Ganesh, K.M., and Ramamurthy, S.S. (2022). Gelucire®-mediated heterometallic AgAu nanohybrid engineering for femtomolar cysteine detection using smartphone-based plasmonics technology. *Mater. Chem. Phys.* 279: 125747.

46 Bhaskar, S. and Ramamurthy, S.S. (2021). High refractive index dielectric TiO_2 and graphene oxide as salient spacers for >300-fold enhancements. In: *In 2021 IEEE International Conference on Nanoelectronics, Nanophotonics, Nanomaterials, Nanobioscience & Nanotechnology (5NANO)*, 1–6. IEEE.

47 Rai, A., Bhaskar, S., Battampara, P. et al. (2022). Integrated photo-plasmonic coupling of bioinspired sharp-edged silver nano-particles with nano-films in extended cavity functional interface for cellphone-aided femtomolar sensing. *Materials Lett.* 316: 132025.

48 Rai, A., Bhaskar, S., Ganesh, K.M., and Ramamurthy, S.S. (2022). Cellphone-based attomolar tyrosine sensing based on Kollidon-mediated bimetallic nanorod in plasmon-coupled directional and polarized emission architecture. *Mater. Chem. Phys.* 285: 126129.

49 Bhaskar, S. and Ramamurthy, S.S. (2019). Mobile phone-based picomolar detection of tannic acid on Nd_2O_3 nanorod–metal thin-film interfaces. *ACS Appl. Nano Mater.* 2 (7): 4613–4625.

50 Bhaskar, S., Jha, P., Subramaniam, C., and Ramamurthy, S.S. (2021). Multifunctional hybrid soret nanoarchitectures for mobile phone-based picomolar Cu^{2+} ion sensing and dye degradation applications. *Physica E* 132: 114764.

51 Rai, A., Bhaskar, S., Reddy, N., and Ramamurthy, S.S. (2021). Cellphone-aided attomolar zinc ion detection using silkworm protein-based nanointerface engineering in a plasmon-coupled dequenched emission platform. *ACS Sustain. Chem. Eng.* 9 (44): 14959–14974.

52 Rathnakumar, S., Bhaskar, S., Rai, A. et al. (2021). Plasmon-coupled silver nanoparticles for mobile phone-based attomolar sensing of Mercury ions. *ACS Appl. Nano Mater.* 4 (8): 8066–8080.

53 Moronshing, M., Bhaskar, S., Mondal, S. et al. (2019). Surface-enhanced Raman scattering platform operating over wide pH range with minimal chemical enhancement effects: test case of tyrosine. *J. Raman Spectrosc.* 50 (6): 826–836.

54 Arathi, P.J., Bhaskar, S., and Ramanathan, V. (2018). The photocatalytic role of electrodeposited copper on pencil graphite. *Phys. Chem. Chem. Phys.* 20 (5): 3430–3432.

55 Arathi, P.J., Seemesh, B., and Ramanathan, V. (2018). Disulphide linkage: to get cleaved or not? Bulk and nano copper based SERS of cystine. *Spectrochim. Acta, Part A* 196: 229–232.

56 Ramachandran, K., Daoudi, K., Columbus, S. et al. (2022). Green production of self-assembled silver nanoarrays on flexible substrate for direct detection and catalytic degradation of organic water pollutants. *Environ. Technol. Innov.* 27: 102409.

57 De, A. and Kalita, D. (2021). Bio-fabricated gold and silver nanoparticle based plasmonic sensors for detection of environmental pollutants: an overview. *Crit. Rev. Anal. Chem.* 53: 1–17.

58 Li, L. and Chin, W.S. (2021). Rapid and sensitive SERS detection of melamine in milk using Ag nanocube array substrate coupled with multivariate analysis. *Food Chem.* 357: 129717.

59 Barveen, N.R., Wang, T.J., and Chang, Y.H. (2020). In-situ deposition of silver nanoparticles on silver nanoflowers for ultrasensitive and simultaneous SERS detection of organic pollutants. *Microchem. J.* 159: 105520.

60 Kumar, G. and Soni, R.K. (2020). Silver nanocube-and nanowire-based SERS substrates for ultra-low detection of PATP and thiram molecules. *Plasmonics* 15 (6): 1577–1589.

61 Ayoob, S. and Gupta, A.K. (2006). Fluoride in drinking water: a review on the status and stress effects. *Crit. Rev. Environ. Sci. Technol.* 36 (6): 433–487.

62 Maria, A. (2003). The costs of water pollution in India. In: *Conference on Market Development of Water and Waste Technologies through Environmental Economics*, CERNA, Ecole Nationale Supérieure des Mines de Paris, Paris, France, pp. 30, 31.

63 Expert Committee, W.H.O. and on Oral Health Status, & Fluoride Use. (1994). Fluorides and oral health. In: *Report of a WHO Expert Committee on Oral Health Status and Fluoride Use*, vol. 846 (ed. W.H.O.E. Committee). Switzerland: World Health Organization.

64 Abe, I., Iwasaki, S., Tokimoto, T. et al. (2004). Adsorption of fluoride ions onto carbonaceous materials. *J. Colloid Interface Sci.* 275 (1): 35–39.

65 Sathish, R.S., Raju, N.S.R., Raju, G.S. et al. (2007). Equilibrium and kinetic studies for fluoride adsorption from water on zirconium impregnated coconut shell carbon. *Sep. Sci. Technol.* 42 (4): 769–788.

66 Sathish, R.S., Sairam, S., Raja, V.G. et al. (2008). Defluoridation of water using zirconium impregnated coconut fiber carbon. *Sep. Sci. Technol.* 43 (14): 3676–3694.

67 Janardhana, C., Rao, G.N., Sathish, R.S. et al. (2007). Study on defluoridation of drinking water using zirconium ion impregnated activated charcoals. *Indian J. Chem. Technol. (IJCT)* 14: 350–354.

68 Janardhana, C., Rao, G.N., Sathish, R.S., and Lakshman, V.S. (2006). Study on defluoridation of drinking water by impregnation of metal ions in activated charcoal. *Indian J. Chem. Technol. (IJCT)* 13: 414–416.

69 Ramamurthy, S.S., Chen, Y., Kalyan, M.K. et al. (2011). Carbon nanotube-zirconium dioxide hybrid for defluoridation of water. *J. Nanosci. Nanotechnol.* 11 (4): 3552–3559.

70 Susheela, A.K. (2003). *Fluorosis Research and Rural Development Foundation, A Treatise on Fluorosis, Revised*, 2e. India: New Delhi.

71 Saha, S. (1993). Treatment of aqueous effluent for fluoride removal. *Water Res.* 27 (8): 1347–1350.

72 Ruiz, T., Persin, F., Hichour, M., and Sandeaux, J. (2003). Modelisation of fluoride removal in Donnan dialysis. *J. Membr. Sci.* 212 (1, 2): 113–121.

73 Kéba Diawara, C., Paugam, L., Pontié, M. et al. (2005). Influence of chloride, nitrate, and sulphate on the removal of fluoride ions by using nanofiltration membranes. *Sep. Sci. Technol.* 40 (16): 3339–3347.

74 Sahli, M.M., Annouar, S., Tahaikt, M. et al. (2007). Fluoride removal for underground brackish water by adsorption on the natural chitosan and by electrodialysis. *Desalination* 212 (1–3): 37–45.

75 Kumar, S. (2018). Environmental risk factor linked to the genesis of lung cancer–a review. *Indian J. Environ. Health* 4 (2): 50.

76 Arullanatham, A.J., Ramakrishna, T.V., and Balasubhramaniam, N. (1992). Study on defluoridation of drinking water by impregnation of metal ions in activated charcoal. *Indian J. Environ. Health* 12 (7): 531.

77 Muthukumaran, K., Balasubramanian, N., and Ramakrishna, T.V. (1995). Removal of fluoride by chemically activated carbon. *Indian J. Environ. Prot.* 15 (7): 514–517.

78 Kamble, S.P., Jagtap, S., Labhsetwar, N.K. et al. (2007). Defluoridation of drinking water using chitin, chitosan and lanthanum-modified chitosan. *Chem. Eng. J.* 129 (1-3): 173–180.

79 Sathish, R.S., Sujith, U., Rao, G.N., and Janardhana, C. (2006). Fluoride ion detection by 8-hydroxyquinoline–Zr (IV)–EDTA complex. *Spectrochimica Acta, Part A* 65 (3, 4): 565–570.

80 Ramos, R.L., Ovalle-Turrubiartes, J., and Sanchez-Castillo, M.A. (1999). Adsorption of fluoride from aqueous solution on aluminum-impregnated carbon. *Carbon* 37 (4): 609–617.

81 Seethapathirao, D. (1964). Defluoridation of water using sulphonated coconut shell carbon. *Indian J. Environ. Health.* 64: 11–12.

82 Hashitani, H., Okumura, M., and Fujinaga, K. (1987). Preconcentration method for phosphate in water using activated carbon loaded with zirconium. *Fresenius' Zeitschrift für analytische Chemie* 326 (6): 540–542.

83 Peräniemi, S. and Ahlgrén, M. (1995). Optimized arsenic, selenium and mercury determinations in aqueous solutions by energy dispersive X-ray fluorescence after preconcentration onto zirconium-loaded activated charcoal. *Anal. Chim. Acta* 302 (1): 89–95.

84 Peräniemi, S. and Ahlgrén, M. (1995). Separation of microgram quantities of Cr (III) and Cr (VI) in aqueous solutions and determination by energy dispersive X-ray fluorescence spectrometry. *Anal. Chim. Acta* 315 (3): 365–370.

85 Gopal, V. and Elango, K.P. (2007). Equilibrium, kinetic and thermodynamic studies of adsorption of fluoride onto plaster of Paris. *J. Hazardous Mater.* 141 (1): 98–105.

86 Quek, S.Y., Wase, D.A.J., and Forster, C.F. (1998). The use of sago waste for the sorption of lead and copper. *Water S.A* 24 (3): 251–256.

87 Ho, Y.S. and McKay, G. (1999). Pseudo-second order model for sorption processes. *Process Biochem.* 34 (5): 451–465.

3

Nanomaterials for Air and Water Purification

Lívia Elisabeth Vasconcellos de Siqueira Brandão Vaz

Federal University of Lavras - Engineering School (EE), Engineering Department, Lavras, MG 37200-900, Brazil

3.1 Introduction

Air and water are essential elements for the occurrence and continuity of life, including humans, plants, and animals. With the increasing needs of modern society and the advancement of technical knowledge from the beginnings to the present day, air and water are increasingly demanded. In the last decades, a lot of focus has been given to the development of technologies, whether in the academic or industrial environment, that allow it to be possible: (i) the use and/or exploitation of such elements in a conscious and sustainable way, not making them unfeasible, mainly because it is essential to survival; and (ii) recover damages already caused by such means, whether through misuse, pollution, accidents, or catastrophes, among others.

One of the concepts of technology is the application of scientific knowledge to production in general. Such applications can take place in the most different sectors and with the most diverse tasks. It is possible that in different everyday situations, the population in general does not perceive or know about the use of some type of technology that enables what is seen, acquired, or used. The simple use of natural resources and the transformation of the environment in their favor, for example, is capable of being considered a technological movement that impacts not only the sector of direct use of the generated product, as well as the entire economy worldwide. At this point, it is worth noting how technological our society is and how much we depend on technologies and natural resources to maintain or advance the standard of living that we enjoy today. Virtually everything produced or used has a certain energy demand level associated with it. This demand is met through the exploitation of natural resources, renewable or not, regardless of whether it requires more or less technology. The fact is that there is a great dependence on natural resources so that we can perform everything from simple actions, such as brushing our teeth, to more elaborate functions such as moving from one city to another. Not to mention our body's energy needs, supplied by food in a natural or industrialized way. In

Nanomaterials for Air and Water Purification, First Edition.
Edited by Deepu A. Gopakumar, Jyotishkumar Parameswaranpillai, Jinu Jacob George, and Midhun Dominic C.D.

other words, currently living implies demanding for the most varied purposes. And demand does not decrease; it only increases.

Therefore, it is necessary that we develop technological ways to meet the needs of modern society and to satisfy the demand for energy, but also that we develop means to conserve, reuse, and renew the resources that allow us to enjoy life itself. This chapter fits into that context.

3.2 Nanoscience

3.2.1 Historical Contextualization of Technological Development

The development of materials engineering came to directly affect the development of technologies that satisfy the high demands of society. When observing the highlights of the scale of technological evolution of materials since the beginning, the Stone Age itself (approximately 10 000 BCE) is considered the first technological milestone since stones, skins, wood, and clay were used as a means of developing tooling to meet basic needs such as heating, feeding, and protection. With the continuity of human development and the addition of a historical and geographical context, the needs were changing and increasing.

When techniques of production and alteration of the properties of the materials were discovered, such as applying heat treatments, or adding other substances, ceramics (from the Greek *keramicos*, which means burnt clay) and metals such as gold, copper, or metal alloys such as bronze, became used. On the scale of technological evolution, this period relates to the Bronze Age (approximately 4000 BCE) and the Iron Age (1 onwards). When entering the Medieval Age, the beginning of the use of presses could be identified, and textile, military technologies and those allied to engineering start to be advanced in function of the accentuated maritime exploration of the time, which culminated in the discovery of other continents and exploitation of the resources typical of each region of the planet.

Another important highlight occurred during the industrial revolution, which led to an explosion in the technological world. During this period, society experienced a period of transition toward new manufacturing processes, production by machines, manufacturing of new chemicals, new iron production processes, greater water energy efficiency, increasing use of steam power, and the replacement of wood and other biofuels with coal. After this phase, new materials were discovered, as well as processes of transformation of materials and products, culminating in a sharp development of the industrial sector and an engineering advance. Studies related to the understanding of nature and of the physical/chemical phenomena that involve matter were also conducted, and scientific-technological knowledge evolved considerably. This technological evolution finds support in the different episodes of history that made possible new discoveries or even new questions, for example, in the needs of facing world wars or even in the new needs of a post-war world. Improvements in the industry of metallic and ceramic materials took place, but concomitantly, the development of a breakthrough in polymer technology emerged. A new industry, with a new type of material, rises and this revolutionizes the entire range of devices hitherto existing.

During the onset of the digital age, with a growing list of materials and innovative production and transformation processes, people began to look for new strategies to meet energy needs, increase efficiency, improve products, and conserve resources. So, nature began to be observed again, and ways of mimetization started to be researched to improve engineering processes. With more powerful instruments for synthesis and analysis of materials, new phenomena and characteristics of matter were revealed. A new industrial sector was settled by working the material on the microscale, which strengthened the effective establishment of the digital age and allowed a rapid leap from giant computers, which occupied an entire room, to portable computers and, in continuity, mobile phones, cell phones, tablets, and smart watches. From the microelectronics industry, with the support of computer science and engineering, advances are made in control and automation, electromechanics, and the possibility of working with materials at the nanoscale (10^{-9} m). This revolutionizes technology once again, introducing a significant milestone in technological development, opening up a whole new universe of work possibilities, ranging from studying the physico chemical changes that occur in materials when they are structured at the nanoscale to using them to the development of prototypes.

Currently, we have a vast application of nanoscience for the development of nanotechnology applied to the most diverse sectors.

3.2.2 Nanomaterials

Nanomaterials or nanostructured materials are those that are organized on a nanometric size scale, that is, whose size ranges from 1 to 100 nm. It is the same size as viral particles, the DNA molecule, or the well-known carbon nanotubes (CNTs). Depending on how nanometric a particle is, that is, if closer to one (1) nanometer in size or one hundred (100) nanometers in size, the physicochemical behavior can vary, since the number of atoms available to establish the different responses to external stimuli is much lower when compared to micro or macrometric scales.

The properties of these materials, whether optical, electrical, magnetic, thermal, mechanical, or degradative, undergo abrupt changes as an effect of the reduction in the number of available atoms, resulting in a high surface area, a high availability of electronic interaction, and, consequently, a high chemical activity. Therefore, materials that previously presented certain physicochemical behaviors, start to present differentiated behaviors when nanostructured [1–9]. One particular example is materials that used to be electrical insulators but, when at the nanoscale, start to present semiconductivity or even become electrical conductors. This is the case with glass, which, when nanostructured and exposed to a strong electric field, exhibits electrical conductivity [10].

Furthermore, it is possible to mention the color variation of materials in function of the nanoparticle size that directly influences optical processes of absorption, transmission, or reflection of visible light [11–13]. And yet, it is also necessary to point out the toxicological response of these particles, which is also significantly influenced by the amount of surface atoms available to establish chemical bonds [14, 15].

However, nanotechnology is not an invention or privilege of actual society. Nanostructured particles have been present in nature for thousands of years, and various phenomena in the universe are due to processes associated with the response of such nanoparticles.

In 1959, the physicist Richard Feynman started what we now call nanoscience, when he proposed the synthesis and processing of matter, atom by atom, in terms of nanostructuration [16]. At a meeting of the American Physical Society, he presented and discussed the principles of particle size reduction at the atomic level, demonstrating that there is no violation of the laws established by physics and indicating tooling processes for manipulating atoms and molecules in order to prepare a material on the desired scale. On that occasion, he also explored important steps to be consolidated for the development of this hitherto unexplored area, as well as the need to have enough powerful "glasses" to allow the monitoring of the development of nanomaterials and adjacent processes.

With the development of analysis instruments that allowed access to the reduced universe of the nanometric scale, the viability of the observation of such small structures, as well as the understanding of natural phenomena were possible, leading to the effective development of nanotechnology in addition to the discovery of new materials. This is the case with the hydrophobicity capacity of a few plants, which is effectively due to the nanostructuring of their surfaces. The famous case of the lotus flower is cited here [17, 18]. Depending on the surface energy, it is even possible to observe plant surfaces that are self-cleaning [19]. This aroused the interest of scientists to mimic such behavior and, through nanotechnology, transform surfaces of previously known materials in order to industrially obtain the known hydrophobic and self-cleaning engineered surfaces. This is already applied to fabrics, glass, wood, cars, and even bricks that do not get wet or dirty. But such applications are not restricted to a small range of materials. It is possible to apply it to a range of materials, regardless of whether they are metallic, ceramic, or polymeric.

It is still possible to cite other examples of nature in which nanostructuration is present without human interference. This is the case of carbon nanoparticles, including fullerenes, CNTs, nanodiamonds, and graphene, which are present in combustion processes that occur when meteoroids enter the Earth's atmosphere, volcanic activity, or even forest fires. It is known that combustion processes are also developed and practiced by a series of industrial sectors, and those particles that exist naturally, end up also existing in a manufactured form. This leads to the occurrence of fine and ultrafine particles in the Earth's atmosphere, whether naturally occurring or, caused by pollution processes, in addition to those related to contamination systems for human health, such as viruses, for example.

Looking at the past, we can say that the artisans conducted the first engineering of nanomaterials responsible for the beautiful stained-glass windows of so many famous cathedrals, churches, and chapels around the world, especially the European ones [11]. Without even imagining what they were doing, they worked distinct colors on such stained-glass windows due to the difference in size of the gold nanoparticles existing in the pigments used in the glasses. Depending on the size of the nanoparticle, it could vary from 18 to 250 nm. For example, the cross section was altered and, therefore, its interaction with light was also modified, emitting a different optical response for each available particle size, resulting in color diversity. This is also the basic principle behind the vast number of colors found in the wings of the different butterflies in existence. In its wings are highly organized nanometric structures that, when interacting with light, absorb and emit different optical wavelengths, resulting in a true range of colors [20].

Another very direct application of nanoscience "practiced" by nature and directly used by industry is the ability of the gecko to attach to surfaces. When observing the legs of this reptile, it is noticed that they are also formed by highly organized nanostructures that allow a quite different adhesion capacity [21–23]. This led to the development of adhesives available on the market, especially the biomimetic adhesive developed by German engineers [24]. An elastic membrane is covered with tiny fibers that mimic the structures of geckos' feet and internally provide the material with a pressure differential that prevents the load from concentrating on the edges of the adhesive.

And there is much more that can be mentioned here as a direct application of nanotechnology, such as the development of smart packaging, technology for soil-free farming, the development of catalysts, the possibility of mechanical self-recovery of surfaces, development of motors, actuators, transistors, smart fabrics, electrically conducting cements, artificial muscles, nanorobots for medical treatments and accurate diagnoses, sensors, medicines, cosmetics, biomaterials, energy generators from sunlight, energy generators from residual heat, in short, an infinity of applications relevant to the most different technological sectors [25–31].

At this moment, a question may arise: with so much to observe from nature yet, with so much already discovered, with so much already used, where are we going to lead when it comes to nanomaterials and applied nanotechnology?

First, it should be clear that nanotechnology is about exploring the potential of nanostructures through an understanding of their structure and how it influences their properties. In a second moment, it is about the development of methods of synthesis, processing, characterization, and application of fabricated or modified nanostructures. As well Disner and Cestari [32]:

> Nanotechnology is technology at the molecular scale.

Within the broad spectrum of technology, with a focus on its application at the nanoscale, much more remains to be developed. We can apply it to further evolve mechatronics, computing, medicine, biology, pharmacy, agriculture, food technology, or even space technologies. However, an extremely important field for research and application of nanotechnology is the recovery and conservation of natural resources such as air, water, or diverse ecosystems.

Using this ability to modify material properties through nanostructuring is one of the contour biases of the various problems of the current planetary environmental situation.

3.2.3 Nanotechnology and Its Potential Risk

Despite the discovery of nanoscience and nanotechnology dating from the 1950s, the development of nanomaterials for technological advancement or for use as technological products is very recent. With the advancement of the discovery of new synthesis techniques and with the correct "glasses" to see the phenomena and nanometric particles, new applications were and still are possible and likely to be established.

Nanotechnology is applied in the most diverse industrial sectors, and, as a result not only of the wide use of these materials, but also of all industrial processes that can generate unwanted particles, in the most diverse sizes, it is inevitable that such particles are in the environment in some way, whether through deliberate release by industries, product washing, or wear and tear during use and disposal of devices.

Nanotechnology opens a universe of possibilities for technological advances and facilitators of modern life today, however, at the same time, it could constitute a potential risk. Inevitably, the air, water, and soil are now full of toxic particles and nanoparticles, being found even in places where they had never been detected before, at levels quite different from those expected [33–38]. These are the so-called contaminants of emerging concerns (CECs), and their detection and existence are of concern not only to researchers in the area, but also to agencies and authorities that monitor and preserve the environment at a local and global level [38]. This is because the amount found increases year after year and, for some of these contaminants, there is still no specific environmental legislation [38, 39]. Depending on the size of the particles detected, their chemical and biological reactivity, which is already high, can be even greater, increasing the risk of toxicity or compromising the environment or ecosystem in which they are inserted [33, 36, 40–44].

What makes this increase in CECs so pronounced? Or is it still only nanotechnologically manufactured products that are responsible for the presence and increase of CECs?

The issue is wide open for debate and divides the scientific society between those researchers who acquit and those who condemn nanotechnology. However, carefully analyzing the situation, the generation and increase in the number of CECs occur due to numerous factors and allow for a series of questions. Directly, we have a problem caused by nanostructured manufactured products because, when used, they can be cleaned and worn down by use, generating contaminants that are continuously dumped in air, water, and soil [34, 35, 37, 38, 43]. But in this case, another important concern is also highlighted: the disposal of the nanodevice or the device that contains a nanomaterial. Depending on its corrosion or degradation process and depending on the type of material, what by-product may be generated and on what scale? Therefore, because it is a recent technology of use, much has been studied not only for the development of new materials and processes, but also for the direct impact from an environmental and human health point of view. Some questions are important to study in this inquiry, such as: what would be the safest way to use nanometric products? What would be the safest way to dispose of them? Does using them directly, depending on the type of material, imply any risk to life, whether for humans, animals, or plants, either through contact with these devices or through water, air, and soil?

In an indirect way, the concern can be even greater, but sometimes it is veiled. Industrial processes that use combustion are generators of nanoparticles that contaminate water, air, and soil continuously. The proper process of producing a nanomaterial, whether on a laboratory or industrial scale, implies unique care with the surfaces and instruments where they will be manipulated. Cleaning these instruments implies disposal of nanoparticles, which can lead to indirect contamination of water and soil. In addition, one can mention the use of nanoencapsulated or even directly nanostructured drugs that will generate a residue indirectly through human and animal excretion products, generating a contaminant of water and soil [36, 40–43].

When it comes to disposal, the concern rises on a much larger scale, as macroscopic or microscopic manufactured products, when they start their degradation process, can generate nanoparticles. And, in addition to the nanoparticle itself being a risk factor, it is known that it has a carrier function. This can increase the degree of contamination, as there is the possibility of carrying heavy metals in the effluents, contaminating fauna and flora, or even carrying other chemical elements that can be dangerous from the point of view of the food chain [45].

Other examples worthy of mention are the processes of energy generation through thermoelectric plants and the burning, seasonal or not.

Finally, the issue is worrying and widely discussed. Unfortunately, there are few studies on the degree of influence of nanoparticles on the environment, and even so, they are conducted in the laboratory using chemical reagents in a size that does not compare to the reality of effluent contamination, for example. Considering the lack of precision in the size and degree of complexity of ecosystems – higher than controlled laboratory parameters – understanding, ascertaining, and defining the impact of nanomaterials on the environment is quite challenging. When it comes to water, inconclusive studies can be cited regarding the toxicological effects related to nanosilver, which has been shown to be highly toxic to bacteria, fungi, and algae; titanium dioxide, which acts directly on microorganism membrane cells; and CNTs, whose reports have demonstrated a collection of different, often contradictory, results [46].

What is paramount and urgent is the need to develop routes to preserve the environment from harmful effects caused by conventional or nanometric contaminants and the development of methods of water, air, and/or soil purification. This is because there is still no global standard for assessing the risk of using nanomanufactured products or even consolidated technical standards for defining the acceptable concentration of nanoparticles in air, water, and/or soil.

3.3 Nanomaterials for Air and Water Purification

Implementing nanotechnology implies both harm and benefits. Using nanoscale matter manipulation to recover and purify water and air is possible because the reactivity of nanoparticles is drastically altered, which improves the sensitive capacity of materials thousands of times. In addition, these characteristics favor dilution and adsorption, and may even have gains in the area of water remediation when using quantum effects and specific magnetic phenomena [45]. However, this does not mean that when using the technology for purification purposes, secondary effects of its use will not be entailed.

Some methodologies have already been developed, and some have even been implemented. Table 3.1 shows an organization of the methods already proposed by the scientific community, as well as references to works that contributed to their development or use. However, it is worth emphasizing that there is an imminent concern when using them, as the risk of causing damage to health and the environment exists, as already discussed.

There are methods for water treatment, wastewater treatment, or even air purification. The most commonly used and studied methods are adsorption, filtration, photocatalysis, and disinfection, and some of these methods allow *in situ* action, which favors the entire procedure [26, 32, 45, 78].

The adsorption method is a superficial physicochemical process used to retain ions, molecules, or atoms. In general, solid surfaces are used for this purpose. Solid adsorbents, whether metallic or organic, are used for purification purpose, attracting organic and inorganic contaminants by chemical or magnetic means to remove them from the water or air and in specific cases, even to cause their degradation. The efficiency of this method depends on the type of adsorbent used, depending on its chemical affinity, which will directly interfere with the selectivity of the contaminant, its size for the purposes of the diffusion process,

Table 3.1 Water and air treatment methodologies using nanotechnology, as well as bibliographic references to methods already proposed by the scientific community related to its use.

Methodology	Nanomaterials
Adsorption	• CNTs [47–49] • Metallic base [50–52] • Dendrimers [53, 54]
Membranes	• Nanofibers [55, 56] • Nanocomposites [57–62] • Zeolites [57, 63] • Membranes with bioreactors [64] • Osmosis [65]
Photocatalysis	• Titanium dioxide [66–68] • Graphene oxide [49, 67]
Disinfection	• Nano-Ag [69–71] • Nano-ZnO [72] • CNTs [47, 48] • Graphene [73]
Monitoring and sensoring	• Carbon based [74] • Metal oxide [75, 76] • Copper [77]

its surface area, which will interfere both with the selectivity and the diffusion process, and the kinetics of the entire adsorption process. Due to these parameters, nanomaterials are extremely attractive, since they have a reduced surface area, which can be controlled, in addition to the possibility of synthesizing chemical conjugates that can contribute even more to the process of attracting unwanted molecules.

There are carbon- and metal-based nanostructured adsorbents [49–54, 79–85]. CNTs are useful for organic and heavy metal removal and are even more efficient than activated carbon. The great advantage of using individualized nanotubes is the use of their external surface as an effective surface area, i.e. high availability of sorption sites. With a surface rich in graphitic compounds, which contribute to a hydrophobic effect, the possibilities of interaction of contaminants/nanotubes are diverse and include polar organic compounds, organic carbon molecules with chain unsaturation, benzene rings, polycyclic aromatic hydrocarbons, compounds with –OH, $-NH_2$ functional groups, positively charged organic compounds, hydrogen bonding, and electrostatic interactions [45]. When used in solution, there is the formation of aggregates, which reduces their surface area, but on the other hand, these aggregates can have imperfections, such as vacancies, for example, which constitute sites of high energy from a thermodynamic point of view, favoring the process of attraction of organic contaminants, specially bulky ones. This ends up increasing their adsorption capacity and makes nanotubes an excellent nanomaterial for the purification process, even superior to activated carbon.

Polymeric nanoadsorbents belong to the class of carbon-based adsorbents and because they allow molecule design, they could purify both organic compounds and heavy metals. They are known as dendrimers and are designed to present a hydrophobic interior, to attract organic compounds, with functionalized external branches to attract heavy metals, amino terminated for example [53, 54, 82]. The sorption effect occurs either by the hydrophobic effect or by electrostatic interaction, as well as by the formation of hydrogen bonds or complexation reactions [45, 54, 82, 86].

Metal-based nanoadsorbents basically work through complexation reactions between dissolved metals and the oxygen present in metal oxides such as iron oxide or titanium dioxide [50–52]. They are efficient precisely as a function of their surface area, which confers greater adsorption capacity and better process kinetics. Like CNTs, they also have available sorption sites on the surface and, due to their reduced size, favor the diffusion process. As they are metallic nanoadsorbents, it is possible that some metallic oxides present particular magnetic behavior, as in the case of iron oxide in specific phases. It is known that the magnetic properties of materials are an interaction response between an external magnetic field and the magnetic dipole moments of the atoms that constitute them. Thus, there is a volumetric dependence, which is extremely important when dealing with nanomaterials. This favors the process of purification and recovery of the adsorbent after cleaning, as the reduced size can directly influence the permanent magnetic moment of the adsorbent particle and, consequently, change its response to an external magnetic field [51, 81].

Another method employed in the water and air purification process is filtration membranes. Membranes are physical barriers or filters that allow the dissolution of salts and removal of contaminants. They are widely used with micropores and when nanotechnology is used, it is interesting that they present nanopores to establish a more aggressive filtration. The use of nanotechnology in this method can be seen in the process of synthesis of these materials as well as in the filtration technology itself, which will be specific for nanocontaminants. It is possible to use a multitude of materials, such as nanoporous membranes, but when it comes to air and water treatment, one parameter becomes essential to achieve a proficient level of efficiency: its permeability. This permeability, which comes precisely from the diffusivity of the required solute, will be the selection criterion to select which contaminants could or could not be retained, either by a surface chemistry process or even by particle size. Currently, nanomaterials have been incorporated into conventional membranes (polymeric or inorganic) or thin film membranes as a way to improving the performance of this material, that is, its performance as a barrier, its mechanical, thermal, and degradative resistance, in addition to improve clogging resistance, a major impasse when using this type of methodology. It is also possible to functionalize such membranes in order to allow them to function as a degrading agent or catalyst in the degradation of contaminants. This is achieved by using, for example, (i) nanoparticles of metallic oxides such as zeolite and alumina; (ii) CNTs; (iii) photocatalytic nanomaterials – known as reactive membranes; and (iv) metallic nanoparticles. The efficiency of functionalization will depend not only on the type of nanoparticle used, but also on the concentration and nanometric range used.

Nanofibers are a particularly useful nanomaterial to function as a filtration membrane and which is increasingly presenting itself as a multifunctional material. Nanofibers have advantages similar to those of nanotubes, as they offer a high surface area, in addition to the particular porosity that will function as an important energetic site in the filtration process. What makes this material attractive for this purpose is the possibility of obtaining them by a simple method, electrospinning, which can be applied to any class of materials. This expands the range for the synthesis of a specific nanofiber and allows the manipulation of natural fibers, such as cellulose [55–57, 87]. In this sense, it is even possible to combine the process of developing nanoporous membranes with sustainable development, since the amount of agricultural waste generated worldwide is quite high, which can be directly used in this sense. In a nanofiber design process, it is possible to order and align them, in addition to controlling diameter, morphology, and even the concentration of existing pores. This is a quite common and useful process for air purification and, currently, widely studied for water purification, since the fibers are able to remove, in addition to organic pollutants, heavy metals. In the literature, there are also works on biological membranes, which use aquaporins in polymeric membranes [88–90]. Aquaporins are a class of integral membrane proteins that function as water channels across cell membranes.

Another possibility would be the use of metallic nanoparticles as a solute in direct osmosis, since a solute with high osmolality and ease of separation from water are parameters necessary to obtain satisfactory direct osmosis [91, 92].

The photocatalysis method is widely studied for energy production, through hydrogen production. However, it is also a useful method for the water and air treatment process. Through a process of photocatalytic oxidation it is possible to remove nonbiodegradable or even dangerous contaminants [68, 93–96]. In addition, when thin films are used on surfaces and the photocatalysis process is activated, it is possible to achieve nanophotocatalysis and obtain a positive response against pathogens such as fungi, bacteria, and viruses, in addition to the possibility of removing odors from the air. The major limiting challenge of this technology is the need to increase the kinetics of the photocatalytic reaction, as well as the photocatalytic activity. Titanium dioxide is a well-known photocatalyst that can be used in both water treatment and air purification [66–68]. Its photoactivity can be altered depending on the size and shape of the particles. The smaller the TiO_2 particle size, the better the interfacial charge transfer due to the reduction of electron/hole recombination, responsible for the formation of reactive oxygen species. However, by decreasing its size, surface recombination increases, and photocatalytic activity ends up decreasing. Therefore, finding the balance between particle size and recombination effect is necessary to obtain a satisfactory result. Among the possibilities of structures for nanostructured TiO_2, titanium dioxide nanotubes are considered more efficient for decomposing organic compounds [71, 97]. This is attributed to the shorter diffusion paths in the tube walls, which generate a faster mass flow of reactants to the nanotube surface and, consequently, higher photocatalytic activity. Another way to improve this activity is through morphology control, as there are specific, more energetic crystallographic planes that better adsorb the reagents, in addition to being able to separate the electron/hole pair, although they decrease during crystal growth. Therefore, using titanium dioxide phases richer in energy planes such as the (001) crystallographic plane, or even some leveling agent or even modifying the surface to improve the adsorption of contaminants by such faces is a means of improving the photocatalysis process by TiO_2 [45].

The disinfection process uses chemical procedures to disintegrate and separate contaminants, including turbid water, which contains microbes, viruses, and heavy metals. However, the great challenge lies in the possibility of generating toxic disinfectant by-products (DBPs) after the disinfection process. Thus, alternative methods to the use of conventional disinfectants or even alternative methods to the use of ultraviolet light become necessary to avoid the formation of DBP. One way to achieve this is to use nanomaterials that do not have a high oxidation rate and will have a lower tendency to form DBP, such as silver nanoparticles, TiO_2, CNTs, graphene, and nanocomposites, for example [69, 72, 73, 98, 99].

The monitoring of air and water regarding the presence of contaminants is another aspect of this extremely important attack, as the influence is direct on public health conditions. Although the process of monitoring these media seems simple and even legible, the procedure itself is hampered by a series of factors, such as: (i) low concentration of contaminants, which sometimes go unnoticed by conventional air and water sensors; (ii) slow detection of pathogens, such as coliform bacteria, protozoa such as *Giardia*, viruses such as hepatitis A, among others; (iii) high complexity of water/sewage matrices. The presence of these microorganisms, especially in potable water, indicates the need for disinfection as soon as possible. Therefore, the development of mechanisms that can guarantee agility, high sensitivity, and effectiveness in the identification process to trigger a disinfection process is very necessary. In this sense, the use of nanomaterials has been growing. Activated pathogen biosensors are currently being developed, consisting of recognition agents, nanomaterials, and a transducer. Nanomaterials, due to their unique properties, can improve not only the selective capacity of the sensor or biosensor as a function of the surface area combined with their different electrochemical properties, but also accelerate the detection process, since they present different optical and magnetic behaviors [74–77, 100–102]. This even allows the detection of whole cells and biomolecules. Combined with a good transducer, a quick response could be obtained in the event of pathogen recognition, which accelerates the beginning of the disinfection process, triggered by the detection of specific microorganisms. Examples of nanomaterials for sensors or biosensors can be cited: (i) quantum dots, due to their electronic characteristics; (ii) magnetic nanoparticles; (iii) nanoparticles doped with luminescent (organic or inorganic) dyes; (iv) CNTs due to electrical conductivity; (v) nanostructured noble metals due to their enhanced and localized surface plasma resonance (LSPR) and due to specific electromagnetic behavior; and (vi) nanocomposites. Research has already been developed and is continuing to be developed in this field of study, since there are still numerous obstacles to obtaining good sensors based on nanomaterials. This is because there is dependence on the size, shape, composition, and level of aggregation of the nanoparticles involved. As much as they present excellent photostability and sensitivity, an aggregation process can change the shape and size of the nanoparticles, and this implies altering the desirable properties. In addition, there is a possibility of starting a nucleation process from contact of nanoparticles with impurities, or even the aggregation can be initiated by thermal favoring. This points to the need for more research in this area.

A new possibility that has emerged is the development of multifunctional devices, that is, the combination of two or more treatment and/or purification methods to optimize properties, response, and direct action to remove contaminants or pathogens. This could be achieved, for example, by functionalizing the membrane's surface with magnetic nanoparticles or doping elements. It is also possible to synthesize of functional multilayer devices [57–61, 67].

3.4 Conclusion

Using nanotechnology for air and water purification or even for treatment and monitoring processes is possible, but it still requires development and research due to innumerable factors. Not only because the technical character still requires optimization, but also because of the difficulty of commercializing the final device due to excessive costs when it comes to nanomaterials and due to the possibility of high environmental and human toxicity.

Currently, there are commercial devices that are intended for water and air purification; however, most of the technology developed has still been evaluated on a laboratory scale. A systematic and prolonged study of the use of this type of device in real conditions of use needs to be developed to check biosafety, real efficiency, cost-benefit ratio, and commercial viability. From a technical point of view, among all the possibilities presented in this review chapter, much is still in the research and development phase, and some have started tests for commercialization. Among these, three methods with potential applications stand out, considering commercial viability and existing infrastructure: adsorption, membranes, and photocatalysis. Other treatment processes can be applied in local, small-scale niches or even in exergetic use systems.

From an economic point of view, when it comes to nanotechnology, the excessive cost resides in the need to use high purity materials. One solution would be the use of nanomaterials that allow low purity without compromising the efficiency of the final device and the development of materials with a long lifetime, that is, that allow reuse without compromising efficiency.

Nanotechnology presents obstacles, but also benefits. Certainly, the bonuses of its use justify a network of collaboration between research institutions, industry, and governmental associations to, in fact, implement an innovative technology that acts quickly and safely in solving critical problems that already exhaust the conventional possibilities.

References

1 Jain, P.K., Huang, X., El-Sayed, I.H., and El-Sayed, M.A. (2008). Noble metals on the nanoscale: optical and photothermal properties and some applications in imaging, sensing, biology, and medicine. *Acc. Chem. Res.* 41: 1578–1586. https://doi.org/10.1021/AR7002804/ASSET/IMAGES/MEDIUM/AR-2007-002804_0001.GIF.

2 Kianinia, M., Xu, Z.Q., Toth, M., and Aharonovich, I. (2022). Quantum emitters in 2D materials: emitter engineering, photophysics, and integration in photonic nanostructures. *Appl. Phys. Rev.* 9: 011306. https://doi.org/10.1063/5.0072091.

3 Cao, L.M., Zhang, J., Zhang, X.F., and He, C.T. (2022). Confinement synthesis in porous molecule-based materials: a new opportunity for ultrafine nanostructures. *Chem. Sci.* 13: 1569–1593. https://doi.org/10.1039/D1SC05983A.

4 Joshi, D.J., Malek, I.N., and Kumar Kailasa, S. (2022). Borophene as a rising star in materials chemistry: synthesis, properties and applications in analytical science and energy devices. *New J. Chem.* 46: 4514–4533. https://doi.org/10.1039/D1NJ05271C.

5 Jergens, E. and Winter, J.O. (2022). Nanoparticles caged with DNA nanostructures. *Curr. Opin. Biotechnol.* 74: 278–284. https://doi.org/10.1016/J.COPBIO.2021.12.010.

6 Vanlandingham, M.R., Dagastine, R.R., Eduljee, R.F. et al. (1999). Characterization of nanoscale property variations in polymer composite systems: 1. Experimental results. *Composites Part A* 30: 75–83. https://doi.org/10.1016/S1359-835X(98)00098-0.

7 Gemma, R., to Baben, M., Pundt, A. et al. (2020). The impact of nanoscale compositional variation on the properties of amorphous alloys. *Sci. Rep.* 10: 1–9. https://doi.org/10.1038/s41598-020-67495-4.

8 Zhang, C., Barnett, R.N., and Landman, U. (2008). Bonding, conductance, and magnetization of oxygenated Au nanowires. *Phys. Rev. Lett.* 100: 046801. https://doi.org/10.1103/PHYSREVLETT.100.046801/FIGURES/4/MEDIUM.

9 Garitaonandia, J.S., Insausti, M., Goikolea, E. et al. (2008). Chemically induced permanent magnetism in Au, Ag, and Cu nanoparticles: localization of the magnetism by element selective techniques. *Nano Lett.* 8: 661–667. https://doi.org/10.1021/NL073129G/SUPPL_FILE/NL073129GSI20071130_124812.PDF.

10 Lee, S., An, R., and Hunt, A.J. (2010). Liquid glass electrodes for nanofluidics. *Nat. Nanotechnol.* 5: 412–416. https://doi.org/10.1038/nnano.2010.81.

11 Farges, F., Etcheverry, M.P., Scheidegger, A., and Grolimund, D. (2006). Speciation and weathering of copper in "copper red ruby" medieval flashed glasses from the Tours cathedral (XIII century). *Appl. Geochem.* 21: 1715–1731. https://doi.org/10.1016/J.APGEOCHEM.2006.07.008.

12 Wagner, F.E., Haslbeck, S., Stievano, L. et al. (2000). Before striking gold in gold-ruby glass. *Nature* 407: 691–692. https://doi.org/10.1038/35037661.

13 Freestone, I., Meeks, N., Sax, M., and Higgitt, C. (2007). The Lycurgus cup – a Roman nanotechnology. *Gold Bull.* 40: 270–277. https://doi.org/10.1007/BF03215599.

14 Martin, L.M.A., Gan, N., Wang, E. et al. (2022). Materials, surfaces, and interfacial phenomena in nanoplastics toxicology research. *Environ. Pollut.* 292: 118442. https://doi.org/10.1016/J.ENVPOL.2021.118442.

15 Gong, J.Y., Holt, M.G., Hoet, P.H.M., and Ghosh, M. (2022). Neurotoxicity of four frequently used nanoparticles: a systematic review to reveal the missing data. *Arch. Toxicol.* 96: 1141–1212. https://doi.org/10.1007/S00204-022-03233-1.

16 (n.d.). The Beginning of Nanotechnology at the 1959 APS Meeting. https://www.aps.org/publications/apsnews/201611/nanotechnology.cfm (accessed 14 July 2023).

17 Gao, N. and Yan, Y. (2009). Modeling superhydrophobic contact angles and wetting transition. *J. Bionic Eng.* 6: 335–340. https://doi.org/10.1016/S1672-6529(08)60135-3.

18 (n.d.). Purity of the sacred lotus, or escape from contamination in biological. https://studylib.net/doc/18757080/purity-of-the-sacred-lotus–or-escape-from-contamination- (accessed 14 July 2023).

19 Koch, K. and Barthlott, W. (2009). Superhydrophobic and superhydrophilic plant surfaces: an inspiration for biomimetic materials. *Philos. Trans. R. Soc. A* 367: 1487–1509. https://doi.org/10.1098/RSTA.2009.0022.

20 (n.d.). X-ray tomography and chemical imaging within butterfly wing scales - CERN Document Server. https://cds.cern.ch/record/1067123?ln=en (accessed 14 July 2023).

21 Autumn, K., Sitti, M., Liang, Y.A. et al. (2002). Evidence for van der Waals adhesion in gecko setae. *Proc. Natl. Acad. Sci. U.S.A.* 99: 12252–12256. https://doi.org/10.1073/PNAS.192252799.

22 Huber, G., Mantz, H., Spolenak, R. et al. (2005). Evidence for capillarity contributions to gecko adhesion from single spatula nanomechanical measurements. *Proc. Natl. Acad. Sci. U.S.A.* 102: 16293–16296. https://doi.org/10.1073/PNAS.0506328102/ASSET/1F393DC9-4779-4711-AF66-99B6D8263A0D/ASSETS/GRAPHIC/ZPQ0430599770004.JPEG.

23 Huber, G., Gorb, S.N., Spolenak, R., and Arzt, E. (2005). Resolving the nanoscale adhesion of individual gecko spatulae by atomic force microscopy. *Biol. Lett.* 1: 2–4. https://doi.org/10.1098/RSBL.2004.0254.

24 Song, S., DiM, D., Majidi, C., and Sitti, M. (2017). Controllable load sharing for soft adhesive interfaces on three-dimensional surfaces. *Proc. Natl. Acad. Sci. U.S.A.* 114: E4344–E4353. https://doi.org/10.1073/PNAS.1620344114/SUPPL_FILE/PNAS.1620344114.SM10.MP4.

25 Singha, K., Regubalan, B., Pandit, P. et al. (2022). Introduction to nanotechnology-enhanced food packaging industry. In: *Nanotechnology-Enhanced Food Packaging*, 1–17. https://doi.org/10.1002/9783527827718.CH1.

26 Mehta, K. (2022). Nanotechnology for energy and the environment. *Nanotechnology* 19–33. https://doi.org/10.1201/9781003220350-2.

27 Zain, M., Yasmeen, H., Yadav, S.S. et al. (2022). Applications of nanotechnology in biological systems and medicine. In: *Nanotechnology for Hematology, Blood Transfusion, and Artificial Blood*, 215–235. https://doi.org/10.1016/B978-0-12-823971-1.00019-2.

28 Shah, M.A., Pirzada, B.M., Price, G. et al. (2022). Applications of nanotechnology in smart textile industry: a critical review. *J. Adv. Res.* https://doi.org/10.1016/J.JARE.2022.01.008.

29 Ghouse, H., Slewa, L., Mahmood, M. et al. (2022). Importance of nanotechnology, various applications in electronic field. In: *Nanotechnology for Electronic Applications*, 1–28. https://doi.org/10.1007/978-981-16-6022-1_1.

30 Dubey, S.K., Dey, A., Singhvi, G. et al. (2022). Emerging trends of nanotechnology in advanced cosmetics. *Colloids Surf., B* 214: 112440. https://doi.org/10.1016/J.COLSURFB.2022.112440.

31 (n.d.). Yuichi Mori (CEO, Mebiol) soil free farming on ted talk - The online Q&A community and marketplace for global farmers. https://oneacreland.com/?/question/280 (accessed 17 April 2022).

32 Disner, G.R. and Cestari, M.M. (2016). Nanociência & Nanotecnologia. *Evidência* 16: 71–76. 10.1016/j.colsurfb.2022.112440.

33 Omari Shekaftik, S., Shirazi, F.H., Yarahmadi, R. et al. (2020). Investigating the relationship between occupational exposure to nanomaterials and symptoms of nanotechnology companies' employees. 77: 209–218. https://doi.org/10.18593/eba.v16i1.10660.

34 Sun, T.Y., Gottschalk, F., Hungerbühler, K., and Nowack, B. (2014). Comprehensive probabilistic modelling of environmental emissions of engineered nanomaterials. *Environ. Pollut.* 185: 69–76. https://doi.org/10.1016/J.ENVPOL.2013.10.004.

35 Keller, A.A., McFerran, S., Lazareva, A., and Suh, S. (2013). Global life cycle releases of engineered nanomaterials. *J. Nanopart. Res.* 15: 1–17. https://doi.org/10.1007/S11051-013-1692-4.

36 Illuminato, Ian; Friends of the Earth - U.S. Tiny ingredients, big risks: Nanomaterials rapidly entering food and farming. 2014. 1–43. https://media-alliance.org/wp-content/uploads/2016/05/TinyIngredients_BigRisks.pdf

37 Oberdörster, G., Oberdörster, E., and Oberdörster, J. (2005). Nanotoxicology: an emerging discipline evolving from studies of ultrafine particles. *Environ. Health Perspect.* 113: 823–839. https://doi.org/10.1289/EHP.7339.

38 United States Environmental Protection Agency (EPA), Nanomaterials – Technical Fact Sheet. 2017, 1–9. https://www.epa.gov/sites/default/files/2014-03/documents/ffrrofactsheet_emergingcontaminant_nanomaterials_jan2014_final.pdf

39 Rani, M., Keshu, Y.J., and Meenu, S.U. (2022). Environmental, legal, health, and safety issues of green nanomaterials. In: *Green Functionalized Nanomaterials for Environmental Applications*, 567–594. https://doi.org/10.1016/B978-0-12-823137-1.00020-8.

40 López, A.D.F., Fabiani, M., Lassalle, V.L. et al. (2022). Critical review of the characteristics, interactions, and toxicity of micro/nanomaterials pollutants in aquatic environments. *Mar. Pollut. Bull.* 174: 113276. https://doi.org/10.1016/J.MARPOLBUL.2021.113276.

41 Ji, Z., Guo, W., Wood, E.L. et al. (2022). Machine Learning Models for Predicting Cytotoxicity of Nanomaterials. *Chem. Res. Toxicol.* 35: 125–139. https://doi.org/10.1021/ACS.CHEMRESTOX.1C00310/ASSET/IMAGES/MEDIUM/TX1C00310_0005.GIF.

42 Liu, Y., Zhu, S., Gu, Z. et al. (2022). Toxicity of manufactured nanomaterials. *Particuology* 69: 31–48. https://doi.org/10.1016/J.PARTIC.2021.11.007.

43 Handy, R.D., Henry, T.B., Scown, T.M. et al. (2008). Manufactured nanoparticles: their uptake and effects on fish – a mechanistic analysis. *Ecotoxicology* 17: 396–409. https://doi.org/10.1007/S10646-008-0205-1.

44 Handy, R.D., Owen, R., and Valsami-Jones, E. (2008). The ecotoxicology of nanoparticles and nanomaterials: current status, knowledge gaps, challenges, and future needs. *Ecotoxicology* 17: 315–325. https://doi.org/10.1007/S10646-008-0206-0.

45 Qu, X., Alvarez, P.J.J., and Li, Q. (2013). Applications of nanotechnology in water and wastewater treatment. *Water Res.* 47: 3931–3946. https://doi.org/10.1016/J.WATRES.2012.09.058.

46 Greßler, S.; Nentwich, M. Nano and Environment – Part II: Hazard potentials and risks. Nano trust Dossiers. Institute of Technology Assessment (ITA); Strohgasse 45/5, A-1030 Vienna (Editor). 2012, 27, 1-5. http://epub.oeaw.ac.at/0xc1aa5576_0x002acc9d.pdf

47 Hussain, N., Bilal, M., and Iqbal, H.M.N. (2022). Carbon-based nanomaterials with multipurpose attributes for water treatment: greening the 21st-century nanostructure materials deployment. *Biomater. Polym. Horizon* 1: 48–58. https://doi.org/10.37819/BPH.001.01.0131.

48 Chung, J.H., Hasyimah, N., and Hussein, N. (2022). Application of carbon nanotubes (CNTs) for remediation of emerging pollutants – a review. *Trop. Aquat.Soil Pollut.* 2: 13–26. https://doi.org/10.53623/TASP.V2I1.27.

49 Wang, Y., Pan, C., Chu, W. et al. (2019). Environmental remediation applications of carbon nanotubes and graphene oxide: adsorption and catalysis. *Nanomaterials* 9. https://doi.org/10.3390/NANO9030439.

50 Beydaghdari, M., Saboor, F.H., Babapoor, A., and Asgari, M. (2022). Recent progress in adsorptive removal of water pollutants by metal-organic frameworks. *ChemNanoMat* 8: e202100400. https://doi.org/10.1002/CNMA.202100400.

51 Ali, S.M., Emran, K.M., and Alrashedee, F.M.M. (2022). Removal of organic pollutants by lanthanide-doped MIL-53 (Fe) metal-organic frameworks: effect of dopant type in magnetite precursor. *J. Rare Earths* https://doi.org/10.1016/J.JRE.2022.01.012.

52 Pauletto, P.S. and Bandosz, T.J. (2022). Activated carbon versus metal-organic frameworks: a review of their PFAS adsorption performance. *J. Hazard. Mater.* 425: 127810. https://doi.org/10.1016/J.JHAZMAT.2021.127810.

53 Tong, Y., Li, S., Wu, Y. et al. (2022). Graphene oxide modified magnetic polyamidoamide dendrimers based magnetic solid phase extraction for sensitive measurement of polycyclic aromatic hydrocarbons. *Chemosphere* 296: 134009. https://doi.org/10.1016/J.CHEMOSPHERE.2022.134009.

54 Tang, B., Wu, Y., Wu, K. et al. (2022). Adsorption performance of silica supported polyamidoamine dendrimers for Cd(II) and Cu(II) in N,N-dimethylformamide. *J. Mol. Liq.* 357: 119098. https://doi.org/10.1016/J.MOLLIQ.2022.119098.

55 Liu, Y., Liu, H., and Shen, Z. (2021). Nanocellulose based filtration membrane in industrial waste water treatment: a review. *Materials* 14: 5398. https://doi.org/10.3390/MA14185398.

56 Deng, Y., Lu, T., Cui, J. et al. (2021). Bio-based electrospun nanofiber as building blocks for a novel eco-friendly air filtration membrane: a review. *Sep. Purif. Technol.* 277: 119623. https://doi.org/10.1016/J.SEPPUR.2021.119623.

57 Sultana, N. and Rahman, R. (2021). Electrospun nanofiber composite membranes based on cellulose acetate/nano-zeolite for the removal of oil from oily wastewater. *Emergent Mater.* 5: 145–153. https://doi.org/10.1007/S42247-021-00326-Y.

58 Li, C., Sun, W., Lu, Z. et al. (2020). Ceramic nanocomposite membranes and membrane fouling: a review. *Water Res.* 175: 115674. https://doi.org/10.1016/J.WATRES.2020.115674.

59 Anadão, P. (2017). Nanocomposite filtration membranes for drinking water purification. In: *Water Purification*, 517–549. https://doi.org/10.1016/B978-0-12-804300-4.00015-0.

60 Bassyouni, M., Abdel-Aziz, M.H., Zoromba, M.S. et al. (2019). A review of polymeric nanocomposite membranes for water purification. *J. Ind. Eng. Chem.* 73: 19–46. https://doi.org/10.1016/J.JIEC.2019.01.045.

61 Wei, Z., Su, Q., Yang, J. et al. (2021). High-performance filter membrane composed of oxidized poly(arylene sulfide sulfone) nanofibers for the high-efficiency air filtration. *J. Hazard. Mater.* 417: 126033. https://doi.org/10.1016/J.JHAZMAT.2021.126033.

62 Zhijiang, C., Ping, X., Cong, Z. et al. (2018). Preparation and characterization of a bi-layered nano-filtration membrane from a chitosan hydrogel and bacterial cellulose nanofiber for dye removal. *Cellulose* 25: 5123–5137. https://doi.org/10.1007/S10570-018-1914-0.

63 Mahdavi Far, R., van der Bruggen, B., Verliefde, A., and Cornelissen, E. (2022). A review of zeolite materials used in membranes for water purification: history, applications, challenges and future trends. *J. Chem. Technol. Biotechnol.* 97: 575–596. https://doi.org/10.1002/JCTB.6963.

64 Kharraz, J.A., Khanzada, N.K., Farid, M.U. et al. (2022). Membrane distillation bioreactor (MDBR) for wastewater treatment, water reuse, and resource recovery: a review. *J. Water Process Eng.* 47: 102687. https://doi.org/10.1016/J.JWPE.2022.102687.

65 Kahrizi, M., Gonzales, R.R., Kong, L. et al. (2022). Significant roles of substrate properties in forward osmosis membrane performance: a review. *Desalination* 528: 115615. https://doi.org/10.1016/J.DESAL.2022.115615.

66 Shafique, M., Mahr, M.S., Yaseen, M., and Bhatti, H.N. (2022). CQD/TiO_2 nanocomposite photocatalyst for efficient visible light-driven purification of wastewater containing methyl orange dye. *Mater. Chem. Phys.* 278: 125583. https://doi.org/10.1016/J.MATCHEMPHYS.2021.125583.

67 Wang, Q., Yu, Z., Liu, Y. et al. (2022). Co-intercalation of TiO_2 and LDH to reduce graphene oxide photocatalytic composite membrane for purification of dye wastewater. *Appl. Clay Sci.* 216: 106359. https://doi.org/10.1016/J.CLAY.2021.106359.

68 Mavrikos, A., Papoulis, D., Todorova, N. et al. (2022). Synthesis of Zn/Cu metal ion modified natural palygorskite clay – TiO_2 nanocomposites for the photocatalytic outdoor and indoor air purification. *J. Photochem. Photobiol., A* 423: 113568. https://doi.org/10.1016/J.JPHOTOCHEM.2021.113568.

69 Woldegebreal, T., Teju, E., Kebede, A. et al. (2022). Water disinfection using Kaolinite supported magnetic silver nanoparticle. *Chem. Data Collect.* 39: 100857. https://doi.org/10.1016/J.CDC.2022.100857.

70 Mo, F., Zhou, Q., and He, Y. (2022). Nano–Ag: environmental applications and perspectives. *Sci. Total Environ.* 829: 154644. https://doi.org/10.1016/J.SCITOTENV.2022.154644.

71 Assadi, A.A., Karoui, S., Trabelsi, K. et al. (2022). Synthesis and characterization of TiO_2 nanotubes (TiO_2-NTs) with Ag silver nanoparticles (Ag-NPs): photocatalytic performance for wastewater treatment under visible light. *Materials* 15: 1463. https://doi.org/10.3390/MA15041463.

72 Misra, A.J., Basu, A., Behera, S.K. et al. (2022). Point-of-use photocatalytic device for water disinfection under visible light using ZnO/Gypsum@alginate beads. *J. Environ. Chem. Eng.* 10: 107520. https://doi.org/10.1016/J.JECE.2022.107520.

73 Omran, B. and Baek, K.H. (2022). Graphene-derived antibacterial nanocomposites for water disinfection: current and future perspectives. *Environ. Pollut.* 298: 118836. https://doi.org/10.1016/J.ENVPOL.2022.118836.

74 Bakhshandeh, B., Sorboni, S.G., Haghighi, D.M. et al. (2022). New analytical methods using carbon-based nanomaterials for detection of Salmonella species as a major food poisoning organism in water and soil resources. *Chemosphere* 287: 132243. https://doi.org/10.1016/J.CHEMOSPHERE.2021.132243.

75 Bulusheva, G., Li, T., Yin, W. et al. (2022). The combination of two-dimensional nanomaterials with metal oxide nanoparticles for gas sensors: a review. *Nanomaterials* 12: 982. https://doi.org/10.3390/NANO12060982.

76 Tyagi, S., Manika Chaudhary, K., Ambedkar, A. et al. (2022). Metal oxide nanomaterial-based sensors for monitoring environmental NO_2 and its impact on the plant ecosystem: a review. *Sensors Diagnostics* 1: 106–129. https://doi.org/10.1039/D1SD00034A.

77 Anh, N.T., Dinh, N.X., van Tuan, H. et al. (2022). Eco-friendly copper nanomaterials-based dual-mode optical nanosensors for ultrasensitive trace determination of amoxicillin antibiotics residue in tap water samples. *Mater. Res. Bull.* 147: 111649. https://doi.org/10.1016/J.MATERRESBULL.2021.111649.

78 Silveira, S.M.B. and Foladori, G. (2016). <b>Nanotecnologia e água no Brasil. *Acta Sci. Human Soc. Sci.* 38: 153. https://doi.org/10.4025/ACTASCIHUMANSOC.V38I2.32668.

79 Shahriyari Far, H., Hasanzadeh, M., Najafi, M., and Rahimi, R. (2022). In-situ self-assembly of mono- and bi-metal organic frameworks onto clay mineral for highly efficient adsorption of pollutants from wastewater. *Chem. Phys. Lett.* 139626. https://doi.org/10.1016/J.CPLETT.2022.139626.

80 Khan, A., Malik, S., Ali, N. et al. (2022). Nanoadsorbents as a green approach for removal of environmental pollutants. In: *Nano-Bioremediation: Fundamentals and Applications*, 435–454. https://doi.org/10.1016/B978-0-12-823962-9.00007-6.

81 Iqbal, Z., Tanweer, M.S., and Alam, M. (2022). Recent advances in adsorptive removal of wastewater pollutants by chemically modified metal oxides: a review. *J. Water Process Eng.* 46: 102641. https://doi.org/10.1016/J.JWPE.2022.102641.

82 He, Z., Wu, F., Liu, L. et al. (2022). Simultaneous removal of pollutants from sub-nanometric to nanometric scales by hierarchical dendrimers modified aerogels. *Sep. Purif. Technol.* 286: 120440. https://doi.org/10.1016/J.SEPPUR.2021.120440.

83 Mokhena, T.C., Matabola, K.P., Mokhothu, T.H. et al. (2022). Electrospun carbon nanofibres: preparation, characterization and application for adsorption of pollutants from water and air. *Sep. Purif. Technol.* 288: 120666. https://doi.org/10.1016/J.SEPPUR.2022.120666.

84 Reynel-Ávila, H.E., Aguayo-Villarreal, I.A., Diaz-Muñoz, L.L. et al. (2022). A review of the modeling of adsorption of organic and inorganic pollutants from water using artificial neural networks. *Adsorpt. Sci. Technol.* 2022: https://doi.org/10.1155/2022/9384871.

85 Veeman, D., Shree, M.V., Sureshkumar, P. et al. (2021). Sustainable development of carbon nanocomposites: synthesis and classification for environmental remediation. *J. Nanomater.* 2021: https://doi.org/10.1155/2021/5840645.

86 Singh, N., Malik, M.A., and Hashmi, A.A. (2021). Nanotechnology and its application in wastewater treatment. In: *Sustainable Practices in the Textile Industry* (ed. L.J. Rather, M. Shabbir, and A. Haji). https://doi.org/10.1002/9781119818915.ch13.

87 Shabafrooz, V., Mozafari, M., Vashaee, D., and Tayebi, L. (2014). Electrospun nanofibers: from filtration membranes to highly specialized tissue engineering scaffolds. *J. Nanosci. Nanotechnol.* 14: 522–534. https://doi.org/10.1166/JNN.2014.9195.

88 Sicard, F. and Yazaydin, A.O. (2022). Biohybrid membrane formation by directed insertion of aquaporin into a solid-state nanopore. *ACS Appl. Mater. Interfaces* 14 (42): 48029–48036. https://doi.org/10.1021/acsami.2c14250.

89 Reddy, A.S., Kalla, S., and Murthy, Z.V.P. (2022). Biomimetic membranes: advancements and applications — a minireview. *Biores. Technol. Rep.* 18: 101047. https://doi.org/10.1016/J.BITEB.2022.101047.

90 Sharma, L., Ye, L., Yong, C. et al. (2022). Aquaporin-based membranes made by interfacial polymerization in hollow fibers: visualization and role of aquaporin in water permeability. *J. Membr. Sci.* 654: 120551. https://doi.org/10.1016/J.MEMSCI.2022.120551.

91 Chen, R., Dong, X., and Ge, Q. (2022). Lithium-based draw solute for forward osmosis to treat wastewater discharged from lithium-ion battery manufacturing. *Front. Chem. Sci. Eng.* 16: 755–763. https://doi.org/10.1007/S11705-022-2137-3.

92 Jain, H., Kumar, A., Rajput, V.D. et al. (2022). Fabrication and characterization of high-performance forward-osmosis membrane by introducing manganese oxide incited graphene quantum dots. *J. Environ. Manage.* 305: 114335. https://doi.org/10.1016/J.JENVMAN.2021.114335.

93 Dalhat, M.H. and Ahmad, A. (2022). Removal of pesticides from water and wastewater by solar-driven photocatalysis. In: *Development in Wastewater Treatment Research and Processes*, 435–458. https://doi.org/10.1016/B978-0-323-85583-9.00012-0.

94 Kato, K. and Shirai, T. (2022). Highly efficient water purification by WO_3-based homo/heterojunction photocatalyst under visible light. *J. Alloys Compd.* 901: 163434. https://doi.org/10.1016/J.JALLCOM.2021.163434.

95 Pang, X., Xue, S., Zhou, T. et al. (2022). 2D/2D nanohybrid of Ti_3C_2 MXene/WO_3 photocatalytic membranes for efficient water purification. *Ceram. Int.* 48: 3659–3668. https://doi.org/10.1016/J.CERAMINT.2021.10.147.

96 Verbruggen, S.W., van Hal, M., Bosserez, T. et al. (2017). Harvesting hydrogen gas from air pollutants with an unbiased gas phase photoelectrochemical cell. *ChemSusChem* 10: 1413–1418. https://doi.org/10.1002/CSSC.201601806.

97 Wang, Q., Zhu, S., Zhao, S. et al. (2022). Construction of Bi-assisted modified CdS/TiO_2 nanotube arrays with ternary S-scheme heterojunction for photocatalytic wastewater treatment and hydrogen production. *Fuel* 322: 124163. https://doi.org/10.1016/J.FUEL.2022.124163.

98 Priya, T., Mishra, B.K., and Prasad, M.N.V. (2020). Physico-chemical techniques for the removal of disinfection by-products precursors from water. In: *Disinfection By-Products in Drinking Water*, 23–58. https://doi.org/10.1016/B978-0-08-102977-0.00002-0.

99 Shi, Y., Ma, J., Chen, Y. et al. (2022). Recent progress of silver-containing photocatalysts for water disinfection under visible light irradiation: a review. *Sci. Total Environ.* 804: 150024. https://doi.org/10.1016/J.SCITOTENV.2021.150024.

100 Wunderlich, L. and Hirsch, T. (2022). Current challenges in nanomaterial-based sensors for online monitoring of drinking water by surface plasmon resonance. *Curr. Opin. Environ. Sci. Health* 26: 100326. https://doi.org/10.1016/J.COESH.2022.100326.

101 Potes-Lesoinne, H.A., Ramirez-Alvarez, F., Perez-Gonzalez, V.H. et al. (2022). Nanomaterials for electrochemical detection of pollutants in water: a review. *Electrophoresis* 43: 249–262. https://doi.org/10.1002/ELPS.202100204.

102 Hou, H., Xu, L., Yang, W., and Wong, W.Y. (2022). Applications of 1D mesoporous inorganic nanomaterials as sensors. *Springer Ser. Mater. Sci.* 321: 165–174. https://doi.org/10.1007/978-3-030-89105-3_10.

4

Green Nanotechnology for Water Purification

Mahmoud A. Abdulhamid[1], *Nawader Alqadhi*[2], *Claudia Osornio*[2], *and Gyorgy Szekely*[2]

[1] *King Fahd University of Petroleum and Minerals (KFUPM), College of Petroleum Engineering and Geosciences (CPG), Sustainable and Resilient Materials Lab, Center for Integrative Petroleum Research (CIPR), Dhahran 31261, Saudi Arabia*
[2] *King Abdullah University of Science and Technology (KAUST), Advanced Membranes and Porous Materials Center, Physical Science and Engineering Division (PSE), Thuwal 23955-6900, Saudi Arabia*

4.1 Introduction

The rapid expansion of cities and world population growth has led to a considerable increase in the demand for fresh and drinkable water, thus driving a shift toward treating contaminated water via sustainable and energy-efficient methods [1]. From 2000 to 2016, the global urban population increased by 5% (from 28% to 33%), with a growth rate of >6% in terms of gross domestic product (GDP) [2]. Due to population increases and the need to fulfill human requirements, large countries such as China and India have increased their chemical, pharmaceutical, and petrochemical production. Moreover, the diminishing water resources in these countries are being contaminated by unprocessed industrial waste, leading to a significant decrease in the amount of drinkable water available to the residents [1]. Accordingly, the Department of Economic and Social Affairs of the United Nations has made clean water and sanitation the pillars of their sustainability aims that must be achieved globally by combined efforts of both the government and citizens [3]. Since water is involved in food, health, industry, and climate change, there are considerable challenges associated with producing clean water. Over the past 50 years alone, there has been an ~83% decrease in the amount of clean water that is available for human use, thus prompting the necessity to develop permanent solutions to recover this loss [1].

Different conventional technologies, such as adsorption and membrane separation, have been employed in water treatment processes. However, among these technologies, adsorption is one of the best techniques to eliminate contaminants from water. Activated carbon (AC) is extensively used to remove contaminants from wastewater streams and drinking water [4]. However, due to its high production cost, researchers started to look for alternative adsorbents. The best alternative to conventional technologies is green nanotechnology. This technology depends

Nanomaterials for Air and Water Purification, First Edition.
Edited by Deepu A. Gopakumar, Jyotishkumar Parameswaranpillai, Jinu Jacob George, and Midhun Dominic C.D.

on the green design of various adsorbents from natural resources with unique surface characteristics and properties. Therefore, different research groups started the development of green bio-based materials that can help in purifying the water efficiently using eco-friendly and efficient adsorbents [5].

Recent studies demonstrated that nanomaterials have potential to be used to resolve issues surrounding water shortages, thus playing an important role in the decontamination of polluted water. For instance, nanochemistry has been used to produce filters to remove pesticides from contaminated drinking water [1]. This technology has proven to be extremely popular and by 2016 there were >7.5 million users of this filter technology, with the result that pesticide levels in water were reduced by >20-fold of the current outlined safety standards (0.5 ppb for all pesticides taken together) [6]. Similarly, restrained nanomaterials have been developed for removing As from wastewater. Currently, this technology makes it possible to provide clean water to approximately one million people in India [7]. Various other methods and materials, such as desalination [8], water harvesting [8], microbial disinfection [9], nanoadsorbents [10], and photocatalysis, have been examined to decontaminate water [11]. Through the use of these methods and materials, there has been an explosion in the development of nanomaterials for water purification, promising results for addressing water shortages. However, the development of these nanomaterials requires the following important points to be addressed. (i) Nanomaterial design and structure: the designed materials should have excellent thermal and chemical stability with a high surface area for high sorption of contaminants. The functional groups of a nanomaterial, porosity, and free space determine its efficiency and reactivity toward specific pollutants. Thus, these important properties should be considered when designing nanomaterials for use in water purification. (ii) Reducing the possibility of generating new contaminants: the developed nanomaterials must remove existing contaminants from water without generating secondary waste via the degradation of materials or possible chemical reactions involving the material being used to clean the water. Moreover, the nanomaterials require that they be fabricated from sustainable resources, involving minimal CO_2 emissions. To summarize, implementing nanotechnology in water purification should not generate toxic gases and increase the already high levels of pollution. (iii) The nanomaterials developed should reduce contamination to the safety standards outlined by the World Health Organization (WHO) and the U.S. Environmental Protection Agency, which have set an upper limit of 10 ppb for As levels in drinking water [12]. (iv) Scalability and low cost: the implementation of nanotechnology for water treatment requires inexpensive materials and ease of scalability. Therefore, to convince governments to implement these technologies, it should be possible to produce these materials on a large scale with minimal costs.

Another method by which contaminated water can be treated involves the usage of AC and its derivatives, which have a high capacity for adsorbing pollutants and act as catalysts to remove organic and inorganic contaminants from wastewater (Figure 4.1) [12]. Moreover, these materials have been employed in membrane-based separation and disinfection [13, 14]. Similarly, metal nanoparticles (NPs) exhibit excellent adsorption capacity and photocatalytic reactivity toward

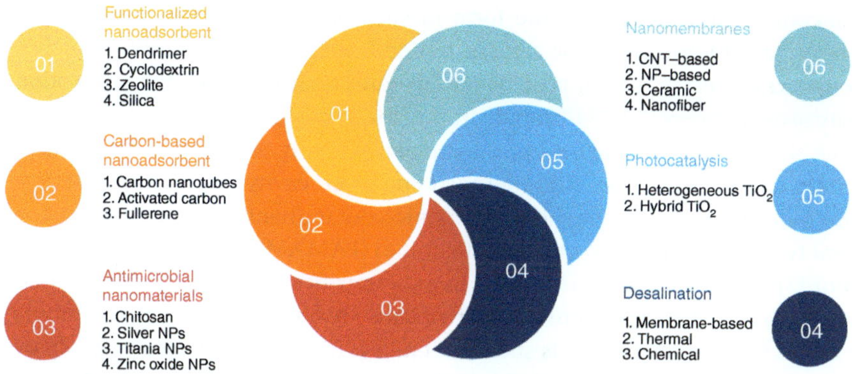

Figure 4.1 The different nanotechnologies used in water treatment applications.

various types of pollutants, making them a good candidate for purifying water [15]. Furthermore, antimicrobial nanomaterials have been used to eliminate and absorb harmful microbes and bacteria while treating wastewater [16]. In addition to the development of materials, research activities have been extended toward developing clean technologies such as thermal and membrane-based desalination processes [17]. For this purpose, solar-powered desalination systems have attracted considerable attention because they represent a clean and sustainable approach to treating wastewater. Additionally, the decomposition of pollutants via photocatalysis presents an efficient and promising pathway toward clean and green water purification [18]. In this chapter, we highlight the involvement of nanotechnology in water treatment applications, with the discussion focused on the excellent performance of this technology to overcome the current challenges for purifying water.

4.2 Functionalized Nanoadsorbents

4.2.1 Dendrimer-Assisted Filtration

Dendritic polymers are three-dimensional (3D) functionalized nanostructured materials featuring a core, branches, and terminal groups [19]. Dendritic polymers are classified based on their molecular structure as dendrimers, star-like, hyper-branched, hyper-grafted, and dendronized polymers [19], among which dendrimers are extensively used for treating water [20]. Because of their functionalized morphology and fine-tuned structures, dendrimers exhibit high reactivity toward the adsorption of metals and organic contaminants and are thus used to remove heavy metals and other pollutants from water [21]. The amphiphilic structure of dendrimers promotes their interaction with organic molecules via their hydrophobic part and with heavy metals via their hydrophilic functionalities, such as amine and hydroxyl groups [22]. Dendrimers interact with external molecules primarily via their terminal groups [23]. The introduction of end groups to dendrimers leads to well-controlled structures in the form of spherical macromolecules saturated with terminal groups. Thus, the surface of dendrimers is covered with branches of groups, resulting in a high specific surface area that increases the adsorption capacity of dendrimers [19, 23]. The adsorption mechanism of dendrimers primarily involves electrostatic interactions and hydrogen bonding with organic pollutants and metals [20]. Moreover, the internal dendrimer cavity allows the selective capture and removal of organic and inorganic contaminants from water. Among different types of dendrimers, poly (amidoamine) (PAMAM) dendrimers have most commonly been employed as adsorbents in water purification [20]. To improve their mechanical properties and increase their surface area, dendrimers are commonly combined with support materials such as titania (TiO_2), silica, chitosan, graphene oxide, and carbon nanotubes (CNTs) [20]. For instance, combining dendrimers with magnetic NPs allows the resultant adsorbent to be magnetically recovered from solution [24]. Dendrimers are then used in membrane-based separation for water purification. Yoo et al. examined the removal of copper(II) (Cu^{2+}) cations from water using polytetrafluoroethylene (PTFE) grafted with PAMAM. The results demonstrated that the grafted membrane adsorbs 1.42 g Cu^{2+}/m^2 with a higher water flux observed compared to that of a non-grafted membrane [25]. Similarly, Han et al. reported the preparation of a mixed matrix membrane by combining polysulfone (PSF) with hyperbranched PAMAM. The resultant membrane demonstrated the efficient removal of cadmium(II) ions (Cd(II)) from water [26]. Dendrimers have been used to remove chromium(IV) (Cr(IV)) metal [27], dyes [28], and polycyclic aromatic hydrocarbons (PAHs) from wastewater [29].

4.2.2 Cyclodextrins

Cyclodextrins are cyclic oligomers obtained from the degradation of starch using the glycosyl transferase of *Bacillus macerans*. Cyclodextrins are low in toxicity and cost, which makes them important candidates for use in medicine [30], food [31], and textiles [32]. Because of their unique tunable cavity [33], host–guest interactions [34], and water disinfection properties [35], these materials have attracted attention for use in treating contaminated water. The molecular structures of cyclodextrins are composed of hydrophobic cavities and an external hydrophilic surface, wherein the cavity accommodates nonpolar molecules, with polar molecules and solvents being adsorbed on the surface of the molecule. The polar hydroxyl groups of cyclodextrin promote its interaction with metal ions to form complexes, from which the metal can be easily removed to treat contaminated water [36]. Liu et al. developed an Ag-functionalized cyclodextrin for the detection and capture of dyes from wastewater [37]. Moreover, Leclercq et al. reported the structural modification of cyclodextrin as an effective approach of modifying the cavity of the molecule to enhance its interactions [38]. Furthermore, mono-functionalized cyclodextrin polymers have been demonstrated to show excellent adsorption of organic pollutants from water [39]. For instance, modifying the cyclodextrin backbone with hexamethylene diisocyanate (HDI) has been demonstrated to produce an adsorbent that adsorbs 70–100% of *p*-nitrophenol and pentachlorophenol (PCP) from contaminated water [39]. Furthermore, polydopamine-modified β-cyclodextrin polymers exhibit excellent adsorption of cationic dyes, such as malachite green (MG) and methylene blue (MB), in addition to significant adsorption of copper(II) (Cu(II)) from water [40]. For instance, the adsorption capacities of modified β-cyclodextrin polymer toward MG, MB, and Cu(II) are 1174.67, 582.95, and 73.64 $mg\,g^{-1}$, respectively [40]. Furthermore, a metal–organic framework (MOF)-based cyclodextrin has been demonstrated to exhibit ultrafast cadmium(II) (Cd(II)) removal from wastewater, adsorbing 90% of Cd(II) in one minute with an adsorption capacity of 140.85 $mg\,g^{-1}$ [41]. Cyclodextrins have been used to fabricate water-resistant fiber mats for removing phenolphthalein and phenanthrene from water [42], which remove 88% of the initial phenanthrene content from water in the first and 70% in the second cycle [42]. Moreover, crosslinked cyclodextrin nanofibrous membranes have been confirmed to demonstrate excellent performance in the removal of micropollutants from water with 99% removal of the pollutant achieved for estradiol [34]. Note that cyclodextrins have been used to prepare membranes to improve their performance in removing pollutants from water. For example, cyclodextrin has been combined with polyamide membranes and graphene oxide to remove salt, heavy metals, and organic contaminants from water [43]. Similarly, β-cyclodextrin has been used to modify graphene oxide membranes toward their removal of bisphenol A from water [44], with the obtained membrane displaying a high flux of 372 $l\,m^{-2}\,h^{-1}\,bar^{-1}$ and a high adsorption capacity of 166.49 $mg\,m^{-2}$ [44]. Furthermore, a thin-film composite (TFC) membrane has been prepared by reacting cyclodextrin that is rich in hydroxyl groups with trimethyl chloride with the resultant material employed as a chlorine-resistant membrane in the purification of water [45]. In terms of the membrane performance, changing the cyclodextrin concentration from 0.1% to 1.8 wt% changed the flux from 104.6 to 197.6 $l\,m^{-2}\,h^{-1}$ with 95.6 and 94.4% rejection recorded for Congo red (CR) and methyl blue, respectively [45].

4.2.3 Zeolites

Zeolites are crystalline aluminosilicate materials that feature uniform micropores with a pore diameter of <2 nm, which are distributed throughout their structure [46]. Zeolites naturally occur in the environment, where they are found in sedimentary rocks [37, 47], but can be synthesized in the laboratory at a low cost [48]. However, naturally occurring zeolites have firm structures with a restricted pore size, which limits their applications. Nevertheless, synthesized and fly ash-based zeolites exhibit tunable pore sizes and channels, making them potential candidates for use in the adsorption of minerals and the trapping of water contaminants [49]. Raw fly ash has demonstrated potential for use in the removal of dyes, heavy metals, and other pollutants from water. However, compared to raw fly ash and natural zeolites, zeolitic fly ash exhibits sixfold to three- to fivefold higher adsorption capacities toward contaminants, respectively [50, 51]. The negatively charged nature of zeolites allows them to readily interact with cationic molecules, thus leading to their surface modification [48]. In terms of their water purification applications, zeolites have attracted considerable attention because of their porous properties and tunable chemical structures, and they have been investigated and studied for applications such as (i) water softening [52], and the removal of (ii) ammonium [53], and (iii) heavy metals from polluted water [54]. Zeolite powder with a high silica content (high-silica zeolites) has been confirmed to be an effective material for adsorbing organic micro-pollutants from water, including industrial chemicals, personal care products, and pharmaceuticals [55, 56]. For instance, high-silica zeolites demonstrated promising results for adsorbing triclosan, phenol, and trichlorophenol from water [57].

Moreover, zinc oxide (ZnO)-modified zeolites have been used to rapidly remove CR from water with the modified material reported to be effective for up to five cycles of this process, reaching equilibrium within 20 minutes [58]. Furthermore, zeolites have been used to remove antibiotics, such as sulfonamides, from water [59]. High-silica zeolite (HSZSM-5), prepared with an average pore size of 0.53–0.58 nm and a SiO_2/Al_2O_3 (mol/mol) ratio of 70–500 has been demonstrated to efficiently remove (90%) sulfamethoxazole and sulfadiazine from an aqueous solution [59]. Moreover, all silica-zeolite Beta is highly selective for removing perfluorinated contaminants from surfaces and wastewater [60]. The selective adsorption of perfluorinated compounds by this zeolite material is attributed to its high surface area and controllable pore size distribution [60]. The use of zeolites in water treatment has been extended to their incorporation in membranes and as fillers in mixed matrix membranes to enhance the performance of materials and allow the selective removal of pollutants from water [61]. For instance, a zeolitic imidazole framework (ZIF-8) with different particle sizes (40, 70, and 100 nm) has been demonstrated to improve the performance of a polyvinylidene fluoride (PVDF) membrane. The 100-nm-sized ZIF-8-based membrane exhibits a water flux of $350\,l\,m^{-2}\,h^{-1}$, which is ten times higher than the performance of the corresponding pristine PVDF membrane, with an ovalbumin flux recovery ratio of 98% observed [61]. Incorporating zeolites in mixed matrix membranes enhances their wettability and molecular sieving properties, thus leading to an improvement in their flux and selectivity [62]. A layer-by-layer assembly approach has been employed to develop a membrane using zeolite NPs, which was then tested against 1 M of $MgCl_2$ or 0.5 M of trisodium citrate in deionized water [63], with the results demonstrating that incorporated zeolite increases the water flux of the membrane by 120% relative to its pristine counterpart without any notable reduction in selectivity [63].

Moreover, zeolites have been incorporated in thin-film nanocomposite membranes [64]. For example, incorporating silicalite-1 nanozeolite in thin-film membranes led to a notable increase in the permeability of the membrane together with an enhancement in its acid and multivalent cation resistance compared to NaA nanozeolite incorporated membranes [65]. The presence of the zeolite was demonstrated to promote better stability of the membrane toward sodium(I) (Na(I)) and calcium(II) (Ca(II)) ions. The zeolite pore size was reported to control the permeability of water across the membrane by providing extra free space and porosity compared to a pristine membrane [65]. The effects that the interfacial polymerization (IP) reaction time and zeolite loading had on the membrane performance were investigated [66], with the results showing that with an increase in the IP reaction time, there were increases in membrane thickness and the degree of crosslinking, a reduction in the water flux, and an enhancement in salt rejection. Moreover, 0.15 wt% zeolite loading was reported to provide the best combination of water flux (1.78 m^3 m^{-2} day^{-1}) and salt rejection (98.8%) [66].

4.2.4 Silica

Because of their high specific surface area, controlled pore size distribution, structural tenability, good thermal stability, and low cost of manufacturing, ordered mesoporous silica materials have demonstrated considerable potential for use in water treatment applications [67]. For this purpose, highly porous silica adsorbents have been explored and examined for the adsorption of organic and inorganic pollutants from water. The presence of functional groups on the silica allows efficient and selective removal of pollutants via interactions such as hydrogen bonding, van der Waals, and electrostatic interactions [67]. For instance, ordered mesoporous silica, prepared from *N,N*-dimethyl dodecyl amine and dodecyl amine, has been used in the adsorption of heavy metals, such as Cd(II), cobalt(II) (Co(II)), Cu(II), and lead(II) (Pb(II)) ions, from aqueous media and has been reported to operate efficiently at low metal ion concentrations [68]. Moreover, grafted mesoporous SBA-15 silica demonstrates excellent adsorption capacity for mercury(II) (Hg(II)) ions and up to 726 mg g^{-1} adsorption observed from aqueous solution, with the ability to simultaneously remove other metal ions from wastewater [69]. The mechanism of contaminant removal depends on certain factors such as ion-pair formation, surface complexation, solid-phase extraction, ligands, and cation exchange [70]. However, reclaiming the aforementioned silica-based adsorbents from water after completion of the treatment is challenging. Thus, magnetic silica-based adsorbents have been developed to allow the easy removal of the adsorbent from the treated aqueous solution using an external magnetic force [67]. For instance, amino-functionalized magnetic mesoporous silica (SBA-15) shows an adsorption capacity of 243.9 mg g^{-1} toward Pb(II) ions. SBA-15 can be used in the treatment before being collected from the aqueous media, treated with acid, and used in another treatment cycle [71]. Moreover, functionalized SBA-15 adsorbents have been examined for removing pesticides, such as PCP, from contaminated water. Compared to nonfunctionalized SBA-15, aminopropyl- and tripolyphosphate-functionalized SBA-15 exhibit 75% higher adsorption capacities. In particular, SBA-15, tripolyphosphate-functionalized SBA-15, and aminopropyl-functionalized SBA-15 exhibit adsorption capacities toward PCP removal of 0.6, 2.5, and 3.3 mg g^{-1}, respectively [72]. Moreover, Si-coated iron oxide magnetic NPs have been used for oil–water separation under both neutral and acidic conditions [73], where the materials hydrophobicity is important in improving the oil adsorption capacity of silica. The hydrophobic part of the material interacts with oil, while the hydrophilic part interacts with water [73]. Similarly, a cellulose-silica sponge has been developed by Liu et al. and tested for oil–water separation [74], where the carbonyl and methyl groups grafted on the silica sponge improve the mechanical properties and enhance its adsorption capacity [74].

Moreover, both pristine and modified silica adsorbents have been used to fabricate composite membranes for water treatment-related applications [75]. For instance, nanofibrous silica membranes have been used for oil–water separation. For this purpose, incorporating silica into polyvinyl alcohol (PVA) results in the preparation of super hydrophilic membranes with a water contact angle of 0° in air and an oil contact angle of 165° [76].

4.2.5 Nanocatalysts and Green Nanoparticles

Various NPs have been utilized for water treatment applications to remove organic, inorganic, and bio-contaminants. However, to ensure a fully green treatment process, bio-based materials should be developed to complement or replace the current non-ecofriendly materials. For this purpose, recent investigations on developing bio-synthesized NPs shed light on the possibility of achieving good performance using green resources. Nanocatalysts were prepared from natural resources by mixing metal salts at various concentrations with microorganisms or plant extracts. Similar to regular synthesis, biosynthesis allows the formation of NPs in different shapes and morphologies such as square, spherical, triangular, hexagonal, and rod [77, 78]. Moreover, the green-fabricated NPs showed potential to remove toxic materials such as arsenic from water. For instance, green-iron NPs demonstrated maximum adsorption capacity of 14.6 mg g^{-1} for arsenic [79]. Likewise, hemin-modified Bi_2WO_6 composite material prepared from green resources displayed excellent catalytic activity, assisted with H_2O_2 to efficiently remove organic pollutants under solar irradiation [80]. Furthermore, these green nanocatalysts and NPs were used to remove pharmaceutical contaminants and employed in composite membranes to treat water [77].

4.3 Carbon-based Nanoadsorbents

Pollution caused by heavy metals presents a severe threat to both the environment and human health. Heavy metals cannot be chemically degraded, therefore indicating that detoxification of metals from the environment should primarily involve their removal. To obtain clean water, the removal of heavy metals from water is important and has attracted considerable attention. For this purpose, many adsorbent-based technologies have been developed as potential methods to remove metals from water. Because of their surface chemistry, chemical stability, structural tenability, and ease of recovery, nanomaterial adsorbents have demonstrated promising results in replacing conventional adsorbents. The mechanism of heavy metal adsorption using nanomaterials involves the capture of metals in the pores of the adsorbent and at the active sites located on the material surface [81].

Certain factors, such as the adsorption capacity, surface area, physical and chemical stability, support materials, and separation of the nanoadsorbent from aqueous solution after completing the water decontamination process, are important when selecting a nanoadsorbent for removing heavy metals from aqueous solutions [82]. Owing to their nontoxicity and high sorption capacity, carbon-based adsorbents are

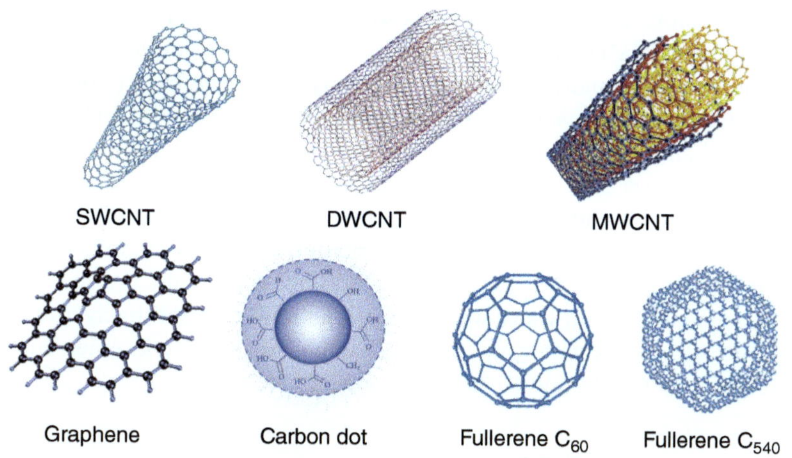

Figure 4.2 Structures of different carbon-based NPs used in treating water. SWCNT = single-walled carbon nanotube, DWCNT = double-walled carbon nanotube, and MWCNT = multiwalled carbon nanotube. Source: Reproduced from Ajith et al. [10] and printed with permission.

attractive materials for remediating water. For this purpose, AC was one of the first adsorbents to be used in filtration and purification processes, in addition to CNTs, graphene, and fullerene nanomaterials (Figure 4.2).

4.3.1 Activated Carbon (AC)

AC has been used as an adsorbent because of its high surface area, porosity, sorption capacity, low cost, low toxicity, and environmental friendliness. AC is a promising industrial sorbent material that is used in various fields, such as the pharmaceutical, agricultural, petroleum, and textile industries, as well as in environmental remediation [83–86]. AC can be derived from natural sources, such as charcoal, husks, wood, coal, coconut shells, and agricultural waste [87]. Furthermore, AC can be fine-tuned via surface modifications to enhance its sorption capacity for the capture of ions and pollutants [88]. Many studies have reported that adsorbent-based AC nanomaterials can be recovered from reaction media, regenerated, and reused for many cycles, indicating that the use of these materials is associated with minimal costs and energy consumption.

Ordinarily, NPs are synthesized via chemical reactions that involve the use of reducing and capping agents, which are considered to be expensive and hazardous chemicals. These chemicals are then discharged into the environment without additional treatment, leaving them free to have a negative impact on both human health and the ecosystem [89]. Therefore, the green synthesis of NPs from plant-based products represents a suitable alternative to conventional chemical synthesis [89]. As reported previously, AC is considered an eco-friendly material that can be used in nanotechnology processes for removing various pollutants from water [90, 91]. For example, a coconut-based AC has been prepared and used as an adsorbent to remove Cr(VI) from aqueous solution. In this study, the sorption capacity was evaluated via batch mode adsorption with the AC exhibiting a sorption capacity of 3.46 $mg\,g^{-1}$. Similarly, green vegetable waste has been used to produce an environmentally friendly adsorbent that was used for the removal of Cu(II) from wastewater, thus exhibiting an adsorption capacity of 75.0 $mg\,g^{-1}$ [92].

An NP-coated AC was developed as a potential adsorbent for the removal of cephalexin from water because this antibiotic is a pollutant that is discharged into the environment by the pharmaceutical, agricultural, and biomanufacturing industries. In laboratory tests conducted under optimal conditions, the removal efficiencies of cephalexin from water were 96.06% and 94.17%, respectively. In the same study, the adsorbents were put to test in actual wastewater samples with removal efficiencies of 69.78% and 66.32% being achieved [89].

4.3.2 Carbon Nanotubes (CNTs)

CNTs are one of the major building blocks used in nanotechnology and are categorized as single-walled carbon nanotubes (SWCNTs) and multiwalled carbon nanotubes (MWCNTs). SWCNTs comprise one layer of carbon atoms and have diameters in the range of 0.3–3 nm, whereas MWCNTs are composed of more than one layer of carbon atoms to form concentric tubes with diameters of ≥100 nm [93]. Because of their excellent properties, such as high specific surface area, light weight, hollowness, layered structure, high tensile strength, and good electrical properties, CNTs have been investigated for their use in separation applications. Moreover, their good biocompatibility, chemical stability, and ability to interact with organic and inorganic molecules make them extremely useful for removing contaminants from solutions [94–97].

Heavy metals are removed from water by CNTs via different mechanisms such as physical adsorption, chemical interaction, electrostatic attraction, and sorption precipitation. The surface of CNTs can be oxidized to functionalize them with different groups (i.e. carboxyl, hydroxyl, and phenol), thus resulting in the activation of sites on their surface. Using surface modification, the absorption capacity of CNTs can be improved because of electrostatic interactions and chemical bonding [98].

As carbon nanomaterials are often synthesized using fossil fuels and hazardous chemicals at high temperatures, the design of eco-friendly materials with high efficiency toward pollutant removal is highly desirable. For this purpose, environmentally friendly alternatives have been explored using natural hydrocarbon sources such as lignin and camphor [93]. For example, a green method has been reported for synthesizing CNTs from camphor. In this method, camphor was vaporized and pyrolyzed over a zeolite support impregnated with Fe–Co at 650 °C to afford 1.62 g of 10-nm-diameter MWNTs in 80% purity, thus representing an increased production efficiency of up to 50% compared to a conventional method [99].

Moreover, lignin has been combined with CNTs to develop a green nanocomposite adsorbent with a high capacity for removing Pb and oil droplets from water. However, the disadvantage of the method used to produce this nanocomposite is that, although the preparation can be considered green, CNTs were initially synthesized from methane at a high temperature of 900 °C [100]. Among the currently used grafting approaches that are used to add functional groups to the surface of CNTs, a plasma technique has been reported [101]. In this plasma technique, ultrasound is used to improve the sorption and desorption kinetics of the CNT-based material with the hazardous solvent traditionally used in such processes being replaced with an alternative green solvent. For example, using this method, the cationic surfactant cetyl trimethyl ammonium bromide was grafted on MWCNTs via ultrasonication in the presence of the green solvent isoamyl alcohol. The efficacy of the produced MWCNT-based material in the removal of fluoride ions from aqueous solution was then assessed with the maximal fluoride uptake determined to be 20.1 $mg\,g^{-1}$. The applicability of this green technique for removing fluoride from groundwater is particularly notable [95]. In the pharmaceutical industry, highly purified products are produced using different methods, requiring expensive equipment and tedious procedures. Molecular imprinting has been proposed as a technique to purify a lysozyme antibacterial agent by synthesizing imprinted materials with recognition sites using carboxyl-functionalized MWCNTs (MWCNTs–COOH) as a support and dopamine (DA) as a functional monomer. Using this material, a high binding capacity of 418 $mg\,g^{-1}$ was observed toward lysozymes [96].

4.3.3 Fullerenes

Fullerenes are carbon molecules that form a hollow closed polyhedral cage of >20 atoms, with the formula ($C_{20} + m$) where m is an integer. The most famous example of a fullerene is C_{60}, comprising 12 pentagonal and 20 hexagonal fused ring structures, which was discovered three decades ago by buckminsterfullerene. Buckyballs are another type of small fullerene structure that has ≤20 atoms [98, 102]. These unique structures have garnered interest in multiple potential industrial applications because of the inner cage of fullerenes, which traps molecules or pollutants. Furthermore, novel fullerene-based technologies have been developed by industries for environmental applications such as water treatment, separation, and semiconductors. Fullerenes have excellent electrochemical and electronic properties, and their hydrophobicity, strong electron affinity, and high surface area play an important role in environmental applications [103]. In addition to their application in environmental remediation, fullerenes are good candidates for use as hydrogen storage materials [104] and in electronics [105].

Interestingly, fullerenes have been incorporated in a polystyrene nanocomposite membrane for removing Cu(II) from wastewater [25], with the results showing that the incorporation of the fullerene enhanced the sorption capacity of the membrane compared to its pristine polystyrene counterpart [106]. Furthermore, fullerenes have the potential to be used in the disinfection of water because of their tendency to produce reactive oxygen species (ROS), which then photodegrade organic pollutants. After the photodegradation process, the fullerene is removed from the water [107].

Recently, C_{60} fullerene has been examined as a magnetic solid-phase extraction (MSPE) sorbent for the removal of heavy metals from water. A soluble eggshell membrane protein (SESMP) composite was prepared by mixing C_{60} with Fe_3O_4 NPs to form a fullerene-based biopolymer–magnetite nanoinorganic composite sorbent for removing arsine (AS) from oil and water. The sorbent was applied to determine the As concentration in samples of crude oil collected from oil field reservoirs and water samples from different sources, including ground, tap, and mineral water, with a high recovery of AS of ~99% from most of the water and oil samples, thus indicating the effectiveness of the C_{60} sorbent [105].

4.4 Antimicrobial Nanomaterials

The global misuse of antibiotics and their preemptive use in animal husbandry have resulted in the emergence of antibiotic-resistant bacteria [108, 109]. Because the development of other drugs is more profitable to pharmaceutical companies than the development of new antibiotics, and because it is impossible to generate antibiotics at the same rate at which microorganisms evolve, it is necessary to identify novel approaches to controlling these organisms [110]. For this purpose, nanotechnology has been used to produce materials that kill or reduce the activity of multiple microorganisms [111]. For example, antimicrobial materials have been developed to which bacteria cannot adhere, and bactericidal surfaces have been prepared that allow the attachment of bacteria, which are subsequently killed after attachment. Generally, for evaluating the antimicrobial properties of novel compounds, two strains of bacteria are primarily used: *Escherichia coli* and *Staphylococcus aureus* [110]. The results show how the compound may interact with other gram-positive and gram-negative bacteria [110].

In the past few years, because of their assumed nontoxicity and environmentally benign nature, there has been considerable interest in using natural products to produce materials to deal with unwanted microorganisms [110]. As microorganisms pose a constant threat to all living organisms, most cells produce antibiotic-like molecules [48]. They are abundantly observed in nature, primarily in plants. Structurally, these molecules are in the form of polysaccharides and their derivatives, peptides, enzymes, essential oils, weak organic acids, and polyphenols [110].

4.4.1 Chitosan and Peptides

Chitin, which is present in exoskeletons of crustaceans, insects, and certain fungi, can be deacetylated to produce chitosan, as shown in Figure 4.3, which is a compound that exhibits antimicrobial properties. The principal source of chitin is as a waste product in shellfish, from which it is extracted [110, 112]. Chitin is a polysaccharide that is composed of β-(1–4)-linked glucosamine units. To extract chitin, waste shellfish are subjected to acid treatment to dissolve their calcium carbonate, followed by extraction in base to dissolve the proteins and decolorize the solution to allow chitosan to be obtained as a colorless product [112]. Chitosan has many properties, including biocompatibility, adsorption ability, biodegradability, nontoxicity, a macromolecular structure, and a low-cost material to obtain. Because of these advantageous properties, chitosan has been applied in various fields,

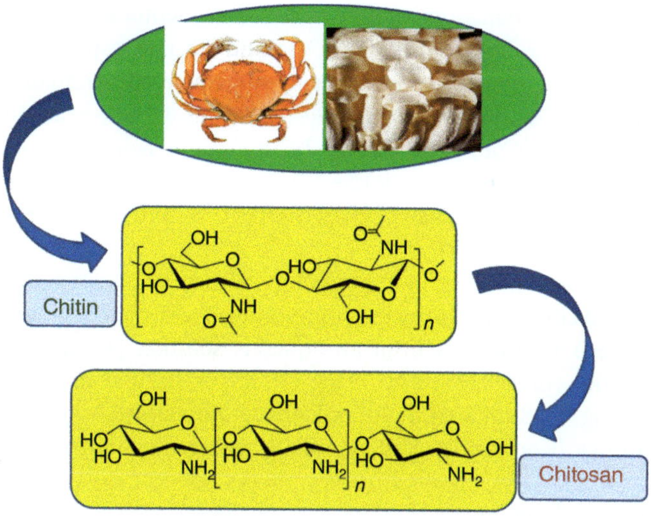

Figure 4.3 Chitosan is produced from chitin extracted from natural resources such as crab shells and mushrooms.

including medicine, membranes, biotechnology, the food industry, and cosmetics [113–115].

Because it is activated at a pH below its pKa of 6.5 to become polycationic, interacts electrostatically, and forms complexes with the anionic surface of microorganisms, chitosan shows a broad spectrum of antimicrobial activities against gram-positive and gram-negative bacteria [116]. After the adhesion of chitosan to cells, the cell membrane is disrupted, triggering the leakage of intracellular components. Moreover, chitosan can penetrate the cell membrane and interact with deoxyribonucleic acid (DNA), thus inhibiting its transcription and protein synthesis [117].

Chitosan is susceptible to structural changes; minor changes in its structure significantly impact its activity. Moreover, its antimicrobial activity is dependent on the pH of the medium, the degree of deacetylation, and its molecular weight [110]. Therefore, the synthetic method for derivatizing chitosan must be carefully selected to produce a material that has desirable characteristics [118]. Many modified chitosan materials have been synthesized that exhibit improved aqueous solubility and antimicrobial bioactivity compared to pristine chitosan with the most common functional groups used to modify chitosan being hydroxyalkyl, carboxyalkyl, guanidinyl, quaternary ammoniumyl, thiol-containing, and other hydrophobic (including substituted phenyl and long) alkyl chains and benzyl rings) groups [119, 120].

Based on their high adsorption capacity, fast kinetics, reusability, and ability to bind and form stable complexes with metal ions, chitosan-based cross-linked polymeric materials have garnered interest for their use in wastewater treatment [121]. Moreover, PVA–chitosan and poly(ethylene glycol) (PEG)–chitosan composites can remove nitrate ions from aqueous solutions. Using these materials for this purpose, Rajeswari et al. reported adsorption capacities for the composites of >35.03 and 50.68 mg g^{-1}, respectively [122]. Furthermore, modified carboxymethyl chitosan and goethite/chitosan nanocomposites have been demonstrated to be selective toward the complexation of Cd(II)/Cr(VI) and Pb(II), respectively [123, 124]. Similarly, chitosan can be used to effectively remediate contaminated water through its adsorption of pesticides with studies showing it being used to remove >90% of oxadiazon (herbicide) [125] and 76.2% of atrazine from aqueous solution. In addition, chitosan/modified sepiolite has been used to remove a herbicide (with a maximum adsorption capacity of 17.92 $mg\,g^{-1}$) from an aqueous solution [126].

Chitosan-based nanomaterials have been reported to adsorb or remove dye molecules solutions. Marrakchi et al. fabricated a crosslinked chitosan/sepiolite composite to remove orange 16 and MB from an aqueous solution [127], with maximum attained adsorptions of 190.97 and 40.99 $mg\,g^{-1}$, respectively, at an initial dye concentration of 100 $mg\,l^{-1}$. Similarly, a glutaraldehyde-crosslinked magnetic chitosan nanocomposite has been reported to eliminate acid red-2 from textile wastewater [128].

Antimicrobial peptides (AMPs) comprise 12–50 amino acid lengths and can be isolated from multiple sources, including plants, animals, or bacteria, with their role being to kill or inhibit the growth of foreign organisms [108, 129]. In the antimicrobial peptide database (APD), there are >3000 entries registered compared to the discovery or synthesis of peptides with antibacterial, antifungal, antiviral, antiparasitic, and other properties. These materials show antimicrobial activity against a broad spectrum of organisms because of their strong interaction with negatively charged membranes, their inhibition of protein, DNA, and RNA synthesis, and their interaction with intracellular targets [112].

Compared to conventional antibiotics, AMPs show considerable advantages, such as a low propensity to induce resistant pathogens, a high ability to discriminate among the host and external cells, and increased activity against an expanded spectrum of microorganisms [110]. However, AMPs have certain limitations, including high production costs, reduced activity when used for surface coating, sensitivity to protease, and extreme pH conditions [130, 131].

Kanchanapally et al. developed an antimicrobial membrane composed of peptide-conjugated graphene oxide with the obtained membrane showing promising results toward the disinfection of water contaminated with methicillin-resistant *S. aureus* [132]. Therefore, this antimicrobial membrane may be efficient in targeting and killing multiple drug-resistant bacteria. Moreover, AMPs can be used to develop functional coatings that are resistant to fouling [129]. Maity et al. designed a tripeptide that self-assembles in an antifouling coating for use on metals, metal oxides, and polymer surfaces [133]. The coating adheres to surfaces via the amino acid 3,4-dihydroxyphenylalanine (DOPA), which is a compound that in the marine world allows mussels to adhere to various surfaces, even the traditionally non-stick coating Teflon. The formation of the coating is driven by two phenylalanine residues that direct the self-assembly of the peptide via π–π stacking. The aromatic rings of phenylalanine were modified using fluorine atoms to form a hydrophobic-fluorinated polymer to prevent the undesirable adhesion of bacteria and proteins, the performance of which was evaluated against *Pseudomonas aeruginosa* and *Escherichia coli*, with a 90% reduction in the adhesion of bacteria observed compared to an uncoated surface. Such coatings can thus identify uses in health care, the marine industry, and water treatment.

4.4.2 Silver NPs

Metallic NPs are another type of antimicrobial compound, with the most commonly known type being silver NPs [133, 134]. NPs can be easily synthesized in two primary ways: via top-down or bottom-up approaches. The former involves the mechanical transformation of a bulk metal into a nanostructure using a stabilizing agent, whereas the latter involves the reduction of metal ions obtained from metal precursors [135]. Depending on the synthetic method and procedure used, the resultant NPs present multiple different types of morphologies [136]. Green chemistry techniques can be used to produce NPs in an eco-friendly manner using plant material extracts, bacteria, fungi, and agricultural waste in their synthesis. The synthesis of NPs from plant extracts produces less waste in the form of by-products, the synthesis is nontoxic, and reaction exhibits higher productivity [137].

In terms of action mechanism, NPs damage microbial cell membrane by inducing structural and physiological changes. For example, changes occur in the membrane potential and permeability because of the accumulation of NPs, the inhibition of respiratory proteins attached to the membrane, alteration of cell homeostasis leading to microbial cell death, and the generation of ROS that induce oxidative stress in the cell and cause severe damage or death [138]. Moreover, the positive charge of silver(I) ions (Ag^+) is important for antimicrobial properties because when Ag^+ ions are internalized, they bind to the sulfhydryl groups of proteins, thus leading to their denaturation [139, 140].

Singh et al. reported the antibacterial activity of Ag NPs against *Acinetobacter baumannii*, a bacterium associated with hospital-acquired infections that leads to high mortality, with a minimal inhibitory concentration (MIC) observed that was much lower (16 $\mu g\,ml^{-1}$) than those of the antibiotics usually used to treat this pathogen, which are ampicillin (4096 $\mu g\,ml^{-1}$), amoxicillin (2048 $\mu g\,ml^{-1}$), and erythromycin (64 $\mu g\,ml^{-1}$) [141]. In another study, Bonde et al. synthesized NPs loaded with *Murraya koenigii*, obtained from the Indian curry leaf tree, and reported that the 40–80-nm-sized NPs produced were more active against *P. aeruginosa*, *E. coli*, and *S. aureus* than NPs loaded with tetracycline and gentamicin [142]. Selim et al. synthesized NPs with an average size of 50 nm and evaluated them in terms of their performance against *Mycobacterium tuberculosis* reference strains, an MDR *M. tuberculosis* strain, and clinical isolates, demonstrating MIC values of 1, 16, and 1–16 $\mu g\,ml^{-1}$, respectively. The aforementioned results thus demonstrate that silver NPs are excellent antimicrobial agents [143].

In addition to their antimicrobial properties, NPs have characteristic physicochemical and biological properties that make them useful for a wide range of applications. For example, they present a large area-to-volume ratio, excellent plasmon resonance, catalytic properties, are cytotoxic toward cancer cells, and exhibit toxicity against pathogens. Silver NPs have excellent antimicrobial properties [128] and are used for a wide range of purposes, including in clothing and food production, water purification, and the diagnosis of diseases [144–146].

In developing countries that usually do not have centralized water treatment facilities, Ag NPs have been used as a point-of-use (POU) water treatment to reduce infections and diarrhea. For example, Tata Swach, a water filter that uses rice husk ash containing silver NPs, can achieve 1 log *E. coli* removal from water [147]. A filter paper coated with Cu and Ag NPs produced by Folia shows >6 log *E. coli* removal using 1.6 mg AgNP g^{-1} coated filter paper [148]. Moreover, Karthik et al. synthesized a reduced graphene oxide–silver NP hybrid nanocomposite using an aqueous extract of *Brassica nigra*, which, besides showing antibacterial activity as a photocatalytic agent, proved to be effective in removing dyes from aqueous solution [149]. Surprisingly, it was found that bioactive silver NPs can be prepared from bacteria *B. cereus* by exposing the strain to different concentrations of silver salt [$AgNO_3$]. The bioactive NPs displayed potential in water treatment applications, but still require further investigation to evaluate its behavior and efficiency in water treatment [150].

4.4.3 TiO_2 NPs

TiO_2 NPs have been identified as attractive photocatalysts that show promising performance in the removal of pollutants from water and air [151, 152]. These NPs are non-toxic, highly stable, low in cost, and easy to produce because their synthesis only requires reagents, a source of light, and oxygen, and can be performed under room temperature and pressure conditions. If the source of light is the Sun, the method for producing these NPs is cost-effective and environmentally friendly [152]. These NPs exhibit activity against many different types of bacteria, viruses, fungi, and protozoa in water sources [153]. Similar to other metallic NPs, the primary mechanism of action of TiO_2 NPs is ROS. The NPs trigger the formation of ROS when in the vicinity of microbial cells [154]. These ROS disrupt the walls of the cells via phospholipid oxidation, resulting in reduced adhesion and ionic imbalance. Inside the cytosol, the NPs inhibit respiratory enzymes and modify their macromolecular structures, damaging cellular integrity and gene expression. Furthermore, phosphate uptake and cellular communication are reduced across the cell. Moreover, ROS generate a combined effect in the bactericidal process [155].

As is the case for silver NPs, the green synthesis of TiO_2 NPs has been explored [156], achieved using a simple precursor salt and a biological extract. Plants are considered to be the most suitable candidates for the synthesis of NPs because they contain a diverse range of compounds that regulate the reduction and stabilization of the bulk metal into its elemental form [157, 158]. For example, a leaf extract of *Calotropis gigantea* has been reported to reduce TiO_2 to NPs in six hours [159]. In another study, spherical NPs were obtained when *Annona squamosa* was added to an aqueous solution of TiO_2 salt [160]. TiO_2 NPs obtained from the leaf extract of *Hibiscus rosa-sinesis* have demonstrated high antimicrobial activity against gram-positive and gram-negative bacteria [161]. TiO_2 NPs doped with Fe can be produced using an aqueous extract of lemongrass (*Cymbopogon citratus*) [162], with the product exhibiting potential in the photocatalytic treatment of wastewater and the degradation of organic pollutants. In another study, Younas et al. Ag-used doped titania NPs to photocatalytically disinfect water, with the 1% Ag–TiO_2 material exhibiting the synergistic effects of the combination of the metals, thus killing 100% bacteria in only 30 minutes [163].

4.4.4 Zinc Oxide NPs

Zinc oxide has one of the highest photocatalytic efficiencies among inorganic photocatalytic materials, exhibiting high selectivity, good durability, heat resistance, and a wide spectrum of activity against *S. aureus* and *E. coli* [164, 165]. Four mechanisms of action have been proposed for the performance of zinc oxide: (i) the production of ROS, such as superoxide anions (O^{2-}), hydroxyl radicals (HO·), and hydrogen peroxide (H_2O_2), which destroy cellular components such as DNA, proteins, and lipids [166, 167]; (ii) the loss of cellular integrity following contact between the ZnO NPs and cell wall; (iii) the release of Zn^{2+} ions; and (iv) the internalization of the ZnO NPs [168].

Velsankar et al. synthesized ZnO NPs using *E. variegata* leaf extract, whose biofilm activity was shown to have efficient antibacterial characteristics. The antibacterial activity of the material was evaluated against *Salmonella typhi* (gram-negative bacteria), and *S. aureus* (gram-positive bacteria) with maximum activities of 93% and 97% and MIC values of 150 and 175 $\mu g\,ml^{-1}$ being observed [169]. Motshekga et al. reported the synthesis of silver and zinc oxide NPs supported on bentonite and independently tested them against *E. coli* and *Enterococcus faecalis*, thus observing bacterial removal efficiencies of >51%. Moreover, a synergistic effect was observed for a silver–zinc oxide bentonite chitosan nanocomposite, which exhibited a bacterial removal efficiency of ≥78%. The parameters that influence the antibacterial activity of the materials were reported to be the amount of material, contact time, and bacterial concentration. After leaching tests, the NPs were demonstrated to be stable and could be effectively used as antibacterial materials for the disinfection of water [170]. Moreover, zinc oxide NPs have been used to catalyze the photodegradation of MB in aqueous streams, where the optimal conditions used to remove 94% of the dye within 20 minutes were 0.45 $mg\,l^{-1}$ of ZnO NPs, pH 11, and a flow rate of 370 $ml\,min^{-1}$ [167].

4.5 Desalination

Access to clean water is a necessity for humanity. However, by 2030, it is estimated that >50% of the global population will experience water shortage [171, 172]. As it is already an enormous challenge to meet the clean water demands of the world population because of water pollution and scarcity, the situation will continue to worsen with the rapid growth of the population, fast industrialization, and continuous improvement in living standards [173, 174]. Moreover, water scarcity impacts both human health and economic growth. The ever-increasing demands for producing clean water have thus been pushed to the limits, with the requirement to process unusual water sources such as stormwater, contaminated water, brackish water, industrial water, and seawater, particularly in arid regions, as well as reuse and recycle water [112]. For this purpose, many arid and coastal areas obtain their safe drinking water via the desalination of seawater, which is the most abundant source of water on Earth [175].

To obtain abundant fresh water from the sea without being limited by energy supplies, it is necessary that desalination technologies be energy-efficient and operate as close to the thermodynamic limit as possible [176]. The thermodynamic limit is the minimum theoretical energy required to desalinate water, determined thermodynamically by the Gibbs free energy of separation, ΔsepG. It is equivalent in magnitude but has the opposite sign to the Gibbs free energy of mixing and is defined as the difference between the Gibbs free energy of the total mixed solution and the sum of the Gibbs free energies for each of the component species [177]. Desalination processes are normalized per unit volume of product water, referred to as the specific energy consumption (SEC). This increases when the desired recovery (the percentage of seawater converted to freshwater) or the salinity of the feedwater increases, e.g. the theoretical thermodynamic minimum energy required to desalinate water with $35\,g\,l^{-1}$ of salts at 50% recovery is ~$1.1\,kWh\,m^{-3}$, indicating that water and salt cannot be separated at a lower energy.

Desalination technologies are pressure-, temperature-, or chemical-driven processes, with the primary representatives of each being reverse osmosis (RO), multistage flash distillation (MSF), and membrane capacitive deionization (MCDI), respectively. The current challenges of these technologies are fouling; scaling because of the precipitation of poorly soluble salts, cleaning of MSF plants and RO membranes, maximizing the flux in RO membranes or optimizing MSF plants, reducing energy losses in MSF; optimizing the design and operation of desalination plants, cost reduction, and increased sustainability of the processes [178, 179]. Figure 4.4. shows the potential performance and commercial viability of nanotechnology-enabled membranes prepared from different nanomaterials.

4.5.1 Membrane-Based Desalination

Membrane materials are considered important for addressing global water scarcity [180]. Filtration membrane-based methods have been extensively developed, introduced to the market, and used on a large scale for water desalination and wastewater filtration. Different processes have been used for water desalination, such as nanofiltration (NF), ultrafiltration (UF), microfiltration (MF), forward osmosis (FO), and RO, with RO being the technology that is the closest to the thermodynamic limit and the most energy-efficient technology known for desalination with optimized plants reaching $2\,kWh\,m^{-3}$ in a controlled pilot-scale study [181, 182]. Over the past 40 years, the energy requirements of this technology have been dramatically reduced because of improvements made to the membranes used, the installation of recovery devices, and more efficient pumps [181].

In terms of the energy required in this method, 47% corresponds to the minimum energy consumption, and the other 53% is for work performed to generate

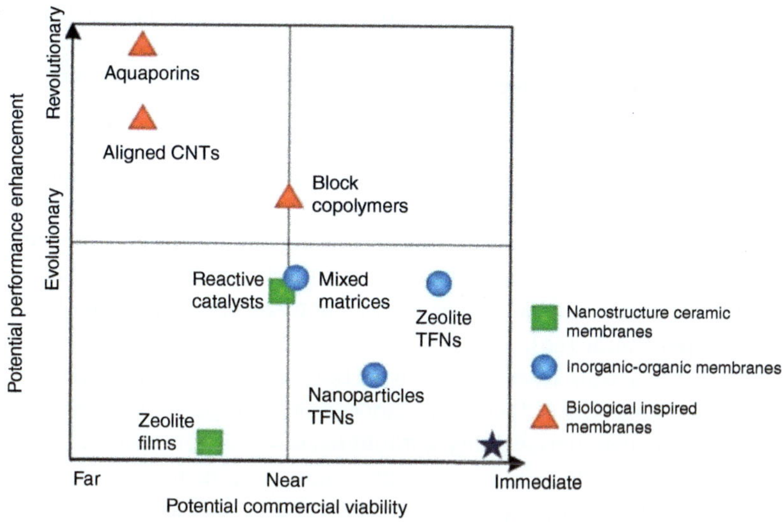

Figure 4.4 Comparison of the potential performance and commercial viability of nanotechnology-enabled membranes. Source: Goh and Ismail [180]/John Wiley & Sons.

membrane flux, overcome module configuration, and overcome friction and concentration polarization losses, as well as overcoming pump, energy recovery device, and system friction losses [183]. Of the required energy, 75% cannot be optimized, with the remaining 25% corresponding to membranes (15%) and modules (10%) [182]. This indicates that, on a large scale, the optimization of membranes does not significantly affect the energy requirements of the process. However, the optimization of membranes will lead to cost savings and sustainability as the lifetime of the membranes can be increased. Certain important areas for future innovations in membranes are as follows:

- Antifouling: fouling is one of the biggest challenges for membranes because of their hydrophobicity. Fouling can be either reversible, indicating that the concentration polarization can be addressed by backwashing or manipulating the flow velocity to remove foulants that are loosely adhered, or irreversible, when foulants cannot be removed. This leads to a decline in flux and a reduction in the life of the membrane in both cases [184]. Antifouling membranes can exhibit high permeability or high operational flux without intensive pretreatment, resulting in savings in terms of capital, energy, and chemical pretreatment [183]. For this purpose, nanocomposite membranes have been developed with hydrophilic, self-cleaning, photocatalytic, and photodegradation properties [184].
- Specific ion rejection: the use of highly selective membranes leads to processes that are more efficient. For example, the use of membranes that exhibit selectivity toward boron, heavy metals, arsenic, or micropollutants could negate the requirement for a second pass [182, 185].
- They require being mechanically robust and resistant to chemicals.

- Improving the membranes used for pretreatment can reduce costs and lower energy requirements.
- Scalable and cost-effective manufacturing.

Currently, the most commonly used membranes for RO are TFC polyamide membranes comprising three layers: a polyester web that acts as a structural support of 120–150 μm in thickness, a microporous interlayer of ~40 μm, and an ultra-thin barrier layer on the upper surface of ~0.2 μm. Decreasing the thickness of the barrier layer minimizes the resistance to permeate transport. A membrane pore size of <0.6 nm generally leads to a salt rejection of >99%, where the selective barrier layer is often made from polyamide [66].

Breakthroughs in the desalination of water have been achieved using polyamide membranes functionalized with NP polymer–nanocomposite membranes (PNCMs) [186], which exhibit excellent permeability, antifouling, salt rejection, and increased energy efficiency [186, 187]. These membranes are produced by incorporating nanomaterials in the form of nanotubes (NTs), nanofibers (NFs), nanosheets (NSs), or NPs in a polymer matrix using different fabrication methods [188]. CNTs, graphene, and zeolites are nanomaterials commonly used for this purpose [186].

Aquaporins are water-channel membrane proteins that allow the passage of water molecules only in their usual biological performance. Recently, these materials have garnered interest because of their high selectivity and permeability and because they have demonstrated good permeability in solute-rejection use for water desalination [189]. Kumar et al. examined the incorporation of bacterial Aquaporin Z into membranes, thus reporting a higher water transport efficiency compared with standard RO membranes, with at least one order of magnitude improvement in permeability [190].

Furthermore, molecular dynamics (MD) simulations demonstrated that CNTs display very similar behavior to aquaporins in terms of their water transport properties [191]. Both types of molecules have a hydrophobic inner cavity that facilitates the formation of one-dimensional hydrogen-bonded water chains and almost frictionless transport of water because of weak interactions [192]. Kalra et al. demonstrated that water moves rapidly inside an NT channel under osmotic pressure [193]. Moreover, because of their antimicrobial properties, CNTs control fouling and are considered to be self-cleaning membranes [194, 195]. CNTs rupture bacterial cells by producing ROS, disrupting their metabolic pathway, and causing oxidative stress [194]. CNTs exhibit good water permeability and salt rejection, which can be improved by functionalizing the tips and pore size of the NTs [195, 196]. A reduction in the pore size and an increase in the number of charged groups at the ends of CNTs significantly increase their solute rejection, but at the cost of a reduction in their water permeability [195].

Moreover, because of their excellent thermal and mechanical stabilities, molecular sieving properties, and competitive adsorption, zeolite membranes have been reported to be suitable for use in different desalination technologies [197, 198]. These membranes can withstand high pressure in RO, as well as high temperature in MDs; the use of them is even preferred for improving the separation efficiency of zeolite membranes [197]. Fathizadeh et al. incorporated a nano-zeolite into polyamide as a thin top layer of a membrane, which made the membrane more thermally stable with a higher water permeability compared to a pristine membrane, but with a lower ion rejection [199]. In another study, MD was used to show that zeolite membranes could theoretically separate ions from aqueous solutions via RO processes [200], separating according to size exclusion and interactions between the hydrated ions. The calculations indicate that a higher valence leads to a greater separation efficiency of the membrane, therefore indicating that it is necessary to control the pore size of the material. For example, to reject Na^+ and Cl^- with hydrated sizes of 0.72 and 0.66 nm, respectively, the intercrystalline interconnection and pore size must be well-controlled to <0.66 nm [187].

Although nanotechnology is leading the way in innovating RO membranes for desalination, these materials are still extremely expensive to use in practical applications. Therefore, to address this issue, the salt rejection must be enhanced and the permeability increased at an affordable cost [190]. Pendergast and Hoek investigated the scalability and compatibility of mixed matrix and thin film nanocomposite (TFN) membranes prepared by dispersing inorganic nanomaterials in a polymer phase [180]. Additionally, hexagonal boron nitride-based (hBN) membranes demonstrated potential in water desalination applications. Interestingly, hBN membranes showed excellent properties against fouling, which is considered a major drawback in other membrane technologies. hBN-based membranes showed good thermal and mechanical stability with fast water transport, which improved their water/ion separation, leading to efficient water desalination. However, these membranes are not heavily explored yet, and further investigation is required to evaluate their performance in water remediation [201].

4.5.2 Thermal Desalination

Membrane-based processes are adequate to use in the desalination of water with a total dissolved solids (TDS) content of <45 000 mg l^{-1}, whereas water with a higher TDS can be treated using conventional thermal techniques. Although it involves additional energy-intensive techniques, thermal desalination technology is considered to be cheap, but has the disadvantage of low productivity [17, 202, 203]. The productivity of this technology can be enhanced by adding NPs, such as in the form of a nanofluid, which improves the optical absorption capability and thermal conductivity of salt water [202, 203]. The visual absorption capability is attributed to the plasmon resonance absorption band ascending in the near-infrared and visible spectrums of the NPs. Moreover, the temperature of the NPs increases and their thermal conductivity decreases [203].

Sharshir et al. examined the effect on the productivity of a single basin solar still upon the addition of copper oxide and graphite microflakes [204, 205], with the results showing an improvement in productivity of 44.91% and 53.95%, respectively, compared to in the absence of a nanofluid. Kabeel et al. examined the effect that cuprous and aluminum oxides have on a single-basin solar cell in the presence and absence of a vacuum inside the still [206]. The results of this study demonstrated that the cuprous oxide–water nanofluid exhibited higher productivity than its aluminum oxide counterpart. Moreover, when an external condenser was integrated in the single basin solar still with the aluminum oxide, the productivity increased by 116%.

Kabeel and El-Said proposed a desalination system using MSF and a nanofluid-based solar collector [207]. In their system, Cu NPs were used to determine the proper NP volume required for maximum freshwater productivity via modeling. The use of solar energy reduces costs and leads to increased water production. Although nanofluids can enhance multiple solar energy applications, they present challenges in terms of scale-up, high production costs, and agglomeration because of their instability [203].

4.5.3 Chemical Desalination

Chemical-driven technologies represent an emerging approach to desalinate water, with the advantages being low maintenance costs (around a third of the costs required for RO, and the long-term cyclability of the electrode materials, which can operate at different levels of ion removal with minimal clogging of their pores). Certain examples of these technologies are ion exchange, liquid–liquid extraction, and precipitation. Among these, ion exchange processes are the most common, such as capacitive deionization (CDI), MCDI, and electrodialysis (ED) [1].

Capacitive deionization follows the principles of the electroadsorption of ions at the surface of electrically charged electrodes [1] and involves the removal of ionic or polarizable species from water [208]. The negative electrode (cathode) attracts positively charged ions (cations), such as calcium (Ca^{2+}), magnesium (Mg^{2+}), and sodium (Na^{+}). However, the positively charged electrode (anode) attracts negative ions (anions), such as chloride (Cl^{-}) and nitrate (NO^{3-}) [209]. After some time, the electrodes become saturated with ions and require to be regenerated. For this purpose, the applied potential is removed such that the ions can be released back into the feed to produce a more concentrated brine stream [209]. Salt removal can be decreased because of the co-ion expulsion effect in this step. Therefore, ion exchange membranes (IEM) can be added to the electrodes to improve their performance and useful life, with an anion exchange membrane (AEM) placed adjacent to the anode and a cation exchange membrane (CEM) placed adjacent to the cathode. Usually, when the cell contains both types of membranes, the process is referred to as MCDI (Figure 4.5) [210]. The membranes in the cell reduce the co-ion expulsion effect

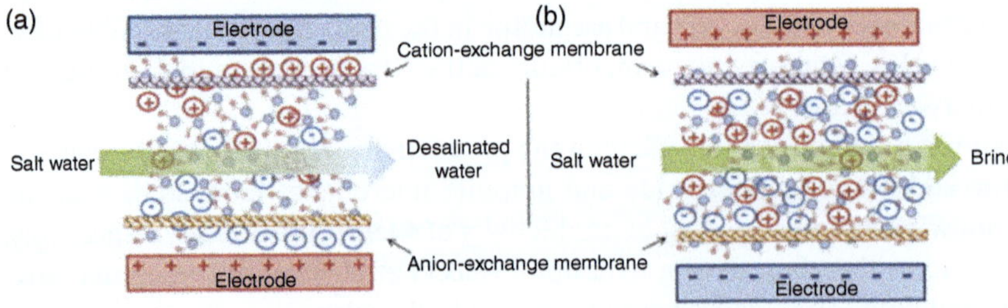

Figure 4.5 (a) Purification and (b) regeneration processes in MCDI.

during the electrosorption process and enhance the overall performance. Using this method, the highest salt adsorption rate and water flux have been achieved, making it a promising desalination technology [211]. MCDI systems have been employed in the processing of power plant wastewater, and it has been reported that the rate of removal of sodium and chloride from water using the MCDI process is 19% higher than the same process achieved using CDI, with a maximum salt removal of 92% and an energy consumption of 1.96 kWh m^{-3} [212].

Over the last 20 years, there has been increasing interest in the area of the desalination of seawater, e.g. with the use of carbon aerogels; carbon cloth; CNTs; and graphite and its composites in this pursuit [1]. The electrode requirements are a high surface area, high porosity, high electrical conductivity, electrochemical stability, bioinertness, fast adsorption–desorption kinetics, good wetting behavior, low cost, and scalability [1]. Usually, carbon-based materials are used as CDI electrodes because of their high specific surface area, high adsorption capacity, reasonable microporous structure, and unique surface reactivity. Among them, CNTs are an ideal material because of their excellent electrosorption performance [213].

In 2011, Wang et al. reported the use of CNT sponge electrodes to desalinate NaCl solutions [214]. These CNT sponges are interconnected CNTs with mesoporous and macroporous structures [215]. These sponges exhibit high conductivity and a high effective surface area, with a maximum adsorption capacity of 40 mg g^{-1}, deduced from a Langmuir isotherm, which is higher than the capacities of other carbon-based materials. Carbon aerogel electrode-based CDI has been demonstrated to be competitive compared to RO at a TDS of <2000 ppm, with a cost of 0.11 \$ m^{-3} for CDI and 0.35 \$ m^{-3} for RO; however CDI demonstrated higher energy saving of 2.03 kWh m^{-3} relative to RO with energy saving of 0.59 kWh m^{-3}. At seawater concentrations, RO remains more effective [216]; however, it is necessary to increase the scalability, salinity, electrode efficiency, and cost-effectiveness of this method to compete with current technologies [217].

4.6 Photocatalysis

4.6.1 Heterogenous TiO_2 Photocatalysis

The technologies currently available for water treatment, such as adsorption, involve the transfer of pollutants to another phase without elimination or destruction [18]. Other conventional technologies, such as sedimentation and filtration, involve high operational costs and lead to the formation of new toxic substances [18]. For instance, although chlorination is one of the most commonly used water disinfection techniques, the products that are generated from the side reactions in this process are considered to be carcinogenic and mutagenic [218]. The limitations of conventional techniques have led to the development of innovative advanced oxidation processes (AOPs) as potential alternative techniques for water treatment [72]. AOPs involve the generation of highly active species, such as

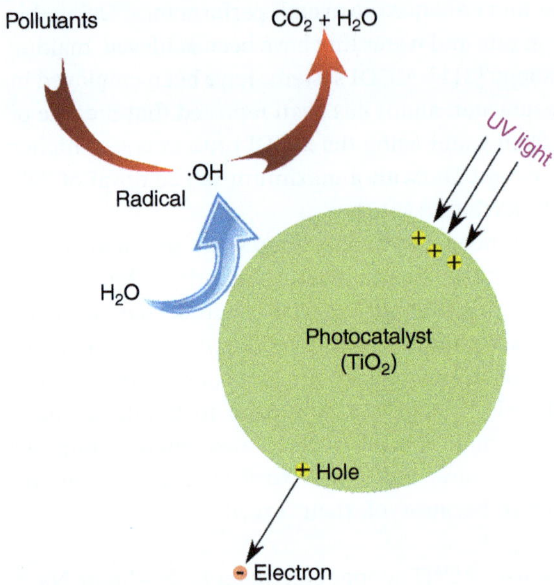

Figure 4.6 Photocatalytic mechanism of TiO_2, which generates active species that assist in the degradation of pollutants.

H_2O_2,. OH, and O_3, which help to eliminate and decompose contaminants in harmless compounds such as CO_2 and H_2O (Figure 4.6). Among the AOP technologies, heterogeneous photocatalysis has demonstrated potential for degrading organic pollutants into biodegradable compounds using semiconductor catalysts such as TiO_2, ZnO, Fe_2O_3, CdS, GaP, and ZnS [18]. Of these, TiO_2 is the most promising semiconductor catalyst because of its high reactivity at a photon energy in the range of 300 nm < λ < 390 nm [219]. TiO_2 has been shown to demonstrate high stability after repeated catalytic cycles, unlike CdS and GaP, which show less resilience and produce toxic products upon their degradation [220]. Moreover, TiO_2 has been extensively used in water treatment because of its high thermal and chemical stability, excellent mechanical properties, operational process at room temperature and pressure, complete mineralization of components without producing extra pollutants, and low operational costs [221]. Furthermore, TiO_2 is produced at low cost in massive quantities industrially and can be used as a powder or thin film in water treatment [222].

Note that the decomposition of organic pollutants is primarily an oxidation process dependent on the valence band (VB) of the photocatalyst [221]. Thus, a photocatalyst with a higher oxidative power improves the oxidation reaction, leading to higher removal efficiency of organic pollutants [221]. The bandgap of TiO_2 is in the range of 3.0–3.2 eV at a wavelength of 400 nm, which allows the oxidation reaction to begin under UV irradiation that has a wavelength of <400 nm [223]. The photon energy of a 400 nm wavelength is equivalent to 30 000 °C of thermal energy, which is a high enough temperature to oxidize all types of materials [224]. However, TiO_2 is efficient under weak light, but inactive under visible light, making it unsuitable for use in mass processing systems [225]. Therefore, to overcome this limitation of TiO_2, metal and non-metal doped TiO_2 materials have been investigated, with hybrid TiO_2 composites of zeolites [226], ceramics [227], glasses [228], and fibers [229] having been prepared and studied.

Other approaches to enhance the photocatalytic activities of TiO_2 particles have been investigated. TiO_2 can take the form of 0D spheres [230], 1D TiO_2 fibers [231], 2D TiO_2 nanosheets [232], and 3D interconnected architectures [233], with the modification of its morphology indicating that structural dimensionality plays an important factor in improving its reactivity. The size and morphology of TiO_2 can be used to maximize its photocatalytic reactivity because they enhance its surface area and pollutant sorption capacity [233]. For instance, upon a decrease in the size of the TiO_2 NPs, there is a decrease in e^-/h^+ recombination because of the enhanced interfacial charge carrier transfer on the TiO_2 surface [234]. Furthermore, TiO_2 NTs have been reported to be more efficient at adsorbing and decomposing non-biodegradable organic compounds than other forms of TiO_2 [235]. This higher efficiency can be attributed to the tube walls of the TiO_2 NTs, which allow short carrier diffusion, fast mass transfer, and high adsorption capacity of organic pollutants [235].

4.6.2 Hybrid TiO_2/Nanocarbon Composites

The high specific area of nano-dimensional TiO_2 promotes charge separation and efficient transfer. Nano-sized TiO_2 shows better performance in terms of higher catalytic activity compared to bulk-sized TiO_2 [236]. However, as a photocatalyst, nano-sized TiO_2 cannot be used directly in water because it tends to aggregate during oxidation processes [237]. Therefore, much research has been conducted in an attempt to overcome this problem, primarily involving the anchoring of TiO_2 onto carbon-based materials [238] to form hybrid composite TiO_2/nanocarbon based on AC NFs [238], CNTs [239], and graphene [240]. This combination between carbon materials and TiO_2 improves the catalytic activity and charge transfer of the resultant photocatalyst. The selection of carbon materials was made because of their excellent chemical and thermal stability, good optical and electrical properties, and structural tenability [239].

Interestingly, combining CNTs with TiO_2 has demonstrated potential for improving the photocatalytic activity of TiO_2 because of the high specific surface area, retardation of e−/h+ recombination, and highly reactive sites of the resultant photocatalyst [221]. In particular, CNTs act as effective electron sinks and as p-type semiconductors that create a Schottky barrier at the composite interface [241]. However, TiO_2 acts as an n-type semiconductor that promotes photo-induced electrons to migrate freely toward the CNTs, with this increase in electron migration improving the catalytic activity on the surface of the TiO_2/CNTs, thus leading to higher efficiency in the decomposition of organic pollutants [242].

Because of their excellent redox properties, graphene sheets have been utilized as a 2D photocatalyst in combination with TiO_2 [243]. High-purity graphene sheets allow the fast transfer of electrons with a mobility rate that exceeds 15 000 $m^2\ V^{-1}\ s^{-1}$ at room temperature.

4.7 Nanomembranes

Due to its low energy consumption, membrane technology plays an important role in improving the separation processes of a range of applications, including water purification, wastewater treatment, desalination, and water reuse processes. Different membrane-based processes, such as the aforementioned RO, NF, MF, and ED, have been developed over the last few decades and used in water treatment applications. However, these membranes are associated with disadvantages such as low flux, fouling, and aging, which affect their long-term use. Therefore, different classes of synthetic materials have been comprehensively integrated into membranes to enhance their permeance and rejection toward improving their performance and resistance. Many different synthetic routes have been reported for fabricating nanopolymer membrane-like materials, such as cellulose acetate, polyacrylonitrile (PAN), polyvinylchloride (PVC), and polydimethylsiloxane (PDMS), representing a significant step toward their diverse applications in producing clean drinking water [244].

4.7.1 CNT and NP Composite Membranes

With their rigid cylindrical nanostructures, CNTs can serve as robust pores in membranes for application in the desalination and decontamination of water. The hollowness of the CNT structure enables the frictionless transport of water molecules, making them suitable for use in high-flux separation techniques. Appropriate pore diameters can result in energy barriers at channel entries in the material, leading to the rejection of salt ions and permitting water transport through the hollows of the NTs. Moreover, the presence of CNTs in the membrane presents scope for additional modification to fine-tune the separation performance. CNTs exhibit antifouling, self-cleaning, and reusable function properties, which, when incorporated in membranes enable the development of stable materials that show enhanced performance compared to membranes without CNTs. The performance of CNT-based membranes is often dependent on the methods used in their processing and fabrication.

Interestingly, CNTs can be incorporated in naturally occurring polymers and their derivatives, such as chitosan and cellulose, to form mixed matrix membranes. A chitosan–MWCNT–polyether block amide/polyethersulfone (PEBA/PES) TFN membrane has been demonstrated to display high water flux performance in a cross-flow cell toward the separation of malachite from solution with 98.7% rejection. Cellulose and chitosan being derivable from renewable sources and having a superior ability to form films and fibers with beneficial characteristics, such as low toxicity, biocompatibility, biodegradability, and economy, make them excellent candidates for use in water purification applications. However, biopolymers often form hydrogen bonds in solution, making them hard to dissolve or melt in addition to their low tensile strength [244]. To overcome this issue, these polymers and MWCNTs can be dissolved or dispersed in ionic liquids, which improves the thermal stability, mechanical properties, and electrical conductivity of the resultant composite materials [245].

Sustainable and renewable solutions using economic and straightforward approaches are at the forefront of in research. For instance, an electrical composite membrane has been produced by incorporating MWCNTs into bacteria cellulose (BC), which is a unique biopolymer gel produced by a genus of acetic acid bacteria that converts ethanol to acetic acid in the presence of oxygen, thus making it useful for various medical applications. This biopolymer is manufactured industrially via microbial fermentation. However, although this cellulose polymer exhibits high mechanical strength and flexibility, its vast size, of ~20 mm, limits its use in electrochemical polymerization. To remedy this, CNTs have been integrated into enzymic biofuel cells by immobilizing laccase (Lac) enzymes as the cathode and glucose oxidase (GOD) enzymes as the anode via a covalent amide link between the MWCTNs and proteins, thus paving the way for a bright future for achieving the purification of water using biological fuel cells [246].

Waste brick powder (WBP) and sodium alginate coated onto PVDF have been developed as an NP-based membrane via a vacuum filtration method. The membrane demonstrates excellent performance in the separation of kerosene, hexane, and petroleum ether in water emulsions, with a separation efficiency of ≤99.8% and a permeating flux of ≤623 $l\,m^2\,h^{-1}$ [247]. Interestingly, the WBP-coated membrane can be used to effectively recover waste from water in wastewater treatment, thus demonstrating good performance in removing crude oil from oily wastewater and thus reducing the levels of environmental pollution.

4.7.2 Ceramic Membranes

Ceramic membranes are known for their chemical and thermal stability and can be produced at a low cost from clay and waste. These membranes attracted considerable attention because of their potential use in treating water and in various other separation applications. A microfiltration ceramic membrane has been developed from a natural waste compound mixture of banana peels and clay via spin coating and used in a cross-flow filtration process with excellent performance in the treatment of wastewater from tanneries and textile effluents, thus removing 96% of the dyes and reducing turbidity [248]. Moreover, a feasible process for desalinating seawater has been developed using a green silica-based ceramic hollow fiber membrane (CHFM), composed of rice husk ash waste via phase inversion and sintering for direct contact membrane distillation (DCMD) (Figure 4.7). CHFM was grafted with a fluoroalkyl silane (FAS) compound to alter its surface hydrophobicity, resulting in a high water flux of 38.2 $kg\,m^2\,h^{-1}$ and a salt rejection performance of ≥99.9% [249].

Similarly, the same process for synthesizing a unique bio-ceramic membrane has been developed using discarded animal bones, which contain hydroxyapatite (HAp), a natural source of calcium phosphates that can act as a promising adsorbent for removal of dyes from water and has ion exchange capacity for heavy metal removal. The use of HAp has been reported in various applications for treating

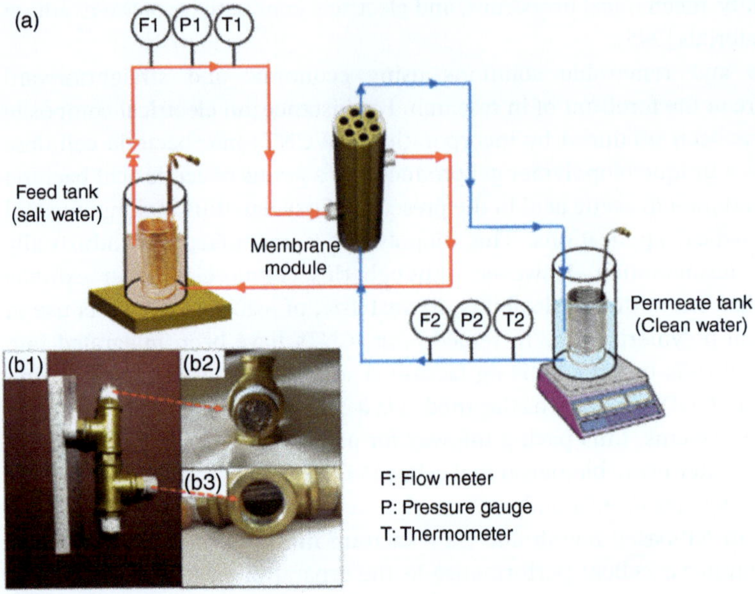

Figure 4.7 (a) Direct contact membrane distillation system, and (b1–b3) the membrane module and related parts.

wastewater from the textile industry. HAp powder has been prepared from waste cow bones via a thermal method to fabricate a HAp-based bio-ceramic hollow fiber membrane. The resulting membrane contains 60% HAp, demonstrating a constant flow of $88.31\,m^2\,h^{-1}$ and 99.9% and 100% dye and turbidity removal, respectively, as well as the exceptional removal of heavy metals from solution [250]. Pristine metakaolin and corn cob ash (CCA) waste have been reported as membrane supports for hollow ceramic fibers for use in oil–water separation using phase inversion and sintering processes. The MK–corn CCA was used to improve pore formation in a ceramic membrane to achieve a high oil removal efficiency (74.73%) toward an oil/water emulsion during the 10 minutes of filtration time. The oil rejection of the material dramatically increased because of the depositing of oil droplets on the surface of the membrane, thus reaching 94.5% after 90 minutes [251].

4.7.3 Nanofiber Membranes for Water Purification

Because of their easy fabrication, pore size tenability, and potential preparation from natural resources, nanofiber membranes have demonstrated great potential for use in water treatment applications. For example, cellulose NFs have been fabricated and functionalized with MOFs to create materials that are highly robust and exhibit high efficiency in their performance in various water purification applications [252]. Similarly, the fabrication of protein-based NFs using soy protein prepared via electrospinning has been reported. The stability of this nanofiber membrane in water was achieved using green oxidized sugars (OS) that involve the oxidant hydrogen peroxide to form cross-links in the material that result in its good resistance to water, thus making it a promising membrane for use in water purification [253]. The resultant membrane has been demonstrated to be an effective adsorbent material for removing crystal violet dye from water, as well as exhibiting high retention ratios against *E. coli* and *B. diminuta*.

Green nanomaterials have been reported to remove other contaminants from water, such as the polyatomic ions fluoride and nitrate. For example, TiO_2 NPs have been successfully synthesized using a *Bacillus licheniformis* Al_2O_3/bio-TiO_2 nanocomposite (ABN), with ABN then incorporated into an electrospun thermoplastic polyurethane nanofibrous membrane (ABN/TPU-NFM). The produced membrane demonstrates sufficient capacity as an adsorbent for reducing nitrate and fluoride concentrations in groundwater [254].

Electrospun nanofibrous membranes have been heavily investigated for water remediation applications. For instance, nanofibrous membranes were developed to remove oil [255, 256], bacteria [257], dyes [258], and heavy metals [259], from contaminated water. Moreover, nanofibrous membranes demonstrated excellent potential for water desalination [260]. The high porosity, which exceeds 80%, and the ability to modify electrospun nanofibrous membranes' surface make them unique candidates for adsorption applications and scavenging contaminants from wastewater. Similarly, two-dimensional materials, such as MXene, have attracted attention for their excellent capability to separate gases and remove pollutants from water. MXene-based membranes were fabricated and used for removal of inorganic salts and dyes and displayed notable performance. However, up to date, only a few publications on membrane-based MXene have been reported for water treatment. Although, due to the high stability of MXene and its high adsorption capacity, MXene seems like a good candidate for water treatment applications, further exploration and investigations are required [261].

4.8 Future Challenges and Prospects

Water contamination is rapidly increasing because of an increase in industrial production and world population. Therefore, developing sustainable technologies for water purification may help reduce contamination and provide better water quality. For this purpose, nanotechnology has become a fundamental tool to purify and treat water via various methodologies and approaches. However, the treatment of water using nanotechnology has to-date been limited to small pilot-scale studies because of multiple challenges presented by its implementation, such as high costs and the possible risk to human health and the environment if the use of nanomaterials is not well controlled. Therefore, to bring nanotechnology into everyday life, safe methodologies must be provided, based on intensive research studies and evaluations. For instance, in the development of nanomaterials for treating water, direct or indirect exposure studies should be conducted to identify the possible risks and toxicity associated with their exposure to living organisms. In this chapter, different methods have been defined, and nanomaterials, such as organic, inorganic, and polymer-based nanomaterials, have been presented in terms of their potential performance in removing various pollutants from water.

New pathways need to be established via which nanomaterials can be synthesized from sustainable resources with minimum energy consumption and low gas emissions. Because low manufacturing costs are important for implementing nanomaterials in the treatment of water, their economic efficiency should be considered in future research, and their related interaction processes in the treatment of water require assessment. Furthermore, it is imperative to evaluate the long-term performance of nanomaterials to determine their stability and efficiency. Note that in addition to the efficiency and performance of the removal of pollutants from water by nanomaterials, their recyclability is important for promoting their use in sustainable water purification processes. Thus, for all types of water treatment nanotechnologies, recyclability should be investigated to enhance the performance of the materials in these processes and minimize the generation of secondary pollutants and waste.

The production of clean water is dependent on clean manufacturing, green synthesis, the intelligent use of materials, and concerns for humanity. It is possible that nanotechnology could play an important role in providing clean water from seawater and wastewater or in the production of water, with one of the most promising approaches being solar-powered desalination systems. Desalination involving the use of membrane separation technology has proven to be efficient in converting seawater into drinkable water. However, the significant drawback of standard desalination techniques is their high energy consumption and emissions. Integrating solar or wind power into desalination processes would therefore address these downfalls by significantly reducing energy consumption. Therefore, more research is needed in this area to optimize the conditions, membrane selection, and power generation routes of these processes toward the development of green nanotechnology for water purification.

References

1 Nagar, A. and Pradeep, T. (2020). *ACS Nano* 14: 6420.
2 World Bank, *Urban Population (Percent, % of Total)*, 2013.
3 UN (2015). United Nations Development Programme. (2015). World leaders adopt sustainable development goals.
4 Crini, G., Lichtfouse, E., Wilson, L.D., and Morin-Crini, N. (2019). *Environ. Chem. Lett.* 17: 195.
5 Singh, P., Yadav, S.K., and Kuddus, M. (2020). *Adv. Struct. Mater.* 126: 227.
6 Parlamento Europeo (2008). *Off. J. Eur. Union* 44: 3–741.
7 Kumar, A.A., Som, A., Longo, P. et al. (2017). *Adv. Mater.* 29: 1604260.
8 Tu, R. and Hwang, Y. (2020). *Energy* 201: 117630.
9 Kokkinos, P., Venieri, D., and Mantzavinos, D. (2021). *Food Environ. Virol.* 13: 283.
10 Ajith, M.P., Aswathi, M., Priyadarshini, E., and Rajamani, P. (2021). *Bioresour. Technol.* 342: 126000.

11 Ren, G., Han, H., Wang, Y. et al. (2021). *Nanomaterials* 11: 1804.
12 WHO (2008). *Guidel. Drink. Qual.* 1: 319–389.
13 Cuthbertson, A.A., Kimura, S.Y., Liberatore, H.K. et al. (2019). *Environ. Sci. Technol.* 53: 5987.
14 Chen, W., Luo, J., Du, X. et al. (2021). *J. Cleaner Prod.* 317: 128224.
15 Sousa, V.S. and Ribau Teixeira, M. (2020). *Sci. Total Environ.* 707: 136077.
16 Yusuf, A., Al Jitan, S., Garlisi, C., and Palmisano, G. (2021). *Chemosphere* 278: 130440.
17 Xiao, G., Wang, X., Ni, M. et al. (2013). *Appl. Energy* 103: 642.
18 Chong, M.N., Jin, B., Chow, C.W.K., and Saint, C. (2010). *Water Res.* 44: 2997.
19 Han, Y.L., Kim, S.Y., Kim, T. et al. (2020). *J. Cleaner Prod.* 250: 119494.
20 Sajid, M., Nazal, M.K., Ihsanullah et al. (2018). *Sep. Purif. Technol.* 191: 400.
21 Ali, Z. and Ahmad, R. (2020). *Nanotechnology for Water Treatment*, 143–163. Cham: Springer.
22 Ali, Z., Khan, A., and Ahmad, R. (2015). *Microporous Mesoporous Mater.* 203: 8.
23 Abbasi, E., Aval, S.F., Akbarzadeh, A. et al. (2014). *Nanoscale Res. Lett.* 9: 1.
24 Kim, K.J. and Park, J.W. (2017). *J. Mater. Sci.* 52: 843.
25 Yoo, H. and Kwak, S.Y. (2013). *J. Membr. Sci.* 448: 125.
26 Han, K.N., Yu, B.Y., and Kwak, S.Y. (2012). *J. Membr. Sci.* 396: 83.
27 Beraa, A., Hajjaji, M., Laurent, R. et al. (2016). *Desalin. Water Treat.* 57: 14290.
28 Hayati, B., Mahmoodi, N.M., Arami, M., and Mazaheri, F. (2011). *CLEAN-Soil, Air, Water* 39: 673.
29 Arkas, M. and Tsiourvas, D. (2009). *J. Hazard. Mater.* 170: 35.
30 Tian, B., Hua, S., and Liu, J. (2020). *Carbohydr. Polym.* 232: 115805.
31 Chen, X., Chen, M., Xu, C., and Yam, K.L. (2019). *Crit. Rev. Food Sci. Nutr.* 59: 2386.
32 Singh, N. and Sahu, O. (2018). *The Impact and Prospects of Green Chemistry for Textile Technology*, 83–105. Woodhead Publishing.
33 Dong, Z., Tagliavini, M., Darmadi, J. et al. (2020). *Adv. Mater. Interfaces* 7: 1902100.
34 Khalil, A.M. and Schäfer, A.I. (2021). *J. Membr. Sci.* 618.
35 Liu, Y., Jia, J., Gao, T. et al. (2020). *J. Colloid Interface Sci.* 576: 302.
36 Barbosa, P.F.P., Cumba, L.R., Andrade, R.D.A., and do Carmo, D.R. (2019). *J. Polym. Environ.* 27: 1352.
37 Liu, J., Zhang, C., Zhang, S. et al. (2020). *Chin. Chem. Lett.* 31: 539.
38 Leclercq, L. and Nardello-Rataj, V. (2020). *Eur. J. Pharm. Sci.* 155: 105559.
39 Mamba, B.B., Krause, R.W., Malefetse, T.J., and Nxumalo, E.N. (2007). *Environ. Chem. Lett.* 5: 79.
40 Chen, H., Zhou, Y., Wang, J. et al. (2020). *J. Hazard. Mater.* 389: 121897.
41 Liu, C., Wang, P., Liu, X. et al. (2018). *Chem. An Asian J.* 14: 261.
42 Celebioglu, A., Topuz, F., and Uyar, T. (2019). *ACS Appl. Polym. Mater.* 1: 54.
43 Badmus, S.O., Oyehan, T.A., and Saleh, T.A. (2021). *J. Mol. Liq.* 340: 116991.
44 Chen, Z.H., Liu, Z., Hu, J.Q. et al. (2020). *J. Membr. Sci.* 595: 117510.
45 Xue, J., Jiao, Z., Bi, R. et al. (2019). *J. Membr. Sci.* 584: 282.
46 Li, Y. and Yu, J. (2014). *Chem. Rev.* 114: 7268.

47 Ramesh, K. and Reddy, D.D. (2011). *Advances in Agronomy*, 219–241.
48 Reeve, P.J. and Fallowfield, H.J. (2018). *J. Environ. Manage.* 205: 253.
49 Koshy, N. and Singh, D.N. (2016). *J. Environ. Chem. Eng.* 4: 1460.
50 Itskos, G., Koukouzas, N., Vasilatos, C. et al. (2010). *J. Hazard. Mater.* 183: 787.
51 Lee, M.G., Yi, G., Ahn, B.J., and Roddick, F. (2000). *Korean J. Chem. Eng.* 17: 325.
52 Wajima, T. (2013). *Anal. Sci.* 29: 139.
53 Guida, S., Potter, C., Jefferson, B., and Soares, A. (2020). *Sci. Rep.* 10: 1.
54 Kobayashi, Y., Ogata, F., Nakamura, T., and Kawasaki, N. (2020). *J. Environ. Chem. Eng.* 8: 103687.
55 Jiang, N., Shang, R., Heijman, S.G.J., and Rietveld, L.C. (2018). *Water Res.* 144: 145.
56 Sarti, E., Chenet, T., Stevanin, C. et al. (2020). *Molecules* 25: 3331.
57 Jiang, N., Shang, R., Heijman, S.G.J., and Rietveld, L.C. (2020). *Sep. Purif. Technol.* 235: 116152.
58 Madan, S., Shaw, R., Tiwari, S., and Tiwari, S.K. (2019). *Appl. Surf. Sci.* 487: 907.
59 Zuo, X., Qian, C., Ma, S. et al. (2019). *Water Sci. Technol.* 80: 507.
60 Van den Bergh, M., Krajnc, A., Voorspoels, S. et al. (2020). *Angew. Chemie-Int. Ed.* 59: 14086.
61 Liu, L.J., Chen, G.E., Mao, H.F. et al. (2021). *High Perform. Polym.* 33: 309.
62 Ursino, C., Castro-Muñoz, R., Drioli, E. et al. (2018). *Membranes* 8: 18.
63 Kang, Y., Emdadi, L., Lee, M.J. et al. (2014). *Environ. Sci. Technol. Lett.* 1: 504.
64 Dong, L.X., Huang, X.C., Wang, Z. et al. (2016). *Sep. Purif. Technol.* 166: 230.
65 Huang, H., Qu, X., Ji, X. et al. (2013). *J. Mater. Chem. A* 1: 11343.
66 Dong, H., Zhao, L., Zhang, L. et al. (2015). *J. Membr. Sci.* 476: 373.
67 Diagboya, P.N.E. and Dikio, E.D. (2018). *Microporous Mesoporous Mater.* 266: 252.
68 Benhamou, A., Baudu, M., Derriche, Z., and Basly, J.P. (2009). *J. Hazard. Mater.* 171: 1001.
69 Zhang, L., Yu, C., Zhao, W. et al. (2007). *J. Non-Cryst. Solids* 353: 4055.
70 Salman, M., Jahan, S., Kanwal, S., and Mansoor, F. (2019). *Environ. Sci. Pollut. Res.* 26: 21065.
71 Wang, S., Wang, K., Dai, C. et al. (2015). *Chem. Eng. J.* 262: 897.
72 Diagboya, P.N., Olu-Owolabi, B.I., and Adebowale, K.O. (2014). *J. Environ. Manage.* 146: 42.
73 Cho, E.C., Chang-Jian, C.W., Chen, H.C. et al. (2017). *Chem. Eng. J.* 314: 347.
74 Liu, B., Zhang, L., Wang, H., and Bian, Z. (2017). *Ind. Eng. Chem. Res.* 56: 5795.
75 Bashambu, L., Singh, R., and Verma, J. (2021). *Materials Today: Proceedings*, 538–545. Elsevier.
76 Zhu, Z., Li, Z., Zhong, L. et al. (2019). *J. Membr. Sci.* 572: 73.
77 Nasrollahzadeh, M., Sajjadi, M., Iravani, S., and Varma, R.S. (2021). *J. Hazard. Mater.* 401: 123401.
78 Singh, P., Kim, Y.J., Zhang, D., and Yang, D.C. (2016). *Trends Biotechnol.* 34: 588.

79 Wu, Z., Su, X., Lin, Z. et al. (2019). *J. Hazard. Mater.* 379: https://doi.org/10.1016/j.jhazmat.2019.120811.

80 Yi, H., Jiang, M., Huang, D. et al. (2018). *J. Taiwan Inst. Chem. Eng.* 93: 184.

81 Parvin, F., Rikta, S.Y., and Tareq, S.M. (2018). *8-Application of Nanomaterials for the Removal of Heavy Metal From Wastewater*, 137–157. Elsevier Inc.

82 Lata, S. and Samadder, S.R. (2016). *J. Environ. Manage.* 166: 387.

83 Januário, E.F.D., Fachina, Y.J., Wernke, G. et al. (2022). *Chemosphere* 289: 133213.

84 Cosgrove, S., Jefferson, B., and Jarvis, P. (2022). *Sci. Total Environ.* 815: 152626.

85 Sullivan, G.L., Prigmore, R.M., Knight, P., and Godfrey, A.R. (2019). *J. Environ. Manage.* 251: 109551.

86 Naushad, M., Alqadami, A.A., Al-Kahtani, A.A. et al. (2019). *J. Mol. Liq.* 296: 112075.

87 Pradeep, T., Anshup, and Bootharaju, M.S. (2014). *Detection and Extraction of Pesticides from Drinking Water Using Nanotechnologies*, 241–270. Elsevier Inc.

88 Azzaza, S., Kumar, R.T., Vijaya, J.J., and Bououdina, M. (2017). *RSC Detection Science*, 139–166. The Royal Society Of Chemistry.

89 Rashtbari, Y., Hazrati, S., Azari, A. et al. (2020). *Adv. Powder Technol.* 31: 1612.

90 Suriyaraj, S.P. and Selvakumar, R. (2016). *RSC Adv.* 6: 10565.

91 Selvi, K., Pattabhi, S., and Kadirvelu, K. (2001). *Bioresour. Technol.* 80: 87.

92 Sabela, M.I., Kunene, K., Kanchi, S. et al. (2019). *Arabian J. Chem.* 12: 4331.

93 Vetrivel, R., Navinselvakumar, C., and Kumar, P.S.S.R. (2018). *Int. J. Mech. Prod. Eng. Res. Dev.* 8: 288.

94 Madima, N., Mishra, S.B., and Mishra, A.K. (2022). *Nanotechnology and Green Materials: Introduction, Fundamentals, and Applications*, 3–19. Elsevier Inc.

95 Sankararamakrishnan, N., Singh, N., and Gupta, A. (2013). *RSC Adv.* 3: 22421.

96 Xu, X., Guo, P., Luo, Z. et al. (2017). *RSC Adv.* 7: 18765.

97 Fosso-Kankeu, E., Mulaba-Bafubiandi, A.F., and Mishra, A.K. (2014). *Application of Nanotechnology in Water Research*, 37–61.

98 Hasdi, N.D. (2020). *Nanoscale Res. Lett.* 14: 1.

99 Kumar, M. and Ando, Y. (2007). *J. Phys. Conf. Ser.* 61: 643.

100 Li, Z., Chen, J., and Ge, Y. (2017). *Chem. Eng. J.* 308: 809.

101 Wang, Q., Wang, X., Chai, Z., and Hu, W. (2013). *Chem. Soc. Rev.* 42: 8821.

102 Kroto, H.W., Heath, J.R., O'Brien, S.C. et al. (1985). *Nature* 318: 162.

103 Prato, M. (1997). *J. Mater. Chem.* 7: 1097.

104 Joseph, J., Sivasankarapillai, V.S., Nikazar, S. et al. (2020). *ChemSusChem* 13: 3754.

105 Shetti, N.P., Mishra, A., Basu, S., and Aminabhavi, T.M. (2021). *Mater. Today Chem.* 20: 100454.

106 Alekseeva, O.V., Bagrovskaya, N.A., and Noskov, A.V. (2016). *Prot. Met. Phys. Chem. Surf.* 52: 443.

107 Arbogast, J.W., Darmanyan, A.P., Foote, C.S. et al. (1991). *J. Phys. Chem.* 95: 11.

108 Bell, B.G., Schellevis, F., Stobberingh, E. et al. (2014). *BMC Infect. Dis.* 14: 13.

109 Hosain, Z., Kabir, S.M.L., and Kamal, M. (2021). *Int. J. One Heal.* 14: 210.

110 Wang, J. and Vermerris, W. (2016). *Materials* 9: 255.

111 Seil, J.T. and Webster, T.J. (2012). *Int. J. Nanomed.* 7: 2767.
112 Boominathan, T. and Sivaramakrishna, A. (2021). *Top. Curr. Chem.* 379: 19.
113 Zhang, J., Xia, W., Liu, P. et al. (2010). *Mar. Drugs* 8: 1962.
114 Venkatesan, J. and Kim, S.K. (2010). *Mar. Drugs* 8: 2252.
115 Li, B., Shan, C.L., Zhou, Q. et al. (2013). *Mar. Drugs* 11: 1534.
116 Hu, Z. and Gänzle, M.G. (2019). *J. Appl. Microbiol.* 126: 1318.
117 Elsabee, M.Z., Naguib, H.F., and Morsi, R.E. (2012). *Mater. Sci. Eng., C* 32: 1711.
118 Rabea, E.I., Badawy, M.E.T., Stevens, C.V. et al. (2003). *Biomacromolecules* 4: 1457.
119 Sahariah, P., Gaware, V.S., Lieder, R. et al. (2014). *Mar. Drugs* 12: 4635.
120 Hu, L., Meng, X., Xing, R. et al. (2016). *Bioorganic Med. Chem. Lett.* 26: 4548.
121 Croisier, F. and Jérôme, C. (2013). *Eur. Polym. J.* 49: 780.
122 Rajeswari, A., Amalraj, A., and Pius, A. (2016). *J. Water Process Eng.* 9: 123.
123 Rahimi, S., Moattari, R.M., Rajabi, L., and Derakhshan, A.A. (2015). *Colloids Surf., A* 484: 216.
124 Medeiros Borsagli, F.G.L., Mansur, A.A.P., Chagas, P. et al. (2015). *React. Funct. Polym.* 97: 37.
125 Arvand, M., Latify, L., Tajmehri, H. et al. (2009). *Anal. Lett.* 42: 856.
126 Liu, H.C., Chen, W., Cui, B., and Liu, C. (2015). *J. Cent. South Univ.* 22: 4168.
127 Marrakchi, F., Khanday, W.A., Asif, M., and Hameed, B.H. (2016). *Int. J. Biol. Macromol.* 93: 1231.
128 Kadam, A.A. and Lee, D.S. (2015). *Bioresour. Technol.* 193: 563.
129 Sakala, G.P. and Reches, M. (2018). *Adv. Mater. Interfaces* 5: 1800073.
130 Bommarius, B., Jenssen, H., Elliott, M. et al. (2010). *Peptides* 31: 1957.
131 Marr, A.K., Gooderham, W.J., and Hancock, R.E. (2006). *Curr. Opin. Pharmacol.* 6: 468.
132 Kanchanapally, R., Nellore, B.P.V., Sinha, S.S. et al. (2015). *RSC Adv.* 5: 18881.
133 Marambio-Jones, C. and Hoek, E.M.V. (2010). *J. Nanopart. Res.* 12: 1531.
134 Zhai, Z., Zhang, F., Chen, X. et al. (2017). *Lab Chip* 17: 1306.
135 De Oliveira, P.F.M., Torresi, R.M., Emmerling, F., and Camargo, P.H.C. (2020). *J. Mater. Chem. A* 8: 16114.
136 Aygün, A., Gülbağça, F., Nas, M.S. et al. (2020). *J. Pharm. Biomed. Anal.* 179: 113012.
137 Velsankar, K., Sudhahar, S., Parvathy, G., and Kaliammal, R. (2020). *Mater. Chem. Phys.* 239: 121976.
138 Rautela, A., Rani, J., and Debnath (Das), M. (2019). *J. Anal. Sci. Technol.* 10: 5.
139 Abbaszadegan, A., Ghahramani, Y., Gholami, A. et al. (2015). *J. Nanomater.* 2015: 1.
140 Rai, M., Yadav, A., and Gade, A. (2009). *Biotechnol. Adv.* 27: 76.
141 Singh, R., Vora, J., Nadhe, S.B. et al. (2017). *J. Nanosci. Nanotechnol.* 18: 3806.
142 Bonde, S.R., Rathod, D.P., Ingle, A.P. et al. (2012). *Nanosci. Methods* 1: 25.
143 Selim, A., Elhaig, M.M., Taha, S.A., and Nasr, E.A. (2018). *OIE Rev. Sci. Tech.* 37: 823.

144 Barkalina, N., Charalambous, C., Jones, C., and Coward, K. (2014). *Nanomed. Nanotechnol. Biol. Med.* 10: e921.

145 Islam, S., Butola, B.S., and Mohammad, F. (2016). *RSC Adv.* 6: 44232.

146 Deshmukh, S.P., Patil, S.M., Mullani, S.B., and Delekar, S.D. (2019). *Mater. Sci. Eng., C* 97: 954.

147 He, D., Ikeda-Ohno, A., Boland, D.D., and Waite, T.D. (2013). *Environ. Sci. Technol.* 47: 5276.

148 Dankovich, T.A., Levine, J.S., Potgieter, N. et al. (2016). *Environ. Sci. Water Res. Technol.* 2: 85.

149 Karthik, C., Swathi, N., and Pandi Prabha, S. (2020). *J. Environ. Chem. Eng.* 8: 103577.

150 Sushma, D. and Richa, S. (2015). *Int. Res. J. Environ. Sci.* 4: 103.

151 Jardak, K., Drogui, P., and Daghrir, R. (2016). *Environ. Sci. Pollut. Res.* 23: 3195.

152 Philippe, K.K., Timmers, R., Van Grieken, R., and Marugan, J. (2016). *Ind. Eng. Chem. Res.* 55: 2952.

153 Foster, H.A., Ditta, I.B., Varghese, S., and Steele, A. (2011). *Appl. Microbiol. Biotechnol.* 90: 1847.

154 Jayaseelan, C., Rahuman, A.A., Roopan, S.M. et al. (2013). *Spectrochim. Acta, Part A* 107: 82.

155 Parangi, T. and Mishra, M.K. (2019). *Comments Inorg. Chem.* 39: 90.

156 Nadeem, M., Tungmunnithum, D., Hano, C. et al. (2018). *Green Chem. Lett. Rev.* 11: 492.

157 Mittal, A.K., Bhaumik, J., Kumar, S., and Banerjee, U.C. (2014). *J. Colloid Interface Sci.* 415: 39.

158 Ilyas, M., Waris, A., Khan, A.U. et al. (2021). *Inorg. Chem. Commun.* 133: 108968.

159 Marimuthu, S., Rahuman, A.A., Jayaseelan, C. et al. (2013). *Asian Pac. J. Trop. Med.* 6: 682.

160 Roopan, S.M., Bharathi, A., Prabhakarn, A. et al. (2012). Spectrochim. *Acta Part A.* 98: 86.

161 Kumar, P.S., Francis, A.P., and Devasena, T. (2014). *J. Environ. Nanotechnol.* 3: 73.

162 Solano, R.A., Herrera, A.P., Maestre, D., and Cremades, A. (2019). *J. Nanotechnol.* 2019: 1.

163 Younas, H., Qazi, I.A., Hashmi, I. et al. (2014). *Environ. Sci. Pollut. Res.* 21: 740.

164 Liu, Y., He, L., Mustapha, A. et al. (2009). *J. Appl. Microbiol.* 107: 1193.

165 Rai, M. (2013). *Cienc. Cult.* 65: 44.

166 Padmavathy, N. and Vijayaraghavan, R. (2008). *Sci. Technol. Adv. Mater.* 9: 35004.

167 Applerot, G., Lellouche, J., Perkas, N. et al. (2012). *RSC Adv.* 2: 2314.

168 Sirelkhatim, A., Mahmud, S., Seeni, A. et al. (2015). *Nano-Micro Lett.* 7: 219.

169 Velsankar, K., Venkatesan, A., Muthumari, P. et al. (2022). *J. Mol. Struct.* 1255: 132420.

170 Motshekga, S.C., Ray, S.S., Onyango, M.S., and Momba, M.N.B. (2015). *Appl. Clay Sci.* 114: 330.

171 OECD (Online service) (2008). *OECD Environmental Outlook to 2030*, OECD.
172 Schwarzenbach, R.P., Egli, T., Hofstetter, T.B. et al. (2010). *Annu. Rev. Environ. Resour.* 35: 109.
173 World Health Organization 2011. Guidelines for drinking-water quality, 4th World Health Organization F. E.-W. chronicle, undefined 2011 (2008) *joinforwater. ngo*, 3–4.
174 Qu, X., Brame, J., Li, Q., and Alvarez, P.J.J. (2013). *Acc. Chem. Res.* 46: 834.
175 Teow, Y.H. and Mohammad, A.W. (2019). *Desalination* 2.
176 Deshmukh, A., Boo, C., Karanikola, V. et al. (2018). *Energy Environ. Sci.* 11: 1177.
177 Wang, L., Violet, C., Duchanois, R.M., and Elimelech, M. (2020). *J. Chem. Educ.* 97: 4361.
178 Wang, Z., Dou, B., Zheng, L. et al. (2012). *Desalination* 299: 96.
179 Humplik, T., Lee, J., O'Hern, S.C. et al. (2011). *Nanotechnology* 22: 292001.
180 Goh, P.S. and Ismail, A.F. (2015). *J. Chem. Technol. Biotechnol.* 90: 971.
181 Elimelech, M. and Phillip, W.A. (2011). *Science* 333: 712.
182 Shrivastava, A. and Stevens, D. (2018). *Sustainable Desalination Handbook: Plant Selection, Design and Implementation*, 25–54. Elsevier.
183 Gude, V.G. (2018). *Sustainable Desalination Handbook: Plant Selection, Design and Implementation*, 437–463. Butterworth-Heinemann.
184 Jhaveri, J.H. and Murthy, Z.V.P. (2016). *Desalination* 379: 137.
185 Werber, J.R., Deshmukh, A., and Elimelech, M. (2016). *Environ. Sci. Technol. Lett.* 3: 112.
186 Amin, M.T., Alazba, A.A., and Manzoor, U. (2014). *Adv. Mater. Sci. Eng.* 2014: 1–24.
187 Oliveira, N.C., Maia, M.T., Noronha, V.T. et al. (2019). *Nanomaterials Applications for Environmental Matrices: Water, Soil and Air*, 227–262. Elsevier.
188 Esfahani, M.R., Aktij, S.A., Dabaghian, Z. et al. (2019). *Sep. Purif. Technol.* 213: 465.
189 Lee, K.P., Arnot, T.C., and Mattia, D. (2011). *J. Membr. Sci.* 370: 1.
190 Kumar, M., Grzelakowski, M., Zilles, J. et al. (2007). *Proc. Natl. Acad. Sci. U. S. A* 104: 20719.
191 Noy, A., Park, H.G., Fornasiero, F. et al. (2007). *Nanofluidics in carbon nanotubes. Nano Today* 2: 22–29.
192 Sui, H., Han, B.G., Lee, J.K. et al. (2001). *Structural basis of water-specific transport through the AQP1 water channel. Nature* 414: 872–878.
193 Kalra, A., Garde, S., and Hummer, G. (2003). *Osmotic water transport through carbon nanotube membranes. Proc. Natl. Acad. Sci.* 100: 10175–10180.
194 Das, R., Ali, M.E., Hamid, S.B.A. et al. (2014). *Desalination* 336: 97.
195 Jain, H. and Garg, M.C. (2021). *Environ. Technol. Innov.* 23: 101561.
196 Ahn, C.H., Baek, Y., Lee, C. et al. (2012). *J. Ind. Eng. Chem.* 18: 1551.
197 Daer, S., Kharraz, J., Giwa, A., and Hasan, S.W. (2015). *Desalination* 367: 37.
198 Swenson, P., Tanchuk, B., Gupta, A. et al. (2012). *Desalination* 285: 68.
199 Fathizadeh, M., Aroujalian, A., and Raisi, A. (2011). *J. Membr. Sci.* 375: 88.
200 Kazemimoghadam, M. (2010). *Desalination* 251: 176.

201 Liu, L., Liu, Y., Qi, Y. et al. (2020). *Sep. Purif. Technol.* 251: 117409.

202 Sharshir, S.W., Elsheikh, A.H., Peng, G. et al. (2017). *Renewable Sustainable Energy Rev.* 73: 521.

203 Elsheikh, A.H., Sharshir, S.W., Mostafa, M.E. et al. (2018). *Renew. Sustain. Energy Rev.* 82: 3483.

204 Sharshir, S.W., Peng, G., Wu, L. et al. (2017). *Appl. Therm. Eng.* 113: 684.

205 Sharshir, S.W., Peng, G., Wu, L. et al. (2017). *Appl. Energy* 191: 358.

206 Kabeel, A.E., Omara, Z.M., and Essa, F.A. (2014). *Energy Convers. Manage.* 86: 268.

207 Kabeel, A.E. and El-Said, E.M.S. (2014). *Desalination* 333: 10.

208 Sen Gupta, S., Islam, M.R., and Pradeep, T. (2018). *Advances in Water Purification Techniques Meeting the Needs of Developed and Developing Countries*, 165–202. Elsevier.

209 AlMarzooqi, F.A., Al Ghaferi, A.A., Saadat, I., and Hilal, N. (2014). *Desalination* 342: 3.

210 Suss, M.E., Porada, S., Sun, X. et al. (2015). *Energy Environ. Sci.* 8: 2296.

211 Zhao, R., Satpradit, O., Rijnaarts, H.H.M. et al. (2013). *Water Res.* 47: 1941.

212 Lee, J.B., Park, K.K., Eum, H.M., and Lee, C.W. (2006). *Desalination* 196: 125.

213 Andelman, M. (2011). *Sep. Purif. Technol.* 80: 262.

214 Wang, L., Wang, M., Huang, Z.H. et al. (2011). *J. Mater. Chem.* 21: 18295.

215 Gui, X., Cao, A., Wei, J. et al. (2010). *ACS Nano* 4: 2320.

216 Welgemoed, T.J. and Schutte, C.F. (2005). *Desalination* 183: 327.

217 Nadakatti, S., Tendulkar, M., and Kadam, M. (2011). *Desalination* 268: 182.

218 Lu, J., Zhang, T., Ma, J., and Chen, Z. (2009). *J. Hazard. Mater.* 162: 140.

219 Raizada, P., Soni, V., Kumar, A. et al. (2021). *J. Mater.* 7: 388.

220 Malato, S., Fernández-Ibáñez, P., Maldonado, M.I. et al. (2009). *Catal. Today* 147: 1.

221 Lee, S.Y. and Park, S.J. (2013). *J. Ind. Eng. Chem.* 19: 1761.

222 Liu, C., Lei, Z., Yang, Y. et al. (2013). *Bioresour. Technol.* 137: 57.

223 Mutzhas, M.F., Hoelzle, E., Hofmann, C., and Plewig, G. (1981). *J. Invest. Dermatol.* 76: 42.

224 Hernández-Alonso, M.D., Fresno, F., Suárez, S., and Coronado, J.M. (2009). *Energy Environ. Sci.* 2: 1231.

225 Malato, S., Blanco, J., Vidal, A., and Richter, C. (2002). *Appl. Catal., B* 37: 1.

226 Nomura, Y., Fukahori, S., and Fujiwara, T. (2020). *Chem. Eng. J.* 380: 122479.

227 Li, C., Sun, W., Lu, Z. et al. (2019). *Chem. Eng. J.* 378: 122138.

228 Pestana, C.J., Portela Noronha, J., Hui, J. et al. (2020). *Sci. Total Environ.* 745: 141154.

229 Odhiambo, V.O., Ongarbayeva, A., Kéri, O. et al. (2020). *Nanomaterials* 10: 882.

230 Gao, P., Liu, Z., Tai, M. et al. (2013). *Appl. Catal., B* 138–139: 17.

231 Kanth, N., Xu, W., Prasad, U. et al. (2020). *Nanomaterials* 10: 1.

232 Wang, Y., Zhou, M., He, Y. et al. (2020). *J. Alloys Compd.* 813: 152184.

233 Cai, J., Cao, J., Tao, H. et al. (2021). *Int. J. Hydrogen Energy* 46: 36201.

234 Yu, W., Zhao, L., Chen, F. et al. (2019). *J. Phys. Chem. Lett.* 10: 3024.

235 Macak, J.M., Zlamal, M., Krysa, J., and Schmuki, P. (2007). *Small* 3: 300.

236 Nagaveni, K., Sivalingam, G., Hegde, M.S., and Madras, G. (2004). *Environ. Sci. Technol.* 38: 1600.

237 Carbajo, J., Tolosana-Moranchel, A., Casas, J.A. et al. (2018). *Appl. Catal., B* 221: 1.

238 Zhang, J., Li, Y., Li, L. et al. (2018). *ACS Sustain. Chem. Eng.* 6: 12893.

239 Awfa, D., Ateia, M., Fujii, M. et al. (2018). *Water Res.* 142: 26.

240 Kusiak-Nejman, E. and Morawski, A.W. (2019). *Appl. Catal., B* 253: 179.

241 Chen, Y., Crittenden, J.C., Hackney, S. et al. (2005). *Environ. Sci. Technol.* 39: 1201.

242 Yan, L., Liu, C., Xia, J. et al. (2020). *J. Cleaner Prod.* 275: 124011.

243 Akhavan, O. and Ghaderi, E. (2009). *J. Phys. Chem. C* 113: 20214.

244 Mousavi, S.R., Asghari, M., and Mahmoodi, N.M. (2020). *Carbohydr. Polym.* 237: 116128.

245 Xiao, W., Wu, T., Peng, J. et al. (2013). *J. Appl. Polym. Sci.* 128: 1193.

246 Lv, P., Feng, Q., Wang, Q. et al. (2016). *Materials* 9: 183.

247 Shi, G., Shen, Y., Mu, P. et al. (2020). *Green Chem.* 22: 1345.

248 Mouiya, M., Bouazizi, A., Abourriche, A. et al. (2019). *Mater. Chem. Phys.* 227: 291.

249 Hubadillah, S.K., Othman, M.H.D., Matsuura, T. et al. (2018). *Sep. Purif. Technol.* 205: 22.

250 Hubadillah, S.K., Othman, M.H.D., Tai, Z.S. et al. (2020). *Chem. Eng. J.* 379: 122396.

251 Kamarudin, N.H., Harun, Z., Othman, M.H.D. et al. (2020). *Ceram. Int.* 46: 1512.

252 Li, J., Bendi, R., Malla, R. et al. (2021). *Emergent Mater.* 4: 1259.

253 Lubasova, D., Mullerova, J., and Netravali, A.N. (2015). *J. Appl. Polym. Sci.* 132: 41852.

254 Tolkou, A.K., Meez, E., Kyzas, G.Z. et al. (2021). *C J. Carbon Res.* 7: 74.

255 Topuz, F., Abdulhamid, M.A., Nunes, S.P., and Szekely, G. (2020). *Environ. Sci. Nano* 7: 1365.

256 Topuz, F., Abdulhamid, M.A., and Szekely, G. (2022). *Chem. Eng. J.* 449: 137821.

257 Fahimirad, S., Fahimirad, Z., and Sillanpää, M. (2021). *Sci. Total Environ.* 751: 141673.

258 Cseri, L., Topuz, F., Abdulhamid, M.A. et al. (2021). *Adv. Mater. Technol.* 6: https://doi.org/10.1002/admt.202000955.

259 Zhu, F., Zheng, Y.M., Zhang, B.G., and Dai, Y.R. (2021). *J. Hazard. Mater.* 401: 123608.

260 Yang, Z., Zhou, Y., Feng, Z. et al. (2019). *Polymers* 11: 1252.

261 Al-Hamadani, Y.A.J., Jun, B.M., Yoon, M. et al. (2020). *Chemosphere* 254: 126821.

5

Electrospun Nanofiber Membranes for Effective Air Filtration

Narendra Reddy and Kavya S. Bhushan

Center for Incubation Innovation Research and Consultancy, Jyothy Institute of Technology, Thataguni Post, Bengaluru 560082, India

5.1 Introduction

Air filtration has become inevitable both at the domestic and industrial levels due to rising pollution, changing lifestyles, the evolution of new diseases such as COVID-19, and consequent health concerns. It is estimated that about eight to nine million people die every year due to air pollution. The current global filtration market is estimated to be worth about US$ 13.8 billion annually across the globe (https://www.fortunebusinessinsights.com/industry-reports/air-filters-market-101676). Filters are also a critical part of heating, ventilation, and air conditioning (HVAC) systems and need regular monitoring, maintenance, and replacement. In addition to domestic and industrial applications, automotive air filters are another crucial and large market. Filters come in different shapes, sizes, densities, compositions, and prices depending on the intended applications [1]. Similarly, air filters are made using renewable and nonrenewable polymers and techniques such as melt blowing, nonwoven air laying and electrospinning. Electrospinning is a preferred choice for producing air filtration membranes because of its ability to control the porosity, ease of production, and relatively environmentally friendly process [2]. However, many polymers do not dissolve in electrospinnable solvents, and the process of electrospinning is slower compared to other techniques used for manufacturing the air filters. In this chapter, we provide an overview of the process, properties, and potential of developing air filtration membranes through electrospinning. An overview of the polymers used and the properties of the membranes/filters obtained is shown in Table 5.1.

Nanomaterials for Air and Water Purification, First Edition.
Edited by Deepu A. Gopakumar, Jyotishkumar Parameswaranpillai, Jinu Jacob George, and Midhun Dominic C.D.

Table 5.1 Polymers used to develop electrospun membranes for filtration applications [3].

Polymer	Solvents	Special features	Characteristics
		—	1. Under the airflow rate of $4.2\,cm\,s^{-1}$, low-pressure drop of 27 Pa, and the efficiency of the PAN filter can easily reach over 99% 2. Similar performance to a HEPA filter with low bulk density
Poly(acrylonitrile) (PAN)	DMF (Dimethyl formamide)	Carbonization	1. Excellent thermal stability to temperatures up to 450 °C, supporting the development of autoclaved and recyclable membranes 2. Adsorbs volatile and biological pollutants
Poly(acrylonitrile) (PAN)	DMF	Surface modification (functional group)	1. $PM_{2.5}$ removal efficiency of 94.02%, pressure drop of 18 Pa 2. Adsorption performance is enhanced
PVDF (Polyvinylidene fluoride)	DMF	—	1. The filtration efficiency is stable for 98.137–96.36% 2. The electrospun PVDF fiber also has a certain hydrophobicity, which can effectively improve the service life of the composite filter 3. Charge storage ability and piezoelectricity
Polyvinyl chloride (PVC)	DMF:THF (1 : 1)	Beaded structure	1. Hydrophobicity 2. The structure improved the filtration efficiency
Polyvinyl chloride (PVC)	DMF:THF (1 : 1)	—	1. Filtration efficiency is 1.022 times higher than commercial PVC filter membrane 2. Lower pressure drop
Nylon	88% Formic acid	—	1. High stability 2. Filtration efficiency reaches 99%
Polyamide 6	99% Formic acid	—	1. The dipole moment of the repeating unit in nylon 6 is 3.67 2. Filtering efficiency is 200% of the original
Polylamide 66	99% Formic acid	—	1. Good light transmittance 2. Filtering efficiency of $PM_{0.3}$ was up to 99.99%

Polyvinyl alcohol (PVOH)	H_2O	—	1. Recyclable up to three times 2. Filtration efficiency was 3.4 times higher compared to nonelectrospun membranes
Poly(lactic acid) (PLA)	DCM/DMF	—	1. Electrostatic charge generated by PLLA nanofibers significantly improves the application of the fibers 2. Filtration efficiency was 99.3% 3. Has piezoelectric properties and degradability
EC (Ethyl acetate)	DMAc/THF	Triboelectricity	1. Friction generates static electricity, which enhances filtering ability 2. Up to 100% filtration efficiency
CA (Cellulose acetate)	Ac:DMAc (3 : 1)	—	1. High performance at low thickness 2. Considerably thinner membranes
PMMA Poly(methyl methacrylate)	DMF	—	1. Stronger charge stability 2. Longer retention time and stronger filtration performance stability
Poly(ethylene terepthalate)	TFA/DCM	—	1. Mechanical property up to 4 Mpa 2. High porosity (96%), high collection efficiency (100%), and low pressure drop (212 Pa)

THF: Tetrahydrofuran, DCM: Dichloromethane, DMF: Dimethylformamide, DMAc: Dimethylacetamide, Ac: Acetate.
Source: Reproduced with permission through open access publication.

5.2 Electrospun Air Filtration Membranes from Biopolymers

5.2.1 Polysaccharide-Based Electrospun Membranes

Common polysaccharides such as cellulose, starch, and chitosan have been made into electrospun membranes for filtration applications. Cellulose in its derivative form, cellulose acetate, which is easily soluble, has been electrospun and made into membranes for air filtration applications [4]. Tree-like nanocellulose membranes were manufactured by modifying cellulose acetate solution using tetrabutylammonium chloride (TBAC). Cellulose acetate was dissolved in acetic acid, and up to 0.15 mol l^{-1} of TBAC was added. The mixture was electrospun at 1 ml h^{-1} at a voltage of 30 to 40 kV. Membranes obtained were subject to deacetylation and later characterized for their properties [5]. The fibers obtained had diameters between 20 and 500 nm, and pore size of the membranes was between 0.65 and 0.69 μm. Membranes having weight of 0.81 g cm^{-2} and filtration efficiency of up to 98.37% at pressure drop of 115 Pa were obtained. In another study, electrospun fibers produced from cellulose acetate had diameters between 0.1 and 1 μm and filter thickness was between 7 and 51 μm. When tested at face velocity of 45 cm s^{-1}, the maximum penetrating particle size was between 40 and 270 nm and mean penetration was 0.03 to 70 [6]. To prepare heat-resistant and efficient air filtration membranes, cellulose acetate was electrospun into fleece form on various substrates such as glassine paper (GP), Lyocell (L), fibrillated Lyocell (LF), and nylon grids to form a multilayered membrane. Average fiber diameters were relatively large varying from 0.32 to 0.58 μm and cellulose acetate membrane thickness was about 65 μm. The membranes were able to remove more than 80% of particles larger than 400 nm at 5.3 cm S^{-1} with a low pressure drop of 40 Pa. Nylon substrates with 100 μm pore size and a quality factor (QF) of 0.080 ± 0.020 Pa at 300 nm, suggesting suitability for HVAC, cigarette filters, and other applications, were obtained [7]. Nanofiber membranes made by combing cellulose acetate with cationic cetylpyridinium bromide (0.5% CPB) and having fiber diameter of 0.2 μm and 7 nm pore size were found to have tensile strength of 2.7 MPa and to be suitable for air filtration applications [8, 9]. Cellulose acetate solutions prepared using *N*,*N*-dimethylacetamide (DMAC) were electrospun into nanofibrous membranes having fibers with diameters between 175 and 890 nm and were found suitable for removing NaCl aerosol particles (4–240 nm in diameter) with up to 99.8% filtration efficiency and a filter quality factor of 0.14 Pa [10].

Several studies have also explored the possibility of developing air filters by modifying cellulose acetate. In one such approach, cellulose acetate was modified using a poly(ionic liquid) i.e. 1-vinyl-3-butylimidazale bromide at different concentrations as shown in Figure 5.1. The average diameter of the fibers varied from 446 to 680 nm, and membranes were able to remove up to 99.65% and 97.94% of PM_{10} and $PM_{2.5}$ particles and also had significant resistance to *Escherichia coli* and *Staphylococcus aureus* [11]. A comparative study was done to evaluate the performance of cellulose acetate membranes for aerosol filtration with membranes made from several synthetic polymers. Nylon (Polyamide 6/6 and 6) was dissolved in formic

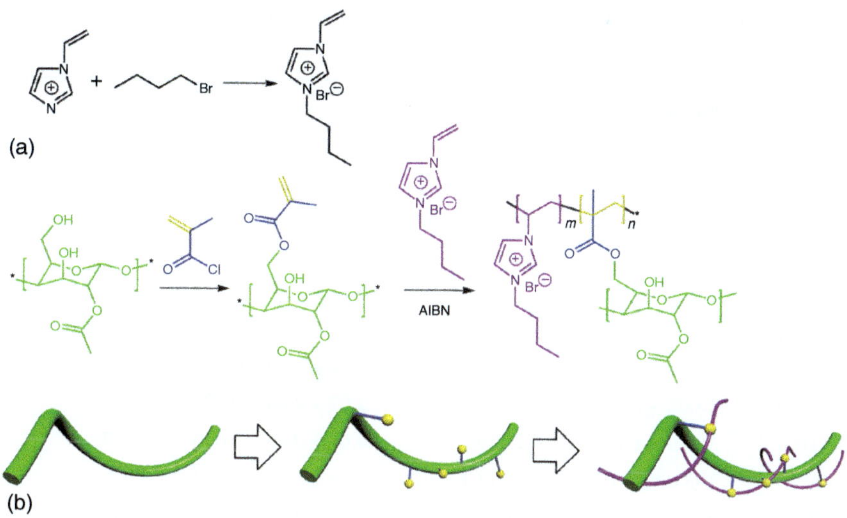

Figure 5.1 (a) Synthesis route of 1-vinyl-3-butylimidazale bromide ([VBIm]Br); (b) The idea of molecular design. [11]. Source: Reproduced with permission from Springer Nature through open access publishing.

acid, Polyvinyl acetate (PVAc) in DMF/acetic acid, and Poly(acrylonitrile) (PAN) in DMF and electrospun into nanofibrous membranes at various electrospinning conditions and polymer concentrations. Fibers with beads and also those with uniform morphology were obtained with diameters ranging from 149 to 437 nm, and membranes formed had a weight per unit area of between 1.36 and 12.0 $g\,m^{-2}$ [12]. Some of the filtration properties against NaCl aerosol obtained by the various nanofibrous membranes are given in Table 5.2.

As seen from Table 5.2, the PVAc and the PAN membranes had better filtration efficiency due to their morphological structures. A three-layered polypropylene-cellulose acetate nanofibers-polypropylene membrane was also found to have a high filtration efficiency of 91%, depending on the thickness of the membranc [13]. A cationic surfactant, CPB, was used to improve the properties of electrospun cellulose acetate nanofibrous membranes for filtration of airborne nanoparticles [8, 9]. Fibers wih an average diameter of 239 nm were obtained with permeability of 10^{-11} m^2 and 100% filtration efficiency, even for carbon black and coronavirus. With a pressure drop of 1.8 kPa at 1.6 $cm\,s^{-1}$ and the ability to retain 90% of carbon black nanoparticles, the membranes were considered to be ideally suited for indoor and outdoor air filtration applications [8, 9]. A summary of the various biopolymers used for developing electrospun air filtration membranes is given in Table 5.3.

A new approach called emulsion electrospinning was used to develop hybrid cellulose acetate nanofibers containing TiO_2 for potential air filtration applications. In this study, cellulose acetate was dissolved using a mixture of *N,N*-dimethylacetamide and acetone in 1 : 2 ratio. TiO_2 was added to the emulsion and electrospun using a voltage of 18 kV and extrusion speed of 1 $ml\,h^{-1}$. Nanofibrous

Table 5.2 Filtration properties of nanofibers/membranes developed using various polymers and their blends [12].

ID	Pressure drop (Pa)	Filtration efficiency at (%)		Quality factor at (Pa^{-1})	
		100 nm	300 nm	100 nm	300 nm
PA6/6_120s	50.77 ± 3.67	62.02 ± 3.17	72.30 ± 2.75	0.019	0.025
PA6/6_120s × 2	94.13 ± 2.74	82.83 ± 2.31	87.6 ± 2.01	0.019	0.022
PA6/6_120s × 3	125.50 ± 3.32	85.36 ± 2.72	90.17 ± 1.89	0.015	0.019
PA6/6_1201	5.10 ± 2.01	24.33 ± 3.34	22.22 ± 3.89	0.055	0.049
PA6/6_1201 × 2	9.60 ± 2.22	26.43 ± 43.05	24.22 ± 3.54	0.032	0.029
PA6/6_1201 × 3	14.02 ± 2.39	32.16 ± 2.56	32.02 ± 2.89	0.028	0.028
PA6/6_1601	18.70 ± 1.70	25.79 ± 0.13	26.69 ± 0.05	0.016	0.017
PA6_300	148.00 ± 17.19	84.73 ± 1.94	89.24 ± 1.38	0.013	0.015
PVAc_30	4.93 ± 1.31	21.51 ± 0.53	21.16 ± 2.18	0.049	0.047
PVAc_30 × 2	8.80 ± 1.98	26.62 ± 1.24	23.61 ± 1.87	0.035	0.031
PVAc_30 × 3	15.20 ± 2.01	36.29 ± 1.67	31.81 ± 1.48	0.030	0.025
PVAc_40	21.68 ± 4.92	39.73 ± 6.51	43.20 ± 6.57	0.023	0.026
PVAc_60	80.57 ± 16.03	98.79 ± 0.37	96.79 ± 0.49	0.055	0.042
PVAc_80	132.83 ± 15.49	99.57 ± 0.31	97.38 ± 0.16	0.041	0.027
PAN_40_6	21.37 ± 1.25	55.94 ± 2.17	59.92 ± 2.30	0.038	0.043
PAN_40_6 × 2	42.40 ± 2.10	80.90 ± 1.45	83.36 ± 1.98	0.039	0.042
PAN_40_6 × 3	60.00 ± 2.45	86.61 ± 1.23	88.27 ± 1.55	0.034	0.036
PAN_40_8	33.77 ± 8.39	77.64 ± 6.27	79.49 ± 6.75	0.044	0.047
PAN_40_8 × 2	64.01 ± 4.21	94.29 ± 2.52	91.89 ± 3.01	0.045	0.039
PAN_40_8 × 3	99.80 ± 2.10	95.91 ± 1.20	96.02 ± 1.34	0.032	0.032
PAN_60	64.60 ± 1.54	94.24 ± 0.90	91.96 ± 0.94	0.044	0.039
PAN_80	90.37 ± 2.57	98.01 ± 0.47	95.83 ± 0.35	0.043	0.035
CA_40	6.57 ± 0.25	24.47 ± 1.77	24.18 ± 2.61	0.043	0.042
CA_40 × 2	13.6 ± 2.43	44.30 ± 1.57	30.74 ± 1.76	0.043	0.027
CA_40 × 3	20.2 ± 2.78	55.51 ± 1.42	50.79 ± 1.36	0.040	0.035
CA_70	49.03 ± 1.21	62.91 ± 0.59	60.22 ± 2.16	0.020	0.019
CA_70 × 2	105.99 ± 2.49	85.83 ± 1.24	84.40 ± 1.80	0.018	0.018
CA_70 × 3	139.90 ± 3.81	88.64 ± 1.00	88.59 ± 0.92	0.016	0.016

Source: Reproduced with permission from Elsevier.

membranes with an average diameter between 200 and 1050 nm were obtained with clear deposits of TiO_2 on the surface of the fibers. The addition of TiO_2 imparted photocatalytic properties to the membranes with ability to capture up to 99.6% of 0.1–0.6 μm particles (Figure 5.2) and degrade volatile organic carbons (VOCs) such as toluene, SO_2, and formaldehyde [14]. Electrospun nanofibrous cellulose acetate

Table 5.3 Some of the biobased polymers used to develop air filtration membranes through electrospinning.

Polymer	Composition	Dia, (nm)	Weight, (g m^{-2})	PM, (μm)	Air velocity (cm s^{-1})	Efficiency, (%)	Pressure drop, (Pa)	*Q*/Pa
Cellulose	CNF/PVP	230	—	0.3	5.3	86.4	170	0.117
	CA	363	23.67	0.3	5.3	60.22	49.03	0.019
	CA/CPB	239	2.17	0.3	1.6	99.99	1000	0.00256
	CA/poly(ionic liquids)	400	—	2.5	—	97.94	7	—
		—	—	—	—	—	115	0.06
	CA/Nylon	400–500	0.81	0.3	—	43	—	0.038
	CA/TBAC	250		0.3	—	90.07	—	—
Chitosan	CS/PVA/P (ADMH-NVF)	100	2.1	0.3	5.3	99.3	183	0.027
		—	—	—	—	—	—	—
	CS/PEO@MOF-	250	—	2.5	—	99.95	44	0.0737
	CS/PVA@SiO_2/ Ag NPs	150	5.64	0.3	—	96.1	295	0.011
		100						
Proteins	Zein	350	—	2.5	5.3	93.5	91	0.03
	Ribbon zein	120	12	2.5	6.2	99.99	107	0.075
	Zein/PVO	230	5.51	2.5	−2.0	99.99	175	0.07
	Zein/PVA	230	—	0.3	−5.3	99.99	—	—
	Zein/PVA/GA	700	—	0.3		97.3	112.5	0.0409
	Z-CoP/ZNF	643	2.31	2.5		99	310	0.029
	SPI/PVA	260	3.74	0.3		99.99	96	0.045
	SF	480	3.4	2.5		96.6		
Novel Bio Matrix	β-CD/PAN	463	1.2	0.3	6	91.66	92	0.07
	β-CD/PVA	500–600	5	2.5	−5.33	99.7	460	0.0125
	ZnO@PVA/KGM	343	—	0.3	1.76	99.9	130	0.053
	LS/PVA	439		2.5		99.44	24.5	0.212
Commercial Mask	PP		26.61	0.3	3	87.26	40	0.052

Source: Reproduced with permission from Elsevier.

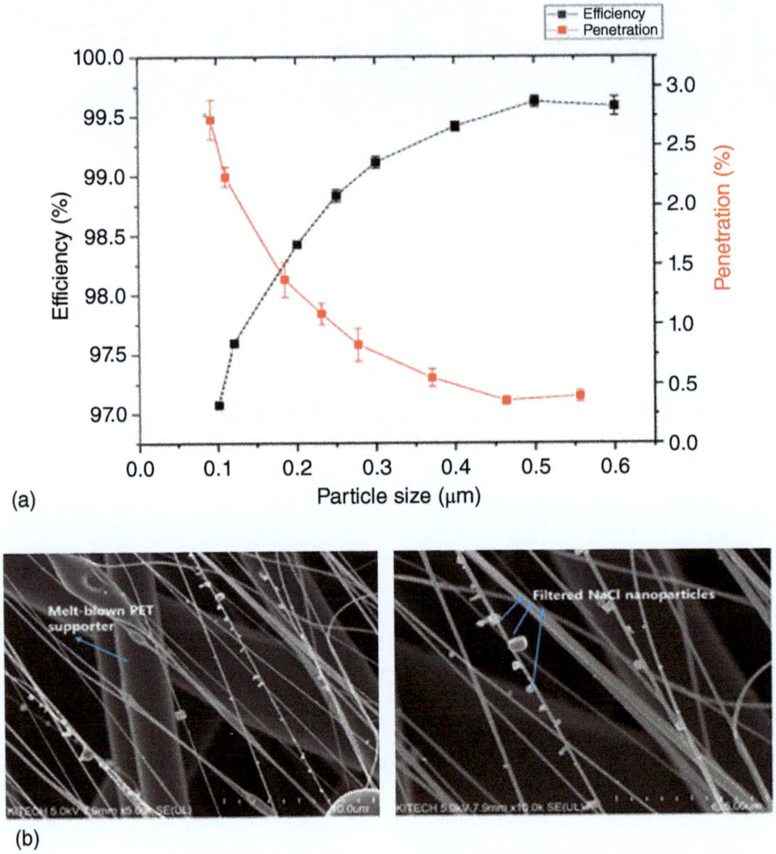

Figure 5.2 (a) Diagrams of the filtration efficiency (%) and air penetration (%) of 10 wt% TiO_2-CA nanofiber webs, and (b) captured NaCl particles (represented as PMs) on TiO_2-CA nanofiber. Source: Kwon et al. [14] Reproduced with permission from MDPI/CC BY 4.0.

was combined with polyvinylidene fluoride (PVDF)-coated polypropylene spunbond layers to develop N95 particulate filtering masks ([15]). Fibers with an average diameter of 319 nm, mean flow pore size of 5.71 μm at 15 minutes with thickness of 0.019 nm were possible when PVDF concentration was 10%. The performance of the membranes satisfied the requirements of NIOSH with a penetration of 1.85% and a pressure drop of 33.87 mm H_2O and was considered suitable for developing masks [15]. Further improvement in the performance of the air filter masks was possible by the addition of activated carbon to TiO_2. Up to 90% removal efficiency was obtained for particle size between 1 and 10 μm and 100% removal efficiency for particles between 5 and 10 μm with low pressure drop of 63–63.8 Pa at face velocity of 0.8 m s^{-1} (Figures 5.3 and 5.4), and hence the filters were considered to be ideally suited for building filtration applications [16]. In addition to particulate matter, indoor airfilters are supposed to remove bad odors and volatile compounds. Electrospun cellulose acetate nanofibers (773 average diameter) were enriched with beta-cyclodextrin to remove odor of hexanol. Up to 85% of cyclodextrin could be

Figure 5.3 Experimental set-up to test the filtration efficiency of the cellulose acetate membranes containing TiO_2 and activated carbon. Source: Orlando et al. [16]/Reproduced with permission from MDPI/CC BY 4.0.

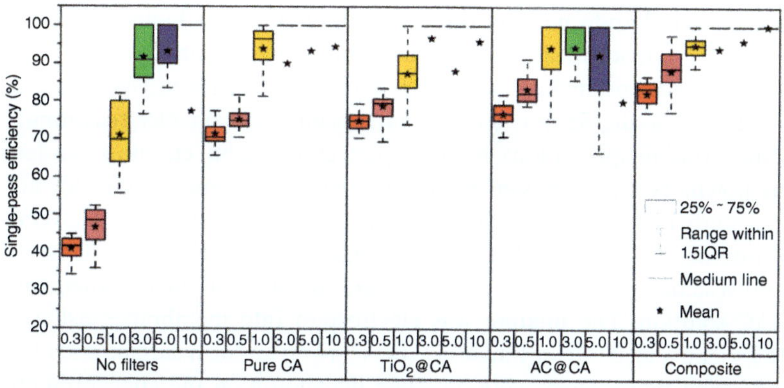

Figure 5.4 Filtration efficiency of pure cellulose acetate, TiO_2 and activated carbon-containing electrospun nanofibrous cellulose acetate membranes. Source: Orlando et al. [16]. Reproduced with permission through open access publishing/MDPI/Public domain.

encapsulated without affecting the structure of cellulose acetate. About 80% of the hexanol present in the air could be entrapped by the electrospun fibers containing β-cyclodextrin suggesting their use for indoor air filtration [17, 18].

(NIOSH: National Institute for Occupational Safety and Health)

5.2.1.1 Starch-Based Filters/Membranes

Starch in rice flour was combined with poly(vinyl alcohol) (PVA) and made into electrospun nanofibrous membranes by applying a voltage of 65 kV. β-cyclodextrin or $AgNO_3$ were also added to impart antimicrobial activity to the filters. Blend fibers had diameters of about 150 nm, and addition of silver nitrate or β-cyclodextrin not only increased bead formation but higher crystallinity was also observed. Improvements in strength and modulus but a marginal decrease in elongation was observed due to the additives. Although β-cyclodextrin affected permeability of the membranes (Table 5.4), considerable increase in absorption of toluene was noticed [19].

Table 5.4 Air permeability of nanofibers on filter paper at varying spinning times.

Sample	Air permeability[a] ($cm^3\ s^{-1}\ cm^{-2}$)		
	15 min	30 min	60 min
PVA	6.12 ± 0.30	4.41 ± 0.22	0.37 ± 0.06
NF	5.87 ± 0.15	4.23 ± 0.27	0.21 ± 0.04
NFS2	6.47 ± 0.21	4.19 ± 0.19	0.22 ± 0.12
NFB2	5.65 ± 0.29	3.92 ± 0.32	0.16 ± 0.09
NFSB	5.79 ± 0.31	4.08 ± 0.21	0.19 ± 0.11

Source: Woranuch et al. [19]. Reproduced with permission through open access publication/Royal Society of chemistry/CC BY 3.0/Public domain.

5.2.1.2 Chitosan-Based Air Filters

Chitosan is a polysaccharide with antimicrobial properties and hence is widely used in the food, pharmaceutical, and medicine industries. In addition, chitosan dissolves in acidic solutions, making it convenient to make electrospun membranes and freeze-dried scaffolds. Several studies have been done on electrospinning chitosan for air filtration applications. Not only pure chitosan, but chitosan blended with various polymers has also been made into electrospun membranes for air filtration. In a study by Li et al. chitosan nanoparticles having a diameter between 100 and 600 nm, degree of deacetylation of 95%, and viscosity of 42 cps was combined with poly(lactic acid) (PLA) (1 : 4 ratio) and dissolved in dichloromethane (DCM)/DMAC solution. The mixture was electrospun into membranes using a voltage of 18 kV and a feed rate of 1 ml h^{-1}. Fibers having uniform surface and also with bead-on-string morphology were obtained depending on the concentration of the solution used. The properties and filtration performance (Table 5.5) of the membranes were also dependent on the ratio of the two polymers and the conditions used during electrospinning. In addition to the good filtration performance, the presence of chitosan imparted antimicrobial properties [20]. A blend of chitosan and polyethylene oxide (PEO) (90/10 w/w ratio) was electrospun to

Table 5.5 Properties and performance of chitosan-PLA blend electrospun membranes.

Samples	Thickness (mm)	Filtration efficiency (%)	Pressure drop (Pa)	Quality factor
FY3107	1.152	89.18	71.67	0.0310
0 wt% CS + 8 wt% PLA	0.215	99.90	335.90	0.0207
1 wt% CS + 8 wt% PLA	0.196	98.10	167.05	0.0237
1.5 wt% CS + 8 wt% PLA	0.198	98.78	162.35	0.0271
2 wt% CS + 8 wt% PLA	0.206	98.26	147.00	0.0276
2.5 wt% CS + 8 wt% PLA	0.202	98.99	147.60	0.0312

Source: Li et al. [20]. Reproduced with permission through open access publishing/MDPI/CC BY 4.0.

form membranes on spun-bonded polypropylene for filtration of particles in air and water. The membranes obtained had relatively weak mechanical properties but were able to remove aerosol particles and polystyrene beads. In addition, the membranes showed good antibacterial properties against gram-positive and gram-negative bacteria [21]. Unique multilayer membranes were developed by combining PVA-chitosan membrane with P(ADMH-NVF), a novel *N*-halamine biopolymer, in a sandwich assembly. A high filtration efficiency of 99.3% for sodium chloride and 99.4% for diisooctyl sebacate aerosol was possible with a low pressure drop of 183 and 2838 Pa for the two particles, respectively. The membranes had high tensile strength of 6.1 MPa and high activity against *E. coli* and *S. aureus* [22, 23]. In another study, chitosan was converted into quaternary ammonium chitosan and combined with PVA. Later, the mixture was electrospun into nanofibrous membranes with good mechanical properties and thermal stability. Membranes formed have porosity of 125 nm, and maximum removal rates were 92, 86, and 82% for PM_{10}, $PM_{2.5}$, and $PM_{1.0}$ particles, respectively. High antibacterial activity of up to 99% was noticed for both gram-positive and gram-negative bacteria [24]. A new approach (Figure 5.5) was used to develop continuous, uniform and stable metal-organic framework-5 (MOF-5) supported by flexible and freestanding chitosan/PEO electrospun membranes for removal of $PM_{2.5}$ particles. As seen from Figure 5.5, the chitosan-PEO membranes were functionalized with amino groups and Zn^{2+}. Modified membranes obtained showed high filtration efficiency of 99.95% and quality factor of 0.0737 making it suitable for filtering $PM_{2.5}$ particles [25].

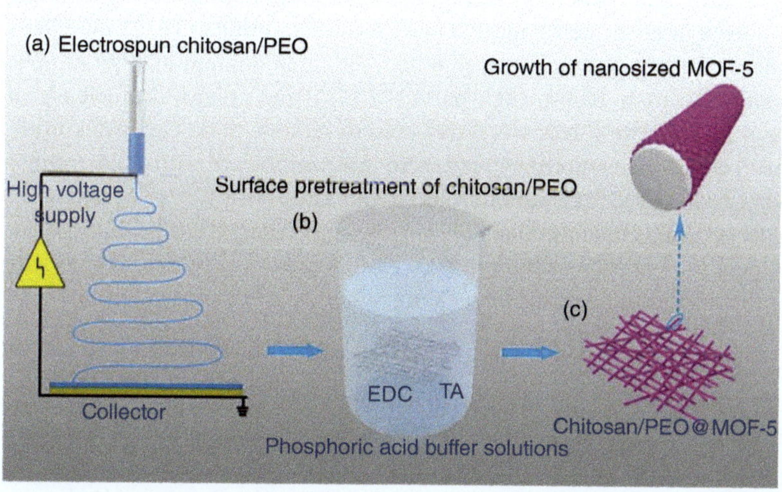

Figure 5.5 Schematic illustration of electrospun chitosan/PEO, used as a skeleton to obtain a self-supported and flexible chitosan/PEO@MOF-5 membrane. (a) Electrospinning of chitosan/PEO nanofibrous membrane, (b) surface pretreatment of chitosan/PEO nanofibers, and (c) growth of nanosized MOF-5 on the chitosan/PEO membrane [25]. Source: Reproduced with permission from Elsevier.

5.2.2 Protein-Based Air Filtration Membranes

Proteins are difficult to dissolve in electrospinnable solvents, and hence limited studies have been done on developing protein based electrospun membranes for filtration applications. However, sericin, a unique water-soluble protein produced by silkworms, has been made into an electrospinable membrane for various applications. Sericin is gum or glue protein that has to be removed before processing the silk fibers. Hence, sericin is considered a byproduct. Water-soluble sericin was combined with PVA and electrospun into membranes for air filtration applications. The filtration efficiency of the fibroin-PVA membranes was affected by the electrospinning time because of the changes in morphology. Up to 100% efficiency for both $PM_{2.5}$ and PM_{10} particulate matter was possible, which was considerably higher than pure PVA membranes at a pressure drop of 50 Pa. Functional groups present in fibroin were suggested to improve the interaction and affinity of the membranes with particles and hence their efficiency [26]. In another study, silk cocoons were treated to extract sericin of about 250 kDa, which was combined with PVA in equal proportions. Clay Closite 30 B (0.1–0.75%) was added to the blend solution, and conductivity was maintained between 0.044 and 1.056 m mho. The blend solution was electrospun into nanofibers under optimized conditions of 27.5 kV voltage, distance of 8 cm and flow rate of 0.8 ml h^{-1}. The nanofibers obtained had diameters between 300 and 400 nm, burst strengths of up to 10 N, and high antimicrobial activity. The membranes were able to filter particulate matter with size range between 1 and 10 μm suggesting suitability for cloth-based air filtration masks [27].

Zein, the protein in corn, dissolves in aqueous ethanol or acetic acid and is hence relatively easy to be electrospun compared to other proteins. Filtration membranes intended to remove both particulate matter and aerosols were prepared by combing zein (7% dissolved in acetic acid) and PEO and electrospinning at 16–21 kV. Fibers with diameters of about 200 nm were obtained with zein content as high as 96%. With pore size less than 10 μm, high particle (0.3–10 μm) removal efficiency of >99.6% was possible. The filters were also able to remove toxic chemicals up to 70% (Figure 5.6) [28]. In another study, zein was combined with PVA (60/40) and electrospun at 14 kV into nanofibrous membranes. The fibers formed were crosslinked by exposure to glutaraldehyde vapors in the presence of trifluoroacetic acid as the catalyst [29]. Crosslinking with glutraldehyde increased the average diameter of the fibers but provided better water stability and increased mechanical properties up to 3.5 MPa. Filtration efficiency of up to 97.3% (0.3 μm particle) was obtained, which was lower than high-efficiency particulate air (HEPA) filters but higher than soyprotein-based filters [29]. In a unique approach, zein was combined with gelatin and electrospun into fibers with diameters of 270 nm for gelatin fibers and 1.12 μm for zein fibers. Considerably high removal efficiencies of 99.67% for $PM_{2.5}$ and 98.80% for $PM_{0.3}$ at low pressure drop of 38 Pa were developed [30].

Similar to zein, soyproteins have also been made into electrospun nanofibrous membranes for filtration applications. Soy protein isolates were dissolved in acetic acid/water solution and treated at 85 °C for four hours. For electrospinning, the soyprotein solution was combined with polyamide 6 (PA6) solution along with the

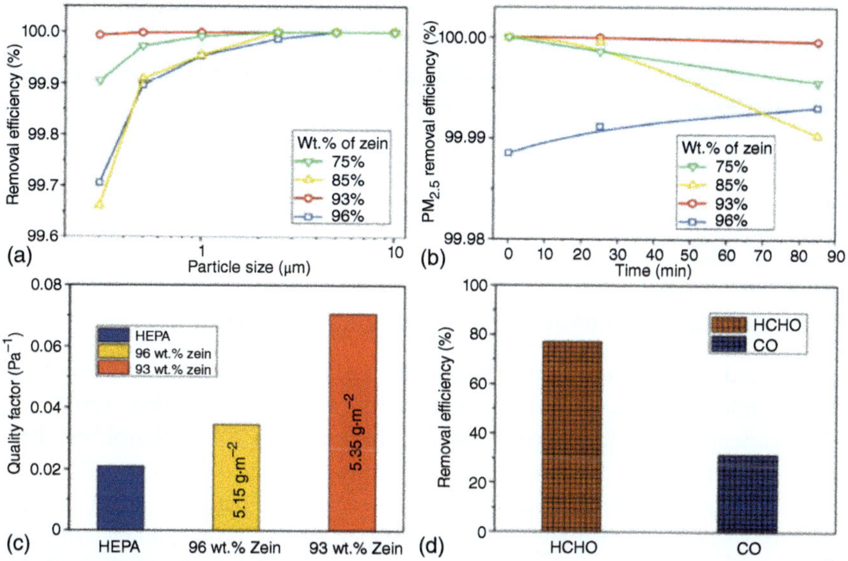

Figure 5.6 Filtration performance of a zein-based nanofilter. (a) Particulate removal efficiencies for PM of various particle sizes. (b) Effect of testing time on $PM_{2.5}$ removal efficiencies. (c) Comparison of the quality factors of the commercial HEPA filter and zein-based nanofabric filters with zein loadings of 96 and 93 wt%. (d) Chemical removal efficiency of zein-based nanofabric filters for formaldehyde (HCHO) and carbon monoxide (CO). Source: Tian et al. [28]. Reproduced with permission from Springer Nature/Springer Nature.

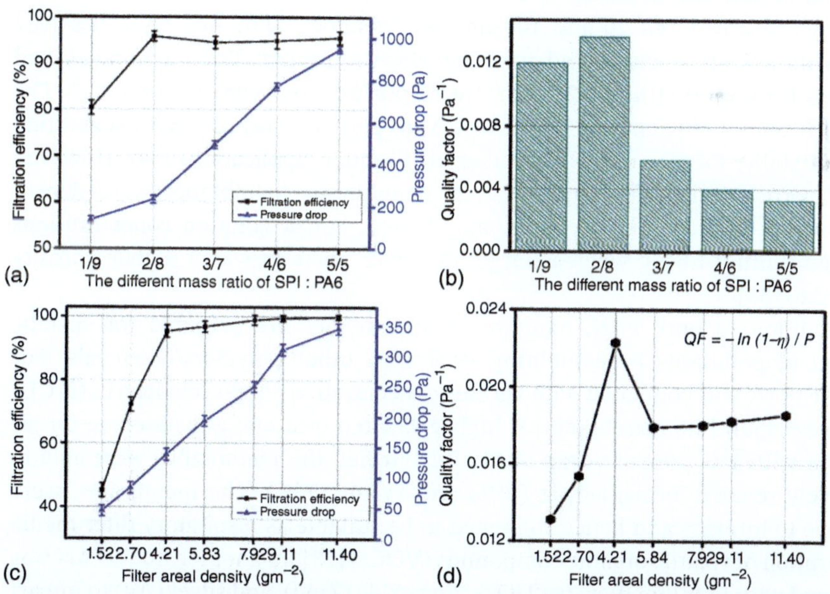

Figure 5.7 (a) Removal efficiency and pressure drop (air-flow resistance) of SPI/PA6 nanofibrous membranes with different mass ratio of SPI: PA6. (b) QF of SPI/PA6 nanofibrous membranes also with different mass ratio of SPI: PA6. (c) Filtration efficiency and pressure drop with an SPI/PA6 mass ratio of 2 : 8 nanofibrous membranes with various areal density. (d) QF of SPI/PA6 with mass ratio of 2 : 8 with various areal density. Source: Jiang et al. [31]. Reproduced with permission from John Wiley and Sons/John Wiley & Sons.

addition of 0.1–0.6 wt% of silver nanoparticles (AgNPs) ([31]). Fibers obtained had average diameters between 225 and 600 nm and membranes had an areal density of 4.21 g m^{-2}. Filtration efficiency was greater than 99%, and the membranes also showed antimicrobial properties due to the presence of AgNPs, and the fibers were also able to sorb VOCs [31] (Figure 5.7).

5.3 Synthetic Polymers

5.3.1 Poly(acrylonitrile) (PAN)-Based Filtration Membranes

PAN was used to develop a three-dimensional (3D) structure composed of nanofibers, microspheres, and thin nanofibers with low filtration resistance through a concept called free surface electrospinning (Gao 2017). Nanofibers had diameters of about 84 nm and the membranes formed showed filtration efficiency of 99.99% and lower pressure drop of 126.7 Pa to aerosol particles when the air velocity was 5.3 cm^{-1}. Even $PM_{2.5}$ filtration efficiency was 99.24%, suggesting the suitability of the membrane for industrial and commercial applications [32]. In another approach, PAN nanofibrous membranes intended for air filtration were functionalized by adding titanium dioxide (TiO_2), zinc oxide (ZnO), and silver (Ag). Filters containing AgNPs showed highest filtration efficiency of 100%, the highest quality factor of about 0.06 Pa^{-1} and permeability coefficients between 1.83×10^{-12} and 9.17×10^{-13}. In addition, the filters were able to provide antibacterial resistance to *E. coli* at concentrations up to 10^8 CFU ml^{-1} [33]. PAN nanofiber membranes with thickness between 20 and 100 μm were obtained with 14% PAN dissolved in *N*,*N*-dimethylformamide and used for electrospinning. The membranes had densities between 0.11 and 0.21 g cm^{-3} with porosity between 80% and 92%. The highest filtration efficiency of 99.7% for 0.3 μm particles was achieved, suggesting their suitability for face masks and other air filtration applications [34]. However, the properties and performance of PAN nanofibrous membranes were dependent on the electrospinning conditions adopted. Lower solution concentrations provided thinner fibers, whereas larger diameter fibers provided higher porosity (Table 5.6) [35].

In addition to pure PAN, nanofibrous membranes are prepared for specific removal of pollutants by combining PAN with other polymers/chemicals. For instance, PAN was combined with various concentrations of β-cyclodextrin (β-CD) and electrospun into membranes. A higher 95% removal rate was observed for air particles with low pressure drop of 112 Pa. Further, the membranes were able to selectively remove formaldehyde (66%) and xylene (73%). The membranes were nontoxic to humans and hence suggested to be suitable as respiratory filter media and sorption of volatile organic compounds (VOCs) [18]. In another study, PAN was combined with titanium dioxide (TiO_2), zinc oxide (ZnO), and silver (Ag) to impart antibacterial properties and improve the removal of air pollutants. TiO_2 containing filters had the lowest diameters (Figure 5.8) and highest filtration efficiency of 100% whereas the silver containing filters had highest quality factor and low pressure

Table 5.6 Effect of electrospinning conditions on the properties of the PAN nanofibrous membranes obtained.

PAN concentration	Voltage (kV)	Tip to collector distance (cm)	Average fiber diameter (nm)	Porosity (%)
16	20	10	263.3	57.87
16	10	15	313.6	40.03
8	20	15	69.05	29.46
8	10	10	46.44	36.53
9.6	15	12.5	27.78	38.15
14.4	15	12.5	113.4	44.73
12	12	12.5	105.5	37.49
12	18	12.5	45.36	43.25
12	15	11.0	43.02	51.17
12	15	14.0	98.50	48.09
12	15	12.5	54.80	44.23
12	15	12.5	68.10	45.69
12	15	12.5	84.36	44.23
12	15	12.5	72.40	45.69
12	15	12.5	52.22	44.23

Source: Dehghan et al. [35]/Taylor and Francis.

drop and very high antibacterial activity better than commercial filters available on the market [33]. PAN electrospun into nanofibrous membranes was further made into microscaffolds having mesh structures by aligning the electrospun membranes into various configurations, followed by solvent vapor annealing. Up to 99.98% filtration efficiency for 0.25 μm particles at a face velocity of 10 cm S^{-1} was achieved using a nine-layer cross ply structure [36]. The scaffolds developed were suggested to be suitable for use as HEPA filters.

5.3.2 Poly(vinyl alcohol)-Based Filters

PVA is a water-soluble polymer and is also easily electrospinnable. Hence, several efforts have been made to develop PVA-based electrospun membranes for various applications, including air filtration. Copper-1,3,5-benzoic acid (Cu-BTC) particles were added into PVA solution and the mixture was electrospun at 15 kV and flow rate of 0.5 ml h^{-1} to obtain electrospun films. Films obtained after modification with Cu-BTC had 8.5 times higher filtration performance than unmodified PAN films [37]. A colloidal graphene solution and in situ synthesized hydroxyapatite nanoparticles were combined with PVA in different ratios and electrospun into membranes using an extrusion speed of 0.1 mm min^{-1} at an applied voltage of 20 kV [38]. The membranes obtained were treated at temperatures between 120 and 200 °C for 30 minutes to crosslink and improve their properties. Heating altered the properties and also the morphology of the fibers, with increasing temperature decreasing fiber diameters from 184 to 104 nm. Inclusion of the graphene and HA was found to promote UV absorption, which is preferable for face mask

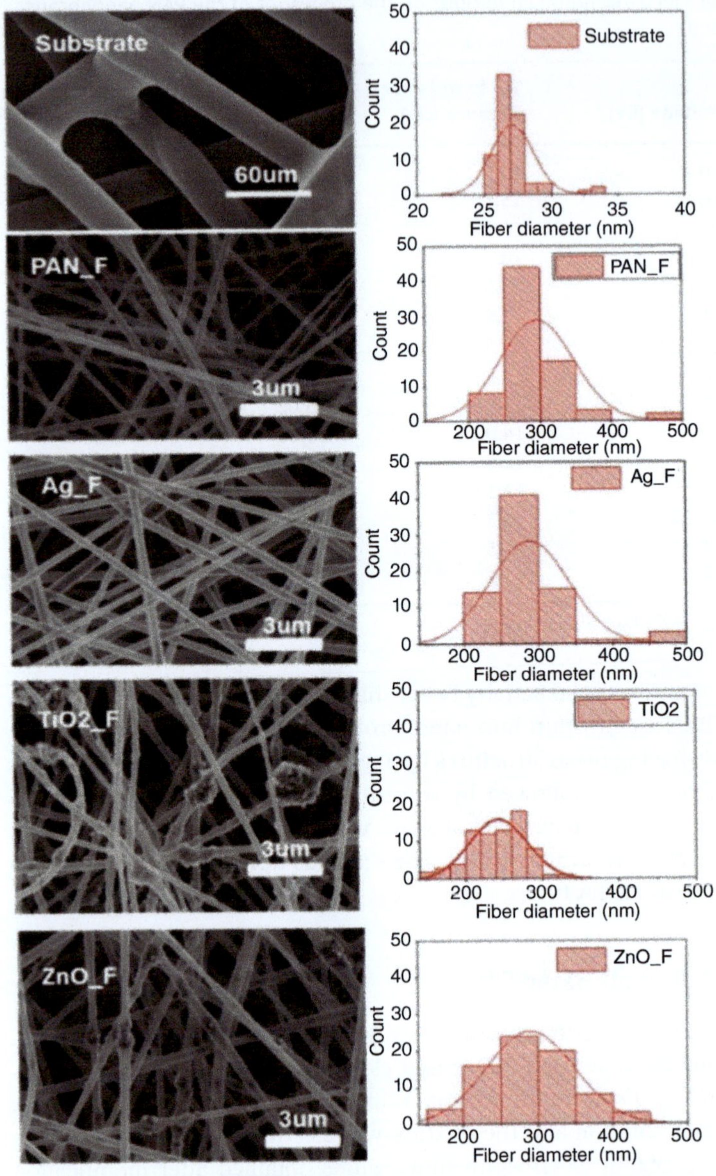

Figure 5.8 Morphology and diameter of the nanofibers obtained by electrospinning PAN with different functional compounds. Source: Bortolassi et al. [33]/Reproduced with permission from MDPI/CC BY 4.0.

applications. The blend membranes also had considerably high tensile strength of 4.2 MPa and also had antifungal and antibacterial properties due to the presence of hydroxyapatite [38]. Water resistant, washable and reusable highly efficient filters were developed by blending PVA and cellulose nanocrystals (CNCs) (5–20 nm in width and 150–200 nm in length). Electrospun PVA-CNC membranes were treated between 80 and 140 °C for five minutes to impart thermal stability and improve the properties of the membranes [22, 23]. Heat treatment increased the % crystallinity of the fibers from 55% to 85% and hence imparted increased tensile strength and water resistance. Membranes (filter) treated at 140 °C showed filtration efficiency of 96% for $PM_{2.5}$ particles and a pressure drop of 60 Pa. Filters that were contaminated (pressure drop of >1000 Pa) retained high efficiency of 95% despite being contaminated and even after repetitive usage, the filters retained an efficiency of 95% even for filtering $PM_{2.5}$ mass concentration higher than 500 $\mu g\,m^{-3}$ [22, 23].

Unique flexible and transparent nanofibrous air filtration membranes were prepared for capture of $PM_{2.5}$ particles through a green electrospinning approach [39]. To develop such a filter, PVA was combined with sodium lignosulfonate (LS), which contains several hydrophilic groups and can form a strong interaction between PVA and LS through sacrificial hydrogen bonds. Pure PVA nanofibers obtained had average diameter of 240 nm compared to 439 nm for the blends. Inclusion of LS not only improved tensile strength of the membranes from 15.6 to 25.8 MPa, and elongation from 86% to 358% but also increased the transparency by 78% with a quality factor of 0.212 Pa. Even after 10 filtration cycles, a high filtration efficiency of 99.38% at a low pressure drop of 25.5 Pa was obtained. With high performance, transparency, and reusability, the filters were considered ideal for face masks and other applications [39]. Functionalized nanofibrous air filters were prepared by adding carbon nanoparticles and tea leaf extract (TLE) to PVA. A PVA solution (8% w/w) was prepared with water, 0.1% nanoparticles, and 1 ml of TLE were added, and the blend was electrospun at 22 kV with a 0.15 $ml\,h^{-1}$ flow rate and a distance of 2 cm. Membranes obtained were functionalized by heating to 600 °C for one hour for stabilization to occur. The addition of the nanoparticles and TLE increased mechanical properties significantly and marginally varied water absorption and desorption (Table 5.7). The PVA/CNP/TLE membranes had a total surface area of 5.24 $m^2\,g^{-1}$, total pore volume of 0.15 $cm^3\,g^{-1}$ and average pore width of 76.9 nm. A high filtration efficiency of 99.25% and 99.29% for $PM_{2.5}$ and $PM_{10-2.5}$, respectively with a low pressure drop of 110 Pa was possible (Figure 5.9). In addition to the excellent filtration performance and mechanical properties, the filters also had antimicrobial properties suitable for face mask applications [40]. Multifunctional

Table 5.7 Properties of PVA-based electrospun nanofiber membranes containing CNPs and TLE.

Samples	Tensile strength (MPa)	Elongation at break (%)	Flexing index (%)	Water absorption (%)	Water desorption (%)
PVA	15.40 ± 0.18	16.52 ± 0.26	5.52 ± 0.35*	32.66 ± 0.65	40.21 ± 0.01*
PVA: CNPs	17.23 ± 0.08*	18.32 ± 0.09	6.15 ± 0.07	34.48 ± 0.37*	42.16 ± 0.03*
PVA: CNPs: TLE	18.31 ± 0.09	18.95 ± 0.03	6.95 ± 0.03*	35.44 ± 0.41	43.12 ± 0.04

*$p < 0.05$. Compared to PVA, using Duncan's multiple range analysis.
Source: Senthil et al. [40]. Reproduced with permission through open access publication/Springer Nature/CC BY 4.0/Public domain.

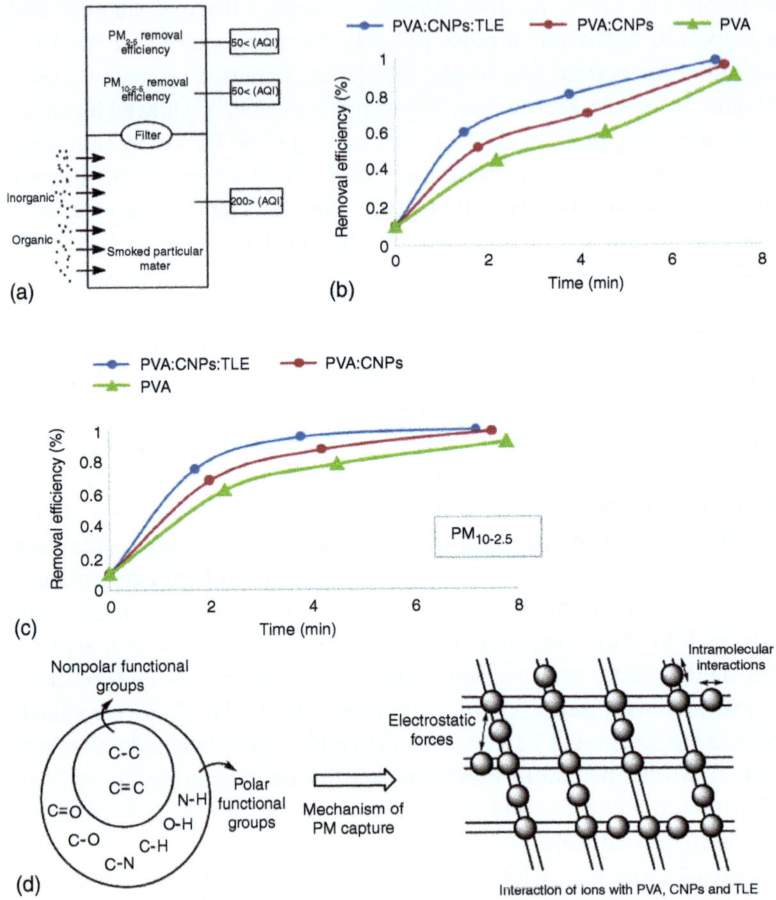

Figure 5.9 (a) The setup designed for testing the efficiency capture of PM2.5 and PM10−2.5 capture (b) PM2.5 capture efficiency of PVA, PVA: CNPs and PVA: CNPs: TLE (c) PM10−2.5 capture efficiency of PVA, PVA: CNPs and PVA: CNPs: TLE (d) Schematic showing the mechanism of PM capture. Source: Senthil et al. [40]. Reproduced with permission through open access publication/Springer Nature/CC BY 4.0/Public domain.

electrospun nanofibrous membranes were made by combining PVA with konjac glucomannan (KGM), citric acid, and ZnO_2 nanoparticles. Citric acid acted as a crosslinker to improve aqueous stability, and ZnO_2 was used to impart antimicrobial activity specifically when exposed to light, leading to photocatalytic degradation (Figure 5.10). Filtration efficiency of the membranes varied considerably based on ZnO concentration and fiber diameters (Table 5.8). The filters obtained were reported to have better efficiency than commercially available HEPA filters, high photocatalytic degradation, and more than 98% antibacterial activity against both gram positive and gram negative bacteria [41, 42]. High filtration efficiency for particles and VOCs such as toluene was possible by adding TiO_2 or activated carbon to PVA, electrospinning into membranes (filters weighing 16.6–67.6 g m^{-2} and specific surface areas between 1.4 and 17.4 m^2 g^{-1}) and subjecting them to photocatalytic

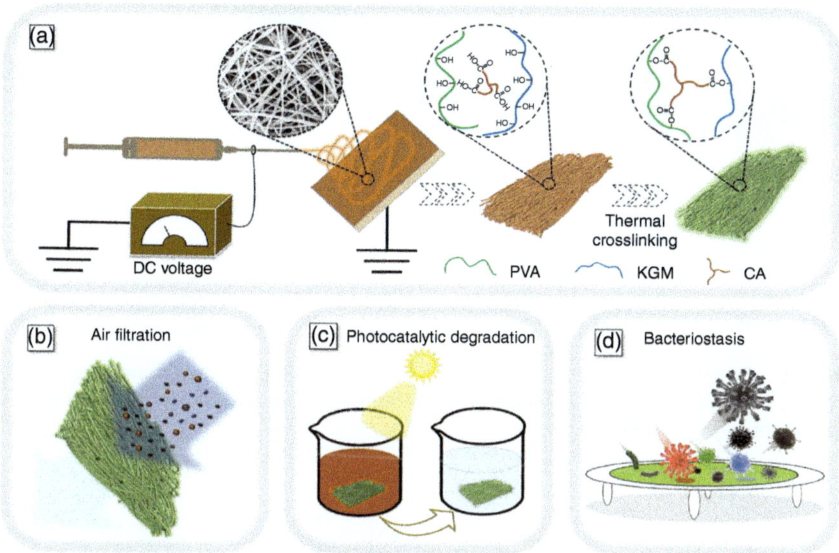

Figure 5.10 Multifunctional ZnO@PVA/KGM electrospun nanofiber membranes. (a) Schematic presentation of the preparation of the membranes by electrospinning and (b) their application for air filtration. (c) Photocatalytic degradation. (d) Antibacterial activity of the membranes [41]. Source: Reproduced with permission through the American Chemical Society.

Table 5.8 Filtration efficiency (%) of the PVA nanofibrous membranes containing ZnO, KGM, and crosslinked with citric acid [41].

ZnO concentration (%)	Particle diameter (μm)						Pressure drop (Pa)
	0.3	0.5	1	3	5	10	
0	97.07	98.59	99.54	99.98	100	100	89.0
0.5	98.56	99.66	100	100	100	100	108.0
1.0	99.99	99.99	100	100	100	100	130.0
1.5	98.85	99.69	99.98	99.99	100	100	139.5
2.0	98.01	98.68	99.69	99.85	100	100	158.0

Source: Reproduced with permission through the American Chemical Society.

oxidation. Removal efficiency higher than 97% for 12–480 nm particles at face velocity of 5.3 cm S^{-1} was possible with low pressure drop between 195 and 2693 Pa [16]. The addition of tannic acid to PVA resulted in marginally finer fibers (470 nm diameter) and electrospun membranes with higher mechanical properties (12.4 MPa) than the commercially available $PM_{1.0}$ filters. PVA-TA membranes had excellent filtration efficiency with quality factor of 0.15 Pa and a pressure drop of 35 Pa, which was 100% higher than pure PVA membranes. Blend membranes produced high filtration efficiency at different flow rates and better performance than surgical mask or dust mask even after repeated usages [39] (Figure 5.11).

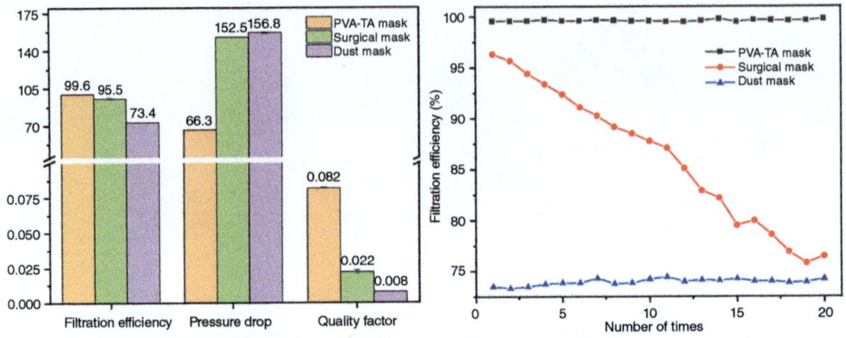

Figure 5.11 Comparison of the performance of PVA-TA nanofibrous membranes compared to commercially available surgical masks and dust masks [39]. Source: Reproduced with permission through Elsevier.

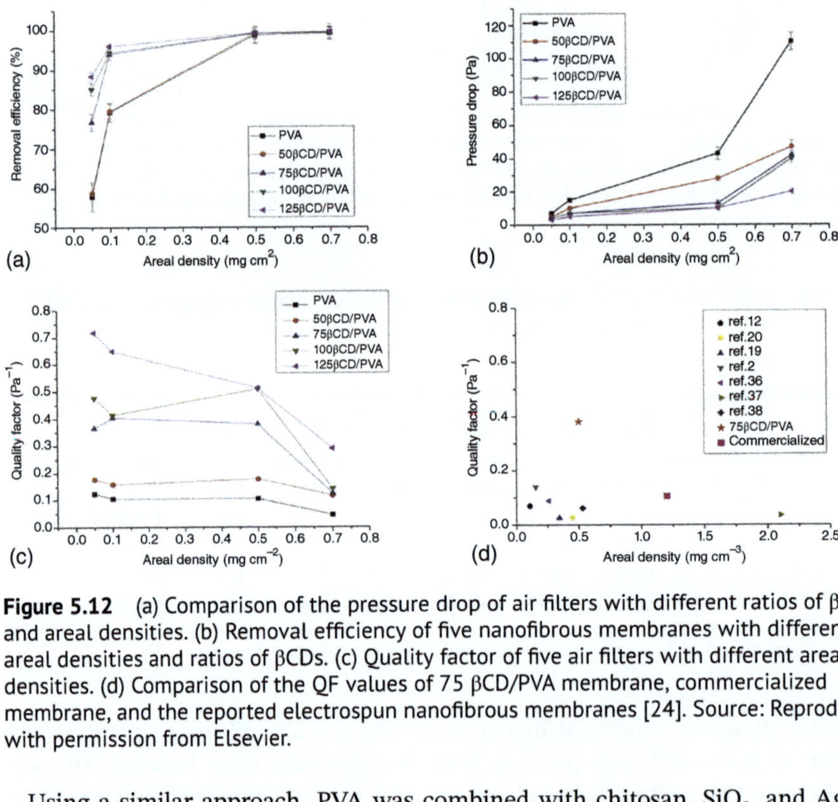

Figure 5.12 (a) Comparison of the pressure drop of air filters with different ratios of βCDs and areal densities. (b) Removal efficiency of five nanofibrous membranes with different areal densities and ratios of βCDs. (c) Quality factor of five air filters with different areal densities. (d) Comparison of the QF values of 75 βCD/PVA membrane, commercialized membrane, and the reported electrospun nanofibrous membranes [24]. Source: Reproduced with permission from Elsevier.

Using a similar approach, PVA was combined with chitosan, SiO_2, and AgNPs and made into nanofibrous membranes. The membranes were treated in UV light to form semi-interpenetrating polymer networks and also to impart antimicrobial properties [43]. Up to 97.73% filtration efficiency for NaCl and 97.3% efficiency for DEHS were obtained. Further, the membranes showed high antimicrobial activity but were non-toxic to mouse fibroblast cells suggesting suitability for personal air filtration [43]. PVA mixed with soyprotein isolate and activated carbon had filtration efficiency of up to 90% for particulate matter and was also able to sorb toxic gases such as formaldehyde [44].

β-Cyclodextrin (50–125% based on weight of PVA) was dissolved in water, and the solution was electrospun into nanofibrous membranes with areal density between 0.05 and 0.7 mg cm^{-1} depending on the amount of β-Cyclodextrin. The addition of β-Cyclodextrin did not affect morphology of the fibers but decreased the diameter of pores from 6.7 to 0.65 μm. The removal efficiency for $PM_{2.5}$ particles and the quality factor varied depending on the amount of cyclodextrin and the areal density (Figure 5.12). Membranes obtained showed significantly higher quality factor compared to commercially available filtration membranes. The Addition of β-Cyclodextrin also enhanced the mechanical properties and increased sorption of SO_2 up to 91% compared to 79% for the pure PVA filter [24].

5.3.3 Air Filtration Membranes from Poly(lactic acid) (PLA)

PLA is one of the common biopolymers that is soluble in electrospinnable solutions such as chloroform, acetone, and dimethyl formamide. Numerous studies (Table 5.9) have been done to develop PLA-based electrospun membranes for various filtration applications. A novel approach to electrospinning PLA onto 3D structures was developed by Pierpaoli et al. 3D PLA structures were developed by extruding at 210 °C onto which PLA (14%) dissolved in acetone was electrospun at 5 ml h^{-1} for two hours to form a membrane on the surface (Figure 5.13). Filtration efficiency obtained by the PLA membranes varied for different particle size but was highest for the membranes electrospun on the 3D printed substrate with ability to retain 95% of the particles with a diameter greater than 230 nm [45]. In addition to pure PLA, blends of PLA and polyhydroxybutyrate (PHB) were made into nanofibrous membranes through electrospinning for removal of NaCl aerosol particles. A quaternary ammonium compound [DDA][NO_3] was added to the membranes to provide antimicrobial properties. Multiple layers of the electrospun membranes were used to develop a filter with high filtration efficiency but low pressure drop. Quality factor of the PLA-PHB filters was between 0.015and 0.018 Pa at 0.3 μm, which is lower compared to commercially available HEPA filters (0.023–0.029 Pa) [46]. Unique bead-on-string nanofibrous membranes were prepared from PLA through electrospinning for air filtration applications [47]. PLA was dissolved in 4–6% concentrations in mixture of *N*,*N*-dimethylacetamide (DMAC), and DCM in different proportions to alter the fiber properties obtained. Nanofibers with unique bead-on-string morphology (Figure 5.14) were obtained depending on the electrospinning conditions used. The bead area and ratio of bead area to membrane varied depending on the conditions used during electrospinning. Weight of the membranes was between 5.18 and 5.31 g m^{-2} and pore size was between 2.1 and 3.97 μm. A high filtration efficiency of 99.97% with low pressure drop of 165.3 Pa at face velocity of 5.8 cm s^{-1} (Figure 5.15) was possible suggesting suitability for commercial air filtration [47]. Further, functionalizing PLA with polysilsesquioxane was found to substantially alter the viscosity of the solution and the bead-on-string morphology and properties of the PLA electrospun membranes [48]. The addition

Table 5.9 A comparison of PLA based electrospun fibers and their filtration efficiencies.

Fiber composition	Experimental parameters (rate, solvent, app. potential, distance)	Fiber diameter	Pressure drop	Filtration efficiency
PLA/CyD	0.5 ml h^{-1}, 8% *w/v* DCM/DMF (8 : 2), 16 kV, 10 cm	350–990 nm	24–34 Pa	97% ($PM_{2.5}$, incense burning)
PLA	1 ml h^{-1}, 10 wt%, a DCM/DMF (7 : 3) solvent +15 (needle) −3 kV (collector)	~500 nm	96–102 Pa (8 cm s^{-1})	96.22–99.1% ($PM_{2.5}$, NaCl)
PLA–PLA/chain extender	DCM/DMF/TFE	From 3 μm to 270 nm	50–1000 Pa (20–948 l m^{-2} s^{-1})	92.7–99% Clay (12.07 μm) carbon black (235 nm) TiO_2 (95.3 nm)
PLA/cotton fabric layer	DCM Needleless electrospinning collector, −30 kV, rotating electrode 60 kV	8 ± 0.2 μm	35.78 Pa cm^{-2} (8 l min^{-1})	97.9% (*S. aureus*)
PLA/PSQ	DMF/DCM (3 : 7 vol) 16 kV, 30 °C, 35–50% (RH), 8–20 cm	Average value from 400 to 580 nm	21.43–37.85 Pa μm^{-1} (5.3 cm s^{-1})	>99.98% NaCl (0.3 μm)
PLA/montmorillonite/keratin	CHL/AC (70/30 *v/v*) and formic Acid, 19 kV, 15 cm, and 5 μl min^{-1}	125–171 nm	2.88–107.92 Pa (0.12–0.24 m s^{-1})	Not evaluated
PLA	CHL/DMSO 15 kV, 1.2 ml h^{-1}, 10 cm	40–100 nm	Not evaluated	Not evaluated
PLA/chitosan	DCM/DMAC 18 kV, collector rotating at 5.5 m min^{-1}, 14 cm	910–1500 nm	148 Pa (14 cm s^{-1})	98.99% (NaCl) 99.4% (*E. coli*) 99.5% (*S. aureus*)

PLA: poly(lactic acid); CyD: cyclodextrins; DMC: dimethyl carbonate; DMF: *N,N*-dimethylformamide; NaCl: sodium chloride; TFE: 2,2,2-trifluoroethanol; TiO_2: titanium dioxide; DCM: dichloromethane; PSQ: poly(epoxy-amino) silsesquioxane; RH: relative humidity; CHL: chloroform; AC: acetone; DMSO: dimethyl sulfoxide; DMAC: dimethylacetamide; *E. coli*: *Escherichia coli*; *S. aureus*: *Staphylococcus aureus*.

Source: Pierpaoli etal. [45]. Reproduced with permission through open access publication/MDPI/CC BY 4.0/Public domain.

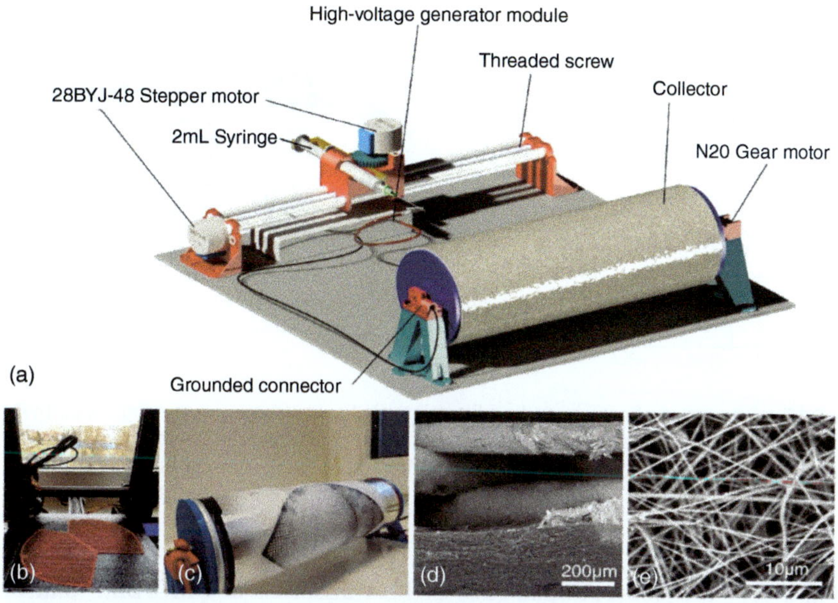

Figure 5.13 (a) Illustration of the 3D-printed simplified electrospinning setup; (b) 3D printing of the filter mask support; (c) electrospinning of the fibers over the 3D-printed PLA support; SEM images (d): cross-section; (e): top-view of electrospun nanofibers. Source: Pierpaoli et al. [45]/Reproduced with permission from MDPI/CC BY 4.0.

Figure 5.14 SEM images of the PLA fibrous membranes fabricated at various concentrations: (a) 4.5, (c) 5.5, (e) 6 wt% PLA solutions in DCM/DMAC (10/1, w/w). Highly magnified images corresponding to the membranes are shown in (b), (d), and (f), respectively. Source: Wang et al. [47]/with permission from Elsevier.

of polysilsesquioxane changed the hydrophobicity and resulted in more dense and uniform beads, which led to a lower pressure drop in the membranes and higher filtration efficiency. A pressure drop of 21.4 Pa μm^{-1} and separation efficiency of 100% was achieved for the functionalized membranes compared to 37.9 Pa μm^{-1} for pure PLA membranes.

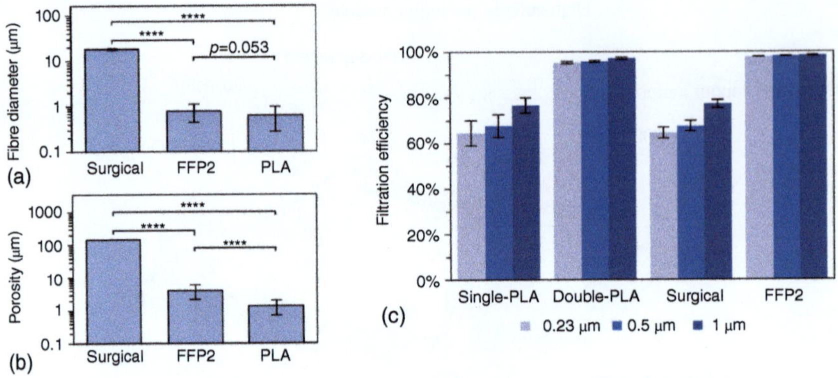

Figure 5.15 Properties of corresponding filtration media, Knudsen number regimes, flow mechanisms, and filtration efficiency: (a) fiber diameter; (b) porosity; (c) percentage filtration efficiency of particles with diameters of 0.23, 0.5, and 1 μm. Significance is expressed by the *p*-value (****$p < 0.005$) [47]. Source: Reproduced with permission through Elsevier.

Using a very low concentration (3%) of PLA dissolved in DMC/DMF and in the presence of TBAC, nanofibers membranes having average fiber diameters of about 37 nm were obtained. The ultrafine fibers obtained were combined to form a multi-scale structure having pore size between 1.83 and 0.26 μm, porosity between 86% and 92% depending on the amount of TBAC and conditions used. Membranes having low weight of $0.91\,g\,m^{-2}$, small pore size of 0.73 μm, and high porosity (91.7%) had filtration efficiency of 99.996% for $PM_{0.3}$ particles with low pressure drop of 104 Pa, which was suggested to be lower than commercially available N95 filters [49]. Some of the properties and proposed mechanisms for sorption of the air particles by the membranes are shown in Figure 5.16. Hierarchical nanofibrous membranes having fibers with nanopores and nanometer-scale protrusions were obtained by adding various concentrations of TiO_2 and varying the relative humidity during electrospinning [50]. Surface area of the membranes varied between 23.4 to $30.5\,m^2\,g^{-1}$ and pore volumes between 0.086 and $0.139\,m^3\,g^{-1}$ were obtained by changing the amount of TiO_2 compared to 21.86 and $0.073\,m^3\,g^{-1}$, respectively. Up to 99.996% filtration efficiency with low pressure drop of 129 Pa and face velocity of $5.3\,cm\,S^{-1}$ was achieved due to the presence of nanoscale pores and protrusions [50]. The presence of TiO_2 also imparted excellent antimicrobial activity of up to 99.5%. It was proposed that electrostatic charges generated from PLA nanofibers could also enhance air filtration and provide high efficiency of 99.3% and better quality factor than commercially available respirator filters [51].

5.3.4 Miscellaneous Polymer Membranes

Several other polymers such as polyamides (Nylon 6 and Nylon 6.6) have also been used to develop electrospun membranes for filtration applications. However, difficulties in dissolving these polymers in electrospinnable solutions limit extensive

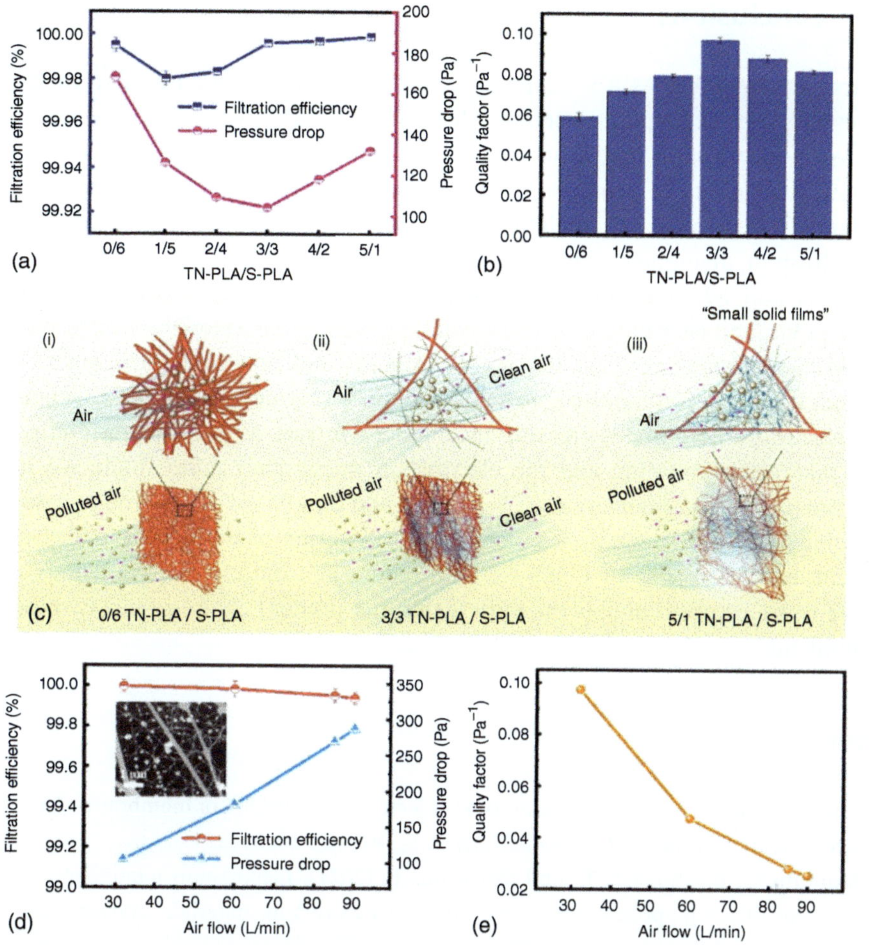

Figure 5.16 (a) Filtration performance and (b) quality factor of multi-scale structured membranes with different numbers of spinning unit ratios of TN-PLA to S-PLA. (c) Schematic diagram of filtration mechanism of MSM-PLA with different ratios of TN-PLA to S-PLA including (i) 0/6, (ii) 3/3, and (iii) 5/1. (d) Filtration performance and (e) quality factor of the multiscale structured membranes (3/3). Source: Wang et al. [49]/Reproduced with permission from Elsevier.

production and usage. Nevertheless, reports on developing novel filtration membranes using polyamide-6 combined with nanoclay and TiO_2 have recently been reported [52]. Nanofibers with diameters between 76 and 136 nm and filters having strength of 0.5–3.5 MPa were obtained. The absorption (CO_2) properties of the membranes increased with increasing amounts of nanoclay and reached up to 43.7%. A blend of poly(ε-caprolactone) (PCL)/zein/AgNP were made into electrospun membranes with the assistance of ultrasonication. Membranes having 1% AgNP showed bead-on-string morphology with air permeability of up to 93.6% and filtration efficiency of >97% for particles between 0.3 and 1.0 μm. The presence of silver resulted in high antimicrobial properties with efficiency of 98% [53].

5.4 Conclusions

Electrospinning provides a versatile approach to developing air filtration membranes with specific properties. Both synthetic and natural polymers have been used to develop electrospun membranes with a wide variety of filtration efficiencies, stability, and ability to separate various types of particles and volatile organic compounds. Although the solubility of the polymers restricts the development of electrospun membranes, the soluble polymers can be blended or additional chemicals and additives added to ensure the required performance. Cellulose acetate, PVA, and PAN have been more commonly used to develop the electrospun air filtration membranes, but the use of proteins and polysaccharides such as chitosan has also been reported. To solubilize more polymers, techniques such as melt electrospinning and increasing the production rate during electrospinning will make electrospinning and the electrospun membranes more conducive to mass production and extensive use. The advent and prevalence of airborne diseases including COVID-9, presents new challenges in terms of developing effective, efficient, and affordable air filters for both indoor and outdoor applications. We believe that electrospinning is the most viable and prudent choice to satisfy these requirements.

References

1 Zhu, M., Han, J., Wang, F. et al. (2017). Electrospun nanofibers membranes for effective air filtration. *Macromol. Mater. Eng.* 302 (1): 1600353.

2 Mamun, A., Blachowicz, T., and Sabantina, L. (2021). Electrospun nanofiber mats for filtering applications—technology, structure and materials. *Polymers* 13 (9): 1368.

3 Zhou, Y., Liu, Y., Zhang, M. et al. (2022). Electrospun nanofiber membranes for air filtration: a review. *Nanomaterials* 12 (7): 1077.

4 Lv, D., Zhu, M., Jiang, Z. et al. (2018). Green electrospun nanofibers and their application in air filtration. *Macromol. Mater. Eng.* 303 (12): 1800336.

5 Zhang, K., Li, Z., Kang, W. et al. (2018). Preparation and characterization of tree-like cellulose nanofiber membranes via the electrospinning method. *Carbohydr. Polym.* 183: 62–69.

6 Chattopadhyay, S., Alan Hatton, T., and Rutledge, G.C. (2016). Aerosol filtration using electrospun cellulose acetate fibers. *J. Mater. Sci.* 51 (1): 204–217.

7 Nicosia, A., Keppler, T., Müller, F.A. et al. (2016). Cellulose acetate nanofiber electrospun on nylon substrate as novel composite matrix for efficient, heat-resistant, air filters. *Chem. Eng. Sci.* 153: 284–294.

8 de Almeida, D.S., Duarte, E.H., Hashimoto, E.M. et al. (2020a). Development and characterization of electrospun cellulose acetate nanofibers modified by cationic surfactant. *Polym. Test.* 81: 106206.

9 de Almeida, D., Sanches, L.D., Martins, E.C. et al. (2020b). Biodegradable CA/CPB electrospun nanofibers for efficient retention of airborne nanoparticles. *Process Saf. Environ. Prot.* 144: 177–185.

10 Ahne, J., Li, Q., Croiset, E., and Tan, Z. (2019). Electrospun cellulose acetate nanofibers for airborne nanoparticle filtration. *Text. Res. J.* 89 (15): 3137–3149.

11 Zhu, M., Cao, Q., Liu, B. et al. (2020). A novel cellulose acetate/poly (ionic liquid) composite air filter. *Cellulose* 27 (7): 3889–3902.

12 Matulevicius, J., Kliucininkas, L., Prasauskas, T. et al. (2016). The comparative study of aerosol filtration by electrospun polyamide, polyvinyl acetate, polyacrylonitrile and cellulose acetate nanofiber media. *J. Aerosol Sci.* 92: 27–37.

13 Omollo, E., Zhang, C., Mwasiagi, J.I., and Ncube, S. (2016). Electrospinning cellulose acetate nanofibers and a study of their possible use in high-efficiency filtration. *J. Ind. Text.* 45 (5): 716–729.

14 Kwon, M., Kim, J., and Kim, J. (2021). Photocatalytic activity and filtration performance of hybrid TiO_2-cellulose acetate nanofibers for air filter applications. *Polymers* 13 (8): 1331.

15 Akduman, C. (2021). Cellulose acetate and polyvinylidene fluoride nanofiber mats for N95 respirators. *J. Ind. Text.* 50 (8): 1239–1261.

16 Orlando, R., Polat, M., Afshari, A. et al. (2021). Electrospun nanofibre air filters for particles and gaseous pollutants. *Sustainability* 13 (12): 6553.

17 Ghorani, B., Kadkhodaee, R., Rajabzadeh, G., and Tucker, N. (2019). Assembly of odour adsorbent nanofilters by incorporating cyclodextrin molecules into electrospun cellulose acetate webs. *React. Funct. Polym.* 134: 121–132.

18 Kadam, V., Truong, Y.B., Easton, C. et al. (2018). Electrospun polyacrylonitrile/β-cyclodextrin composite membranes for simultaneous air filtration and adsorption of volatile organic compounds. *ACS Appl. Nano Mater.* 1 (8): 4268–4277.

19 Woranuch, S., Pangon, A., Puagsuntia, K. et al. (2017). Starch-based and multi-purpose nanofibrous membrane for high efficiency nanofiltration. *RSC Adv.* 7 (56): 35368–35375.

20 Li, H., Wang, Z., Zhang, H., and Pan, Z. (2018). Nanoporous PLA/(chitosan nanoparticle) composite fibrous membranes with excellent air filtration and antibacterial performance. *Polymers* 10: 1085.

21 Desai, K., Kit, K., Jiajie Li, P. et al. (2009). Nanofibrous chitosan non-wovens for filtration applications. *Polymer* 50 (15): 3661–3669.

22 Zhang, L., Li, L., Wang, L. et al. (2020a). Multilayer electrospun nanofibrous membranes with antibacterial property for air filtration. *Appl. Surf. Sci.* 515: 145962.

23 Zhang, Q., Li, Q., Zhang, L. et al. (2020b). Preparation of electrospun nanofibrous poly (vinyl alcohol)/cellulose nanocrystals air filter for efficient particulate matter removal with repetitive usage capability via facile heat treatment. *Chem. Eng. J.* 399: 125768.

24 Wang, L., Kang, Y., Xing, C.-Y. et al. (2019). β-cyclodextrin based air filter for high-efficiency filtration of pollution sources. *J. Hazard. Mater.* 373: 197–203.

25 Pan, W., Wang, J.-P., Sun, X.-B. et al. (2021). Ultra uniform metal–organic framework-5 loading along electrospun chitosan/polyethylene oxide membrane fibers for efficient $PM_{2.5}$ removal. *J. Cleaner Prod.* 291: 125270.

26 Bian, Y., Wang, R., Ting, S.H. et al. (2018). Electrospun SF/PVA nanofiber filters for highly efficient PM 2.5 capture. *IEEE Trans. Nanotechnol.* 17 (5): 934–939.

27 Purwar, R., Sai Goutham, K., and Srivastava, C.M. (2016). Electrospun sericin/PVA/clay nanofibrous mats for antimicrobial air filtration mask. *Fibers Polym.* 17 (8): 1206–1216.

28 Tian, H., Xuewei, F., Min Zheng, Y. et al. (2018). Natural polypeptides treat pollution complex: moisture-resistant multi-functional protein nanofabrics for sustainable air filtration. *Nano Res.* 11 (8): 4265–4277.

29 Yu, X., Li, C., Tian, H. et al. (2020). Hydrophobic cross-linked zein-based nanofibers with efficient air filtration and improved moisture stability. *Chem. Eng. J.* 396: 125373.

30 Fu, X., Liu, J., Ding, C. et al. (2021). Building bimodal structures by a wettability difference-driven strategy for high-performance protein air-filters. *J. Hazard. Mater.* 415: 125742.

31 Jiang, Z., Zhang, H., Zhu, M. et al. (2018). Electrospun soy-protein-based nanofibrous membranes for effective antimicrobial air filtration. *J. Appl. Polym. Sci.* 135 (8): 45766.

32 Gao, H., Yang, Y., Akampumuza, O. et al. (2017). A low filtration resistance three-dimensional composite membrane fabricated via free surface electrospinning for effective PM 2.5 capture. *Environ. Sci. Nano* 4 (4): 864–875.

33 Bortolassi, C., Cláudia, A., Guerra, V.G. et al. (2019). Composites based on nanoparticle and pan electrospun nanofiber membranes for air filtration and bacterial removal. *Nanomaterials* 9 (12): 1740.

34 Naragund, V.S. and Panda, P.K. (2021). Electrospun polyacrylonitrile nanofiber membranes for air filtration application. *Int. J. Environ. Sci. Technol.* 19: 10233–10244.

35 Dehghan, S.F., Golbabaei, F., Maddah, B. et al. (2016). Optimization of electrospinning parameters for polyacrylonitrile-MgO nanofibers applied in air filtration. *J. Air Waste Manag. Assoc.* 66 (9): 912–921.

36 Kao, T.-H., Shuenn-Kung, S., Ching-Iuan, S. et al. (2016). Polyacrylonitrile microscaffolds assembled from mesh structures of aligned electrospun nanofibers as high-efficiency particulate air filters. *Aerosol Sci. Technol.* 50 (6): 615–625.

37 Yang, C.-F., Tian, D., and He, J.-H. (2021). Preparation of a Cu-Btc/Pan electrospun film with a good air filtration performance. *Therm. Sci.* 25: 1469–1475.

38 Mukherjee, S., Kumar, S., Sahu, R.K., and Nayar, S. (2020). PVA-graphene-hydroxyapatite electrospun fibres as air-filters. *Mater. Res. Express* 6 (12): 125366.

39 Cui, J., Wang, Y., Tao, L. et al. (2021). High performance, environmentally friendly and sustainable nanofiber membrane filter for removal of particulate matter 1.0. *J. Colloid Interface Sci.* 597: 48–55.

40 Senthil, R., Sumathi, V., Tamilselvi, A. et al. (2022). Functionalized electrospun nanofibers for high efficiency removal of particulate matter. *Sci. Rep.* 12 (1): 1–14.

41 Lv, D., Wang, R., Tang, G. et al. (2019). Ecofriendly electrospun membranes loaded with visible-light-responding nanoparticles for multifunctional usages: highly efficient air filtration, dye scavenging, and bactericidal activity. *ACS Appl. Mater. Interfaces* 11 (13): 12880–12889.

42 Geetha, K., Sivasangari, D., Kim, H.S. et al. (2022). Electrospun nanofibrous ZnO/PVA/PVP composite films for efficient antimicrobial face masks. *Ceram. Int.* 48 (19): 29197–29204.

43 Zhu, M., Xiong, R., and Huang, C. (2019). Bio-based and photocrosslinked electrospun antibacteril nanofibrous membranes for air filtration. *Carbohydr. Polym.* 205: 55–62.

44 Nallathambi, G., Robert, B., Esmeralda, S.P. et al. (2020). Development of SPI/AC/PVA nano-composite for air-filtration and purification. *Res. J. Text. Apparel* 24 (1): 72–83.

45 Pierpaoli, M., Giosuè, C., Czerwińska, N. et al. (2021). Characterization and filtration efficiency of sustainable PLA fibers obtained via a hybrid 3D-printed/electrospinning technique. *Materials* 14 (22): 6766.

46 Nicosia, A., Gieparda, W., Foksowicz-Flaczyk, J. et al. (2015). Air filtration and antimicrobial capabilities of electrospun PLA/PHB containing ionic liquid. *Sep. Purif. Technol.* 154: 154–160.

47 Wang, Z., Zhao, C., and Pan, Z. (2015). Porous bead-on-string poly (lactic acid) fibrous membranes for air filtration. *J. Colloid Interface Sci.* 441: 121–129.

48 Cao, Q., Meng, X., Tan, S. et al. (2020). Electrospun bead-in-string fibrous membrane prepared from polysilsesquioxane-immobilising poly(lactic acid) with low filtration resistance for air filtration. *J. Polym. Res.* 27: 1–3.

49 Wang, L., Gao, Y., Xiong, J. et al. (2022). Biodegradable and high-performance multiscale structured nanofiber membrane as mask filter media via poly (lactic acid) electrospinning. *J. Colloid Interface Sci.* 606: 961–970.

50 Wang, Z., Pan, Z., Wang, J., and Zhao, R. (2016). A novel hierarchical structured poly (lactic acid)/titania fibrous membrane with excellent antibacterial activity and air filtration performance. *J. Nanomater.* 2016: 39–39.

51 Zhang, J., Gong, S., Wang, C. et al. (2019). Biodegradable electrospun poly(lactic acid) nanofibers for effective $PM_{2.5}$ removal. *Macromol. Mater. Eng.* 304 (10): 1900259.

52 Aydin-Aytekin, D., Gezmis-Yavuz, E., Esra Buyukada-Kesici, C. et al. (2022). Fabrication and characterization of multifunctional nanoclay and TiO_2 embedded polyamide electrospun nanofibers and their applications at indoor air filtration. *Mater. Sci. Eng., B* 279: 115675.

53 Liu, Y., Li, S., Wenting Lan, M. et al. (2021). Electrospun antibacterial and antiviral poly (ε-caprolactone)/zein/Ag bead-on-string membranes and its application in air filtration. *Mater. Today Adv.* 12: 100173.

6

Nanocatalysts and Redox Nanoparticles for Water Purification

Kannappan Lakshmi, Arunasalam Sridhar, and Rajmohan Rangasamy

Guru Nanak College (Autonomous), Affiliated to University of Madras, Department of Chemistry, Velachery, Chennai 600042, India

Nanomaterials play a vital role in catalysis owing to their greater surface area-to-volume ratio. Nanoparticles were utilized as an efficient and effective catalyst in many fields such as organic transformations, water remediation, dye degradation, and heavy metal removal. Due to the increased human population, the demand for drinking water has increased. Since the source of drinking water is very limited and, due to rapid industrialization, the drinking water is polluted to a greater extent. Usually, water gets contaminated because of pharmaceutical products, organic dyes, personal care products, industrial effluents, etc. [1]. Primarily, ozonation and chlorination methods are used to disinfect municipal drinking water by disinfection [2]. However, chlorination results in carcinogenic by-products on the other hand ozone, which is expensive, often forms toxic bromate in water [3]. The primary contaminants in fresh water are pathogens, toxins and waste from pharmaceutical, plastic and textile industries; disinfection methods are not suitable for treating industrial contaminants and many toxins [4]. The rapid growth of human population and climate change are the major reasons for diminishing fresh water resources. The strong bond between human health and water quality necessitates the development of effective sustainable water treatments to meet the global need for pure drinking water [5, 6].

Therefore, researchers are making efforts for water purification through various methods such as adsorption, ion exchange, reverse osmosis, catalysis, and UV light treatment. Owing to their smaller size and high surface area, nanomaterials demonstrate high absorption capacity and enhanced chemical activity. Nanomaterials in the form of metal nanoparticles, metal oxide nanoparticles, nanofilters, nanomembranes, and nanoadsorbents are some of the protocols used for water remediation. Nanomembranes have several advantages, including high permeability and catalytic reactivity, which results in effective disinfection of pathogens at low costs. Two-dimensional (2D) materials like zeolite, graphene, metal organic frameworks (MOFs), and carbon nitride (C_3N_4) are some of the most widely used nanomembranes [7]. Redox processes are an appealing method for the treatment of

Nanomaterials for Air and Water Purification, First Edition.
Edited by Deepu A. Gopakumar, Jyotishkumar Parameswaranpillai, Jinu Jacob George, and Midhun Dominic C.D.

polluted water; harmful high valence heavy metals are reduced to less or nontoxic forms; organic pollutants and persistent chemical contaminants are either reduced to nontoxic forms or subjected to oxidative degradation in the presence of redox nanomaterials [8, 9].

Nanomaterials utilized in these strategies must be recovered and recycled in order to avoid further disposal in the environment, which increases pollution. Hence, nanomaterials as catalysts can be proficiently involved in water purification, especially for the treatment of organic pollutants, as they remove the pollutant and convert the contaminant into nontoxic or useful products. Some of the common organic pollutants found in water are nitro compounds, chloroethanes, organic dyes, pesticides, etc., that cause adverse effects on humans and the environment [10–12]. Nanocatalysts as oxidants and reductants have a significant role in the water purification processes. Also, some nanomaterials act as photocatalyst in the presence of various light sources, enhancing the reactivity of the catalyst for the purification of water. Even though numerous nanomaterials were reported, only a few nanomaterials have catalytic activity for water purification. Among the available nanomaterials, carbon-based nanomaterials, graphene-based nanomaterials, titanium oxide-based nanomaterials, metal oxide-based nanomaterials, transition metals, and noble metal nanoparticles play a vital role in the catalytic activity of the nanomaterial for water purification. This chapter deals with the nanocatalyst and redox nanomaterials for water purification.

6.1 Fenton and Photo-Fenton Processes for the Treatment of Contaminated Water

Advanced oxidation processes (AOPs) have emerged as an efficient technique for water treatment processes to degrade organic pollutants. Generally, AOPs involves the generation of various reactive oxygen species in industrial and municipal wastewater and subsequent degradation of its recalcitrant chemical constituents under ambient conditions. AOPs proceeds through a series of reactions that produce hydroxyl radicals, in which ozone and hydrogen peroxides act as oxidants with the aid of light, a catalyst (e.g. Fe^{2+}, Fe^{3+}, and TiO_2) and there are several combinations such as homogeneous Fenton (H_2O_2/Fe^{2+}), heterogeneous Fenton (H_2O_2/Fe-support), photo-Fenton ($H_2O_2/UV/Fe^{2+}$, Fe^{3+}), direct photolysis (H_2O_2/UV), peroxone system (O_3/H_2O_2), and photocatalytic peroxone system $O_3/TiO_2/H_2O_2$. Compared to direct oxidation reactions, AOPs operate with a lower energy requirement. However, in the case of AOPs using ozone, the products obtained by the ozonation include ketones, aldehydes, and phenols, which are quite toxic, and ozonation processes are expensive. Another important issue related to ozone is that bromides present in the contaminants are oxidized to bromates, which are carcinogenic. Similarly, direct photolysis of hydrogen peroxide requires high energy requirements to generate reactive oxygen species; moreover, the molar absorption coefficient of H_2O_2 is very low. On the other hand, Fenton and photo-Fenton processes are energy-efficient, cost-effective, and less tedious methods for the treatment of polluted water. Special equipment and expensive reagents are not required for these processes, expanding their applications to large-scale water treatment [13].

The Fenton reaction was first reported by H.J. Fenton (1894). In a typical reaction, hydrogen peroxide is used as an oxidant with the help of a transition metal to promote the generation of hydroxyl radicals, which are effective oxidizing agents for the oxidative degradation and transformation of organic pollutants. The Fenton reaction is simply represented as follows:

$$M^{n+} + H_2O_2 \rightarrow x02794; M^{(n+1)+} + OH^{\bullet -} + OH^{\bullet}$$

where M is a transition metal (Fe, Cu), acts as a catalyst, undergoes oxidation in the first step, then again reduced by hydrogen peroxide as follows:

$$M^{(n+1)+} + H_2O_2 \rightarrow M^{n+} + {}^{\bullet}OH_2 + H^+$$

The above two reactions create a catalytic cycle in Fenton processes, and hence, less concentration chemicals are needed for the reaction. Iron is widely used as a catalyst in Fenton reactions; other transition metals such as cobalt and copper are also employed as catalysts at a lower oxidation state. pH plays an important role in the Fenton reactions, Fe^{2+} ion exists in solution form up to a pH level of 8, whereas, Fe^{3+} ion is soluble only when pH level is 3. Hence, Fenton processes need to be carried out at an acidic pH usually between 2.5 and 3.5 and the overall Fenton process involves.

$$Fe^{2+} + H_2O_2 \rightarrow Fe^{3+} + OH^- + HO^{\bullet}$$

$$Fe^{3+} + H_2O_2 \rightarrow Fe^{2+} + H^+ + HOO^{\bullet}$$

$$Fe^{2+} + HO^{\bullet} \rightarrow Fe^{3+} + HO^-$$

$$HO^{\bullet} + H_2O_2 \rightarrow HOO^{\bullet} + HOO^-$$

$$Fe^{2+} + HO_2{}^{\bullet} \rightarrow Fe^{3+} + HOO^-$$

$$Fe^{3+} + HO_2{}^{\bullet} \rightarrow Fe^{2+} + H^+ + O_2$$

$$HO^{\bullet} + HO^{\bullet} \rightarrow H_2O_2$$

$$\text{Organic pollutant} + HO^{\bullet} \rightarrow \text{Degraded products}$$

As mentioned earlier, the Fenton reaction produces a hydroxyl radical, which is proved by many techniques, including electron spin resonance (ESR) spectroscopy. The rate-determining step of the Fenton reaction involves the reduction of Fe^{3+} to Fe^{2+} when the reaction is performed under dark conditions. Under UV irradiation, photochemical regeneration of ferrous ions by the photoreduction of ferric ions occurs ($\xi > 250$ nm) as follows:

$$Fe^{3+} + H_2O + h\nu \rightarrow Fe^{2+} + {}^{\bullet}OH + H^+$$

The newly generated Fe^{2+} ions react with H_2O_2 generating a new hydroxyl radical and a Fe^{3+} ion, and the cycle continues. Moreover, it is well known that UV light irradiation can also enhance the homolytic cleavage of hydrogen peroxide [14, 15].

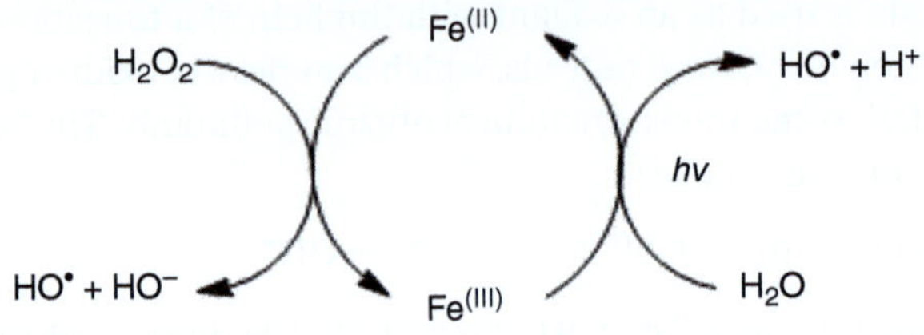

Concentration of H_2O_2 also plays an important role in Fenton processes; a higher concentration of H_2O_2 leads to a side reaction between •OH and H_2O_2 resulting in less reactive $•OH_2$ radicals. Similarly, a lower concentration of H_2O_2 also leads to unwanted side reactions. Hence, it is important to identify the optimal concentration of H_2O_2 for Fenton and photo-Fenton processes.

Some of the drawbacks of using homogeneous Fenton processes are that the reaction has to be performed at lower pH to avoid the precipitation of ferric hydroxide, formation of large quantities of ferric sludge, and the catalysts are isolated. These difficulties can be overcome by using heterogeneous Fenton processes. Fenton AOPs have achieved new heights with the concepts of nanotechnology. Nanoparticles of magnetite (Fe_3O_4), hematite (Fe_2O_3), goethite (α-FeOOH), and lepidocrocite (γ-FeOOH) have been used as nanocatalysts. These nanocatalyst facilitates easier separation and help to overcome the operative pH range limitations. Nanoparticles of iron minerals are anchored on various materials such as clays, zeolites, carbon-based materials, MOFs, and silica. Apart from these Fenton processes, they are enhanced by combining them with electrochemical processes (electro-Fenton processes), ultrasound (sono-Fenton processes) and other wastewater treatment technologies [16].

6.2 Carbon-Based Nanocatalysts for Water Purification

Carbon-based nanomaterials embedded with unique properties act as a building blocks in catalysis especially in the treatment of water. Carbon materials, especially graphene oxide, act as an oxidation catalyst for the degradation of pollutants in water. Activated carbon is widely used in the global market for water purification. Carbon nanotubes (CNTs) with superior adsorption capacity can uptake heavy metal ions present in the water. CNTs have increased their utility for water treatment such as adsorption of pollutants, catalyst support, and removal of microorganisms. Even though CNT can undergo easy surface modification, recollecting the CNTs after use is a challenging task. Hence, CNT supported on alumina and silica carbide can improve their reusability, but lesser proportion of CNT compared to

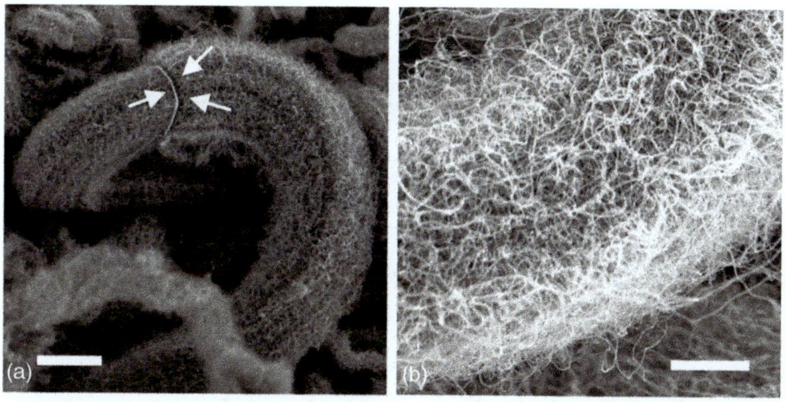

Figure 6.1 Carbon nanotube ponytails. (a, b) Scanning electron micrographs (layered double- oxide discs marked by arrows).

support has decreased their activity. Therefore, a hierarchical structure has to be developed in such a way that the nanomaterial contains higher concentrations of CNT. Carbon nanotube ponytails (CNPs) were synthesized possessing hundreds of CNT in micrometers of length without using any solid supports (Figure 6.1). CNP nanoparticles have better catalytic efficiency in the degradation of methylene blue dye and the reduction of nitro compounds than CNT [17].

Nitro compounds are common organic pollutants in water, which on adsorption remains as contaminant, but reduction of nitro compounds into amino compounds is an important intermediate for various pharmaceutical compounds. A metal-free catalyst is of great interest nowadays, hence a well-defined 3D nitrogen-based graphene nanoparticle was synthesized for the reduction of nitro compounds. Nitrogen-doped graphene oxide nanoparticles were synthesized by ultrasonication method and this nanomaterial was confirmed through various analytical techniques. Reduction of 4-nitrophenol was carried out using nitrogen-doped graphene oxide nanoparticles in the presence of $NaBH_4$, and the catalyst showed better catalytic activity due to the synergistic effect of nitrogen over graphene oxide nanoparticles [18]. Palladium nanoparticles stabilized on carbon nanofibers (CNFs) were synthesized for the first time through one pot electrospinning method. Pd/CNF was also doped on multiwalled carbon nanotubes (MWCNTs) by electrospinning method and the nanomaterials were characterized through various techniques. Pd/CNF has a diameter of 217 ± 55 nm, whereas MWCNT-doped Pd/CNF nanomaterial showed curled nanofibers with reduced diameter of 64 ± 24 nm. The palladium nanoparticle was observed to have a spherical morphology with a size of about 8 nm and be uniformly dispersed (Figure 6.2). The nanomaterials Pd/CNF and MWCNT-loaded Pd/CNF (0–0.25%) was explored as catalysts in the nitrite reduction. Pd nanoparticle on the nanofibrous support showed enhanced catalytic activity in the reduction of waterborne pollutants. Pd/CNFs-2.5% MWCNTs resulted in superior catalytic activity in nitrite reduction than other Pd-doped carbon nanofibrous nanomaterials [19].

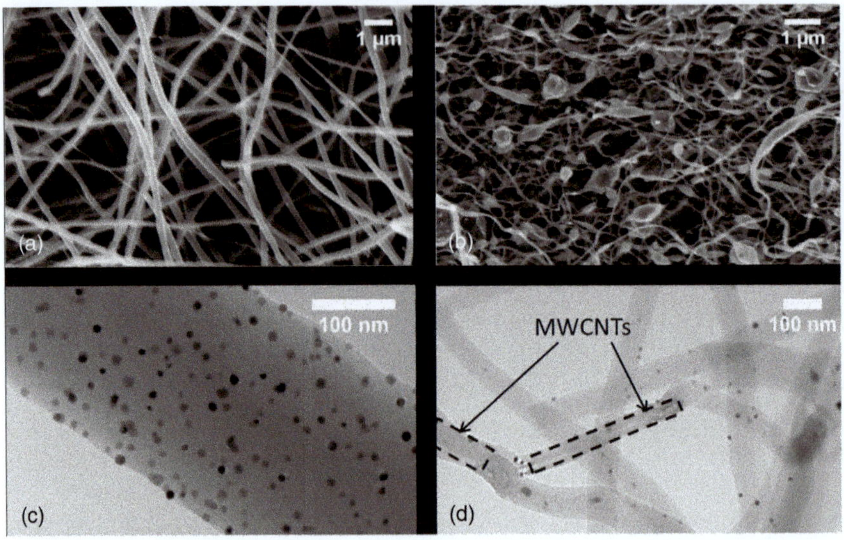

Figure 6.2 Electron microscopic images of Pd/CNFs (a and c), and Pd/CNFs-2.5% MWCNTs (b and d). The presence of MWCNTs in CNFs is highlighted in (d).

(a) (b)

Figure 6.3 (a) FESEM image (b) TEM image of TiO_2 nanosheets.

Among various carbon-based nanomaterials, graphene oxide plays an important role as a support material that improves stability and enhances the catalytic activity of nanomaterials. TiO_2 nanosheets (TNSs) supported on graphene oxide were utilized in purification of water and photodegradation of contaminants. TNSs were synthesized by solvothermal methods, and it was observed as individual square nanosheets from HRTEM images. TNS were grafted onto the surface of graphene oxide (Figure 6.3).

Hydroxyl groups on TiO_2 facilitate the adsorption of Congo red dye, which, on UV irradiation, enables electron transfer between TiO_2 and GO, enhances the degradation of dye molecules present in water. Photocatalytic degradation of the dye increased as the loading capacity of TNSs on GO increases [20].

6.3 Metal Oxide-Based Nanocatalysts

Water purification has been essential due to increased urbanization and increased industrial developments, especially in sectors such as paper, textile, rubber, and cosmetics. In order to reduce the deterioration of water, several methods were adapted, such as adsorption, catalysis, and disinfectant. The catalytic activity of the metal oxides involves purification of water and also degrades the pollutants present in it, resulting in pollutant-free water. Among various catalytic materials, metal oxides such as vanadates, alumina, and ferrites TiO_2 find remarkable applications in water remediation. Rare earth metal vanadates are more significant; in particular, neodymium vanadates are considered as important catalyst in the degradation of organic pollutants present in water bodies. Neodymium vanadate is synthesized from neodymium nitrate and ammonium vanadate using solid-state reaction in the presence of air and Schiff's base. The size and morphology of the nanocatalyst were modified based on the concentration of the ligand and the solvent. The photocatalytic efficiency of the nanocatalyst was investigated for the degradation of Eriochrome Black T dye under UV light at pH 11. Neodymium vanadate acts as an efficient photocatalyst in the degradation of dye contaminants in water during purification process [21].

Similarly. Holonium vandate nanocatalysts were also synthesized via sonochemical method using triethylenetetramine as capping agent. Microscopic images confirm the spindle-shaped spherical morphology of the nanocatalyst, which has an average uniform size of 36 nm. Holonium vanadate nanocatalyst eliminated methylene violet dye, pollutant in water, by photocatalytic degradation under UV light. These nanocatalysts showed better stability and could be regenerated and reused up to three catalytic cycles [22].

Heterogeneous photocatalysis facilitates degradation of pollutants via a fundamental redox mechanism. Usually, semiconductor materials are widely used as photocatalyst, which absorb light in the UV–Visible range depending on the material's band gap resulting in the formation of a conduction band electron ($e_{cb}{}^-$) and valence band hole ($h_{vh}{}^+$) pair. The generated electron–hole pair undergoes a redox reaction in an aqueous solution produces hydroxyl radical, which is believed to be an important oxidant for the purification of polluted water. Metal oxides such as titanium oxide, zinc oxide, and composites made of metal-doped semiconductors can act as photocatalysts due to their ability to absorb light and generate reactive oxygen species, which are the key steps for the degradation of polluted water. The band gaps of TiO_2 (anatase) and zinc oxide are 3.2 and 3.4 eV, respectively; hence, these compounds absorb light in the ultraviolet region, and a series of reactions take place as follows [23].

$$TiO_2 + h\nu\,(< 380\,\text{nm}) \rightarrow e_{cb}{}^- + h_{vb}{}^+$$

$$h_{vb}{}^+ + H_2O \rightarrow {}^\bullet OH + H^+ \text{ (Oxidation)}$$

$$e_{cb}{}^- + O_2 \rightarrow O_2{}^\bullet \text{ (Reduction)}$$

The major drawback of TiO_2 photocatalysis has been the absorption of light in the ultraviolet region and lower quantum yields. The addition of dopants such as Fe, Co, Cu, Ag, and nonmetals decreases the band gap energy. It eases the electron transfer from valence band to the conductance band, which also improves the performance of TiO_2 from the ultraviolet region to the visible region. Various factors, such as dopant concentration, nature, and method of doping influence the properties of photocatalyst [24].

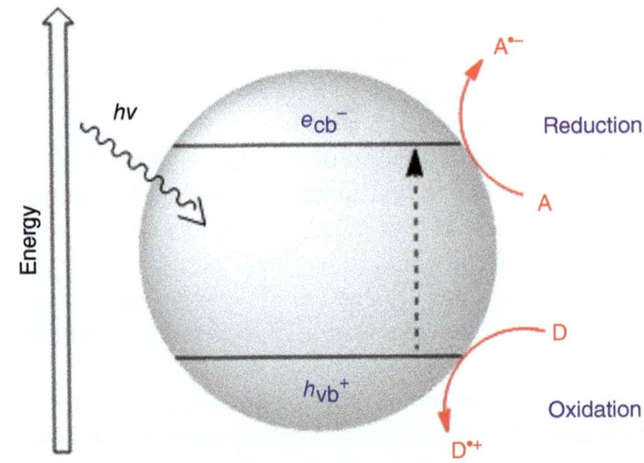

Reaction mechanism of photocatalysis

Nonmetals, especially nitrogen-doped TiO_2 has been extensively studied as photocatalyst for water treatment. Since the nonmetal is doped in the form of an isolated atom, the inherent properties of the photocatalyst remain unaltered. Moreover, the band gap energy of N-doped TiO_2 is around 2.5 eV, leading to the absorption of light in the visible region. Yang and coworkers prepared N-doped TiO_2 by heating the commercially available P25-TiO_2 in flowing NH_3. ESR studies showed triplet g values (g-1.987, 2.004, and 2.024), which are correlated to the formation of single-electron-trapped oxygen-vacancy (SETOV). The visible light photocatalysis of N-doped TiO_2 is attributed to the formation of SETOV on TiO_2 matrix and existence of nitrogen on the surface of the TiO_2 [25].

Ata et al. immobilized N-doped TiO_2 on the surface of the polystyrene by using solvent-cast method. The efficiency of the resulting photocatalyst was evaluated by the decolorization of methylene blue and the inactivation of *Escherichia coli* (*E. coli*) strain selected from urban wastewater. Higher photocatalytic efficiency was observed for the N-doped TiO_2 photocatalyst compared to commercially available TiO_2 powders under direct solar light [26]. Wantala et al. utilized diamide as a nitrogen source for the synthesis of N-TiO_2 and utilized for the photodegradation of ciprofloxacin under UV light. Studies revealed that increasing nitrogen content increased the photocatalytic ability of N-TiO_2 [27].

Carbon doping has also been widely studied due to its low cost, variety of sources, and ability to narrow the band gap energy, leading to the absorption of light in the visible region. Lal and coworkers synthesized TiO_2 and carbon nanoflakes via a hydrothermal route. Graphene oxide and reduced graphene oxides were used as carbon sources. The photocatalytic ability was evaluated by the degradation of methylene blue and the bacterial inactivation of gram-positive and gram-negative bacteria under sunlight [28].

Falaras and coworkers synthesized mesoporous N-F, C, and S-doped TiO_2 using a surfactant-assisted sol–gel method. Hydrocarbon surfactants were utilized that controlled the extent of hydrolysis and condensation rates of the titanium precursor, resulting in the formation of porous TiO_2. Surfactants, as well as combinations with suitable reagents, were used as sources for the anions. Photocatalytic efficiency was explored in the degradation of an aqueous solution of hazardous microcytin LR (MC-LR) toxin under visible light [29].

Dye effluents from textile, printing, and photographic industries drastically affect drinking water sources even in low concentrations. Zhang et al. reported the hydrothermal synthesis of $Bi_{12}TiO_{20}$ nanostructures using $Bi(NO_3)_3$ and $Ti(SO_4)_2$. The reported band gap energy for the synthesized nanomaterial is about 2.7 eV, which facilitated the photodegradation of acid orange 7 under visible light. Results show that $Bi_{12}TiO_{20}$ has higher photocatalytic activity than N-doped TiO_2 [30].

Li and colleagues reported a comparative study of the photodegradation of dye over TiO_2 and ZnO, where the microscale ZnO showed better efficiency than TiO_2, due to the higher efficiency of electron transfer from an excited organic dye to the conduction band of the microscale ZnO than that of TiO_2. TiO_2-modified ZnO was developed as a novel light-to electricity conversion device and has proven to be a capable candidate for the photocatalytic degradation of dye pollutants, as well as a renewable energy source [31].

Titanium oxide nanosheeets supported g-C_3N_4 acts as an ideal and promising nanocatalyst in the treatment of water. The nanocatalyst was synthesized by hydrogen treatment, which generated more active defects and also produced oxygen vacancies. The photocatalyst accelerates the degradation of Congo red effectively by interfacial charge transfer [32].

Zirconium-doped TiO_2 nanothin film shows enhanced photocatalytic activity in the degradation of organic pollutants present in water. Zr-modified TiO_2 thin films were synthesized by varying Zr contents. The thickness of Zr–TiO_2 (05–500) nm was observed to have a thickness of 340 nm. Zr–TiO_2 thin film exhibit superior photodegradation of organic pollutants under Xe lamp radiation. The photocatalytic activity of Zr-doped TiO_2 thin films decreased with increase in Zr content. This nanocatalyst has enhanced photocatalytic activity in the degradation of contaminants such as phenol, chloridazon, and 4- chlorophenol. Also, Zr-grafted anatase crystals show increased catalyst activity than pure TiO_2 thin films in the photodegradation of the pollutants [33].

Pesticides have extensive use in agricultural lands in developed countries, which results in a major contaminant in groundwater. Hence, removal of pesticide contaminants from water is of great concern, as even trace amount is more toxic and cause adverse effects in birds, animal, and humans. Nano TiO_2 nanoparticles were proven to undergo UV-induced photodegradation of mixed pesticides such as lindane,

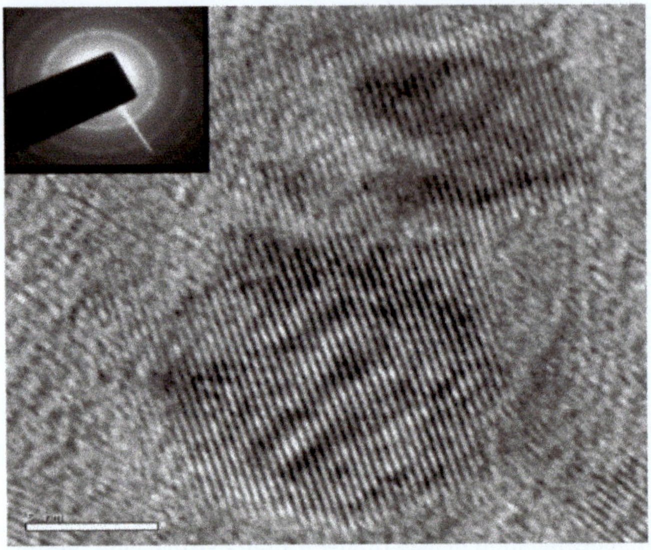

Figure 6.4 TEM analysis of surfactant stabilized TiO_2 nanoparticle.

methyl parathion, and dichlorvos. Nano TiO_2 nanoparticles were synthesized by several methods such as acid stabilized, alcohol stabilized, and surfactant stabilized nanoparticles. Surfactant stabilized TiO_2 nanoparticles of about 12 nm have superior photocatalytic activity than the commercially available TiO_2 (Figure 6.4). Furthermore, nano TiO_2 degrades mixed pesticides at faster rate than single pesticides [34].

Cystamine functionalized $Ni_{0.5}Zn_{0.5}Fe_2O_4$ (NZF) nanoferrite is an eminent photocatalyst for the degradation of organic dye in water purification. $Ni_{0.5}Zn_{0.5}Fe_2O_4$ nanoferrrite was synthesized by co-precipitation method, resulting in nanospinels, followed by cystamine functionalization of the nanoferrites. The nanocatalyst was confirmed through various analytical techniques such as X-ray photoelectron spectroscopy (XPS), HRTEM, and energy dispersive X-ray (EDX), which confirmed the phase purity, and also no secondary phase was observed. The superparamagnetic behavior of the nanoferrite was proved by VSM analysis. Nanocatalyst acts as an efficient photocatalyst in the degradation of contaminant Brilliant Blue G and Bisphenol A under visible light during water purification. Also, the mechanism confirms the photocatalytic degradation of the dyes rather a mere adsorption of the dye on the catalytic surface. The nanocatalysts were isolated by magnetic separation and the regeneration nanocatalyst retained its catalytic efficiency up to four cycles [35].

6.4 Nano-Zero-Valent Iron (nZVI)

Nano-zerovalent-iron (nZVI) is one of the most studied nanomaterial for water treatment due to its good reducing ability, reactivity, and excellent adsorption properties.

Moreover, it has antibacterial and antimicrobial properties. In polluted water, nZVI can be oxidized to Fe^{2+} and hydrogen due to corrosion, which are good reducing agents for pollutants present in the water [36].

$$Fe + 2H_2O \rightarrow Fe^{2+} + H_2 + 2OH^-$$

nZVI has been widely used for the reductive dehalogenation of chlorinated organic compounds and chlorinated organic solvents. nZVI and dissolved aqueous Fe^{2+} forms a redox couple with the reduction potential of −0.44 V, which is sufficient to reduce many organic pollutants. In reductive halogenation of organic compounds in the presence of nZVI, water acts as a weak proton donor [37].

$$Fe^0 + RX + H+ \rightarrow Fe^{2+} + RH + X^-$$

Corrosion of nZVI occurs in water in the presence of dissolved oxygen during reactions 1 and 2, resulting in the release of Fe^{2+}. Consequently, the presence of Fe^{2+} and OH^- ions increases the pH of aqueous solution. With the increase of Fe^{2+} and OH^- ion concentration, precipitation of ferrous hydroxide $Fe(OH)_2$, and ferric hydroxide ($Fe(OH)_3$), which is an excellent flocculent also occurs. Fe^{3+} is formed as a result of oxidation of Fe^{2+} in the presence of pollutants. Since corrosion is faster under aerobic conditions, the formation of Fe^{3+} is also faster in the presence of dissolved oxygen [38].

In addition, nZVI can enhance the precipitation, which helps the removal of heavy metal pollutants from the contaminated water. Chromium(VI), a toxic and harmful heavy metal present in the water, can be readily adsorbed by nZVI due to its high surface area. Since nZVI is reactive, it instantaneously reduces Cr(VI) to $Cr(OH)_3$ due to the oxidation of nZVI to $Fe(OH)_3$. As mentioned earlier, $Fe(OH)_3$ is an excellent flocculant and facilitates the precipitation of Cr(III) in water. Moreover, the toxicity of Cr(III) is much lesser than that of Cr(VI).

$$Cr_2O_7{}^{2-} + 2Fe + 14H^+ \rightarrow 2Cr^{3+}\ 2Fe^{3+} + 7H_2O$$

$$2Cr^{3+}\ 2Fe^{3+} + 6OH^- \rightarrow Cr(OH)_3 + Fe(OH)_3$$

nZVI can efficiently oxidize a series of organic pollutants, such as textile industry dyes and phenols, in the presence of dissolved oxygen. nZVI, in combination with dissolved oxygen, produces hydroxyl radicals, which possess strong oxidizing tendency toward variety of organic pollutants [39].

$$Fe^0 + O_2 + 2H^+ \rightarrow Fe^{2+} + H_2O_2$$

$$Fe^0 + H_2O_2 + 2H^+ \rightarrow Fe^{2+} + 2H_2O$$

$$Fe^{2+} + H_2O_2 \rightarrow Fe^{3+} + {}^{\bullet}OH + {}^{-}OH$$

6.5 Metal-Based Nanocatalyst for Water Purification

Metal nanoparticles are of great importance in the water purification process, as metal nanoparticles are capable of determining the ultralow concentration of organic pollutants in water. Transition metal nanoparticles, noble metal nanoparticles and bimetallic nanoparticles, show enhanced catalytic activity in the degradation of contaminants in water. Noble metal nanoparticle Au was utilized as an effective catalyst in the degradation process. Au nanoparticles were synthesized by a green chemical method using polydopamine as a reductant and stabilizer. As synthesized Au nanoparticles were deposited on g-C_3N_4, a support for Au nanoparticles, which enhanced the catalytic activity of the nanoparticles. The spherical morphology of the nanocatalyst with an average size of 24 nm was proven from HRSEM and FESEM images. The nanocatalyst shows superior catalytic activity in the reduction of the pollutant *p*-nitrocompounds with a rate constant of 0.0514 s^{-1} and TOF 545.60 h^{-1}. Furthermore, polydopamine-stabilized Au nanoparticle on g-C_3N_4 support proved to be an efficient catalyst in the reduction of other contaminants in water during the water purification process. This nanocatalyst proved to be an excellent catalyst in degradation of the pollutant from Congo red, Eriochrome Black T, 2,4-dinitrophenol, 2-nitrophenol to methyl orange. Also, this nanocatalyst was effectively reused for ten catalytic cycles in the treatment of water purification [40].

Fe nanoparticles synthesized by a green chemical method using green tea (GT) extracts has enhanced catalytic activity in the degradation of methylene blue and methyl orange dye pollutants during water purification. Clusters of GT–Fe nanoparticles were observed as irregular clusters with average particle sizes of 40 and 60 nm through TEM analysis. The nanoparticles were also characterized using scanning electron microscopy (SEM), energy dispersive X-ray spectroscopy (EDX), X-Ray photoelectron spectroscopy (XPS), X-ray diffraction (XRD), and Fourier- transform infrared spectroscopy (FTIR) techniques. This Fe nanoparticle degraded methylene blue dye and methyl orange dye under UV–Visible light radiation, and it was utilized as Fenton-like catalyst for the degradation of dyes. In addition, GT-Fe nanoparticle has better catalytic activity in the degradation reaction than Fe nanoparticles synthesized by borohydride reduction. Further, this catalyst completely removed the dye pollutant with similar activity to the Fenton catalyst [41].

Due to the increase in pharmaceutical industries, contamination of ground water with antibiotics, painkillers, analgesics, antidepressants, impotent drugs, and tranquilizers have been increasing and needs more attention for remediation. Among various techniques, catalytic degradation of these contaminants plays a significant role as it converts the toxic materials into nontoxic materials. Fe nanoparticles can effectively degrade the antibiotic contaminants amoxicillin and ampicillin. The removal of the antibiotics using Fe nanoparticles may involve reduction, adsorption, or coprecipitation methods. Fe nanoparticles were prepared through borohydride reduction method, which eliminates antibiotics in batch reduction experiments. The mechanism of degradation of amoxicillin and ampicillin involves the reduction of β-lactam ring, which completely makes the antibiotic disappear [42].

Chlorinated compounds remain toxic pollutants in drinking water, which may have adverse effects on humans and the environment. Iron nanoparticles are an enhancing nanocatalyst in the degradation of chlorinated methanes, ethanes, and ethenes. The chlorinated compound showed an exceptional rate of degradation, and the rate increased with a decrease in size of the nanoparticle. However, dichloromethane does not show an exceptional degradation rate compared to other chlorinated compounds. This served as an efficient in situ nanocatalyst for the remediation of groundwater contamination [43].

Pd/Fe bimetallic nanoparticles can dechlorinate compounds such as 1,2,4 trichlorobenzene, dichlorobenzene, and monochlorobenzene. Zero-valent iron nanoparticles were synthesized by reduction; subsequently palladium nanoparticles were synthesized resulting in Pd/Fe bimetallic nanoparticles. Pd/Fe nanoparticles were capable of reducing chlorinated benzene compounds completely into benzene, and the reaction followed pseudo-first-order kinetics. The rate of the reaction increased in order as follows: Trichlorobenzene, Dichlorobenzene, and monochlorobenzene. Dechlorination was not observed in the absence of palladium nanoparticles. Furthermore, the activity of aged Pd/Fe nanoparticles was also observed in dechlorination reaction and they found that the reactivity of the nanoparticle decreased due to dislodgment of Pd. The catalytic activity of the aged Pd/Fe nanoparticles can be restored by treatment with HCl [44].

Trichloroethene a contaminant present in water and soil can be degraded using Fe–Pd bimetallic nanoparticle stabilized on carboxymethyl cellulose (CMC). CMC stabilizes the zero-valent iron oxide nanoparticles by reduction using $NaBH_4$, followed by the reduction of palladium ions to obtain CMC stabilized Fe–Pd nanoparticles. The nanoparticles were analyzed through various analytical techniques such as TEM, FTIR, and DLS confirms the size of the nanoparticles in the range of 17 nm. Degradation of trichloroethene using CMC stabilized Fe–Pd nanoparticles was found to be 17 times faster than unstabilized Fe–Pd bimetallic nanoparticles. The degradation process follows a pseudo-first-order rate constant. Hence, purification of water through the removal of chlorinated compounds can be carried out efficiently using in situ-generated CMC-stabilized Fe–Pd nanoparticles [45].

Ag and CuO nanoparticles decorated on Cu composites act as a superior catalyst in the degradation of pollutants such as dye and the reduction of nitrocompounds in water. In situ synthesized Ag and CuO nanoparticles were immobilized on Cu nanoparticles and the synthesized nanocomposite was confirmed through various techniques (Scheme 6.1). The size of Cu nanoparticles, CuO shell, and Ag nanoparticles were found to be 100, 10, and 5 nm, respectively using TEM analysis (Figure 6.5). XPS and HAADF-STEM confirmed the presence of constiuent elements in the nanocomposite.

Nitro compounds are considered to be the most toxic and persistent organic pollutant, and degradation seems to be very difficult. Cu/CuO-Ag nanocomposite reduced 4-nitrophenol efficiently in the presence of $NaBH_4$ and the nanocomposite was reused for three catalytic cycles without any significant loss in activity. In addition to catalytic reduction of nitrophenol, Cu/CuO-Ag nanocomposites degraded

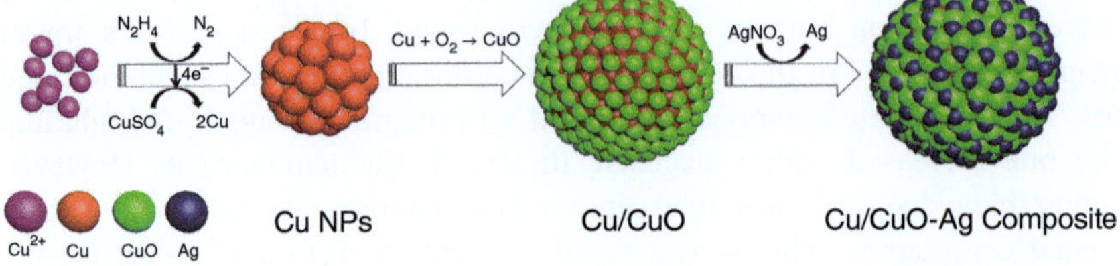

Scheme 6.1 Schematic illustration for the synthesis of Cu/CuO-Ag composite.

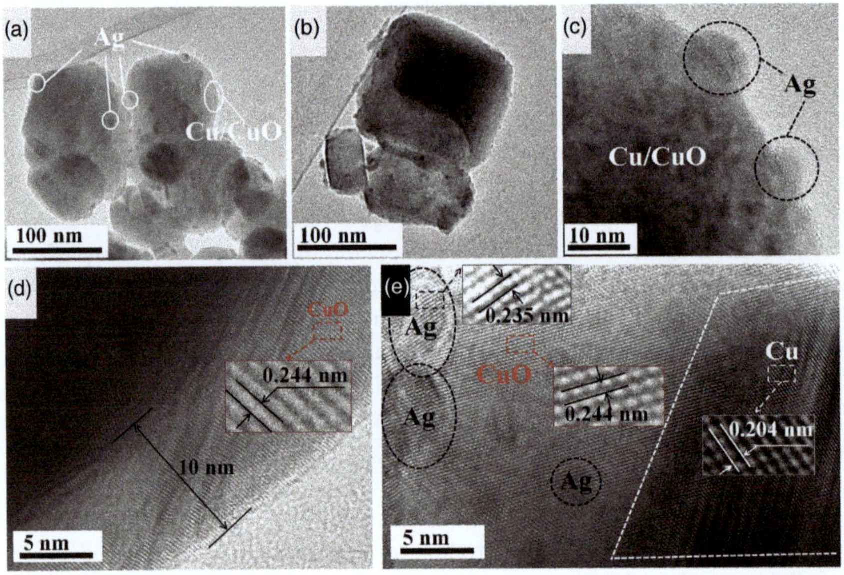

Figure 6.5 TEM (a, b, c) and the HRTEM (d, e) image of Cu/CuO-Ag composite at different viewing angle and magnification.

methylene blue dye in presence of hydrogen peroxide at room temperature. Hence, Cu/CuO-Ag nanocomposites serve as an efficient nanocatalysts in the reduction and degradation of organic pollutants during the process of water purification [46].

Gold nanoparticles grafted on diamond nanoparticulate act as an efficient catalyst in the degradation of pollutants present in water. Au nanoparticles were deposited through the precipitation method on diamond nanoparticles, and the size of Au nanoparticle was found to be 2 nm (Figure 6.6). Au decorated diamond nanoparticles in the presence of hydrogen peroxide generate highly active hydroxyl radicals, like Fenton catalysts, which can competently degrade phenol contaminants in water into alcohols. It has been found to have 79% catalytic activity like Fenton catalyst in the conversion of phenols at room temperature along with the stoichiometric use of hydrogen peroxide. Au on diamond nanoparticle support is highly stable, selective, and reusable [47].

Figure 6.6 TEM images of Au on diamond nanoparticles.

6.6 Conclusion

In conclusion, the application of nanocatalysts and redox nanoparticles in the decontamination of industrial wastewater obtained from pharmaceuticals, organic dyes, personal care products, detergent industries, etc., has been described. Redox nanoparticles and nanocatalysts are more prominent nanomaterials than nanomembranes and nanoabsorbents, as they convert the toxic pollutants in water into nontoxic and useful products. Furthermore, the final target is to develop a sustainable nanocatalyst for water purification method, which is cost-effective, nontoxic, regenerated, reused, and eco-friendly.

References

1 Lu, F. and Astruc, D. (2020). Nanocatalysts and other nanomaterials for water remediation from organic pollutants. *Coord. Chem. Rev.* 408: 213180.

2 Ahmed, J., Thakur, A., and Goyal, A. (2021). Chapter 1 Industrial wastewater and its toxic effects. In: *Biological Treatment of Industrial Wastewater* (ed. M.P. Shah), 1–14. The Royal Society of Chemistry.

3 Li, X.F. and Mitch, W.A. (2018). Drinking water disinfection byproducts (DBPs) and human health effects: multidisciplinary challenges and opportunities. *Environ. Sci. Technol.* 52 (4): 1681–1689.

4 Patel, M., Kumar, R., Kishor, K. et al. (2019). Pharmaceuticals of emerging concern in aquatic systems: Chemistry, occurrence, effects, and removal methods. *Chem. Rev.* 119 (6): 3510–3673.

5 Boretti, A. and Rosa, L. (2019). Reassessing the projections of the World Water Development Report. *npj Clean Water* 2 (1): https://doi.org/10.1038/s41545-019-0039-9.

6 O'Shea, K.E. and Dionysiou, D.D. (2012). Advanced oxidation processes for water treatment. *J. Phys. Chem. Lett.* 3 (15): 2112–2113.

7 Vaiano, V., Sannino, D., and Sacco, O. (2020). *The Use of Nanocatalysts (and Nanoparticles) for Water and Wastewater Treatment by Means of Advanced Oxidation Processes*. Elsevier Inc.

8 Kurian, M. (2021). Advanced oxidation processes and nanomaterials – a review. *Clean. Eng. Technol.* 2 (2020): 100090.

9 Simeonidis, K., Martinez-Boubeta, C., Zamora-Pérez, P. et al. (2019). Implementing nanoparticles for competitive drinking water purification. *Environ. Chem. Lett.* 17 (2): 705–719.

10 Pradeep, T. and Anshup (2009). Noble metal nanoparticles for water purification: a critical review. *Thin Solid Films* 517 (24): 6441–6478.

11 Sharma, R. (2021). Nanotechnology: an approach for water purification-review. *IOP Conf. Ser. Mater. Sci. Eng.* 1116 (1): 012007.

12 Nasrollahzadeh, M., Sajjadi, M., Iravani, S., and Varma, R.S. (2021). Green-synthesized nanocatalysts and nanomaterials for water treatment: current challenges and future perspectives. *J. Hazard. Mater.* 401 (2020): 123401.

13 Divyapriya, G., Nambi, I.M., and Senthilnathan, J. (2016). Nanocatalysts in fenton based advanced oxidation process for water and wastewater treatment. *J. Bionanosci.* 10 (5): 356–368.

14 Ameta, R., Chohadia, A.K., Jain, A., and Punjabi, P.B. (2018). Chapter 3 - Fenton and Photo-Fenton Processes. In: *Advanced Oxidation Processes for Waste Water Treatment Emerging Green Chemical Technology*, 49–87. https://doi.org/10.1016/B978-0-12-810499-6.00003-6.

15 Du, S., Lian, J., and Zhang, F. (2022). Visible light - responsive N - doped - TiO_2 photocatalysis: synthesis, characterizations, and applications. *Trans. Tianjin Univ.* 28 (1): 33–52.

16 Vorontsov, A.V. (2019). Advancing Fenton and photo-Fenton water treatment through the catalyst design. *J. Hazard. Mater.* 103–112. https://doi.org/10.1016/j.jhazmat.2018.04.033.

17 Wang, H., Ma, H., Zheng, W. et al. (2014). Multifunctional and recollectable carbon nanotube ponytails for water purification. *ACS Appl. Mater. Interfaces* 6 (12): 9426–9434.

18 Liu, J., Yan, X., Wang, L. et al. (2017). Three-dimensional nitrogen-doped graphene foam as metal-free catalyst for the hydrogenation reduction of p-nitrophenol. *J. Colloid Interface Sci.* 497: 102–107.

19 Ye, T., Durkin, D.P., Hu, M. et al. (2016). Enhancement of nitrite reduction kinetics on electrospun Pd-carbon nanomaterial catalysts for water purification. *ACS Appl. Mater. Interfaces* 8 (28): 17739–17744.

20 Nair, A.K. and Jagadeesh, J.B. (2017). TiO_2 nanosheet-graphene oxide based photocatalytic hierarchical membrane for water purification. *Surf. Coat. Technol.* 320: 259–262.

21 Monsef, R., Ghiyasiyan-Arani, M., and Salavati-Niasari, M. (2019). Utilizing of neodymium vanadate nanoparticles as an efficient catalyst to boost the photocatalytic water purification. *J. Environ. Manage.* 230 (2018): 266–281.

22 Khorasanizadeh, M.H., Monsef, R., Amiri, O. et al. (2019). Sonochemical-assisted route for synthesis of spherical shaped holmium vanadate nanocatalyst for polluted waste water treatment. *Ultrason. Sonochem.* 58 (July): 104686.

23 Dong, S., Feng, J., Fan, M. et al. (2015). Recent developments in heterogeneous photocatalytic water treatment using visible light-responsive photocatalysts: a review. *RSC Adv.* 5: 14610–14630.

24 Nan, M., Jin, B., Chow, C.W.K., and Saint, C. (2010). Recent developments in photocatalytic water treatment technology: a review. *Water Res.* 44 (10): 2997–3027.

25 Wang, Y., Feng, C., Zhang, M. et al. (2010). Applied catalysis B: environmental enhanced visible light photocatalytic activity of N-doped TiO_2 in relation to single-electron-trapped oxygen vacancy and doped-nitrogen. *Applied Catal. B, Environ.* 100 (1, 2): 84–90.

26 Ata, R., Sacco, O., Vaiano, V. et al. (2017). Visible light active N-doped TiO_2 immobilized on polystyrene as efficient system for wastewater treatment. *J. Photochem. Photobiol. A Chem.* 348 (1): 255–262. https://doi.org/10.1016/j.jphotochem.2017.08.054.

27 Suwannaruang, T., Kidkhunthod, P., Chanlek, N. et al. (2019). High anatase purity of nitrogen-doped TiO_2 nanorice particles for the photocatalytic treatment activity of pharmaceutical wastewater. *Appl. Surf. Sci.* 478 (2018): 1–14. https://doi.org/10.1016/j.apsusc.2019.01.158.

28 Ghumro, S.S., Lal, B., and Pirzada, T. (2022). Visible-light-driven carbon-doped TiO_2 - based nanocatalysts for enhanced activity toward microbes and removal of dye. *ACS Omega* 7 (5): 4333–4341. https://doi.org/10.1021/acsomega.1c06112.

29 Likodimos, V., Han, C., Pelaez, M. et al. (2013). Anion-doped TiO_2 nanocatalysts for water purification under visible light. *Ind. Eng. Chem. Res.* 52 (39): 13957–13964. https://doi.org/10.1021/ie3034575.

30 Zhu, X., Zhang, J., and Chen, F. (2010). Hydrothermal synthesis of nanostructures Bi_{12} TiO_{20} and their photocatalytic activity on acid orange 7 under visible light. *Chemosphere* 78 (11): 1350–1355.

31 Li, Y., Xie, W., Hu, X. et al. (2010). Comparison of dye photodegradation and its coupling with light-to-electricity conversion over TiO_2 and ZnO. *Langmuir* 26 (1): 591–597. https://doi.org/10.1021/la902117c.

32 Shen, S., Fu, J.J., and Wang, H.B. (2021). Unravelling the favorable photocatalytic effect of hydrogenation process on the novel g-C_3N_4-TiO_2 catalysts for water purification. *Diam. Relat. Mater.* 114 (February): 108292.

33 Mbiri, A., Taffa, D.H., Gatebe, E., and Wark, M. (2019). Zirconium doped mesoporous TiO_2 multilayer thin films: influence of the zirconium content on the photodegradation of organic pollutants. *Catal. Today* 328 (2018): 71–78.

34 Senthilnathan, J. and Philip, L. (2010). Removal of mixed pesticides from drinking water system using surfactant-assisted nano-TiO_2. *Water. Air. Soil Pollut.* 210 (1–4, 154): 143.

35 Tanwar, R., Kumari, R., Mandal, U.K., and Kar, K.K. (2020). Catalytic activity of surface-functionalized nanoscale nickel zinc multiferrites: potential vector for water purification. *J. Chem. Technol. Biotechnol.* 95 (3): 739–750.

36 Xu, M., Wu, C., and Zhou, Y. (2020). Advancements in the Fenton lwastewater treatment. In: *Advanced Oxidation Processes - Applications, Trends, and Prospects.* https://doi.org/10.5772/intechopen.90256.

37 Fu, F., Dionysiou, D.D., and Liu, H. (2014). The use of zero-valent iron for groundwater remediation and wastewater treatment: a review. *J. Hazard. Mater.* 267: 194–205.

38 Hartmann, M., Kullmann, S., and Keller, H. (2010). Wastewater treatment with heterogeneous Fenton-type catalysts based on porous materials. *J. Mater. Chem.* 20 (41): 9002–9017.

39 Raman, C.D. and Kanmani, S. (2016). Textile dye degradation using nano zero valent iron: a review. *J. Environ. Manage.* 177: 341–355.

40 Qin, L., Huang, D., Xu, P. et al. (2019). In-situ deposition of gold nanoparticles onto polydopamine-decorated g-C_3N_4 for highly efficient reduction of nitroaromatics in environmental water purification. *J. Colloid Interface Sci.* 534: 357–369.

41 Shahwan, T., Abu Sirriah, S., Nairat, M. et al. (2011). Green synthesis of iron nanoparticles and their application as a Fenton-like catalyst for the degradation of aqueous cationic and anionic dyes. *Chem. Eng. J.* 172 (1): 258–266.

42 Ghauch, A., Tuqan, A., and Assi, H.A. (2009). Antibiotic removal from water: Elimination of amoxicillin and ampicillin by microscale and nanoscale iron particles. *Environ. Pollut.* 157 (5): 1626–1635.

43 Gillham, O. (1994). Enhanced degradation of halogenated aliphatics. *Groundwater* 32 (6): 958–967.

44 Zhu, B.W. and Lim, T.T. (2007). Catalytic reduction of chlorobenzenes with Pd/Fe nanoparticles: Reactive sites, catalyst stability, particle aging, and regeneration. *Environ. Sci. Technol.* 41 (21): 7523–7529.

45 He, F., Zhao, D., Liu, J., and Roberts, C.B. (2007). Stabilization of Fe-Pd nanoparticles with sodium carboxymethyl cellulose for enhanced transport and dechlorination of trichloroethylene in soil and groundwater. *Ind. Eng. Chem. Res.* 46 (1): 29–34.

46 Liang, Y., Chen, Z., Yao, W. et al. (2017). Decorating of Ag and CuO on Cu nanoparticles for enhanced high catalytic activity to the degradation of organic pollutants. *Langmuir* 33 (31): 7606–7614.

47 Navalon, S., Martin, R., Alvaro, M., and Garcia, H. (2010). Gold on diamond nanoparticles as a highly efficient Fenton catalyst. *Angew. Chemie Int. Ed.* 49 (45): 8403–8407.

7

Nanomaterials and Water Purification: Opportunities for Improving and Protecting Water

Natalia Scheverin, Aura B. Patiño, Mariel Díaz, and Maria Fernanda Horst

INQUISUR, Departamento de Química, Universidad Nacional del Sur (UNS)-CONICET, Av. Alem 1253, 8000 Bahía Blanca, Argentina

7.1 Introduction

In the last few decades, nanotechnology has been studied and applied in different areas by researchers. The great variety of applications of nanotechnology gives this discipline great potential for the economic transformation of countries and the modification of the production processes of their industries. Hence the interest of government agencies and also the private sector, which is promoting research centers, the launch of $I + D + i$ programs for the development of nanomaterials (NMs) for multidisciplinary uses. In this sense, in the last few years, the continuous industry growth and population growth all over the world have impacted water supplies. Since the increase of all human activities the quality of water sources has been declining continuously. So, nanotechnology and the implementation of novel NMs and/or nanoparticles (NPs) have emerged as good tools for water purification in conditions suitable for consumption [1]. Nanotechnological strategies can be applied both for the remediation of natural water sources and for anthropogenically polluted waters. Figure 7.1 exemplifies the various types of contaminants that can be found in water resources and the extent to which the different nanotechnological methodologies are efficient in their remediation.

When we are talking about NMs, we refer to materials that include natural or manufactured particles with at least one dimension of 100 nm or less. In contrast, nanoparticles (NPs) include those with at least two dimensions between 1 and 100 nm [2]. One of the most important characteristics of NMs is related to their surface. In this sense, the NMs and nanoparticles display a higher surface/volume ratio, which allows improvement in their applications such as nano-sensors, nano-adsorbents, among others. This special characteristic gives the NMs used in environmental remediation the affinity to retain in their surface or structure the pollutants found in water media by a different mechanism of interaction. In a review of NMs and the environment, it was informed that in the period between 2006

Nanomaterials for Air and Water Purification, First Edition.
Edited by Deepu A. Gopakumar, Jyotishkumar Parameswaranpillai, Jinu Jacob George, and Midhun Dominic C.D.

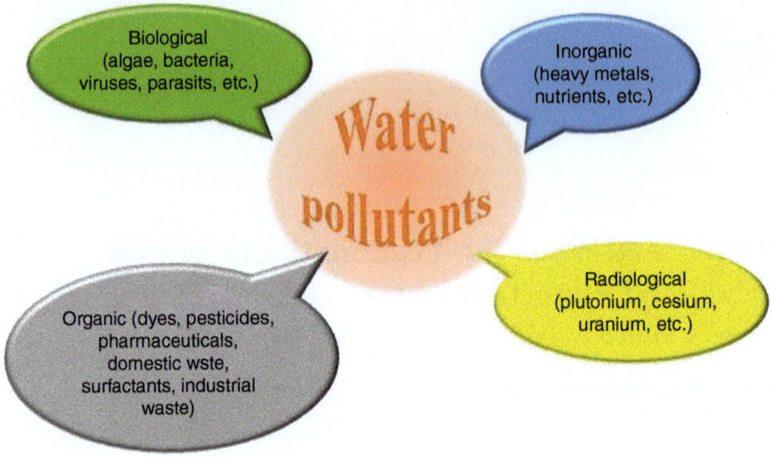

Figure 7.1 Classifications of different pollutants found in water sources.

and 2009, according to the Project and Engineering Nanotechnologies, 212 new nanotechnology-based products have been added to the consumer product list [3].

The investigation is continuously growing around NMs and the researchers have dedicated their work to developing novel nanoparticles for the application, especially in water remediation. So, we can overview the nanoparticles that have been studied: (i) inorganic nanoparticles, (ii) metal-organic frameworks (MOFs), (iii) nanomembranes, and (iv) organic polymer-based NMs. Among all the aforementioned nanoparticles employed for prepared NMs, special attention is given in this chapter to inorganic nanoparticles. This selection was done based on the fact that the methodologies of synthesis of these inorganic nanoparticles involve available reagents, and the use of environmentally friendly solvents; the techniques employed are usually easier to implement and the control of size and surface structure at the nanoscale could be done by modifying variables of synthesis. Table 7.1 shows the most inorganic nanoparticles used for water purification and their principal characteristics considering the application.

The aforementioned NMs and nanoparticles have been extensively studied in recent years. In particular, the researchers focus their work on the development of novel nanomaterials that could be applied to water and wastewater remediation or purification. In this context, the Section 7.2 presents the most relevant nanoparticles and NMs synthesized for this application. The description was especially focused on the material stabilization in the media, the performance of pollutants removal, and a critical point of view about the possibility of scalability of these materials and the development of a system of purification applied in the pollutant sites. In this sense, zero-valent metal-based NMs; silver nanoparticles (Ag-NPs); metal-oxide NMs; nanosized-zeolite; polymer carbon nanocomposites and derivatives were described below.

Table 7.1 Characteristics of Inorganic Nanoparticles for water purification.

		Inorganic nanomaterials			
Carbon-based nanomaterials	**Characteristic**	**Transition metal nanoparticles**	**Characteristic**	**Transition metal oxides nanoparticles**	**Characteristic**
CNT	Easy synthesis; high external surface area; high stability in vigorous low or high pH conditions	Fe(0)NPs	– Cheapest of transition metals, – high reactivity	TiO_2 NPs	Different catalytic reactivity, chemical behavior due to different surface planes
G, GO, and rGO	High sorption capacity; easy surface modification; sufficient functional groups; large surface area; easy modification	Cu NPs, Au NPs	– good performance	Fe_2O_3 NPs, Fe_2O_4 NPs	– active sites for reaction with pollutants, – particle size controllable, – large surface area for interaction
CNT: Carbon nanotubes G: Graphene GO: Graphene oxide rGO: Reduced graphene oxide NPs: nanoparticles		Bimetallic NPs	Higher removal efficiency than monometallic NPs	ZnO NPs	– eco-friendly material, – cheap, – simple way methodology of synthesis

7.2 Zero-Valent Metal-Based Nanomaterials

In recent decades, researchers from various areas have extensively studied zero-valent metal-based nanomaterials (nZVM), with a particular focus on water purification. The use of different nZVM-based materials has been reported for their application in removing different contaminants, with most of the studies dedicated to the elimination of heavy metals and organic pollutants from water. For example, Ag^0 nanoparticles have been extensively studied for their application in water disinfection, owing to their known antimicrobial properties [4]; Zn^0 nanoparticles have been widely reported for the degradation of numerous organic pollutants [5, 6]. On the other hand, nanosized zero-valent iron (nZVI) has been successfully used for the elimination of a wide range of pollutants, including aromatic compounds [7], heavy metals [8], and inorganic anions [9].

Generally, nanoparticles of zero-valent metals are obtained from the reduction of their metal salts in an aqueous solution. The reducing agent and the synthesis conditions used (e.g. stabilizing agents, temperature, and solvents) determine the size and shape of the nanoparticles. In this sense, the higher the surface area/volume ratio, the higher their reactivity and thus their efficiency in removing contaminants. However, the increase in reactivity, associated with the increase in surface energy, leads to aggregation and surface oxidation of the nZVMs, reducing their photocatalytic properties or adsorption capacity. Additionally, the decrease in size at nanometric scales can decrease the stability of these compounds in aqueous media, causing the leaching of metals into the medium.

The main challenges associated with applying these materials lie in their potential toxicity; rapid surface oxidation, which limits their storage; and their tendency to agglomerate, which significantly decreases their efficiency. Additionally, the separation of nZVMs from the medium is a complex process, and their direct application in columns generates very high internal pressures. The Section 7.2.1 will detail the uses, types of mechanisms, and applicability of ZVM-based NMs in water and wastewater treatment.

7.2.1 Nanosized Zero-Valent Iron

nZVI is an emerging material widely studied for water remediation. Its moderate reduction potential gives it stability in an aqueous medium, making it an ideal material for removing many redox-sensitive contaminants, such as halogenated hydrocarbons, organic dyes, antibiotics, and heavy metal ions [10]. Also, due to its small size and thus large specific surface area, nZVI presents excellent adsorption, precipitation, and oxidation properties. In this sense, nZVIs are estimated to be between 10 and 1000 times more reactive than granular zero-valent iron. However, the reactivity of nanoparticles does not depend only on their size but also on factors such as their shape, oxide layer, or surface coatings, among others. The synthesis methodology determines these factors. So then, the synthesis methodology used will depend on the field of application of nZVI.

Nowadays, numerous synthetic methodologies have been developed to obtain nZVI, including chemical co-precipitation, ball milling, liquid-phase reduction, carbothermal method, and green synthesis [11]. Among them, the liquid-phase reduction method has been widely reported in the literature. Normally, iron salts are reduced with sodium borohydride in an aqueous medium under anaerobic conditions (inert atmosphere). However, this type of methodology requires large amounts of expensive reagents, such as $NaBH_4$, that hamper their large-scale industrial application [11].

Bare nZVIs are highly reactive in an oxygen environment. When the nZVI is exposed to air, a thin iron oxide layer forms on its surface. Thus, nZVIs are considered a composite consisting of a core of Fe^0 and a layer of iron (hydr)oxides, as shown in Figure 7.2. In water, this layer appears to be a Fe(0)/Fe(II)/Fe(III) mixed-phase with a surface stoichiometry close to FeOOH [12], whose main crystalline phase would be lepidocrocite (γ-FeOOH) [13]. If the surface oxide layer is thin enough to allow electron transfer, its presence is advantageous from a practical point of view, as it gives the material greater stability than bare zero-valent iron nanoparticles. Besides, this core–shell structure gives the material adsorption and reduction functions for removing and transforming various contaminants.

The use of nZVI with a typical core–shell structure have been successfully applied for the elimination of numerous water contaminants: heavy metal ions, including lead, cadmium, copper, nickel, zinc, and silver [14]; metalloids, such as arsenic; inorganic anions, like nitrates [15]; dyes [16]; and organic compounds, such as phenol and halogenated organic compounds [17].

It is important to note that the removal mechanisms will depend essentially on the pollutant but also on the conditions of the reaction medium, such as pH. For example, the interaction mechanism for the removal of heavy metals will depend on the standard potential E^0 of the metal pollutant. For metals whose E^0 is very close to or more negative than the potential of $Fe^{2+}/^{3+}/Fe^0$, such as Zn and Cd,

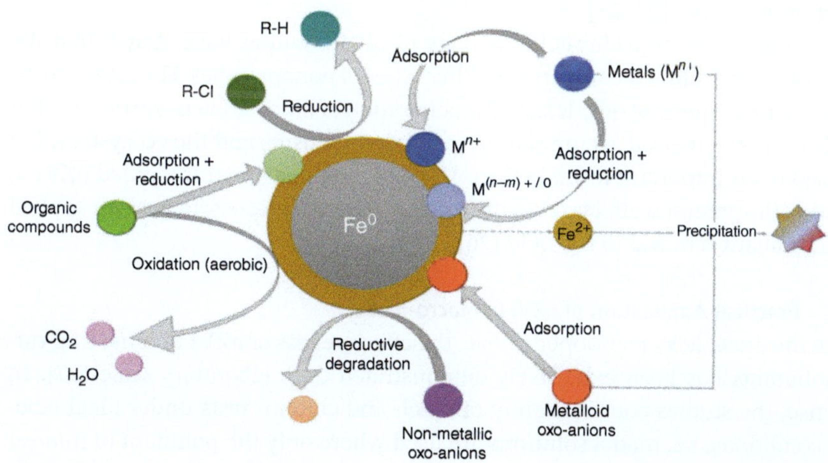

Figure 7.2 Schematic model of reaction mechanism of ZVI with different contaminants.

the predominant mechanisms will be adsorption and surface complexation without reduction. Meanwhile, ion metals whose E^0 is far more positive than Fe^0, such as Cu, As, and Hg, are typically reduced, and the reduced form can precipitate on or close to the nZVI surface. Finally, metals whose standard reduction potentials are slightly higher than E^0_{Fe}, such as Ni and Pb, could be immobilized on the surface by both adsorption and reduction processes [10]. The removal of inorganic oxo-anions (NO_3^-, ClO_4^-, and BrO_3^-) is generally by reduction mechanism.

On the other hand, the removal of organic pollutants is based on the adsorption and reduction capacity of nZVI. nZVI nanoparticles are able to reduce and modify the functional groups of organic compounds, transforming them into derivatives with lower toxicity and even adsorbing them on their surface. Additionally, nZVI can be used as catalysts in the Fenton reaction, degrading the pollutants to mineralization.

The application of nZVI has excellent potential for the remediation of various water sources, such as groundwater and wastewater. However, its use has some drawbacks: (i) the easy oxidation of their surface prevents their long-term storage; (ii) the agglomeration of the nanoparticles produces a notable decrease in their removal efficiency; (iii) difficult separation from the remediation medium; and (iv) leaching of iron into the medium, as a result of the redox processes involved in the removal mechanisms.

To overcome these drawbacks, many researchers have proposed the modification of nanoparticles. Modification strategies mainly include doping with other metals, surface coating, encapsulation in the matrix, and deposition of nZVI on a carrier or support [18]. The approaches used depend on the objective to be achieved. Doping with other metals, such as Pd and Pt, aims to increase their reactivity [19]. Both surface coating [20] and deposition of nZVI on a carrier [21] are usually used to prevent aggregation and enhance dispersibility. Also, the deposition of nZVI on a support and the encapsulation in a matrix are intended to reduce the agglomeration of nanoparticles, retard oxidation, and provide a higher surface area favoring the adsorption processes [22–25].

Another aspect to consider is the toxicity of nZVI. Studies have shown that the toxicity of nZVI is limited compared to other types of nanoparticles. However, modifications of nZVI may not only lead to the generation of more efficient adsorbents but may also cause increased accumulation in living organisms and the ecosystem. For this reason, it is important to study the toxicity of modified and unmodified nZVI to determine the potential effects on living organisms and the ecosystem when applied for contaminant removal in the field [26].

7.2.1.1 Practical Application of nZVI in Macro-Scale

Despite the drawbacks mentioned above, the effectiveness of nZVI in removing various pollutants has been extensively demonstrated on a laboratory scale [27]. In this sense, the studies consist mainly of batch and column tests under ideal laboratory conditions, i.e. model solutions are used where only the pollutant of interest is present, usually at very high concentrations, and pH values that in many cases do not match the values that occur in a real medium. For this reason, the results of laboratory experiments do not always reflect the effectiveness of nanoparticles in natural water or wastewater conditions. In order to know the feasibility of the practical application of nZVI, it is necessary to carry out experiments under real conditions, previously considering the advantages and limitations inherent to the application of this material [18].

The first step to applying any adsorbent on a large scale is conducting pilot-scale tests. The main objective of these tests is the study of different variables to maximize the efficiency of the pollutant removal process. In addition to studying typical variables, such as contact time, water injection flow rates, and dosage, pilot-scale tests also allow a cost study to be carried out. Consequently, with all this information, it is possible to find the combination of variables that will allow a process to have the best cost/efficiency ratio.

Numerous studies of the application of nZVI on a pilot scale can be found in the literature. Li et al. used nZVI in a pilot-scale process to remediate wastewater from the metallurgical industry [8]. The process consisted of a first adsorption step in a two-stage reactor, followed by a polishing treatment. Each nZVI reactor consisted of a reaction zone, a gravity separator, and a recirculation system. The treated water contained high concentrations of heavy metals (Au, Cu, Co, Ni, Tl, Zn, and Pb) and metalloids (As and Se). The treated water matrix had extremely high concentrations of Cu (103 mg l^{-1}) and As (110 mg l^{-1}). Long-term monitoring of the system (>12 months) showed that it achieved >99.5% removal of Cu, As, and other toxic contaminants.

Another pilot-scale study was conducted by Wang et al. to remove Zn and Pb from metallurgical wastewater [28]. A continuous, sequential process was used, consisting of the first stage of treatment with lime ($Ca(OH)_2$) as adsorbent, followed by the second stage of treatment with nZVI. Figure 7.3 shows the flow diagram of the process. The first stage aimed to reduce the concentration levels of the pollutants, which, after treatment with nZVI, were brought into compliance with the established standards for their elimination. The results of this combined process show that the concentrations of Pb and Zn were reduced from 610 and 195 mg l^{-1} to values below 0.8 and 0.5mg l^{-1} after treatment with nZVI.

This type of ex situ treatment for groundwater is costly and slow. So then, the use of in situ groundwater treatment technology is increasing [29]. The literature shows that many in-situ applications have been carried out, both pilot and full-scale [30]. This method mainly consists of the injection of nZVI, which degrades or adsorbs the pollutants as the groundwater comes into contact with the material.

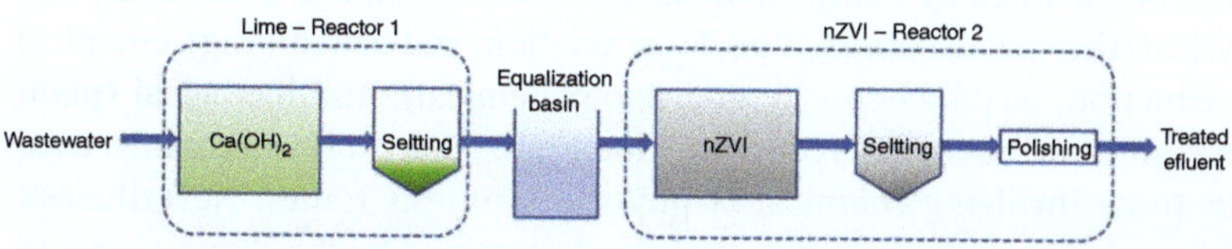

Figure 7.3 Schematic diagram of the remediation process used. Source: Wang et al. [28]/with permission of Elsevier.

Germany was the first European country to use this technology to remove chlorinated organic pollutants on a large scale in the town of Borheim. The site was contaminated with politetrachloroethylene (PCE) from a former industrial laundry [31]. Initially, the contaminated area was treated with another type of technology, such as ex-situ treatment and soil vapor extraction. However, these treatments were not efficient due to the high cost and time required. The injection of nZVI and micro-sized ZVI was able to remove more than 90% of the contaminant at a significantly lower cost than previous treatments used [30]. It is important to note that the success of the treatment cannot be attributed to nZVI alone, as micro-sized ZVIs were also used.

The Czech Republic has also successfully applied this technology on a large scale in Horice and Pisecna [32]. In particular, the area to be treated in Horice was approximately 7.2 km^2 in size, with a depth of 3–10 m. The area was contaminated with different organochlorine compounds, reaching concentrations above 70 $mg\,l^{-1}$. About 82 injection wells were used, through which 600 kg of nZVI were injected in two stages. The reduction of the different pollutants ranged from 60% to 90% after treatment. Similar results were obtained in the locality of Pisecna, where the extent of the contaminated area was 2 km^2 [30].

7.2.2 Silver Nanoparticles

Ag-NPs have been mainly used for the disinfection of contaminated water. Among the various NMs with disinfectant capacity, Ag-NPs are the most widely used due to their high antimicrobial and antibacterial activity and low toxicity in humans [33]. It has been reported for use against a wide range of microorganisms, including viruses [34], bacteria [35], and fungi [36]. The biocidal mechanism of Ag-NPs is still under discussion. To explain this process, researchers have proposed many action mechanisms, which can be summarized in three action pathways: (i) alteration of membrane properties due to the adhesion of Ag-NPs to cell walls, increasing their permeability (ii) irreversible damage to the cell membrane by the generation of free radicals (iii) release of the Ag^+ ion in the cellular environment, capable of destroying the bonds that involve the S and P atoms of vital enzymes and/or DNA/RNA [1].

The antimicrobial properties of Ag-NPs increase with decreasing size and depend on the shape of the nanoparticles. In this regard, it has been shown that, at equal concentrations, the antibacterial activity of Ag-NPs is considerably higher than that of dissolved Ag^+ and micro-sized Ag [35, 37]. Moreover, the efficacy of the bactericidal properties of Ag-NPs depends on their size and shape [38]. The size and shape of Ag-NPs depend on the synthesis methodology used.

Ag-NPs can be obtained by many methods that can be classified into three main groups: physical (thermal decomposition, laser ablation, and ball milling), chemical (chemical reduction, electrochemical, and photochemical), and biological (plant extract reduction and microbial reduction) methods. The most commonly used methods are those involving chemical or physical synthesis routes. Nevertheless, the choice of synthesis method will primarily be determined by the shape, size, and stability of the nanoparticles required but will also be affected by factors such as cost and complexity.

The direct application of Ag-NPs has some associated disadvantages, such as their tendency to aggregate, surface oxidation, and the release of Ag^+ ions into the medium, reducing their reactivity toward their different applications. On the other hand, uncontrolled release of Ag^+ ions into the environment can have eco- and cytotoxicological effects. Ag-NPs are usually stabilized or supported in different polymeric and ceramic matrices to overcome these drawbacks.

Ag-NPs impregnated in porous ceramic filters for point-of-use (POU) water treatment are a promising option for expanding access to clean water, especially in rural areas. Several studies have demonstrated the effectiveness of this technology in filtering and sterilizing water [39–42]. It has been shown that the addition of Ag-NPs to ceramic filters using different methodologies (painting and immersion) improves the performance of the filters and provides them with the ability to remove *Escherichia coli* in the range of 97–100% [42]. Mellor et al. [43] studied the efficacy of these long-term treatments compared to the traditional method of sterilization, which involves the use of sodium hypochlorite. The researchers found no statistically significant differences in water quality over time between the sterilization methods tested. Additionally, there are no cost differences in applying both technologies [44]; however, evidence of overchlorination has been found when using the traditional methodology. It is important to note that commercial treatment devices already use this type of system, such as Aquapure® and MARATHON®.

Other low-cost POU water disinfection systems can be generated from the immobilization of Ag-NPs on the cellulose fibers of a filter paper. Several studies showed that filter papers with deposited Ag-NPs showed excellent antibacterial properties against *E. coli* and *Enterococcus faecalis* [45–47]. Therefore, this type of system has great potential to be used as a personal or household water treatment method.

In addition to using Ag-NPs as disinfecting agents, recent publications have reported their potential application as sensors for aqueous pollutants, such as heavy metals and various organic pollutants. Additionally, the use of Ag-NPs for the oxidative degradation of pollutants has been studied on a laboratory scale.

7.3 Metal Oxide-Based Nanomaterials

Ideally, adsorbent materials used in water treatment should be effective, reusable, multifunctional, stable, nontoxic, economically viable, and environmentally friendly. In this sense, metal oxide nanoparticles (nMO) are excellent candidates, since they have a good capacity for adsorption and reuse against many contaminants. Due to the abundance of their precursor materials, the ease of synthesis, and the simple possibility of surface functionalization, the nMO represent an economically viable option. In addition, as they are compounds naturally present in the earth's crust, their toxicity is considered very low or null. A wide range of nMO have been used as adsorbents. Among them, the most common nMOs used as adsorbents are mainly oxides of iron (Fe), titanium (Ti), aluminum (Al), silicon (Si), and zinc (Zn).

The nMOs have proven effective in removing heavy metals, metalloids, and various organic contaminants, even at low concentrations. Factors such as pH, initial concentration, and the presence of coexisting species govern the adsorption processes, determining the removal efficiency of the materials. However, most applications of metal oxide-based NMs have been studied under laboratory conditions, using model solutions of contaminants or simulated water. Furthermore, nMOs separation currently remains a critical challenge for its large-scale application. The Section 7.3.1 will detail the uses, and types of removal mechanisms, focusing on the practical applicability of nMOs in water and wastewater treatment.

7.3.1 Iron Oxide Nanoparticles

The separation and recovery of adsorbents from water remain a major challenge with an important prospect for research and development in the area of water science and technology. Due to their superparamagnetic behavior, iron oxide nanoparticles (IONPs), such as magnetite (Fe_3O_4) and maghemite (γ-Fe_2O_3), have a great advantage over other adsorbents as they can be easily separated from the medium by applying an external magnetic field. Here, magnetic nanoadsorbents appear promising not only because they can be easily separated but also recovered and thus reused.

The IONPs can be synthesized by physical, chemical, and biological methods. Chemical-based synthesis methods are the most widely used due to their low production cost and high yield [48]. This method includes a wide range of techniques, such as hydrothermal, co-precipitation, solvothermal, and sol–gel synthesis. Each methodology will have its advantages and disadvantages, depending on the purpose and application required.

IONPs are widely used as adsorbent materials due to their fast kinetics and exceptional removal efficiency of various pollutants. Their application has been reported for the adsorption and degradation of heavy metals and metalloids, such as As, Pb, Cu and Ni; non-metallic oxo-anions, NO_3^- and PO_4^{3-}; and various organic micropollutants, such as dyes, pesticides and pharmaceuticals. Due to the enormous variety of contaminants, the removal capacity of IONPs has been explained by several types of mechanisms, including adsorption, complexation, ion exchange, electrostatic attraction, chemical reduction, precipitation, and co-precipitation processes mediated by the surface of the oxides.

IONPs can remove several species simultaneously, which significantly reduces their efficiency compared to empirical model tests. For this reason, it is essential to study the performance of these materials in a real-water environment to have a more realistic perspective on the possibility of their practical use in groundwater or wastewater treatment. Castelo-Grande et al. studied the application of micro-sized magnetite for wastewater remediation [49]. The adsorption performance was investigated at three different points in a wastewater treatment facility. In general terms, the magnetite particles showed good efficiency in removing the vast majority of the pollutants analyzed (heavy metals, detergents, and chemical oxygen demand [COD]). The influence of the type of wastewater sample (i.e. sampling point) proved to be an important parameter, especially in the case of high contaminant concentration.

Despite the promising results of laboratory-scale trials regarding magnetic IONPs for real water treatment, several limitations still delay their industrial-scale application in water purification systems [50]. These limitations mainly involve their chemical stability, toxicity, regeneration, and reusability [51]. Some of the most commonly used strategies to improve these aspects are coating nanoparticles with polymeric materials or generating nanocomposites based on IONPs. The generation of these types of materials will be further discussed in Sections 7.3.2 and 7.3.3.

On the other hand, the separation of magnetic IONPs by large-scale magnetic decantation is still a challenge to overcome. Ideally, it is intended that these technologies can be integrated into existing wastewater treatment facilities. Figure 7.4 shows the synergistic combination of both technologies schematically to improve process efficiency. Thus, after the first treatment stage, the effluents are treated with IONPs to remove residual pollutants. In the second stage, the IONPs, loaded with various pollutants, are separated from the effluent using a magnetic settling system in a second stage. The IONPs can then be regenerated and recovered for subsequent reuse.

Under this concept, Powell et al. developed and operated a device called MagNERD to separate and reuse magnetite nanoparticles from treated water under online and continuous flow, in relevant conditions at pilot scale [52]. MagNERD removed 94% of arsenic in a simulated drinking water medium. Furthermore, the MagNERD device efficiently separated magnetite nanoparticles (>95%) in different water matrices (e.g. ultrapure water and brackish water). Recovery of the adsorbent was performed by hydraulic backwashing.

Drenkova-Tuhtan et al. conducted pilot-scale tests for phosphate removal from wastewater using a magnetic $ZnFeZr@Fe_3O_4/SiO_2$ nanocomposite as an adsorbent [53]. Almost total removal was obtained after the first cycle, which included adsorption, magnetic separation, desorption, and reuse of the nanocomposite. However, a gradual decrease in removal efficiency was recorded in the following cycles. The

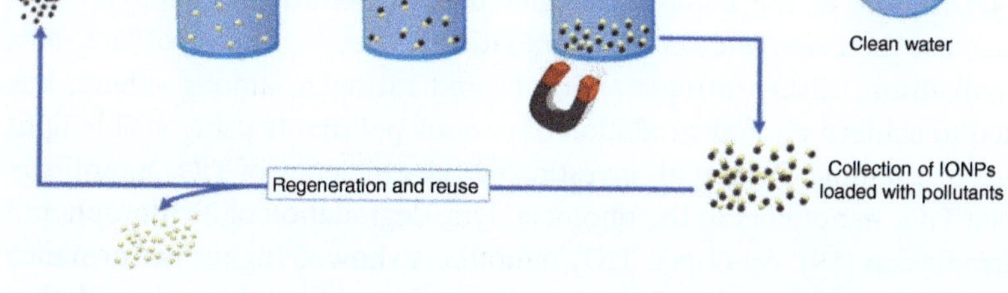

Figure 7.4 Schematic representation of the synergistic combination of adsorption and magnetic settling technologies with existing water treatment plants.

authors reported that this decrease in efficiency was consistent with the loss of adsorbent particles throughout each cycle.

7.3.2 Titanium Dioxide Nanoparticles

Titanium dioxide nanoparticles (TiO_2-NPs) have been widely reported for water purification due to their low cost and toxicity. In virtue of its photocatalytic properties, TiO_2-NPs have been studied mainly to degrade a wide variety of organic pollutants, such as pesticides, herbicides, dyes, and products from the pharmaceutical industry [54]. The photocatalytic degradation reaction is initiated because light irradiation of TiO_2-NPs induces charge separation, causing them to become highly unstable and react with water, generating free radicals. These free radicals are mainly responsible for the degradation process. On the other hand, its application in the removal of heavy metals and metalloids has also been demonstrated [55]. Also, TiO_2-NPs serve as disinfectant against a broad spectrum of microorganisms [56].

The efficiency of using TiO_2-NPs has been tested both at the laboratory and pilot scales. In Argentina, scientists designed an economic POU-type system for water purification in critical areas where resources are very limited [57]. The systems evaluated consisted of the immobilization of TiO_2-NPs inside PET soft drink bottles by different procedures. Different methodologies for the immobilization of the nanoparticles were studied. The efficiency of these devices was tested in different simulated and groundwater samples. The results showed a good photocatalytic efficiency of the tested devices, significantly reducing the organic matter present. However, TiO_2 particles were released into the medium when the bottles were shaken.

A wide variety of reactor designs have been reported, varying in factors such as the irradiation source and how the TiO_2-NPs are used (powder or immobilized). In this regard, Benotti et al. evaluated the performance of a pilot system using UV/TiO_2 photocatalysis for a total of 32 contaminants of pharmaceutical origin in enriched lake water samples [58]. The results showed that 29 of the pollutants were removed by 70%, while the remaining three pollutants were degraded by 50%. These results show the great potential of these technologies for water treatment.

Because TiO_2-NPs require ultraviolet radiation to overcome their bandgap, their photocatalytic capacity is limited under ambient conditions. Only 5% of the solar energy reaching the earth is within this range. The doping of TiO_2-NPs with metals and nonmetals is one of the widest strategies used to reduce the bandgap and, therefore, increase their efficiency in the degradation processes. Doping of TiO_2-NPs with iron, palladium, silver, nitrogen, carbon, and nitrogen, among others, has been reported to achieve photodegradation of various pollutants using visible light successfully. For example, Park et al. investigated the efficiency of TiO_2 nanofibers and Ag-doped TiO_2 nanofibers in the photocatalytic degradation of 2-chlorophenol under UV irradiation [59]. Ag-doped TiO_2 nanofibers showed higher performance in pollutant photodegradation than TiO_2 nanofibers. In this regard, the degradation rate constant of Ag-doped TiO_2 nanofibers was almost 50% higher than that of its undoped counterpart.

Although the application of TiO_2-NPs is mainly focused on the photocatalytic degradation of organic pollutants, they can also be used as adsorbents for various types of inorganic pollutants, such as heavy metals, metalloids and nonmetallic oxo-anions. One of the most studied lines is the use of TiO_2-NPs and its derivatives for arsenic removal. In this sense, the adsorption mechanisms of TiO_2-NPs can be divided into two types: in the presence of UV/Visible light, TiO_2-NPs function as adsorbents and as photocatalysts, promoting the oxidation of As(III)/organic arsenic to its less toxic As(V) species; while in the absence of irradiation, TiO_2-NPs function only as adsorbents. In addition to TiO_2-based materials, the application of titanates and TiO_2-based materials for water removal has been reported. However, most studies were performed at a laboratory scale under empirical model test conditions.

7.3.3 Zinc Oxide Nanoparticles

The use of zinc oxide nanoparticles (ZnO-NPs) in water treatment has been widely reported. Their good adsorptive and photocatalytic properties give them great potential for application in water purification processes. Due to their adsorptive capacity, ZnO-NPs have been mainly used to remove heavy metals such as Cd, Cr, Hg, and Pb [60]. Some research shows that ZnO-NPs have better heavy metal removal efficiency than TiO_2 [61]. Their application as adsorbents for methyl orange and amaranth has been reported recently [62].

Like titanium dioxide, due to their photocatalytic properties, ZnO-NPs can be used to degrade a wide variety of organic compounds, such as dyes, phenolic compounds, fungicides, pesticides, and pharmaceuticals. Several investigations have focused on comparing the photocatalytic activity of ZnO-NPs with that of TiO_2-NPs. Sakthivel et al. compared the photocatalytic activity of ZnO-NPs and TiO_2-NPs to degrade acid brown 14 [63]. From comparing the performance results of the different semiconducting materials, it emerged that ZnO-NPs have higher photocatalytic activity for the degradation of acid brown 14 using sunlight as the energy source. The authors explained this result were obtained because zinc oxide can absorb a large fraction of the solar spectrum, which is not the case for titanium oxide. Similar results for the degradation of naproxen and methylene blue by Štrbac et al. [64].

7.4 Nanosized Zeolite as Adsorbent and Support Material

Zeolites are low-cost, highly porous aluminosilicates with great potential to be used as adsorbents for various pollutants. Over other inorganic adsorbents, zeolites stand out for their large surface area and exceptional ion exchange capacity, which gives them high adsorption capacity. Due to their high stability and intrinsically low toxicity, these materials have been studied for different applications. Nanosized zeolite can be obtained both by synthetic methods, from their chemical precursors, and by physical methods, from milling the naturally present mineral in the earth's crust.

Because zeolites in aqueous media have a negative surface charge, these materials have been studied mainly to remove cationic pollutants, such as heavy metals and cationic dyes [65]. However, their surface modification using different multivalent metal cations has been reported to apply them as adsorbents of anionic species, such as fluorides [66, 67]. On the other hand, the scientific community has used zeolites as support materials to stabilize and immobilize various nanoadsorbents. For example, nZVI nanoparticles have been supported on zeolites to avoid agglomeration and increase the removal efficiency of pollutants such as As, Cd, and Pb [68]. Similarly, it has been reported that the generation of zeolite-magnetite nanoparticle nanocomposites leads to an increase in stability, in terms of iron release to the medium, concerning the application of bare magnetite nanoparticles [69].

The application of nanosized zeolites generates excessive pressure drops when used directly in fixed bed and continuous flow systems. For this reason, bulk zeolites rather than nanosized zeolites are often used for this type of application.

7.5 Nanocomposites

The abovementioned nanoscale materials have exceptional physicochemical properties; nevertheless, some exhibit potential risks to human and animal organisms and hazards to the environment. They could be introduced to human and animal organisms by inhalation, ingestion, and dermal penetration through potable water, air, food, and personal products [70]. When NMs are discharged into the environment, they are chemically and structurally modified, increasing, or decreasing their toxicity. Their dynamic in the environment is complex due to several factors that must be considered: temperature, pressure, humidity, and pH. The toxicity of NMs is directly dependent on their size, shape, surface area, and surface chemistry [71]. The size of the NM is one of the significant physicochemical properties that should be assessed since a small size (comparable to some biomolecules) allows fast absorption in the organism. As the size decreases, the surface area increases, enhancing reactivity and interaction.

The NMs small size facilitates their entrance into the cell membranes and other biological barriers; thus, they could spread through the whole body, reaching biological tissues, and accumulating mainly in the liver, kidney, and spleen [72]. Furthermore, some NMs induce reactive oxygen species (ROS) production, resulting in oxidative stress [73]. For example, Ag nanoparticles are a well-known type of metal nanoparticle widely used as efficient antimicrobial, antibacterial, and antiprotozoal agents [74]. Toxicological studies of Ag^0 have demonstrated that they exhibit cytotoxic and genotoxic effects (oxidative stress, DNA damage, and modulation of cytokine production) against several cell lines [75]. TiO_2-NPs are another favorite type of nanoparticle that have been employed in the preparation of sunscreens, toothpaste, and nutritional supplements, also as pollutants degradation agents in water remediation. Therefore, it is a compound that has been gaining considerable attention; therefore, it is pivotal to study its interaction with biological systems. Recent studies on rats and mice have shown that skin, pulmonary, and oral absorption could be exposed. The results suggested that the tiny TiO_2 nanoparticles of <0.1 μm show pulmonary inflammatory responses. Affections in different organs like kidneys, brain, spleen, and liver have been impacted by NMs uncontrolled use [76].

On the other hand, NMs tend to aggregate due to their high surface energy and large surface area [77]. Furthermore, NMs such as iron oxides (magnetite, maghemite, and hematite) could be easily oxidized by air and alter their physicochemical properties. For these reasons, newly developed materials with low or nontoxicity, oxidation resistance, and low or nonaggregation have been synthesized. As a result, nanocomposites have been widely reported in the literature, and diverse classifications have been suggested depending on various factors.

The classification of carbonaceous or polymer nanocomposites has been intensely investigated. Polymers can be natural or synthetic, including cellulose, starch, chitosan, polyethylene, polyester, Teflon, and epoxy [78]. On the other hand, carbonaceous matrixes or coatings could be obtained from natural sources or residues through a thermal treatment followed by an activation process.

As seen in Figure 7.5, the carbon or polymer could act as a (i) matrix where NMs are introduced or supported into the carbonaceous/polymeric matrix (it acts as binding reinforcement) or as an (ii) coating where the carbon/polymer coats the inorganic core. The hybridization between NMs and carbon/polymer matrix or coating could improve essential properties, including colloidal, chemical, and mechanical stability [79]. Colloidal stability is a crucial factor that can be enhanced by introducing a carbonaceous/polymeric coating or matrix that could, in some cases, decrease the zeta potential to a negative magnitude, which could stimulate the steric and electrostatic (repulsion between particles) stabilization [80]. Regarding matrix nanocomposites, the NMs are considered fillers that could achieve different shapes and sizes depending on the interaction between the NMs and the matrix. Nonetheless, sometimes the matrix does not guarantee to prevent

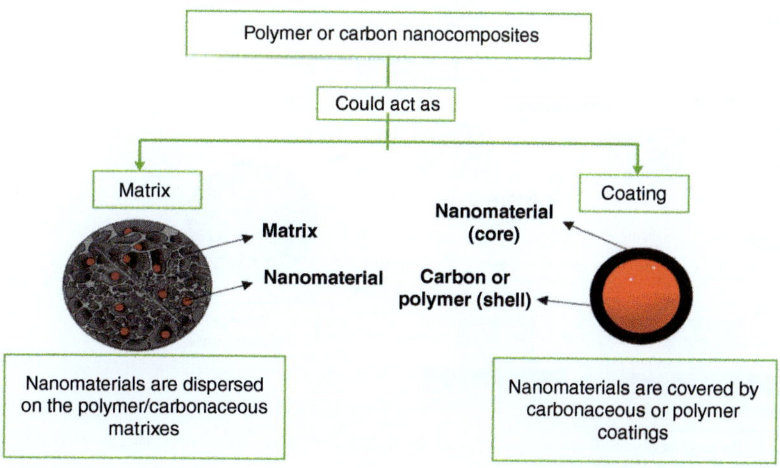

Figure 7.5 Types of carbonaceous/polymeric nanocomposites.

agglomeration of NMs. Sometimes, Van Der Waals forces contribute to creating nanoparticle aggregates in the matrix. This issue can be solved by employing coatings that will functionalize and modify the surface of the NMs.

In coatings nanocomposites, the system is deemed a core–shell structure where NMs are in the core part, and the carbon/polymer acts as a shell where protection of the nanocore is induced. The nanosized fillers and the carbon/polymer matrixes or coatings could interact by physical (electrostatic and London dispersion forces) or chemical forces (covalent bonds), depending on the porous structure and chemical nature (exposed functional groups), or a combination of both. Core–shell structures are systems applied in diverse fields; however, the focal point is biomedicine (especially in controlled drug delivery systems and hyperthermia treatment) and water remediation [81]. For example, carbon-coated magnetite nanoparticles have been widely applied in hydrometallurgy under very aggressive operating conditions (low pH and high temperatures). In addition, water remediation in acid mine sources is another well-known application. These acid conditions (pH <3) could disaggregate some metal nanoparticles (iron oxide) and change their properties by altering their structure [82]. These inconveniences could be solved by employing nanocomposites. The mentioned drawbacks of NMs have led to the search of new alternatives to biocompatible and nontoxic coatings or supporters to avoid the toxic effects of NMs and chemical perturbance.

Many synthesis methods of carbonaceous/polymeric nanocomposites, such as chemical, biological, and physical, have been investigated. However, core–shell systems are more complex to obtain. Usually, their preparation consists of a two-step procedure where NMs are obtained via bottom-up or top-down approaches. After that, each nanocomposite could be obtained by chemical vapor deposition, solution plasma processing, hydrothermal methods, spray pyrolysis, solid-phase, or slow pyrolysis treatment [83–85]. On the other side, as shown in Figure 7.6, NMs

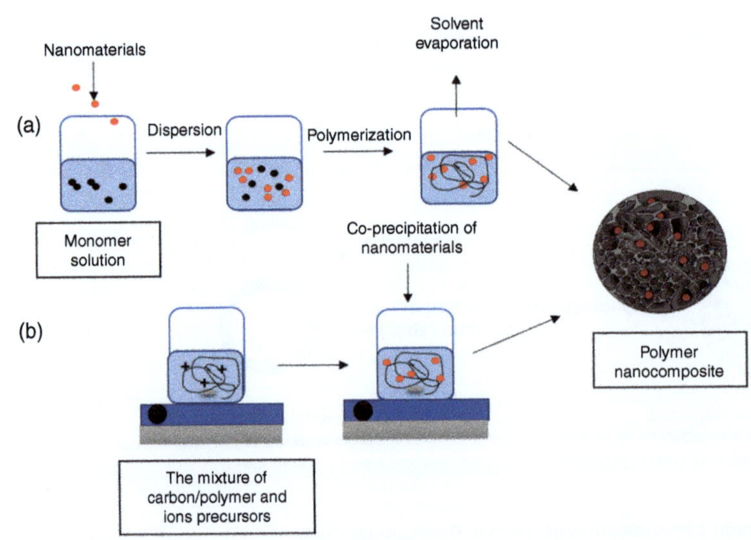

Figure 7.6 Two common approaches to obtaining carbon/polymer nanocomposites; (a) in situ polymerization and (b) mixing method.

supported on carbonaceous/polymeric matrixes can be prepared by two main methods: in situ NMs deposition into the matrix by (i) the in situ polymerization or (ii) mixing method [86, 87].

7.6 Classification of Carbon/Polymer Nanocomposites

7.6.1 Metal/Polymer or Carbon-Based Nanocomposites

7.6.1.1 nZVI (Nanosized Zero-Valent Iron) Nanocomposites

Water treatment technologies have been increasing due to the requirement to follow the accepted limits on maximum permitted concentrations of pollutants in water. However, the constant search for efficient technologies for the degradation or removal of pollutants: metalloids, heavy metals, and organic pollutants has been challenging. One of the most explored elements is iron, the fourth most abundant element in the earth's crust. Therefore, magnetic carbonaceous nanocomposites are being researched because of their vast range of applications. They are usually composed of nZVI (nanoscale zero-valent iron) or iron oxides (magnetite, maghemite, and hematite) supported or coated on/by various polymers or carbon solids.

Furthermore, IONPs are nontoxic, biocompatible, and exhibit a substantial magnetization property, which gives them an added value property that will be advantageous when removing the material from a particular water source by applying an external magnetic field. One of the most commonly used NMs at laboratory and industrial scales are nZVI. Its immobilization should be comprehended to prevent the loss of nZVI and avoid iron pollution. When supporting nZVI, it could achieve different shapes highlighting the beads, which show remarkable potential to be implemented in continuous flow systems, filter columns, domestic filters, and fixed bed reactors [88]. To take into consideration, during the NMs immobilization process, they could suffer oxidation. Thereby, it is advised to carry out the synthesis in anaerobic conditions.

Application of immobilized nZVI focuses on adsorption and degradation (employed as reductants in heterogeneous Fenton oxidation processes) of organic and inorganic pollutants. nZVI has been supported or coated on/by different matrixes, standing out the activated carbon (a popular adsorbent material used in water treatment plants), biochars, hydrochars, diatomite, PEG (Polyethylene), and PVP (Polyvinylpyrrolidone) [89]. nZVI/carbon core–shell and nZVI/diatomite/organosilane nanocomposites have been employed to degrade organic compounds such as phenol (aromatic organic compounds), bisphenol A (BPA), and pharmaceuticals (amoxicillin, metronidazole, tetracycline) by a Fenton-type reaction [90, 91]. Fenton reactions are suggested to reach high percent removal efficiencies and were used to promote the oxidation rate of organic compounds in the presence of some radicals such as $^{\cdot}OH$ and $^{\cdot}O_2$. In some cases, both degradation and adsorption co-occur where the degradation process of the pollutants leaves functional groups in the aqueous medium, and the nanocomposite could adsorb the degradation products [92].

Another application of these nZVI nanocomposites is removing toxic, non-biodegradable organic dyes such as methylene blue, malachite green, and methyl orange [93]. They are used in the textile and leather industries. The dyes impart color to the effluents and remain stable in water sources due to their stabilization owing to the auxotrophic and chromophoric groups. Removing dyes could involve degradation, adsorption, reduction, or precipitation [94]. Cellulose, montmorillonite, and bentonites are other popular coatings and supporters of nZVI to remove dyes. Moreover, cellulose could be used in the form of cellulose nanocrystals, and it enhances the amount of active and reactive sites that will facilitate the interaction with nZVI [95]. Clays are other types of supporters or coatings that have been attractive due to their outstanding cation exchange capacity (CEC), surface hydrophilicity, and surface electronegativity [96].

On the other side, heavy metals and metalloids are some of the most dangerous pollutants released into the environment due to industrial activity. Therefore, they have great potential to accumulate in living organisms. One of the widespread nanocomposites is carbon/polymer. The interaction between nZVI carbon/polymer nanocomposites and the heavy metals could be through reduction, adsorption, oxidation/reoxidation, co-precipitation, and precipitation of Cr, As, Cu, Pb, Se, Pd, Hg, and Ag ions [97]. Some of the processes mentioned previously can be combined depending on the redox potential of the system and the capping or support agent of the nZVI. The carbon supporters are gaining considerable attention, highlighting biochars (carbon-rich solids obtained from materials that are considered residues) that possess great potential as water remediation agents. Biochars could be obtained by thermochemical processes, including pyrolysis (slow, fast, and flash), torrefaction, or gasification [98]. They are converted into carbon-rich solids through cellulose, hemicellulose, and lignin chemical transformations [99]. The main functional groups present in the lignocellulosic framework are mostly oxygen-containing ones (C–O, C=O, and OH) that are capable of interacting with inorganic or organic molecules [100]. Depending on the pH and the point zero charge (PZC), those functional groups can change into their ionized forms: COO^- and OH^- forming strong electrostatic bindings with cationic organic or inorganic molecules. nZVI/biochar nanocomposites can also be used in the removal of Cd(II), Co(II), As(V), and Cr(VI) [101]. From these heavy metals, Cr(VI), a pollutant found in industrial effluents, appears in various chemical forms (trivalent and hexavalent) [102]. It exhibits carcinogenic effects on organisms; therefore, depuration technologies, including chemical precipitation, membrane process, electrodialysis, and ion exchange, have been applied [103]. Nevertheless, they are time- and energy-consuming. The use of nZVI nanocomposite as a reduction agent is preferred because of its efficient degradation through chemical reactions of electrochemical corrosion [104].

7.6.1.2 Silver Nanocomposites (Ag^0)

Silver has been used since ancient times as an antibacterial agent. Its demand has increased over time due to its unique optical, chemical, physical, antibacterial, conductive, and catalytic properties [105]. It has been estimated that 320 tons/year are produced worldwide [106]. Nowadays, AgNPs carbon/polymer nanocomposites have been employed in engineering products, clothing, sprays, and detergents [107]. The use of carbon or polymer reduces the toxic risks that AgNPs intrinsically have and stabilize the thermodynamically unstable AgNPs. It has been compared AgNPs crystal surfaces to other nanoparticles, and it has been noticed that AgNPs possess higher atom density. This characteristic permits easy adhesion to cell membranes [108]. The size of AgNPs influences the antibacterial effect. AgNPs have broad-spectrum effects against gram-positive and gram-negative bacteria by releasing silver ions [109]. As mentioned in the introduction to the nanocomposite section, AgNPs could be harmful at high concentrations, altering live organisms (oral, reproductive, neurotoxic, and immunotoxic) and the environment [110]. Subsequently, AgNPs nanocomposites have been developed to control and slow silver ions release into the aqueous medium.

Chitosan, a polysaccharide obtained naturally by deacetylation of chitin, has gained high interest due to its distinct chemical and physical properties. Chitosan has two main functional groups within its structure corresponding to amine (NH_2) and hydroxyl (OH) that promote the formation of intra and intermolecular interactions [111]. Chitosan, by itself, was found to have a poor surface area [112]. To enhance and activate its surface area, several nanostructures have been introduced into chitosan, including nanoparticles, nanosheets, nanorods, and nanowires [113]. Ag-NPs can be added to their chitosan matrix due to the high affinity for the chitosan functional groups [114]. Different authors have implemented AgNPs chitosan nanocomposite in the remediation of metal ions by ultrafiltration, dyes, antibiotics, and pesticides. The excellent heavy metal adsorption is related to its high chemical reactivity due to active functional groups and a flexible polymer structure [115].

AgNPs cellulose is another type of nanocomposite considered an eco-safe, versatile, and efficient water treatment agent. Cellulose is a polysaccharide composed of chains of glucose monomers [116]. Its structure is rich in hydroxyl groups which actively interact with AgNPs. This nanocomposite has been reported to serve as a plasmonic sensor of heavy metals and organic compounds [117]. Furthermore, it is also employed as a photocatalyst for oxidative degradation [118]. Besides their utilization as an adsorbent and degradation agent, AgNPs cellulose nanocomposite has shown high antibacterial activity by inactivating *E. coli* and *E. faecalis* [119]. Another water treatment system that is gaining much attention is membranes. For example, AgNPs have been supported on polyethersulfone (PES) microfiltration membranes, showing solid antimicrobial properties and prolonged silver release, enhancing the lifetime of water treatment membranes [120].

7.7 Metal Oxides/Polymer or Carbon-Based Nanocomposites

7.7.1 Iron Oxide (Magnetite) Nanocomposites

IONPs, especially Fe_3O_4 (magnetite), exhibit high adsorption capacities against many pollutants, including heavy metals, metalloids, dyes, pesticides, and care products. IONPs are gaining significant attention due to their remarkable physicochemical properties, including nontoxicity, biocompatibility, superparamagnetism, and high reactivity [121]. It is crucial to point out their substantial magnetization property due to their easy recovery from the applied aqueous medium by an external magnetic field [122]. Plenty of matrixes or coatings are available; however, carbonaceous ones have been employed because of their simplicity and availability. The carbonaceous matrix or coating can be obtained through chemical, physical, or biological treatments. Nevertheless, thermochemical treatments, including pyrolysis, hydrothermal carbonization, torrefaction, and gasification, have been widely used due to their high efficiency and yields, low cost (capital, operational, and maintenance), and accessibility [123].

Nowadays, the circular economy plays an essential role in industry and our daily lives [124]. For this reason, the reuse or recycling of materials generated throughout an agro-industrial process could be employed as a feedstock to prepare added-value products. This is the case with lignocellulosic residues. Some of the most common agricultural wastes are husks: rice, sunflower, and coffee [125]. Lignocellulosic residues are considered agricultural wastes that can be transformed into carbonaceous materials by thermochemical treatments. The transformation employing a specific temperature, pressure, and activating agent will condition the final porous characteristics. Activated carbons and biochars usually possess well-developed pores and a large surface.

IONPs supported on carbonaceous matrixes or coated by a carbon layer have been investigated. They have been applied to remove different pollutants such as heavy metals, dyes, and pharmaceuticals [126]. The removal efficiency is dependent on the medium pH, the initial dose of adsorbent, ionic strength, the initial concentration of the pollutant, temperature, and stirring rate. The metal oxide carbonaceous nanocomposites reveal synergism of properties. For example, magnetic biochars are well-investigated nanocomposites to remove arsenic, a toxic metalloid found in groundwater aquifers mainly in two anionic forms: As(III) and As(V). Biochar has a negative surface charge, and the interaction with the anionic forms of As provokes a low sorption efficiency. For this reason, modified biochar with magnetite could induce a stronger interaction between the sorbate and sorbent [127].

Another well-known application of magnetite nanoparticles supported on a biochar matrix is the removal of Cr(VI). The presence of the biochar matrix enhances adsorption and reduction, which results in an efficient, reusable, and cost-beneficial nanocomposite [128]. Besides, other nanocomposites involve two phases and three or more phases, generating trifunctional or multifunctional nanocomposites, such as Fe_3O_4-Ag nanocomposite, which has catalytic antibacterial activity effects, and adsorption capacity [129]. Furthermore, other nanocomposites involve microorganisms, such as a developed iron oxide biochar loaded with photosynthetic bacteria. Each component interacted synergically and played an essential role in the nanocomposite; biochar and IONPs were helpful in the adsorption and catalysis, and on the other side, the photosynthetic bacteria were employed as biodegradation agents to control COD, ammonia (NH_4^+), and phosphate (PO_3^{4-}) in an aqueous medium [130].

7.7.2 Titanium Oxide (TiO_2) Nanocomposites

As early mentioned, TiO_2 is involved in the photocatalytic degradation of various pollutants and can damage the function of various gram-positive and gram-negative cell lines [131]. Both effects can occur under UV light, which promotes the generation of OH radicals, which are the ones in charge of degrading arsenic, heavy metals, pesticides, dyes, and polycyclic aromatic hydrocarbons (PAHs). These processes are called advanced oxidative processes (AOPs) and are efficient tools utilized in the last years [132].

Heterogeneous photocatalysis is when the degradation of pollutants is carried out using a nanocomposite (composed of a semiconductor TiO_2 and a carrier) and a light source [133]. TiO_2 has been supported on clays, which are minerals including montmorillonite and bentonite made up of silica, alumina, water, and weathered rock [134]. TiO_2 immobilization enhances its activity by increasing active surface sites. In addition, it prevents the agglomeration and release of TiO_2 into the environment. Another example could be the preparation of TiO_2 into bacterial cellulose, which is a biocompatible matrix. It is characterized as having high mechanical strength, crystallinity, water retention, and adsorption capacity. The use of bacterial cellulose as a matrix helps to prepare a well-organized nanostructure due to the exposure of hydroxyl functional groups [135].

7.7.3 Zinc Oxide (ZnO) Nanocomposites

ZnO nanoparticles are interesting due to their strong oxidation ability and photocatalytic property. Nevertheless, they present almost the same disadvantage as TiO_2 due to their considerable bandgap energy that demands the use of UV light. Compared to TiO_2, ZnO is less toxic and cost-effective [136]. However, ZnO undergoes photocorrosion; consequently, new strategies are being searched for, such as metal doping or immobilization of ZnO in polymers to enhance the photodegradation of pollutants. Metal doping and ZnO immobilization are two strategies that have been applied. Starch, cellulose, and chitosan are examples of biopolymers employed as carriers of ZnO [137]. In some studies, two or three compounds are added to enhance photocatalytic activity. ZnO supported on chitosan/polypyrrole (conductive polymers) has been reported to exhibit great potential as a photodegradation agent of different dyes, including orange-16, methylene blue, and brilliant blue R-250. The presence of amino groups in chitosan could interact with the anionic part of the dyes improving the degradation of pollutants [138].

Moreover, other conductive polymers are embedded with ZnO, such as polyaniline, polyphenylendiamine, and poly(1-naphthylamine) [139]. The energy levels of the conductive polymers match those of ZnO, thus increasing the sensitiveness of absorption of visible light [140]. The use of sunlight is preferred over ultraviolet radiation due to the harmful effects caused by the artificial source. On the other side, ZnO could also act as membrane antifouling material. To cite, polyvinylidene fluoride (PVDF)–ZnO composite membranes are prepared. The addition of ZnO into the membrane promotes smaller contact water angles and enhances antifouling characteristics [141].

7.8 Silicate/Polymer or Carbon-Based Nanocomposites

7.8.1 Nanosized Zeolites Composites

Nanosized zeolite (a microporous aluminosilicate material) is well-known for being an excellent ion-exchange agent [142]. Owing to its characteristic negative surface charge on the framework, it possesses a high affinity for cationic molecules or cationic metal ions. Adding zeolites to carbonaceous or polymer matrixes will improve stabilization in an aqueous solution, easier reusability, and high adsorption capacities. There has been investigation related to finding supporters such as natural and synthetic polymers and carbonaceous matrixes, including activated carbons, cellulose, chitosan, alginate, polyvinyl chloride (PVC), polycaprolactone (PCL), polyacrylamide (PAA), polyvinyl alcohol (PVA), polypropylene (PP), polypyrrole (PPy), PVP, polyamide (PA), polyacrylonitrile (PAN), and polysulfone (PSF) [143]. Zeolite nanocomposites have been prepared in various forms, including beads, films, and hydrogels. One of the most compelling applications of nanocomposites is the removal of radioactive ions from water sources. For example, magnetic zeolite nanocomposite (nZVI-zeolite) was used as strontium (Sr^{2+}) sorbent in a real seawater sample [144]. This is quite challenging since real water sources present a high amount of ions and compounds that could compete for the active sites of the nanocomposite. Other similar work used a natural zeolite supported on PAA to remove UO_2^{2+}, Tl^{+}, Pb^{2+}, Ra^{2+}, Bi^{3+}, and Ac^{3+} [145].

References

1 Qu, X., Alvarez, P.J.J., and Li, Q. (2013). Applications of nanotechnology in water and wastewater treatment. *Water Res.* 47 (12): 3931–3946.

2 Andy, R.I.D.H., Yon, D.E.Y.L., Ahendra, S.H.M. et al. (2008). Nanomaterials in the environment: behavior, fate, bioavailability, and effects. *Environ. Toxicol. Chem.* 27 (9): 1825–1851.

3 Peralta-Videa, J.R., Zhao, L., Lopez-Moreno, M.L. et al. (2011). Nanomaterials and the environment: a review for the biennium 2008-2010. *J. Hazard. Mater.* 186 (1): 1–15.

4 Deshmukh, S.P., Patil, S.M., Mullani, S.B., and Delekar, S.D. (2019). Silver nanoparticles as an effective disinfectant: a review. *Mater. Sci. Eng., C* 97: 954–965.

5 N. Boronina T, Lagadic I, B. Sergeev G, J. Klabunde K. (1998). Activated and nonactivated forms of zinc powder: reactivity toward chlorocarbons in water and AFM studies of surface morphologies. *Environ. Sci. Technol.* 32 (17): 2614–2622.

6 Bokare, V., Jung, J., and lim, Chang YY, Chang YS. (2013). Reductive dechlorination of octachlorodibenzo-p-dioxin by nanosized zero-valent zinc: modeling of rate kinetics and congener profile. *J. Hazard. Mater.* 250, 251: 397–402.

7 Li, Q., Chen, Z., Wang, H. et al. (2021). Removal of organic compounds by nanoscale zero-valent iron and its composites. *Sci. Total Environ.* 792: 148546.

8 Li, S., Wang, W., Liang, F., and Zhang, W.X. (2017). Heavy metal removal using nanoscale zero-valent iron (nZVI): theory and application. *J. Hazard. Mater.* 322: 163–171.

9 El-Lateef, H., Khalaf, M., Al-Fengary, A., and Elrouby, M. (2022). Removal of the harmful nitrate anions from potable water using different methods and materials, including zero-valent iron. *Molecules.* 27 (8): 1–22.

10 Pasinszki, T. and Krebsz, M. (2020). Synthesis and application of zero-valent iron nanoparticles in water treatment, environmental remediation, catalysis, and their biological effects. *Nanomaterials.* 10 (5): 917.

11 Wu, Y., Pang, H., Liu, Y. et al. (2019). Environmental remediation of heavy metal ions by novel-nanomaterials: a review. *Environ. Pollut.* 246: 608–620.

12 Ling, L., Pan, B., Zhang, W., and xian. (2015). Removal of selenium from water with nanoscale zero-valent iron: mechanisms of intraparticle reduction of Se(IV). *Water Res.* 71: 274–281.

13 Liu, A., Liu, J., Pan, B., and Zhang, W.X. (2014). Formation of lepidocrocite (γ-FeOOH) from oxidation of nanoscale zero-valent iron (nZVI) in oxygenated water. *RSC Adv.* 4 (101): 57377–57382.

14 Yang, J., Hou, B., Wang, J. et al. (2019). Nanomaterials for the removal of heavy metals from wastewater. *Nanomaterials.* 9 (3): 424. 1–39.

15 Diao, Z.H., Qian, W., Lei, Z.X. et al. (2019). Insights on the nitrate reduction and norfloxacin oxidation over a novel nanoscale zero valent iron particle: reactivity, products, and mechanism. *Sci. Total Environ.* 660: 541–549.

16 Bhatti, H.N., Iram, Z., Iqbal, M. et al. (2020). Facile synthesis of zero valent iron and photocatalytic application for the degradation of dyes. *Mater. Res. Express.* 7 (1): 015802. 1–16.

17 Wei, Y.T., Wu, S.C., Chou, C.M. et al. (2010). Influence of nanoscale zero-valent iron on geochemical properties of groundwater and vinyl chloride degradation: a field case study. *Water Res.* 44 (1): 131–140.

18 Stefaniuk, M., Oleszczuk, P., and Ok, Y.S. (2016). Review on nano zerovalent iron (nZVI): from synthesis to environmental applications. *Chem. Eng. J.* 287: 618–632.

19 Wang, X., Zhu, M., Liu, H. et al. (2013). Modification of Pd–Fe nanoparticles for catalytic dechlorination of 2,4-dichlorophenol. *Sci. Total Environ.* 449: 157–167.

20 Wang, X., Yang, J., and Zhu, M. (2014). Effects of PMMA/anisole hybrid coatings on discoloration performance of nano zerovalent iron toward organic dyes. *J. Taiwan Inst. Chem. Eng.* 45 (3): 937–946.

21 Yang, J., Wang, S., Xu, N. et al. (2021). Synthesis of montmorillonite-supported nano-zero-valent iron via green tea extract: enhanced transport and application for hexavalent chromium removal from water and soil. *J. Hazard. Mater.* 419: 126461.

22 Wang, M., Cheng, W., Wan, T. et al. (2019). Mechanistic investigation of U(VI) sequestration by zero-valent iron/activated carbon composites. *Chem. Eng. J.* 362: 99–106.

23 Shaibu, S.E., Adekola, F.A., Adegoke, H.I., and Ayanda, O.S. (2014). A comparative study of the adsorption of methylene blue onto synthesized nanoscale zero-valent iron-bamboo and manganese-bamboo composites. *Materials.* 7: 4493–4507.

24 Pereira, C.A.A., Nava, M.R., Walter, J.B. et al. (2021). Application of zero valent iron (ZVI) immobilized in Ca-Alginate beads for C.I. Reactive Red 195 catalytic degradation in an air lift reactor operated with ozone. *J. Hazard. Mater.* 401: 123275.

25 Bezbaruah, A.N., Kalita, H., Talal, C.A. et al. (2014). Ca-alginate-entrapped nanoscale iron: arsenic treatability and mechanism studies. *J. Nanopart. Res.* 16: 2175.

26 Chen, X., Ji, D., Wang, X., and Zang, L. (2017). Review on Nano zerovalent Iron (nZVI): from modification to environmental applications. *IOP Conf. Ser. Earth Environ. Sci.* 51 (1): 1–7.

27 Mueller, N.C. and Nowack, B. (2010). Nanoparticles for remediation: solving big problems with little particles. *Elements.* 6 (6): 395–400.

28 Wang, W., Hua, Y., Li, S. et al. (2016). Removal of Pb(II) and Zn(II) using lime and nanoscale zero-valent iron (nZVI): a comparative study. *Chem. Eng. J.* 304: 79–88.

29 Karn, B., Kuiken, T., and Otto, M. (2009). Nanotechnology and in situ remediation: a review of the benefits and potential risks. *Environ. Health Perspect.* 117 (12): 1823–1831.

30 Mueller, N.C., Braun, J., Bruns, J. et al. (2012). Application of nanoscale zero valent iron (NZVI) for groundwater remediation in Europe. *Environ. Sci. Pollut. Res.* 19 (2): 550–558.

31 (2007). North Rhine-Westphalia Contaminated Site Remediation and Processing Association. AAV Annual Report: Contaminated Site Remediation and Urban Development.

32 Černík, M. (2010). *Chemicky podporované in situ sanační technologie*. Vydavatelství VŠCHT.

33 Xiu, Z. and ming, Zhang Q bo, L. Puppala H, L. Colvin V, J. J. Alvarez P. (2012). Negligible particle-specific antibacterial activity of silver nanoparticles. *Nano Lett.* 12 (8): 4271–4275.

34 Dung, T.T.N., Nam, V.N., Nhan, T.T. et al. (2019). Silver nanoparticles as potential antiviral agents against African swine fever virus. *Mater. Res. Express.* 6 (12): 1250g9. 1–11.

35 Liang, Z., Das, A., and Hu, Z. (2010). Bacterial response to a shock load of nanosilver in an activated sludge treatment system. *Water Res.* 44 (18): 5432–5438.

36 Krishnaraj, C., Ramachandran, R., Mohan, K., and Kalaichelvan, P.T. (2012). Optimization for rapid synthesis of silver nanoparticles and its effect on phytopathogenic fungi. *Spectrochim. Acta, Part A* 93: 95–99.

37 Choi, O., Deng, K.K., Kim, N.J. et al. (2008). The inhibitory effects of silver nanoparticles, silver ions, and silver chloride colloids on microbial growth. *Water Res.* 42 (12): 3066–3074.

38 Samberg, M.E., Orndorff, P.E., and Monteiro-Riviere, N.A. (2011). Antibacterial efficacy of silver nanoparticles of different sizes, surface conditions and synthesis methods. *Nanotoxicology.* 5 (2): 244–253.

39 Ren, D. and M. Colosi L, A. Smith J. (2013). Evaluating the sustainability of ceramic filters for point-of-use drinking water treatment. *Environ. Sci. Technol.* 47 (19): 11206–11213.

40 Moosa, A.A. and Muhsen, M.F. (2017). Ceramic filters impregnated with silver nanoparticles for household drinking water treatment. *Am. J. Mater. Sci.* 7 (6): 232–239.

41 Abebe, L.S., Smith, J.A., Narkiewicz, S. et al. (2014). Ceramic water filters impregnated with silver nanoparticles as a point-of-use water-treatment intervention for HIV-positive individuals in Limpopo Province, South Africa: a pilot study of technological performance and human health benefits. *J. Water Health* 12 (2): 288–300.

42 Kallman, E.N., Oyanedel-Craver, V.A., and Smith, J.A. (2011). Ceramic filters impregnated with silver nanoparticles for point-of-use water treatment in rural Guatemala. *J. Environ. Eng.* 137 (6): 407–415.

43 Mellor, J.E., Kallman, E., Oyanedel-Craver, V., and Smith, J.A. (2015). Comparison of three household water treatment technologies in San Mateo Ixtatán, Guatemala. *J. Environ. Eng.* 141 (5): 04014085.

44 Adeleye, A.S., Conway, J.R., Garner, K. et al. (2016). Engineered nanomaterials for water treatment and remediation: costs, benefits, and applicability. *Chem. Eng. J.* 286: 640–662.

45 A. Dankovich T, Gray DG. (2011). Bactericidal paper impregnated with silver nanoparticles for point-of-use water treatment. *Environ. Sci. Technol.* 45 (5): 1992–1998.

46 Dankovich, T.A. (2014). Microwave-assisted incorporation of silver nanoparticles in paper for point-of-use water purification. *Environ. Sci. Nano.* 1 (4): 367–378.

47 Mohammed, A.N. (2019). Resistance of bacterial pathogens to calcium hypochlorite disinfectant and evaluation of the usability of treated filter paper impregnated with nanosilver composite for drinking water purification. *J. Global Antimicrobial Resistance.* 16: 28–35.

48 Ali, A., Zafar, H., Zia, M. et al. (2016). Synthesis, characterization, applications, and challenges of iron oxide nanoparticles. *Nanotechnol. Sci. Appl.* 9: 49–67.

49 Castelo-Grande, T., Augusto, P.A., Rico, J. et al. (2021). Magnetic water treatment in a wastewater treatment plant: Part I - sorption and magnetic particles. *J. Environ. Manage.* 281: 111872.

50 Mudhoo, A. and Sillanpää, M. (2021). Magnetic nanoadsorbents for micropollutant removal in real water treatment: a review. *Environ. Chem. Lett.* 19 (6): 4393–4413.

51 You, J., Wang, L., Zhao, Y., and Bao, W. (2021). A review of amino-functionalized magnetic nanoparticles for water treatment: features and prospects. *J. Cleaner Prod.* 281: 124668.

52 Powell, C.D., Atkinson, A.J., Ma, Y. et al. (2020). Magnetic nanoparticle recovery device (MagNERD) enables application of iron oxide nanoparticles for water treatment. *J. Nanopart. Res.* 22 (48): 48. 1–11.

53 Drenkova-Tuhtan, A., Schneider, M., Franzreb, M. et al. (2017). Pilot-scale removal and recovery of dissolved phosphate from secondary wastewater effluents with reusable ZnFeZr adsorbent @ Fe_3O_4/SiO_2 particles with magnetic harvesting. *Water Res.* 109: 77–87.

54 Lazar, M.A., Varghese, S., and Nair, S.S. (2012). Photocatalytic water treatment by titanium dioxide: recent updates. *Catalysts.* 2: 572–601.

55 Gupta, K., Joshi, P., Gusain, R., and Khatri, O.P. (2021). Recent advances in adsorptive removal of heavy metal and metalloid ions by metal oxide-based nanomaterials. *Coord. Chem. Rev.* 445: 214100.

56 Saravanan, A., Kumar, P.S., Jeevanantham, S. et al. (2021). Photocatalytic disinfection of micro-organisms: mechanisms and applications. *Environ. Technol. Innov.* 24: 101909.

57 Lin, H.J., Meichtry, J.M., Piperata, G. et al. (2004). Results of the application of FH technology to make drinking water in the Argentine Republic. In: *Advances in low-cost technologies for disinfection, decontamination and arsenic removal in waters from rural communities of Latin America (HP and SORAS methods) - ASO Project AE 141, Argentina – Brasil – Chile – México – Perú – Trinidad & Tobago* (ed. M.I. Litter and A.J. González), 1–7.

58 Benotti, M.J., Stanford, B.D., Wert, E.C., and Snyder, S.A. (2009). Evaluation of a photocatalytic reactor membrane pilot system for the removal of pharmaceuticals and endocrine disrupting compounds from water. *Water Res.* 43 (6): 1513–1522.

59 Park, J.Y. and Lee, I.H. (2014). Photocatalytic degradation of 2-chlorophenol using Ag-doped TiO_2 nanofibers and a near-UV light-emitting diode system. *J. Nanomater.* 2014: 250803, 6 pages.

60 Sheela, T., Nayaka, Y.A., Viswanatha, R. et al. (2012). Kinetics and thermodynamics studies on the adsorption of Zn(II), Cd(II) and Hg(II) from aqueous solution using zinc oxide nanoparticles. *Powder Technol.* 217: 163–170.

61 Bhatia, M., Satish Babu, R., Sonawane, S.H. et al. (2017). Application of nanoadsorbents for removal of lead from water. *Int. J. Environ. Sci. Technol.* 14 (5): 1135–1154.

62 Zafar, M.N., Dar, Q., Nawaz, F. et al. (2019). Effective adsorptive removal of azo dyes over spherical ZnO nanoparticles. *J. Mater. Res. Technol.* 8 (1): 713–725.

63 Sakthivel, S., Neppolian, B., Shankar, M.V. et al. (2003). Solar photocatalytic degradation of azo dye: comparison of photocatalytic efficiency of ZnO and TiO_2. *Sol. Energy Mater. Sol. Cells* 77 (1): 65–82.

64 Štrbac, D., Aggelopoulos, C.A., Štrbac, G. et al. (2018). Photocatalytic degradation of Naproxen and methylene blue: comparison between ZnO, TiO_2 and their mixture. *Process Saf. Environ. Prot.* 113: 174–183.

65 Hernández-Montoya, V., Pérez-Cruz, M.A., Mendoza-Castillo, D.I. et al. (2013). Competitive adsorption of dyes and heavy metals on zeolitic structures. *J. Environ. Manage.* 116: 213–221.

66 Rahmani, A., Nouri, J., Kamal Ghadiri, S. et al. (2010). Adsorption of fluoride from water by Al^{3+} and Fe^{3+} pretreated natural Iranian zeolites. *Int. J. Environ. Res.* 4 (4): 607–614.

67 Scheverin, V.N., Russo, A.V., Horst, M.F. et al. (2021). Design of magnetic nanotechnological devices for the removal of fluoride from groundwater. *Cleaner Eng. Technol.* 3 (November 2020): 100097.

68 Suazo-Hernández, J., Sepúlveda, P., Manquián-Cerda, K. et al. (2019). Synthesis and characterization of zeolite-based composites functionalized with nanoscale zero-valent iron for removing arsenic in the presence of selenium from water. *J. Hazard. Mater.* 373: 810–819.

69 Scheverin, V.N., Russo, A., Grünhut, M. et al. (2022). Novel iron-based nanocomposites for arsenic removal in groundwater: insights from their synthesis to implementation for real groundwater remediation. *Environ. Earth Sci.* 81 (7): 188. 1–15.

70 Saleh, T.A. (2020). Nanomaterials: classification, properties, and environmental toxicities. *Environ. Technol. Innov.* 20: 101067. 1–16.

71 Sahu, S.C. and Hayes, A.W. (2017). Toxicity of nanomaterials found in human environment. *Toxicol. Res. Appl.* 1: 239784731772635.

72 Sohaebuddin, S.K., Thevenot, P.T., Baker, D. (2010). Nanomaterial cytotoxicity is composition, size, and cell type dependent. http://www.particleandfibretoxicology.com/content/7/1/22.

73 AshaRani, P.V., GLK, M., Hande, M.P., and Valiyaveettil, S. (2009). Cytotoxicity and genotoxicity of silver nanoparticles in human cells. *ACS Nano.* 3 (2): 279–290.

74 Yin, I.X., Zhang, J., Zhao, I.S. et al. (2020). The antibacterial mechanism of silver nanoparticles and its application in dentistry. *Int. J. Nanomed.*Dove Medical Press Ltd. 15: 2555–2562.

75 Stensberg, M.C., Wei, Q., ES, M.L. et al. (2011). Toxicological studies on silver nanoparticles: challenges and opportunities in assessment, monitoring and imaging. *Nanomedicine* 6: 879–898.

76 Shakeel, M., Jabeen, F., Shabbir, S. et al. (2016). Toxicity of nano-titanium dioxide (TiO_2-NP) through various routes of exposure: a review. *Biol. Trace Elem. Res.* Humana Press Inc. 172: 1–36.

77 Shaji, A. and Zachariah, A.K. (2017). Surface area analysis of nanomaterials. In: *Thermal and Rheological Measurement Techniques for Nanomaterials Characterization* (ed. S. Thomas, R. Thomas, A.K. Zachariah, and R.K. Mishra), 197–231. Amsterdam, Netherlands: Elsevier.

78 Bhatia, S. (2016). Natural polymers vs synthetic polymer. In: *Natural Polymer Drug Delivery Systems* (ed. S. Bhatia), 95–118. India: Springer International Publishing.

79 Beyene, H.D. and Ambaye, T.G. (2019). Application of sustainable nanocomposites for water purification process. In: *Sustainable Polymer Composites* *and Nanocomposites* (ed. A. Inamuddin, S. Thomas, R. Kumar Mishra, and A. Asiri), 387–412. India: Springer International Publishing.

80 Tan, N.P.B., Lee, C.H., and Li, P. (2016). Green synthesis of smart metal/polymer nanocomposite particles and their tuneable catalytic activities. *Polymers (Basel).* 8 (4): 105. 1–14.

81 El-Gendy, A.A. (2018). Core/shell magnetic nanoparticles for biomedical applications. In: *Magnetic Nanostructured Materials: From Lab to Fab* (ed. A.A. El-Gendy, J.M. Barandiarán, and R.L. Hadimani), 41–58. United States: Elsevier.

82 Álvarez, M.L., Méndez, A., Rodríguez-Pacheco, R. et al. (2021). Recovery of zinc and copper from mine tailings by acid leaching solutions combined with carbon-based materials. *Appl. Sci. (Switzerland).* 11 (11): 5166. 1–11.

83 Kang, J., Kim, Y., Kim, H.M. et al. (2016). In-situ one-step synthesis of carbon-encapsulated naked magnetic metal nanoparticles conducted without additional reductants and agents. *Sci. Rep.* 6: 38652. 1–9.

84 Lee, J.S., Song, Y.J., Hsu, H.S. et al. (2019). Magnetic enhancement of carbon-encapsulated magnetite nanoparticles. *J. Alloys Compd.* 790: 716–722.

85 Yu, F., Wang, J.N., Sheng, Z.M., and Su, L.F. (2005). Synthesis of carbon-encapsulated magnetic nanoparticles by spray pyrolysis of iron carbonyl and ethanol. *Carbon N Y.* 43 (14): 3018–3021.

86 Darwish, M.S.A., Mostafa, M.H., and Al-Harbi, L.M. (2022). Polymeric nanocomposites for environmental and industrial applications. *Int. J. Mol. Sci.*MDPI 23: 1023. 1–32.

87 Porel, S., Venkatram, N., Rao, D.N., and Radhakrishnan, T.P. (2007). In situ synthesis of metal nanoparticles in polymer matrix and their optical limiting applications. *J. Nanosci. Nanotechnol.* 7 (6): 1887–1892.

88 Pasinszki, T. and Krebsz, M. (2020). Synthesis and application of zero-valent iron nanoparticles in water treatment, environmental remediation, catalysis, and their biological effects. *Nanomaterials.* MDPI AG 10 (5): 917. 1–37.

89 Chen, X., Li, X., Xu, D. et al. (2020). Application of nanoscale zero-valent iron in hexavalent chromium-contaminated soil: a review. *Nanotechnol. Rev.* De Gruyter Open Ltd 9: 736–750.

90 Ma, L., He, H., Zhu, R. et al. (2016). Bisphenol A degradation by a new acidic nano zero-valent iron diatomite composite. *Catal. Sci. Technol.* 6 (15): 6066–6075.

91 He, D., Niu, H., He, S. et al. (2019). Strengthened Fenton degradation of phenol catalyzed by core/shell Fe–Pd@C nanocomposites derived from mechanochemically synthesized Fe-metal organic frameworks. *Water Res.* 162: 151–160.

92 Karim, S., Bae, S., Greenwood, D. et al. (2017). Degradation of 17α-ethinylestradiol by nano zero valent iron under different pH and dissolved oxygen levels. *Water Res.* 125: 32–41.

93 Sabouri, M.R., Sohrabi, M.R., and Moghaddam, A.Z. (2020). A novel and efficient dyes degradation using bentonite supported zero-valent iron-based nanocomposites. *ChemistrySelect* 5 (1): 369–378.

94 Sun, X., Kurokawa, T., Suzuki, M. et al. (2015). Removal of cationic dye methylene blue by zero-valent iron: effects of pH and dissolved oxygen on removal mechanisms. *J. Environ. Sci. Health. Part A Toxic/Hazardous Subst. Environ. Eng.* 50 (10): 1057–1071.

95 Bossa, N., Carpenter, A.W., Kumar, N. et al. (2017). Cellulose nanocrystal zero-valent iron nanocomposites for groundwater remediation. *Environ. Sci. Nano.* 4 (6): 1294–1303.

96 Son, Y.H., Lee, J.K., Soong, Y. et al. (2012). Heterostructured zero valent iron-montmorillonite nanohybrid and their catalytic efficacy. *Appl. Clay Sci.* 62, 63: 21–26.

97 O, Carroll, D., Sleep, B., Krol, M. et al. (2013). Nanoscale zero valent iron and bimetallic particles for contaminated site remediation. *Adv. Water Resour.* 51: 104–122.

98 Gwenzi, W., Chaukura, N., Noubactep, C., and Mukome, F.N.D. (2017). Biochar-based water treatment systems as a potential low-cost and sustainable technology for clean water provision. *J. Environ. Manage.* Academic Press 197: 732–749.

99 Conte, P., Bertani, R., Sgarbossa, P. et al. (2021). Recent developments in understanding biochar's physical–chemistry. *Agronomy.* MDPI AG 11 (4): 615. 1–41.

100 Li, H., Dong, X., da Silva, E.B. et al. (2017). Mechanisms of metal sorption by biochars: biochar characteristics and modifications. *Chemosphere*, Elsevier Ltd 178: 466–478.

101 Zarime, N.A., Yaacob, W.Z.W., Jamil, H. (2018). Removal of heavy metals using bentonite supported nano-zero valent iron particles. In: *AIP Conference Proceedings.* American Institute of Physics Inc., E3S Web of Conferences 101, 04002 (2019) https://doi.org/10.1051/e3sconf/201910104002. ICESD 2019

102 Shu, Y., Ji, B., Cui, B. et al. (2020). Almond shell-derived, biochar-supported, nano-zero-valent iron composite for aqueous hexavalent chromium removal: Performance and mechanisms. *Nanomaterials.* 10 (2): 198. 1–14.

103 Sharma, S.K., Petrusevski, B., and Amy, G. (2008). Chromium removal from water: a review. *J. Water Supply Res. Technol. AQUA* 57: 541–553.

104 Yin, Y., Shen, C., Bi, X., and Li, T. (2020). Removal of hexavalent chromium from aqueous solution by fabricating novel heteroaggregates of montmorillonite microparticles with nanoscale zero-valent iron. *Sci. Rep.* 10 (1): 12137. 1–12.

105 Franci, G., Falanga, A., Galdiero, S. et al. (2015). Silver nanoparticles as potential antibacterial agents. *Molecules.* MDPI AG 20: 8856–8874.

106 Ferdous, Z. and Nemmar, A. (2020). Health impact of silver nanoparticles: a review of the biodistribution and toxicity following various routes of exposure. *Int. J. Mol. Sci.* MDPI AG 21: 2375. 1–31.

107 Muñoz, M., Muraviev, D., Macanás, J. et al. (2013). Metal nanoparticles for microbiological processes View project Catalytic polymeric membrane for process intensification and green chemistry View project Berta Domènech Osram Opto Semiconductors Polymer-Silver Nanocomposites as Antibacterial materials. https://www.researchgate.net/publication/259304795.

108 Jaswal, T. and Gupta, J. (2023). A review on the toxicity of silver nanoparticles on human health. *Mater. Today Proc.* 81 (Part 2): 859–863.

109 Abbaszadegan, A., Ghahramani, Y., Gholami, A. et al. (2015). The effect of charge at the surface of silver nanoparticles on antimicrobial activity against gram-positive and gram-negative bacteria: a preliminary study. *J. Nanomater.* 2015: 720654. 1–9.

110 Akter, M., Sikder, M.T., Rahman, M.M. et al. (2018). A systematic review on silver nanoparticles-induced cytotoxicity: physicochemical properties and perspectives. *J. Adv. Res.*Elsevier B.V. 9: 1–16.

111 Silva, A.O., Cunha, R.S., Hotza, D., and Machado, R.A.F. (2021). Chitosan as a matrix of nanocomposites: a review on nanostructures, processes, properties, and applications. *Carbohydr. Polym.*Elsevier Ltd 272: 118472. 1–13.

112 da Silva Alves, D.C., Healy, B., de Almeida Pinto, L.A. et al. (2021). Recent developments in chitosan-based adsorbents for the removal of pollutants from aqueous environments. *Molecules.* MDPI AG 26: 594. 1–45.

113 Islam, S., Bhuiyan, M.A.R., and Islam, M.N. (2017). Chitin and chitosan: structure, properties and applications in biomedical engineering. *J. Polym. Environ.*Springer New York LLC 25: 854–866.

114 Vrček, I.V., Žuntar, I., Petlevski, R. et al. (2016). Comparison of in vitro toxicity of silver ions and silver nanoparticles on human hepatoma cells. *Environ. Toxicol.* 31 (6): 679–692.

115 Attatsi, I.K. and Nsiah, F. (2020). Application of silver nanoparticles toward Co(II) and Pb(II) ions contaminant removal in groundwater. *Appl. Water Sci.* 10 (6): 152. 1–13.

116 Brigham, C. (2017). Biopolymers: biodegradable alternatives to traditional plastics. In: *Green Chemistry: An Inclusive Approach* (ed. B. Török and T. Dransfield), 753–770. United States: Elsevier Inc.

117 Yu, Z., Wang, W., Kong, F. et al. (2019). Cellulose nanofibril/silver nanoparticle composite as an active food packaging system and its toxicity to human colon cells. *Int. J. Biol. Macromol.* 129: 887–894.

118 Albukhari, S.M., Ismail, M., Akhtar, K., and Danish, E.Y. (2019). Catalytic reduction of nitrophenols and dyes using silver nanoparticles @ cellulose polymer paper for the resolution of waste water treatment challenges. *Colloids Surf., A* 577: 548–561.

119 Xu, Y., Li, S., Yue, X., and Lu, W. (2018). Nanosilver-cellulose antibacterials. *BioResources.* 13 (1): 2150–2170.

120 Haider, M.S., Shao, G.N., Imran, S.M. et al. (2016). Aminated polyethersulfone-silver nanoparticles (AgNPs-APES) composite membranes with controlled silver ion release for antibacterial and water treatment applications. *Mater. Sci. Eng., C* 62: 732–745.

121 Mahmoudi, M., Sant, S., Wang, B. et al. (2011). Superparamagnetic iron oxide nanoparticles (SPIONs): development, surface modification and applications in chemotherapy. *Adv. Drug Delivery Rev.* 63: 24–46.

122 Ponomar, V.P. (2018). Synthesis and magnetic properties of magnetite prepared by chemical reduction from hematite of various particle sizes. *J. Alloys Compd.* 741: 28–34.

123 Escudero-Curiel, S., Acevedo-García, V., Sanromán, M.Á., and Pazos, M. (2021). Eco-approach for pharmaceutical removal: thermochemical waste valorisation, biochar adsorption and electro-assisted regeneration. *Electrochim. Acta* 389: 138694. 1–12.

124 Hu, Q., Jung, J., Chen, D. et al. (2021). Biochar industry to circular economy. *Sci. Total Environ.* 757: 143820. 1–15.

125 Bong, C.P.C., Lim, L.Y., Lee, C.T. et al. (2020). Lignocellulosic biomass and food waste for biochar production and application: a review. *Chem. Eng. Trans. Italian Association of Chemical Engineering - AIDIC* 81: 427–432.

126 Li, X., Wang, C., Zhang, J. et al. (2020). Preparation and application of magnetic biochar in water treatment: a critical review. *Sci. Total Environ.* 711: 134847. 1–14.

127 Wang, S., Gao, B., Zimmerman, A.R. et al. (2015). Removal of arsenic by magnetic biochar prepared from pinewood and natural hematite. *Bioresour. Technol.* 175: 391–395.

128 Shang, J., Pi, J., Zong, M. et al. (2016). Chromium removal using magnetic biochar derived from herb-residue. *J. Taiwan Inst. Chem. Eng.* 68: 289–294.

129 Biswal, S.K., Panigrahi, G.K., and Sahoo, S.K. (2020). Green synthesis of Fe_2O_3-Ag nanocomposite using Psidium guajava leaf extract: an eco-friendly and recyclable adsorbent for remediation of Cr(VI) from aqueous media. *Biophys. Chem.* 263: 106392. 1–8.

130 He, S., Zhong, L., Duan, J. et al. (2017). Bioremediation of wastewater by iron oxide-biochar nanocomposites loaded with photosynthetic bacteria. *Front. Microbiol.* 8 (May): 1–10.

131 Lee, S.Y. and Park, S.J. (2013). TiO2 photocatalyst for water treatment applications. *J. Ind. Eng. Chem.* 19: 1761–1769.

132 Fernandes, A., Makoś, P., Wang, Z., and Boczkaj, G. (2020). Synergistic effect of TiO_2 photocatalytic advanced oxidation processes in the treatment of refinery effluents. *Chem. Eng. J.* 391: 123488. 1–14.

133 Serpone, N. (2018). Heterogeneous photocatalysis and prospects of TiO_2-based photocatalytic DeNOxing the atmospheric environment. *Catalysts*. MDPI 8 (11): 553. 1–99.

134 Celis, R., Hermosín, M.C., Carrizosa, M.J., and Cornejo, J. (2002). Inorganic and organic clays as carriers for controlled release of the herbicide hexazinone. *J. Agric. Food. Chem.* 50 (8): 2324–2330.

135 Khan, S., Ul-Islam, M., Khattak, W.A. et al. (2015). Bacterial cellulose-titanium dioxide nanocomposites: nanostructural characteristics, antibacterial mechanism, and biocompatibility. *Cellulose.* 22 (1): 565–579.

136 Mondal, K., Sharma, A. Photocatalytic oxidation of pollutant dyes in wastewater by TiO_2 and ZnO nano-materials-a mini-review non gassing osmotic pump view project liquid drop impact on liquid bath view project photocatalytic oxidation of pollutant dyes in wastewater by TiO_2 and ZnO nano-materials-a mini-review. https://www.researchgate.net/publication/271512331.

137 Sun, X., Yin, L., Zhu, H. et al. (2022). Enhanced antimicrobial cellulose/chitosan/ZnO biodegradable composite membrane. *Membranes (Basel).* 12 (2): 239. 1–13.

138 Ahmad, N., Sultana, S., Faisal, S.M. et al. (2019). Zinc oxide-decorated polypyrrole/chitosan bionanocomposites with enhanced photocatalytic, antibacterial and anticancer performance. *RSC Adv.* 9 (70): 41135–41150.

139 Jadoun, S., Verma, A., Ashraf, S.M., and Riaz, U. (2017). A short review on the synthesis, characterization, and application studies of poly(1-naphthylamine): a seldom explored polyaniline derivative. *Colloid Polym. Sci.*Springer Verlag 295: 1443–1453.

140 Máynez-Navarro, O.D. and Sánchez-Salas, J.L. (2017). Focus on zinc oxide as a photocatalytic material for water treatment. *Int. J. Biodegrad.* 106: https://www.researchgate.net/publication/329160918.

141 Yong, H.E., Krishnamoorthy, K., Hyun, K.T., and Kim, S.J. (2015). Preparation of ZnO nanopaint for marine antifouling applications. *J. Ind. Eng. Chem.* 29: 39–42.

142 Mintova, S., Gilson, J.P., and Valtchev, V. (2013). Advances in nanosized zeolites. *Nanoscale.* 5 (15): 6693–6703.

143 Mintova, S., Jaber, M., and Valtchev, V. (2015). Nanosized microporous crystals: emerging ap-plications. *R. Soc. Chem.* 44 (20): 7207–7233. https://hal.archives-ouvertes.fr/hal-03154304.

144 Karmaker, S.C., Eljamal, O., Saha, B.B. et al. (2021). Optimization of strontium removal process from contaminated water using zeolite nanocomposites optimization of strontium removal process from contaminated water using. https://doi.org/10.21203/rs.3.rs-309368/v1.

145 Baybaş, D. and Ulusoy, U. (2011). Polyacrylamide-clinoptilolite/Y-zeolite composites: characterization and adsorptive features for terbium. *J. Hazard. Mater.* 187 (1–3): 241–249.

8

The Use of Sol–Gel Technology to Produce Nanomaterials for Air and Water Purification

Andrei Rex and João Henrique Zimnoch dos Santos

Universidade Federal do Rio Grande do Sul, Instituto de Química, Departamento de Química Inorgânica, Av. Bento Gonçalves, Porto Alegre CEP 91501-970, RS, Brazil

8.1 Introduction

For environmental conservation, a set of mandatory interdisciplinary actions is mandatory, including the management of purification agents that may act as optimal substances that can remove or transform pollutants by degrading them and adsorbing or submitting them to any other process. In this context, nanotechnology has provided alternative approaches in the context of environmental management. An optimal substance may be a nanomaterial that must be as efficient as possible in terms of the relationship between the purified volume and the quantity of nanomaterials. In addition, it is expected to produce little to no toxic residues (including itself) and have low costs. In other words, because the purification agent is friendly to large-scale production, it will also be environmentally friendly, i.e. industrial production of nanomaterials for purification may imply a green process and the use of available materials. With this demand, the sol–gel process couples the need for an inert abundant material, which usually involves silica-based materials and their precursors, for support material and/or enables specific nanomaterial synthesis and enhances its usage by improving thermal and stability properties using the sol–gel process.

As seen through decades, the development and application of sol–gel processes have been inexpensive and versatile alternatives in the field of materials synthesis, since they favor complex stoichiometry material synthesis [1], and they produce environmentally friendly composites while using a silica matrix for the support material [2]. Such an approach provides an outstanding support with excellent quality material that, e.g. as a core–shell structure, can be further modified using other organosilanes [3] and even inserting metals to react with other contaminants [4] beyond the target one. As a support material, it enhances the catalyst activity and/or selectivity for determined compounds [5] with different conformations of the silica structure as support materials, e.g. core–shell structures, hollow

Nanomaterials for Air and Water Purification, First Edition.
Edited by Deepu A. Gopakumar, Jyotishkumar Parameswaranpillai, Jinu Jacob George, and Midhun Dominic C.D.

shell nanospheres (HSNs), microcapsules, and other immobilization techniques. These porous structures can also prevent the self-sintering of a nanomaterial [6]. Such approaches favor the synthesis of metal–oxide frameworks (MOFs) [7]; in addition, the immobilization of sensitive biocatalysts to degrade biological contaminants [8] provides selective oxidation of organic drinking water pollutants such as diclofenac, carbamazepine, and paracetamol [9] and adds magnetic particles to better recover purification agents, especially in water bodies [10]. The sol–gel process also promotes control over the porosity of the final material and particle size [11]; hence, it is linked to a more adjusted purification process due to the ease of characterization of the resulting materials.

In the matter of application in air and water purification, as previously mentioned, the sol–gel process can be applied to even complex nanomaterial synthesis. In the last three years, it has been largely applied to photocatalysis, specifically in water bodies for purification. Hence, many applications of sol–gel in water and air purification will be shown in this chapter, starting with the use of sol–gel to synthesize photocatalysts for water purification, followed by air purification, and finally other purification of both media for other applications, e.g. detection, filtration and separation.

Over the years, heterogeneous photocatalysts have shown great potential for water purification or treatment [12]. Such catalysts commonly use ultraviolet radiation by absorbing photons to produce free radicals. Many free radicals, such as hydroxyl ($HO^{\cdot}$) are highly reactive due to unpaired electrons. A common semiconductor for heterogeneous photocatalysis is TiO_2, which forms electron–hole pairs on its electronic structure, i.e. it forms oxidation and reduction sites on its structure and produces radicals such as hydroxyl and $O_2^{\cdot-}$.

$$TiO_2 \xrightarrow{h\nu} e^- + h^+ \tag{8.1}$$

Afterwards, the covalent band electrons and holes (h^+) are entrapped. Then, oxygen is oxidized:

$$O_2 + e^- \rightarrow O_2^- \tag{8.2}$$

$$h^+ + OH^- \rightarrow OH^{\cdot} \tag{8.3}$$

The photodegradation with the oxidized hydroxyls is described as follows:

$$R - H + OH^{\cdot} \rightarrow R'^{\cdot} + H_2O \tag{8.4}$$

Finally, the holes and electrons that act on the dye pollutants are described:

$$\text{Dye} + h^+ \rightarrow \text{Oxidation products} \tag{8.5}$$

$$\text{Dye} + e^- \rightarrow \text{Reduction products} \tag{8.6}$$

Recent reports using TiO_2 are highlighted and detailed in this chapter to provide a better understanding of each synthesized material and its structure. This solid heterogeneous catalyst is highly used due to its abundance, easy TiO_2 synthesis, low cost, and highly appreciable photocatalytic activity. Moreover, titanium dioxide generates H_2, which can be used for self-cleaning and applied to ceramics and solar cells.

This catalytic activity is directly related to its morphology in terms of its surface area, crystanility, particle size, gap energy, and absorption spectrum. Consequently, many studies aim to study its nanostructure using different processes for its synthesis [13]. In this aspect, the sol–gel method is commonly used to enhance the activity and control of these aspects; thus, many authors prepared TiO_2 using the sol–gel method to purify air and water.

8.2 Water Purification

The urgent need for more efficient methods to degrade water contaminants has stimulated research on the use of TiO_2. Recently, Rawal et al. [14] introduced this material into thin films prepared via sol–gel to degrade acetaminophen in water. The nontoxic and cost-efficient material is used as a faster and more efficient method of purification as an alternative to typical water filtration. The materials of multiple layers were prepared via the sol–gel method, deposited onto glass slides, and inserted into a UV-photocatalytic reactor using aluminum slides to determine their activity toward acetaminophen under different pH conditions: acidic, basic, and neutral. As a result, the higher surface area of the thin TiO_2 films, which implies more active sites, provided a greater rate of acetaminophen degradation and an optimized result at neutral pH with 10 layers of TiO_2 catalysts on the film. However, the authors noticed that an appreciable result depended on the initial concentration of the contaminant, the maximum quantity of catalyst deposited on the slides, and the irradiation sources. The authors affirmed that a potential enhancement in the usage of such films was doping a suitable metal with TiO_2, reducing the band gap and absorbing more visible light. For further investigations, the authors aim to eliminate the low quantum efficiency during the band gap recombination of the electron–hole. Thus, other semiconductors may have synergy with TiO_2, such as ZnO and CdS.

As photocatalysts, TiO_2 nanostructures are versatile materials for air and water purification; they enable modifications with other materials for an even more efficient reaction. For instance, Liu and coworkers [15] added graphene oxide (GO) and GO composite materials to evaluate their photocatalytic activity. Titanium dioxide was prepared through sol–gel using titanium butoxide as a precursor and subsequently combined with GO via hydrothermal synthesis. The resulting titanium dioxide on the surface of GO was shown to be uniformly dispersed. As a result, the titania covers a larger area than the GO surface, so this material has excellent adsorption capacities and does not decrease its photocatalytic activity; it even enhances its photocatalytic activity due to the greater area coverage, as tested for sewage treatment. The catalyst was only tested to degrade rhodamine B (RhB), which is a common dye pollutant that can develop toxicity [16, 17] and even be carcinogenic [17, 18]. Nonetheless, it displays the importance of the development of nanomaterials for water treatment of dye pollutants that are far more diverse than RhB, which requires more selective materials added to TiO_2. It illustrates the applicability of sol–gel for materials with a larger range of actuation and uniformity of active sites on the surfaces of these purifying agents.

As urea (nonmetal doping), nitrogen (N) doping increases the TiO_2 range of photoresponse, as stated by Liu et al. [15]. Complementarily, Birlik and Dagdelen [19] used N doping, manganese (Mn) doping, and co-doping of N and Mn to increase the TiO_2 photocatalytic efficiency. The authors used doping to overcome a common problem with this specific photocatalyst, which can be avoided with morphological modifications. The problem consists of the recombination of the cited photogenerated electron–hole pairs, which is partly solved with an increase in surface area or chemical modifications through other components; the latter is the option used by Birlik and Dagdelen. As a result, under UV radiation, the nanoparticles (NPs) of co-doped Mn-N TiO_2, which were synthetized via sol–gel with $TiCl_4$ as a precursor, demonstrated 99% efficiency toward blue methylene, which is better than undoped titanium dioxide with 79.4% efficiency. The Mn-doped catalyst showed 86.3% efficiency. The results prove the hypothesis that adding an acceptor and donor dopant of Mn substitutes Ti atoms and nitrogen for O atoms, respectively, which are n-type and p-type dopants and result in different orbital energies in the titanium dioxide structure.

As seen before, a larger surface and greater photoactivity in UV or visible light are indeed desired for water purification through photocatalysis. Seo et al. presented the synthesis of thin films of titania through the sol–gel method using polystyrene spheres (PSs) [20]. The studied nanosized material, i.e. PS-patterned TiO_2 films, is applied to malachite green (MG) degradation through photocatalysis measured by UV–visible absorptiometry. In the synthesis, first, the nanospheres are prepared; then, sol–gel catalysis is conducted using titanium hydroxide oxide as a precursor mixed with hydrochloric acid (HCl); finally, PS-patterned and nonpatterned TiO_2 thin films are synthesized by adding titania sol–gel in the PS-patterned silicon wafers. The morphology of the final material with different diameters of nanospheres is visualized in Figure 8.1, where the authors used FE-SEM and stated that the 400-nm samples had higher surface charges. The resultant patterned material displayed appreciable results, where the PS-patterned TiO_2 removed 98% of the MG solution, and as the authors demonstrated, a 20-fold faster decomposition was obtained. Seo and coworkers concluded that the catalyst formed by the 400-nm

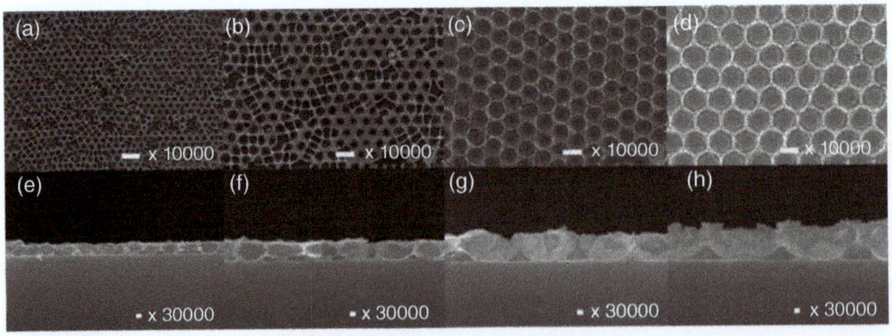

Figure 8.1 Morphology of the samples through FE-SEM, displaying TiO_2 films with PS nanospheres of 400 nm (a, e), 700 nm (b, f), 1000 nm (c, g), and 1300 nm (d, h). Source: Seo et al. [20]/MDPI/CC BY 4.0.

PS spheres, which had the optimal result, was an excellent candidate in the field of water purification through photocatalysts. In addition, it was a cheap process with low synthesis complexity and removed 50% of phenol from water in the first two hours of the process using the same catalyst.

A simple and smart strategy to selectively and efficiently photodegrade pesticides using only TiO_2 as the photocatalytic material through sol–gel was developed by Fiorenza et al. [21]. The method dispensed solvents, which are typical of molecular imprinting (MI) methods without added support reactants. The target organic contaminants are the herbicide 2,4D and insecticide imidacloprid. The synthesis followed a molar ratio of 5 : 1 for TiO_2:pesticide (for one pesticide) and was dissolved in an acid-catalyzed sol–gel process. During this experimental section, through a calcination process in air that used a heating ramp of 2 °C/min up to 500 °C in six hours, the authors separated the pesticide from TiO_2 and obtained different templated materials for the two different pesticides. The authors also synthesized the materials without removing the organic components and bare used TiO_2 for comparison. The mechanism of the MI materials for water purification from pesticides is shown in Figure 8.2. The MI samples displayed superior degradation of their respective templated pesticides and were tested on contaminants of similar structures, such as 1,4-dichlorobenzene and *o*-phenylphenol. 1,4-dichlorobenzene is similar to 2,4D, and *o*-phenylphenol is different from both. Thus, the MI samples displayed similar behavior to bare TiO_2. Moreover, the authors tested the toxicity of titania powders using a zebrafish embryotoxicity test, which proved the non-toxicity of the described materials and provided a glimpse of future applications of environmentally friendly and selective molecular imprinted photocatalysts, whose templating with the target organic component may not be limited to the studied pesticides due to the size and morphology control of the sol–gel process.

Mesoporous silica via the sol–gel process is a relevant support material for heterogeneous catalysts by controlling its size and pore diameters. It can be synthesized

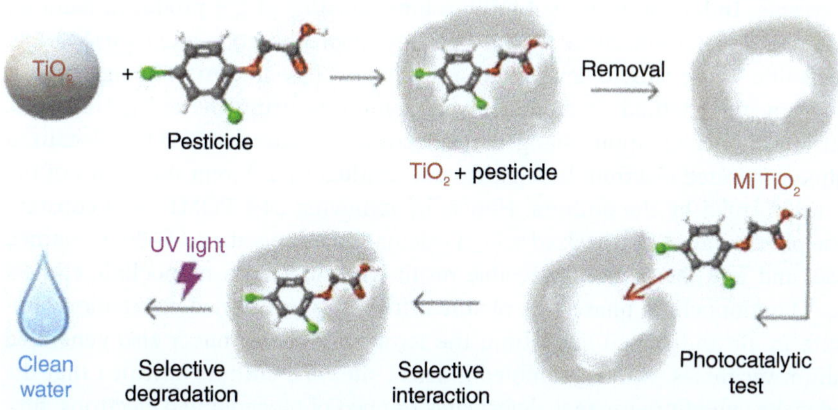

Figure 8.2 A graphical representation of the molecularly imprinted pesticide synthesized titanium dioxide for increased selectivity over the pesticide. Source: Fiorenza et al. [21]/with permission of Elsevier.

in the format of microspheres; in addition to its compact format, it is a support for nanosized materials, provides resistance to external influences, such as enzymes that are very sensitive and need well-adjusted procedures for their usage, and gains thermal stability and the mentioned recyclability of applied materials. From this format of mesoporous silica perspective, Kalu et al. used titanium dioxide-coated SBA-16-type (an ordered mesoporous silica structure) mesoporous silica microspheres and tested the degradation of crystal violet, another dye that can be found in water [22]. The SBA-16 usage is investigated in regard to its 3D cubic body-centered format arrangement of its mesoporous structure, which forms a structure similar to a cage in the matter of its pores, so it is a crystalized system. The synthesis used tetraethyl orthosilicate (TEOS), which is a common sol–gel precursor, and Pluronic F127, which is a structure-directing agent, as a template. Titanium dioxide nanocrystals were used in the surface modification of the synthesized SBA-16-type silica, which resulted in TiO_2-coated SBA-16 silica microspheres. The resulting material presented a relatively high surface area, an ordered porous nature and anatase – one of the three mineral forms of TiO_2 – nanocrystals in the coat with a proportion of 15 Si to 1 TiO_2. A combination of the anatase crystals, which are known for their greater effectiveness in photocatalysis, and high surface area resulting from the mesoscopic structure guaranteed accelerated N-demethylation reactions. Hence, this method presented appreciable and superior results in the photodegradation of crystal violet due to this organic component's sensitivity to light and offered a method of suspending particles for water purification.

As previously mentioned, the material prepared by sol–gel displays a safe method for film synthesis. However, more inorganic materials can be prepared via sol–gel to produce a material that is not limited to ultraviolet radiation for photocatalysis but also operates with UV-visible light radiation. In this context, Saputera et al. synthesized bismuth vanadate (V) ($BiVO_4$) to degrade palm oil mill effluent (POME) [23], a common and highly polluting waste from the palm oil industry and a product with low production costs and a higher yield in the measure of millions of tons (for example, India consumes 9.4 million tons and 19% of the produced palm oil worldwide). The authors chose the sol–gel above coprecipitation and hydrothermal methods due to the simple equipment, low cost, and desired morphology control provided by the method. Nonetheless, the synthesis using the sol–gel method, specifically after calcination, changed the electronic structure of $BiVO_4$, delocalized the photogenerated electron–hole pairs, and resulted in different mobilities of the pairs, as affirmed by the authors. Hence, by removing 24% POME on a constant tax, the photocatalyst synthesized with a calcination temperature of 450 °C, named BV-450, and had the most appreciable result while having a monoclinic clinical phase. The monoclinic phase was obtained from the mentioned calcination temperature in air and transitioned from the tetragonal phase, but it also generated vanadium vacancies, which promoted a defect site. The authors affirmed that the high-density vanadium vacancy defect sites trapped photogenerated electrons, promoted efficient charge separation, and increased the photocatalytic performance. This catalyst was prepared through sol–gel and may provide a method and design for potential future $BiVO_4$ photocatalysts. The sol–gel method provided an alternative to the traditional powder method and showed the optimization of the materials during calcination and changes in optical properties; thus, the morphology and phase transformation control are highlighted by the sol–gel method.

The detection of water pollutants is also important before applying the purification agent, but developments in this subject benefit the removal or degradation of nanomaterials, since the precise and inexpensive detection helps comprehend the species of pollutants in water bodies and their quantity.

Tajik et al. noted that electrochemical sensor platforms based on nanomaterials made significant advancements in the detection of toxic and poisonous pollutants and investigated many reports regarding polymer nanocomposites applied in electrochemical sensors [24]. Briefly, electrochemical sensors generally use redox reactions for a given analyte in an electrolyte; therefore, they exchange electric signals. However, to surpass the nonreproducibility of surface behavior of these sensors and slower kinetics of the mentioned electrochemical reactions in the surface and overpotential requirements, the authors added nanomaterials to the surface of these sensors; more specifically, polymer nanocomposites were used to surpass the surface issues and bring their superb transducer properties as electroanalytical sensors. Among the methods of polymer nanocomposite synthesis, the sol–gel had several reports cited by the authors, and the typical route for the desired composite is displayed in Figure 8.3.

To detect heavy metals in the environment, Khan et al. apud Tajik et al. [24] used the sol–gel method to prepare poly(2-methylaniline) (P2MA)-$Ce_2(WO_4)$ immobilized on carbon nanotubes (CNTs) to detect cadmium ions (Cd^{2+}) on a Nafion-bonded glassy carbon electrode (GCE). The nanocomposite-based electrochemical sensor displayed a sensitivity of approximately 5.138 μM^{-1} cm^{-2} along with a higher selectivity to Cd^{2+} and a linear dynamic range from 1.0 to 0.1 M, according to the authors. The sol–gel method was also reported by Faisal et al. apud Tajik et al. [24] to prepare polythiophene (PTh) on zinc oxide (ZnO) (PTh/ZnO) modified in GCE and applied to hydrazine detection. There is a synergistic effect between metal oxide and conducting polymer for catalysis; the catalyzed reaction generates an electric signal; hydrazine is detected in concentrations of 0.5–4.8 μM with an

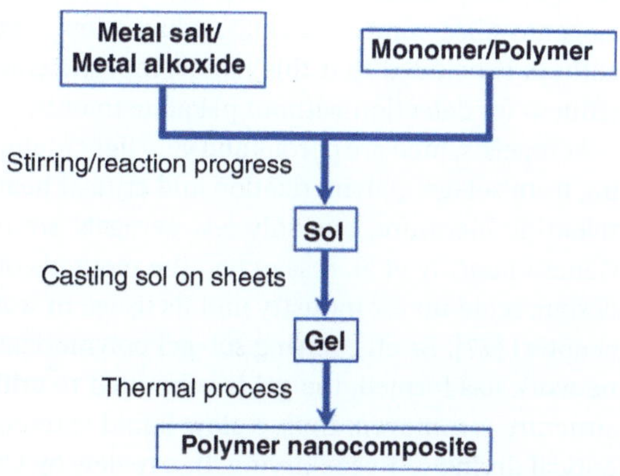

Figure 8.3 Route used by Tajik et al. to obtain the polymer nanocomposite through a sol–gel process. Source: Tajik et al. [24]/American Chemical Society.

appreciable sensitivity of 1.22 $\mu A\ \mu M^{-1}\ cm^{-2}$, a lower detection limit of 0.207 μM and a response faster than five seconds with superb selectivity and long-term stability.

Stieberova et al. studied NPs that positively impact the environment, i.e. the well-cited TiO_2, ZnO, ZrO_2, and $LiFePO_4$, for comparison of their production methods and applicability [25]. The scope of their research went toward the comprehension of continuous-flow hydrothermal synthesis (CFHS) in techno-economic and life cycle environmental evaluation. Thus, the sol–gel method to prepare NPs was presented for comparison. However, the sol–gel was used to synthesize TiO_2, and the authors highlighted its self-cleaning capacities as a photocatalyst. The method provided a similar range of NPs as the CFHS results. The authors appointed sol–gel and other technologies as being equal to CFHS in terms of quality parameters due to their higher environmental impact and low production costs but insufficient productivity on an industrial scale.

As seen above, improvements in detection methods and materials through electrochemical signals are appealing, and innovation in this scope of water purification may be in electrochemical oxidation methods. Nsubuga et al. reported a degradation method to detect beta-blockers (propranolol and acebutolol) using a heterogeneous electro-Fenton reactor [26]. The authors studied the detection of β-blockers due to the emerging presence of these ecotoxicological pollutants in drinking, ground, waste, and surface water that affected plants and aquatic organisms; for the latter, electrochemical advanced oxidation processes (EAOPs) are appealing to remedy these organisms. Moreover, EAOPs are environmentally friendly and have a high degradation efficiency, very low toxicity, and process complexity. Hence, Nsubuga and coworkers developed a biogenic iron-carbon nanocomposite to detect β-blockers in hospital wastewater using a droplet flow-assisted heterogeneous electro-Fenton reactor (DFEF) system. Electro-Fenton processes are new and efficient EAOPs for water treatment that completely oxidize refractory organic pollutants. The Fe–C nanocomposite was synthesized using the sol–gel method and rice husk silica. The resultant sol–gel solutions were unmodified or iron-modified nanocomposites in sol–gel solutions and subsequently aged until gelation. The supported iron–carbon composite catalysts resulted in fast kinetics for H_2O_2 production; they could perform the EF process at neutral pH and operate on an optimal process at minimal experimental runs; among these advantages, the biogenic silica composite catalyst had a low cost and good reusability. Finally, the authors concluded that this nanocomposite could be used in any real wastewater effluent for detection without pH adjustments.

Aerogels, which are porous and very light materials consisting of 90–99% air resulting from sol–gel polymerization and critical heating, are well known in the sol–gel scientific literature, but only few aerogels are on the market. With this premise, Ganesamoorthy et al. researched the methods of aerogel preparation, the reactors' design, scale-up for industry and its usage in water treatment (for the scope of this chapter) [27]. Briefly, during sol–gel polymerization, after the tridimensional silica network has formed, the gel is submitted to critical heating, and the liquid on the structure is removed; thus, a slow liquid extraction can be performed using supercritical drying or freeze-drying. In a review by Ganesamoorthy et al. aerogels based on alginate, biochar, cellulose, chitosan, gelatin, carbon, MOF, polymer, and silica were investigated for water treatment.

Among the reported works by Ganesamoorthy, alginate-based hybrid aerogels are highlighted. These materials were synthetized from sodium alginate combined with TiO_2 (AA-1) and sodiumalginate-meso-2,3-dimercaptosuccinic acid (AA-2) and demonstrated appreciable applicability for water treatment. The first material, AA-1, which was reported by Dai et al. and Ganesamoorthy et al., displayed a photocatalytic degradation of 85% after six cycles for methyl orange, and AA-2, which was investigated by Wang et al. and Ganesamoorthy et al. demonstrated more than 90% heavy metal ion removal and was highly selective for Pb^{2+} and Cd^{2+}. Other water treatment aerogel materials were also reported, highlighting cellulose-based aerogels from bacterial cellulose (BC); hybrid BC and Ag_2O NPs for photocatalytic and oil removal applications; hydrophilic raw cellulose-based aerogels, of which application for water treatment depends on hydrophobic modifications. Finally, the authors made the required effort to apply aerogels to water purification residues in reactor designs for a greater scale of production, stability, and recyclability of these materials.

8.2.1 Air Purification

Although titanium dioxide NPs demonstrate appreciable results for photocatalysis, they are limited to their difficulty of separation after their reactions with the pollutants, i.e. aggregation in solution. Ansari et al. explored the usage of TiO_2 as nanofibers (NFs) through electrospinning (ES) for antibacterial and antibiofilm against gram-positive and gram-negative bacteria, methicillin-resistant *Staphylococcus aureus* (MRSA) and *Pseudomonas aeruginosa*, respectively [28]. The authors affirmed that TiO_2 NFs were an alternative to NP aggregation. However, even with the benefits of a mesoporous structure and high surface, stability in solution, and very low aggregation, they still have low efficiency in photocatalysis without UV light. This occurs due to the large band gap energy in their structure and the low-ordered crystalline structure, as opposed to the previously demonstrated efficient anatase structure [22]. Hence, the authors explored the TiO_2 NF application for bactericide application, since the NP counterpart displayed superb features on the matter using UV light, and NFs have rarely been explored in the past. The NFs were prepared through sol–gel and ES. First, the prepared TiO_2 solution was prepared using titanium isopropoxide (IV) as a precursor. Then, the solution was mixed with PVP to obtain a TiO_2/PVP sol–gel, which was subsequently set for the ES experiment on a plastic syringe, resulted in electrospun TiO_2 NFs and finally calcinated in a mix of 25% air/75% argon. The mix displayed greater antibacterial and antibiofilm activity with minimum values of inhibitory concentration vs. minimum bactericidal concentration of 3 and 6 $mg\,ml^{-1}$, respectively. For the gram-positive strain, the concentrations were 6 and 12 $mg\,ml^{-1}$ in the previous comparison. Moreover, the mixture displayed biofilm formation inhibition of 75.2% for MRSA and 72.3% for *P. aeruginosa* using 2 $mg\,ml^{-1}$, as the authors stated. Finally, the report concluded that TiO_2 NFs showed potential application for food packing and coating for diverse inanimate objects, wastewater treatment and purification (as demonstrated), which prevented bacterial proliferation and biofilm formation.

A well-known process of air purification consists of photocatalytic oxidation (PCO) to remove many air pollutants, particularly volatile organic compounds (VOCs). Motivated by improving the air quality in many spheres, such as well-insulated building air cleaning, Mamaghani et al. studied the TiO_2 synthesis using sol–gel, flame hydrolysis, water-in-oil microemulsion, chemical vapor deposition, solvothermal and hydrothermal treatments [29]. Although hydrothermal and solvothermal methods were their focus, the sol–gel method was applied and compared, but the resultant material was also formed of NPs. The sol–gel method was cited by the authors as part of the reaction medium for isopropanol as the solvent, with ethanol also being present. Their work references previous research from Maira et al. [30], which diverged in particle size growth and affirmed that increasing the isopropanol in the reactor increased the primary particle size. Maira and coworkers provided important data regarding the sol–gel synthesis of anatase TiO_2 catalysts, synthesis adjustments, and relation between primary and secondary TiO_2 particles. Briefly, from [30], the maximum degradation for trichloroethylene (TCE) using anatase TiO_2 from gel spheres after controlled thermal and hydrothermal treatments was seen using primary particles with a size of 7 nm, and larger crystals with smaller particles provided a higher surface area and greater degradation of TCE. Moreover, after transmission electron microscopy (TEM) images, Mamaghani et al. noticed that the presence of water during sol–gel synthesis rapidly decreased the secondary particle sizes. As a conclusion after characterization and comparison to other previous reports, the authors highlighted that as potential suggestions for future research, crystalline and structure properties could be mediated by the reaction medium's pH, water-alcohol usage as solvent and calcination temperature. Finally, titanium precursors in the presence of fluoride ions during hydrothermal treatment enable the fabrication of hollow spheres, and the authors suggest that for higher adsorption ability, charge carrier lifetime and pollutant removal, the NPs can be coupled with graphene for greater efficiency.

8.2.2 Air and Water Purification

In a wider range of applications, comprising both air and water purification, Obregón and Rodríguez-González presented an overview of deposition techniques based on sol–gel methods to prepare TiO_2 films and coatings for applications in both mediums [31]. The authors explored the fundamental sol–gel methodologies, including dip-coating, spin coating, spray hydrolysis, and electrophoretic sol–gel deposition, where the resultant material was applied to purify water and air from many pollutants, especially in megacities and public transport and public spaces. In addition to the cited advantages of the sol–gel process by itself, i.e. its complexity, homogeneity control of surface and porosity, and low costs, the authors highlighted the low temperature during synthesis and possibility of large-scale production.

Furthermore, Obregón and Rodríguez-González cited TiO_2 thin films and coatings applied in the degradation of water and air pollutants but also conducted research on heavy metal treatment through photocatalytic treatment. The treatment chemically reduces heavy metal ions into a zero-valent state through photogenerated electrons, but this process is not well studied, with Cr(VI) or Cr(III) being the most researched. From their review, Kajitvichyanukul et al. apud Obregón and Rodríguez-González [31] demonstrated that anatase TiO_2 thin films displayed the most appreciable photocatalytic behavior while reducing Cr(IV) ions. In addition, the heavy metal reduction through photocatalysis depends on the pH of the medium, since heavy metal species distribute themselves differently according to pH. Other applications, such as hydrogen production via water splitting using titanium dioxide – anatase's bottom edge conduction band, are more negative than the reduction potential of hydrogen, photoreduction of CO_2, gas sensing, and inactivation of harmful microorganisms. Figure 8.4 demonstrates the specificity of their research.

Moreover, nanostructures of TiO_2 as photocatalysts are versatile materials for air and water purification since they enable modifications with other materials for an even more efficient reaction. Such versatility may be acquired by adding other metals. For instance, Tegze et al. modified a titanium dioxide coating prepared through sol–gel with silver, so they prepared different pore structures from the mesoporous TiO_2 coat [32]. The photoactivity was measured on rhodamine 6G and methylene blue dye molecule degradation under UV and visible light at liquid–solid and air-solid interfaces, i.e. an application for both water and air purification. These reaction conditions were also modeled and enabled the self-cleaning of the TiO_2 coatings. The authors used $AgNO_3$ as the silver source and studied how it modified the mesoporous TiO_2 sol–gel coatings and their functionalities using two methods: mixing and impregnation. As a mix, the method consisted of adding $AgNO_3$ to the precursor sol, and the impregnation method consisted of immersing the samples in 0.03-M and 1-M $AgNO_3$ solutions. The titanium dioxide coating synthesis on glass

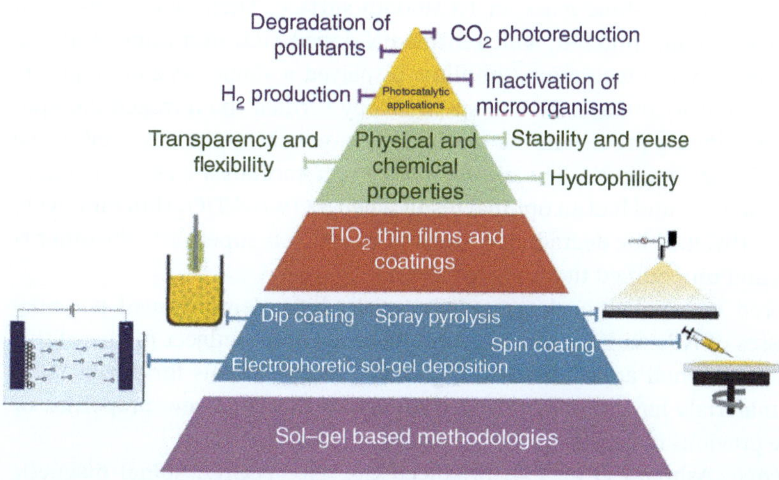

Figure 8.4 A representation of studies on titanium dioxide thin films or coatings synthesis from the sol–gel to the study of its most import characteristics (regarding surface properties and stability) and applications. Source: Obregón et al. [31]/Springer Nature.

was performed in quartz glass, and the sol–gel spin-coated Si substrates were first coated with an SiO_2 barrier layer, which inhibited Na^+ diffusion and prevented a decrease in photoactivity. Each sample was prepared using a different surfactant, Pluronic P123 and cetrimonium bromide (CTAB); then, the resultant titanium dioxide samples were named p-TiO_2 and c-TiO_2, respectively. The surfactants generated different pore structures from these pore-forming agents and were applied to degrade rhodamine 6G and blue methylene. The authors also prepared silver-modified TiO_2 coatings using the impregnation method. As a result, the silver modification increased the photoactivity at the liquid–solid interface, which synergized with titania. However, in an air-solid interface, specifically for methylene blue, silver content was presumably less important, and due to slower mobility in the air medium in contrast to liquid medium, the titanium dioxide content on the surface was insufficient for dye degradation. Coatings prepared through impregnation displayed superior functional properties among the other prepared samples, given their significantly lower Ag content. Finally, during the impregnation method, the quantity of Ag in the pores depends on the pore sizes, and the dye association is also influenced by the pore structure of the coatings. The authors presume that dye-pore wall interactions are inhibited by the Ag on the surface, which increases the dye's mobility during irradiation and their tendency to associate.

A report of nitrogen-doped TiO_2 thin-film synthesis for methylene blue degradation through the sol–gel method was made by Vallejo et al. using urea as the nitrogen source [33]. During the synthesis, titanium tetraisopropoxide and urea were the reactants in the sol–gel method: the former for TiO_2 and the latter for the N source. The authors prepared samples using different loads of urea, i.e. 1%, 3%, and 5% (w/w), and another unmodified TiO_2 powder. The samples were deposited using the referred doctor blade technique: a ceramic powder slurry, which is the N-doped TiO_2 powder in this case, with solvent and additives was cast on a smooth surface. The authors chose soda-lime glass for its smooth surface. Then, it was heated to remove the solvent and sintered, which formed N-doped TiO_2 thin films. With the X-ray diffraction (XRD) technique, the films displayed a single crystalline phase, which corresponds to anatase. Raman spectroscopy verified the nitrogen incorporation after the doping process in the crystal lattice, which can be observed in the light shift in E_{g1} in Figure 8.5. The nitrogen-doped photocatalyst was compared to the nonmodified one and had an optimal result when 5% (w/w) TiO_2 thin films were doped: the methylene blue degradation was 26.5%, which is superior to the other N percentages and unmodified titanium dioxide.

Nanotailored materials, i.e. decorated materials, have demonstrated potential for new classes of NPs with enhanced properties and, in the subject matter of this chapter, environmental applications of highly efficient materials for purification. Decorated materials have other NPs on the surface that add new properties or improve the previous material.

In this scope, Ashtiani et al. [34] produced $Co_{0.5}Ni_{0.5}FeCrO_4$ spinel magnetic nanoparticles (SMNPs) decorated with a nanocomposite, which was synthesized from GO, oxidized single-walled carbon nanotubes (O-SWCNTs) and the organic frameworks (MOF) of UiO-66-NH_2 and UiO-66-COOH; they tested them in the

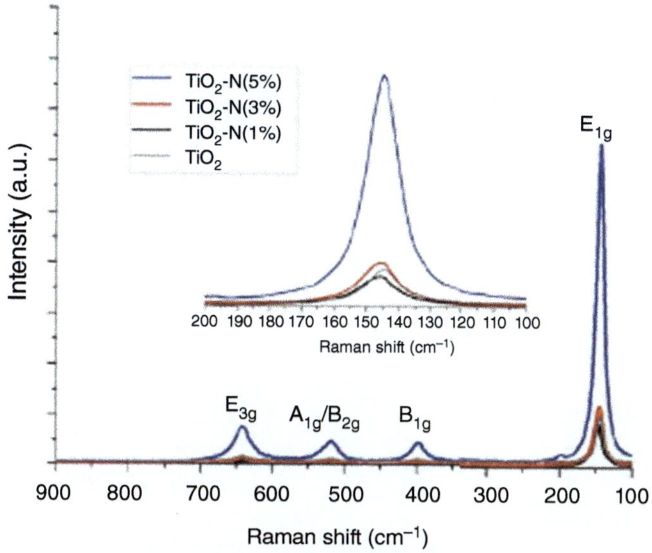

Figure 8.5 Raman spectrum where the anatase form of TiO_2 can be visualized by its typical signals at E_{1g}, B_{1g}, A_{1g}/B_{1g}, and E_3 and an inside figure with the signal intensity of different N-doped TiO_2 thin films. Source: Vallejo et al. [33]/IOP /CC BY 3.0 /Public domain.

reduction of 4-nitrophenol to nontoxic 4-aminophenol. Spinel NPs are widely employed as catalysts for water pollutant purification and have magnetic, optical, electrical, and catalytic properties of interest [35]. These magnetic NPs had interesting properties, but Ashtiani et al. hybridized the surface of the first-time synthesized, sol–gel synthesized$Co_{0.5}Ni_{0.5}FeCrO_4$ SMNPs using MOFs, which expanded them and increased the number of active sites. The authors hybridized the surface through a hydrothermal method to apply it to the GO and O-SWCNT platforms, which can be more clearly observed in Figure 8.6.

As a result, the material improved in interfacial properties and NP physicochemical properties, was effective in reducing 4-aminophenol, increased in catalytic activity due to the higher surface area with SMNP magnetic properties, which enabled facile separation. Thus, it provided more catalytic reaction cycles. Ashtiani and coworkers state that this novel material exhibits potential for environmental applications, specifically in water treatment and gas separation.

Electrosupun ceramic NFs have many potential applications, e.g. electrospun TiO_2–SiO_2 mats can curve to a radius of curvature of 1.3–3.4 mm after heat treatment, and Esfahani et al. reported that bulk ceramics were brittle after the treatment [36]. These materials are prepared by ES, which has been the most promising approach to prepare such materials in a continuous form and control the diameter of the fibers on a scale from microns to nanometers, according to Brown et al. apud Esfahani et al.

Kim et al. studied a combination of the sol–gel method and ES to prepare yttria-stabilized zirconia/silica NFs for water purification, and they studied the effect of silica on the structure [37]. More specifically, they studied the ceramic

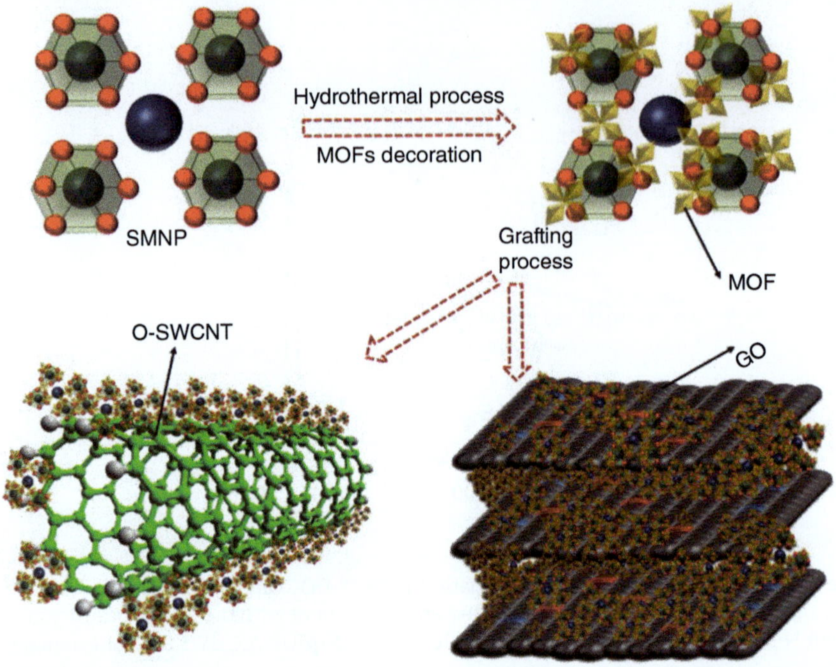

Figure 8.6 The concept proposed by the authors by using the SMNPs decorated with MOFs grafted in the platforms of GO and O-SWCNT. Source: Ashtiani et al. [34]/with permission from Elsevier.

precursors TEOS and zirconium(IV) propoxide (ZrP), which positively affected the flexibility of YSZ/silica NFs. During the synthesis, two parameters were analyzed: grain size and fiber diameter. The samples were prepared in two forms: constant TEOS concentration with variable ZrP concentration and constant ZrP concentration with variable TEOS concentration. Altogether, the research concluded that a smaller grain size and a larger fiber diameter could increase the flexibility of YSZ/silica NFs. Finally, according to the authors, as membranes, these flexible YSZ/silica NFs had appreciable results for microfiltration with high water permeability and rejected a great part of the polymeric NPs.

Another investigation to solve the low surface area of titania was made by Ullah et al. using a hyperbranched polyamide (HBPA) reinforced with titania [38]. The titania-reinforced polymer nanocomposite combines both positive aspects from titania and HBPA: high photocatalytic activity and high surface area. After hyperbranched polyamine had been prepared using a one-step uncatalyzed route, three nanocomposites were synthetized following a sol–gel route with (3-aminopropyl)triethoxysilane as the coupling agent and titanium tetraisopropoxide as the precursor. The three samples were prepared with 6, 12, and 18 wt% titania NP contents and 5 wt% HCl as the acid route catalyst for the sol–gel process. The samples with 6 and 12 wt% TiO_2 had finely distributed dioxide without agglomeration. The authors expect great potential from this combination for water treatment and gas separation using the photocatalytic properties of titania and its surface properties.

8.3 Final Remarks

Categorically, photocatalysis was the prominent reaction for water purification using nanomaterials prepared through the sol–gel process and its modifications. Titanium dioxide is indisputably the most commonly used nanomaterial for treatment, where modifications to the prepared material, such as doping, are used to enhance its photocatalytic and surface properties. Other particular modifications were highlighted, such as MI to increase the nanomaterial selectivity for pesticides and applying NPs to a larger platform, such as in GO and O-SWCNT, for a larger surface area, which created more active sites. Moreover, the anatase form of TiO_2 has the most efficient photocatalytic activity for many pollutants presentin water bodies. For air purification, biofilm formation and gas separation through interfacial properties were highlighted; however, air purification through interfaces had low development throughout the time period of this chapter.

Titania-based nanomaterials for water purification presented issues related to the synthesized materials' low surface area, which can be further investigated, such as applying a larger platform or increasing its surface while synthetizing TiO_2. Through doping methods, adding synergic elements such as manganese or simply nitrogen has proven to be effective in increasing the photoresponse of TiO_2. Bismuth vanadate-based catalysts are also highlighted for their potential application for palm oil treatment, but their synthesis also demonstrates low cost through the sol–gel method and morphological control. Detection through nanomaterials has shown potential through electrochemical sensor platforms, where polymer nanocomposites on CNT surfaces display affinity to detect heavy metals through electric signals with the synergy with metal oxides, and the conducting polymers are not sensor platforms.

Hopefully, this chapter provides a big picture of many paths that one can choose to follow in the development of nanomaterials prepared by the sol–gel method for air and water purification and the challenges of this area of research. Finally, for a greater comprehension of this field's status, a graphic representation using bibliography data from the Scopus database was made through the software VOSviewer in Figures 8.7 and 8.8 with the title "sol–gel air and water purification," which resulted in a 50-different-keyword connection map with the words repeating at least three times.

In Figure 8.7, the most intense colors and texts display the most investigated themes, and the application of these nanomaterials and parameters, e.g. titanium dioxide and sol–gel processes are close and confirms that they are related to the air cleansing and water treatment of dyes and organic pollutants, and parameters such as calcination and surface area are related. Figure 8.8 gives a more detailed map of

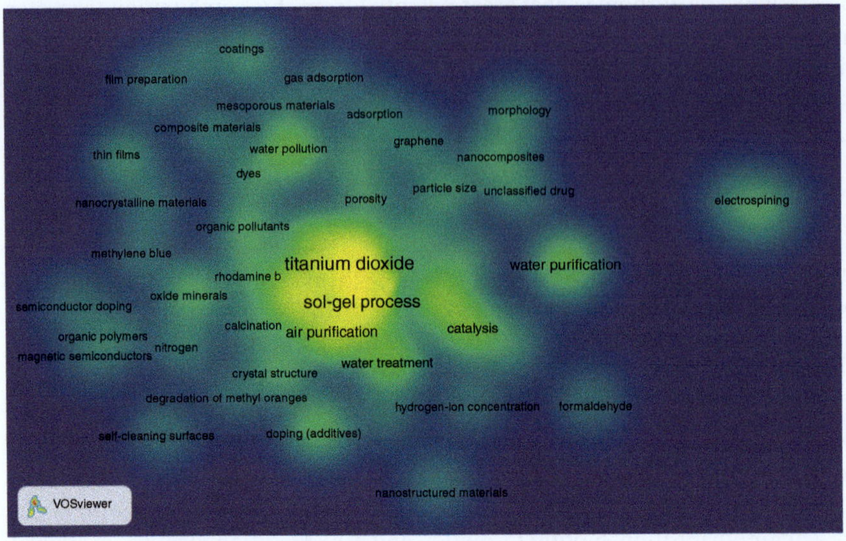

Figure 8.7 Density visualization by keywords of publications in the Scopus database of sol–gel air and water purification. The sol–gel process is strongly correlated with TiO_2 synthesis and applied to air purification and water treatment, with modifications such as doping for water treatment and application on subjects adjacent to the mentioned items.

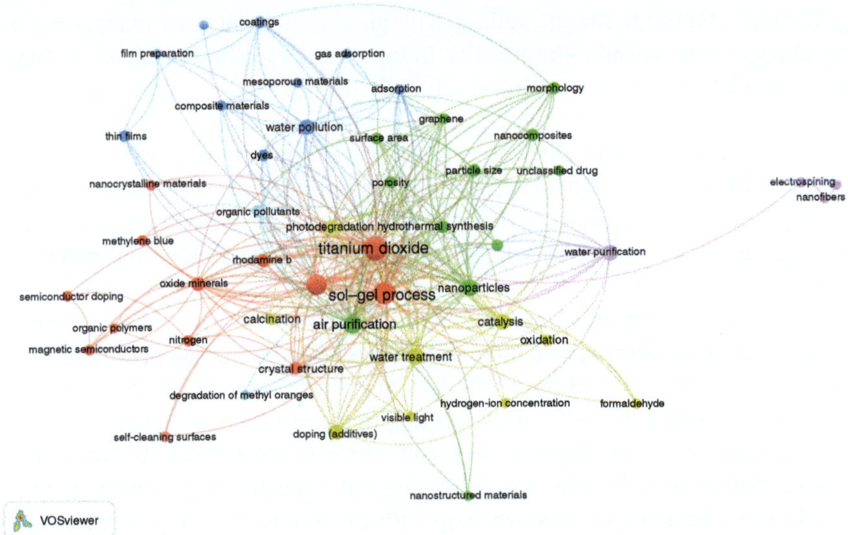

Figure 8.8 A more detailed visualization of how each reaction medium (air or water) is connected to each pollutant or structure modification and a specific studied parameter.

connections of these themes and groups them into clusters of the same colors to relate their coincident investigations and relation to each theme using the lines of connections.

References

1 Hüsing, N. and Schubert, U. (2004). *Synthesis of Inorganic Materials*, 2e. Vienna: Wiley-VCH.

2 Dragnea, D., Gudovan, D., Zahara, E., and Bildea, C.S. (2018). The treatment of aluminium pigments with inorganic polymers for environment-friendly applications. *Bucharest, Revista de Chimie (Rev. Chim.)* 69 (12): 3353–3360. https://doi.org/10.37358/RC.18.12.6748.

3 Lee, C.H., Park, S.H., Chung, W. et al. (2011). Preparation and characterization of surface modified silica nanoparticles with organo-silane compounds. *Colloids Surf., A* 384 (1–3): 318–322. ISSN 0927-7757. https://doi.org/10.1016/j.colsurfa.2011.04.010.

4 Ullmann, M.A. (2019). Síntese Sol-gel Non-hydrolytic Sol-gel Synthesis in the Heterogenization of Metallocenes in Hybrid Silicas for Polyethylene Production. Tese (Doutorado em Química) - Programa de Pós-Graduação em Química da Universidade Federal do Rio Grande do Sul, Porto Alegre (Avaiable in Portuguese).

5 Otor, H.O., Steiner, J.B., García-Sancho, C., and Alba-Rubio, A.C. (2020). Encapsulation methods for control of catalyst deactivation: a review. *ACS Catal.* 10 (14): 7630–7656. https://doi.org/10.1021/acscatal.0c01569.

6 Gunduz-Meric, G., Kaytakoglu, S., and Degirmenci, L. (2020). Catalytic performance of silica covered bimetallic nickel-iron encapsulated core-shell microspheres for hydrogen production. *Int. J. Hydrogen Energy* 45 (60): 34547–34556.

7 Ye, R.-P., Lin, L., Chen, C.-C., and (...), Zhou, Z., Yao, Y.-G. (2018). Synthesis of robust MOF-derived Cu/SiO_2 catalyst with low copper loading via sol-gel method for the dimethyl oxalate hydrogenation reaction. *ACS Catal.* 8 (4): 3382–3394.

8 Tananaiko, O., Trofimchuk, A., Nechiporuk, Y. et al. (2018). Planar electrodes modified by nanodiamonds and biocomposite silica-choline oxidase film for choline determination. In: *IEEE 38th International Conference on Electronics and Nanotechnology, ELNANO 2018 – Proceedings 8477433*, IEEE. Kyiv, Ukraine, pp. 411–416.

9 Pylypchuk, I.V., Daniel, G., Kessler, V.G., and Seisenbaeva, G.A. (2020). Removal of diclofenac, paracetamol, and carbamazepine from model aqueous solutions by magnetic sol–gel encapsulated horseradish peroxidase and lignin peroxidase composites. *Nanomaterials* 10 (2): 282.

10 Wilson, M., Cheng, C.Y.C., Oswald, G., and (...), Beaumont, S.K., Badyal, J.P.S. (2017). Magnetic recyclable microcomposite silica-steel core with TiO_2 nanocomposite shell photocatalysts for sustainable water purification. *Colloids Surf., A* 523: 27–37.

11 Caresani, J.R.F., Dallegrave, A., and Dos Santos, J.H.Z. (2021). Amylases encapsulated in organosilane-modified silicas prepared by sol–gel: evaluation of starch saccharification. *J. Sol-Gel Sci. Technol.* https://doi.org/10.1007/s10971-020-05446-1.

12 Ahmed, S.N. and Haider, W. (2018). Heterogeneous photocatalysis and its potential applications in water and wastewater treatment: a review. *Nanotechnology* 29: 342001. https://doi.org/10.1088/1361-6528/aac6ea.

13 Kataoka, F.P. (2011). Study of photocatalytic efficiency as a function of the morphology of TiO2 nanostructures synthesized by hydrothermal method. Dissertação (Mestrado em Ciência e Tecnologia de Materiais). Universidade Estadual Paulista, Bauru (Available in Portuguese).

Water

14 Rawal, S., Buer, S.H., Hawkins, W. et al. (2022). Photocatalytic degradation of acetaminophen from water solutions via thin films part I: preparation, characterisation, and analysis of titanium dioxide thin films. *Int. J. Chem. Reactor Eng.* 20 (1): 97–112. https://doi.org/10.1515/ijcre-2021-0039.

15 Liu, S., Su, Z.-L., Liu, Y. et al. (1915). Mechanism and purification effect of photocatalytic wastewater treatment using graphene oxide-doped titanium dioxide composite nanomaterials. *Water* 2021: 13. https://doi.org/10.3390/w13141915.

16 Hood, R.D., Jones, C.L., and Ranganathan, S. (1989). Comparative developmental toxicity of cationic and neutral rhodamines in mice. *Teratology* 40: 143–150.

17 Qingguo, L., Gao, W., Jingjing, D. et al. (2012). Discovery of environmental rhodamine b contamination in paprika during the vegetation process. *J. Agric. Food Chem.* 60 (19): 4773–4778. https://doi.org/10.1021/jf300067z.

18 Sweatman, T.W., Seshadri, R., and Israel, M. (1990). Metabolism and elimination of rhodamine 123 in the rat. *Cancer Chemother. Pharmacol.* 27: 205–210. https://doi.org/10.1007/BF00685714.

19 Birlik, I. and Dagdelen, D. (2020). Synergistic effect of manganese and nitrogen codoping on photocatalytic properties of titania nanoparticles. *Bull. Mater. Sci.* 43: 85. https://doi.org/10.1007/s12034-020-2041-8.

20 Seo, H.J., Lee, J.W., Na, Y.H., and Boo, J.-H. (2020). Enhancement of photocatalytic activities with nanosized polystyrene spheres patterned titanium dioxide films for water purification. *Catalysts* 10: 886. https://doi.org/10.3390/catal10080886.

21 Fiorenza, R., Di Mauro, A., Cantarella, M. et al. (2020). Preferential removal of pesticides from water by molecular imprinting on TiO_2 photocatalysts. *Chem. Eng. J.* 379: 122309. ISSN 1385-8947. https://doi.org/10.1016/j.cej.2019.122309.

22 Kalu, O.I., Subramanian, B., MacLean, B.J., and Saha, G.C. (2019). A novel approach to the sol–gel synthesis of titanium dioxide-coated SBA-16 type silica mesoporous microspheres for water purification. *Materialia* 5: 100237. ISSN 2589-1529. https://doi.org/10.1016/j.mtla.2019.100237.

23 Saputera, W.H., Amri, A.F., Mukti, R.R. et al. (2021). Photocatalytic degradation of palm oil mill effluent (POME) waste using $BiVO_4$ based catalysts. *Molecules* 26: 6225. https://doi.org/10.3390/molecules26206225.

24 Tajik, S., Beitollahi, H., Nejad, F.G. et al. (2021). Recent developments in polymer nanocomposite-based electrochemical sensors for detecting environmental pollutants. *Ind. Eng. Chem. Res.* 60 (3): 1112–1136. https://doi.org/10.1021/acs.iecr.0c04952.

25 Stieberova, B., Zilka, M., Ticha, M. et al. (2019). Sustainability assessment of continuous-flow hydrothermal synthesis of nanomaterials in the context of other production technologies. *J. Cleaner Prod.* 241: 118325. ISSN 0959-6526. https://doi.org/10.1016/j.jclepro.2019.118325.

26 Nsubuga, H., Basheer, C., Jalilov, A. et al. (2019). Droplet flow-assisted heterogeneous electro-Fenton reactor for degradation of beta-blockers: response surface optimization, and mechanism elucidation. *Environ. Sci. Pollut. Res.* 26: 14313–14327. https://doi.org/10.1007/s11356-019-04551-1.

27 Ganesamoorthy, R., Vadivel, V.K., Kumar, R. et al. (2021). Aerogels for water treatment: a review. *J. Cleaner Prod.* 329: 129713. ISSN 0959-6526. https://doi.org/10.1016/j.jclepro.2021.129713.

Air

28 Ansari, M.A., Albetran, H.M., Alheshibri, M.H. et al. (2020). Synthesis of electrospun TiO_2 nanofibers and characterization of their antibacterial and antibiofilm potential against gram-positive and gram-negative bacteria. *Antibiotics* 9: 572. https://doi.org/10.3390/antibiotics9090572.

29 Mamaghani, A.H., Haghighat, F., and Lee, C.-S. (2019). Hydrothermal/solvothermal synthesis and treatment of TiO_2 for photocatalytic degradation of air pollutants: preparation, characterization, properties, and performance. *Chemosphere* 219: 804–825. ISSN 0045-6535. https://doi.org/10.1016/j.chemosphere.2018.12.029.

30 Maira, A.J., Yeung, K.L., Lee, C.Y. et al. (2000). Size effects in gas-phase photo-oxidation of trichloroethylene using nanometer-sized TiO_2 catalysts. *J. Catal.* 192 (1): 185–196. ISSN 0021-9517. https://doi.org/10.1006/jcat.2000.2838.

Air and Water

31 Obregón, S. and Rodríguez-González, V. (2021). Photocatalytic TiO_2 thin films and coatings prepared by sol–gel processing: a brief review. *J. Sol-Gel Sci. Technol.* https://doi.org/10.1007/s10971-021-05628-5.

32 Tegze, B., Albert, E., Dikó, B. et al. (2021). Effect of silver modification on the photoactivity of titania coatings with different pore structures. *Nanomaterials* 11: 2240. https://doi.org/10.3390/nano11092240.

33 Vallejo, W., Calderon, D., Díaz-Uribe, C. et al. (2020). Photocatalitic activity of N-doping TiO_2 thin films deposited by the sol-gel technique. *J. Phys.: Conf. Ser.* 1541: 012016.

34 Ashtiani, S., Khoshnamvand, M., Shaliutina-Kolešová, A. et al. (2020). $Co_{0.5}Ni_{0.5}FeCrO_4$ spinel nanoparticles decorated with UiO-66-based metal-organic frameworks grafted onto GO and O-SWCNT for gas adsorption and water purification. *Chemosphere* 255: 126966. ISSN 0045-6535. https://doi.org/10.1016/j.chemosphere.2020.126966.

35 Liaskovska, M., Tatarchuk, T., Bououdina, M., and Mironyuk, I. (2019). Green synthesis of magnetic spinel nanoparticles. In: *Nanophotonics, Nanooptics, Nanobiotechnology, and Their Applications. NANO 2018. Springer Proceedings in Physics*, vol. 222 (ed. O. Fesenko and L. Yatsenko). Cham: Springer https://doi.org/10.1007/978-3-030-17755-3_25.

36 Esfahani, H., Jose, R., and Ramakrishna, S. (2017). Electrospun ceramic nanofiber mats today: synthesis, properties, and applications. *Materials (Basel).* 10 (11): 1238. https://doi.org/10.3390/ma10111238. PMID: 29077074; PMCID: PMC5706185.

37 Kim, J., Lee, J., Ha, J.-H., and Song, I.-H. (2019). Effect of silica on flexibility of yttria-stabilized zirconia nanofibers for developing water purification membranes. *Ceram. Int.* 45 (14): 17696–17704. ISSN 0272-8842. https://doi.org/10.1016/j.ceramint.2019.05.337.

38 Ullah, A., Shabbir, S., and Kausar, H. (2019). In situ synthesis of hyperbranched polyamide/titania nanocomposites via sol-gel process. In: *16th International Bhurban Conference on Applied Sciences and Technology (IBCAST)*, IEEE. Islamabad, Pakistan, pp. 80–85. https://doi.org/10.1109/IBCAST.2019.8667250.

9

Graphitic Carbon Nitride ($g\text{-}C_3N_4$)-based Sustainable Photocatalysts for Purification of Antibiotic Containing Wastewater

Dhruti S. Pattanayak[1], Chandrakant Thakur[1], Dhananjay Singh[2], and Dharm Pal[1]

[1] *National Institute of Technology Raipur, Department of Chemical Engineering, Raipur 492010, Chhattisgarh, India*
[2] *Institute of Engineering & Technology, Department of Chemical Engineering, Lucknow 226021, Uttar Pradesh, India*

9.1 Introduction

Among the numerous issues confronting the modern world, maintaining our environment is emerging as the most difficult. Over one billion people globally do not have access to pure water, resulting in a high number of waterborne infections and fatalities. The World Health Organization (WHO) predicts that by 2025, nearly half of all people on Earth will be affected by water scarcity [1]. Water contaminated by developing contaminants such as pharmaceutical products is hazardous to both the environment and living creatures. Among the different pharmaceuticals, the increasing use of antibiotics in people and animals has resulted in widespread pollution of the environment, deeming it a possible risk and biohazard [2]. Numerous antibiotics have been created and utilized in clinical settings since 1928, when Fleming discovered penicillin to control a variety of outbreaks, such as bacterial illnesses. Between the 1940s and the 1980s, when antibiotics were widely used, The "Golden Age" of antibiotics began for humans [3]. The introduction of antibiotics in the environment causes the formation of antimicrobial resistance (AR) (UN Environment 2017), which the UN Environment has acknowledged as one of the most serious worldwide public health issues [4]. The problem has been made worse by the rising use of various antibiotics and improper medicine disposal by hospitals, pharmaceutical companies, and homes [5]. Even at low levels, antibiotics can have negative effects on the flora and fauna of the ecosystem, including vulture renal dysfunction, fish reproduction issues, and growth retardation and inhibition in some aquatic animals. Highly soluble and persistent antibiotic compounds can act as powerful carcinogens [6]. Although new antibiotic research may not even be able to track the appearance and spread of new superbugs, overuse or incorrect use of antibiotics has resulted in an increase in antibiotic resistance. As a result, in the twenty-first century, the threat that antibiotic pollutants pose to the integrity of our water

Nanomaterials for Air and Water Purification, First Edition.
Edited by Deepu A. Gopakumar, Jyotishkumar Parameswaranpillai, Jinu Jacob George, and Midhun Dominic C.D.

supplies is considered to be one of the most severe environmental issues globally, not only owing to environmental degradation but also because of the inherent human health risks [7]. As a result, there is a push for green, sustainable technology and legislation to reduce or eliminate antibiotic residues in water [2]. Photocatalysis is a promising clean and sustainable energy technique for converting solar energy into chemical energy and degrading emerging contaminants [8, 9]. Nanoparticles (NPs) have unique properties due to their size, shape, huge surface-to-volume ratio, and composition, which make them effective for pollutant degradation as catalysts [10, 11]. Due to the advantages of its well-known basic physical characteristics, g-C_3N_4 has immediately gained substantial attention as a metal-free material. g-C_3N_4 is a great candidate to absorb some visible light due to its fair bandgap of 2.7 eV, suitable conduction band (CB) of −1.3 eV, and suitable valence band (VB) of 1.4 eV [12]. Additionally, g-C_3N_4 has a structural similarity to graphene and is composed of two extremely common earth elements (carbon and nitrogen), which lowers the cost of production while enhancing physicochemical stability. Despite its strong thermal and chemical stabilities and outstanding photoresponse to visible light, pure g-C_3N_4 demonstrates the limits of low charge carrier mobility and quick photo-induced electron–hole pair recombination that are present in many other photocatalysts [13]. To satisfy the needs of a broader range of applications, it is necessary to significantly enhance photocatalytic g-C_3N_4's performance. Some modification techniques have been explored to achieve this goal: basal engineering, element defects and element doping, co-catalyst loading, and hybrid semiconductor engineering composites [14]. To boost the activity of g-C_3N_4, two techniques are mainly proposed: (i) introduction of specific defects and local lattices to the photocatalyst semiconductor to narrow its band gap and improve visible light absorption; and (ii) isolatation of enough excitons to stop electron–hole recombination (which occurs throughout the catalytic process). Several methodologies for obtaining modified g-C_3N_4 have been published, including novel synthesis methods, nanostructure creation, and electrical structure tuning. To date, researchers have reported various strategies to modify bare g-C_3N_4 using semiconductor coupling, noble metal loading, insertion of carbonaceous materials, and elemental doping [15]. It was found that the appealing characteristics of g-C_3N_4 made it easy for a variety of inorganic, organic, and metal NPs to attach on its surface, producing hybrid nanocomposites that improved its efficiency [16]. Other methods for boosting the effectiveness of g-C_3N_4 synthesis include the use of templates, hydrothermal, solvothermal, cold polymerization, freeze drying, interactions between nitrogen lone pairs, copolymerization, dye sensitization, and other processes. In view of the tremendous advances regarding g-C_3N_4-based photocatalysts, we have provided extensive and up-to-date information on g-C_3N_4, which is developing as a hot research issue not only in photocatalysis but also in other domains. Nanostructure design and its various synthetic processes have been studied in terms of modification tactics. To achieve the goal of presenting every feature of g-C_3N_4, each application is described in depth, including the most current and widely utilized g-C_3N_4-based photocatalyst for antibiotic degradation. In this chapter, we try to discuss various components of g-C_3N_4 as well as its diverse spectrum of applications in order to entice multidisciplinary research [17]. The manufacturing techniques, physicochemical characteristics, catalytic activity, and reaction processes of modified carbon nitride-based photocatalysts in antibiotic-wastewater treatment are briefly summarized.

9.2 Structure and Preparation of g-C_3N_4

A polymer without metals that resembles graphene in structure is called g-$C_3N_{4,}$ which forms aromatic C–N heterocycles with six members via sp^2 hybridization of atoms C and N [18]. The electron conjugation process in its network structure is two-dimensional (2D), which is substantially delocalized. g-C_3N_4 has stacked 2D layered structure, and the Van der Waals connections that are weak connect the layers. g-C_3N_4 has extraordinarily high thermal and chemical stability due to its high degree of condensation, stiff network of C–N heterocycles, and strongly conjugated system. The capacity to withstand heat g-C_3N_4 is the greatest of any organic substance, reaching 600 °C [19]. Additionally, it is not soluble in most acids and bases, water, or organic solvents. Two different g-C_3N_4 structural isomers exist: one made of *s*-triazine and another made of tri-*s*-triazine [20]. Planar structures are formed by connecting different repeating units with tertiary amino groups. The magnitude of irregular gaps in the lattice, as well as the N atom's electrical environment, can result in varied energy stabilities. The more stable tri-*s*-triazine is commonly regarded as the foundational component of g-C_3N_4 calculated using first-principles density functional theory (DFT). Although g-C_3N_4 does not exist in nature, it can be easily synthesized. Urea, melamine, melamine hydrochloride, cyanamide, thiourea, and dicyandiamide are examples of N-rich chemical precursors that are widely available and inexpensive [15, 21]. In general, each approach has its own set of benefits and drawbacks, and many associated research developments have previously been documented in several instructive studies elsewhere. The primary and most prevalent approach is thermal polycondensation. It is inexpensive, simple to use, and suitable for preparation on a large scale. It uses short organic compounds N-rich as precursors, and at temperatures between 450 and 650 °C, these precursors easily calcinate to form g-C_3N_4 [15]. The product, like graphene, has a layered 2D structure. The conjugated aromatic system stacks to produce the largest X-ray powder diffraction (XRD) peak, which appears at 27.4 and is equal to 0.326 nm. In addition, all products include some hydrogen residue [22]. The product features of thermal polycondensation, such as the molar ratio C/N, functional groups, crystallinity structure, specific bandgap, surface area, and carrier migration velocity of condensation, are clearly influenced by the precursor choice and pretreatment as well as the reaction circumstances. These features can also influence g-C_3N_4's heterogeneous photocatalytic activity and practical use. However, virgin g-C_3N_4 generated directly through thermal polycondensation has condensation and polymerization taking place as a result of the bulk process being inhomogeneous, resulting in a restricted specific surface area [23]. The hydrothermal approach, is ideal for producing g-C_3N_4 on a large scale, is a straightforward and controlled process without dangerous solvents, templates, or

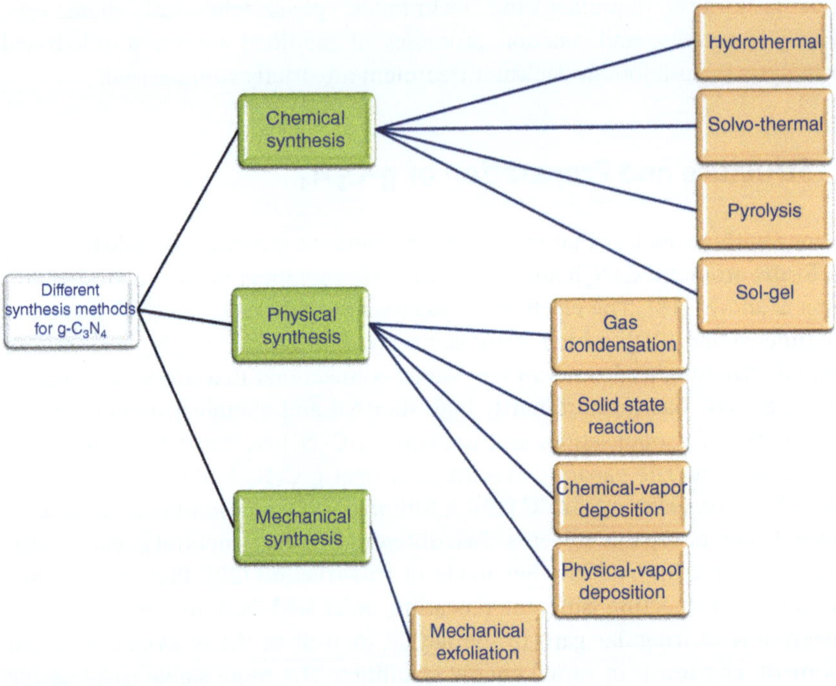

Figure 9.1 Different techniques for synthesis of g-C_3N_4-based photocatalysts.

expensive chemicals and with comparatively mild reaction conditions. Over the past ten years, numerous new techniques for creating g-C_3N_4-based nanomaterials [17], which are depicted in Figure 9.1, have undergone substantial development in addition to the general direction of green, controllable, and refined.

9.2.1 Modification of g-C_3N_4

A novel form of polymeric semiconducting material is graphitic carbon nitride (g-C_3N_4). It is thought to be the most stable allotrope of carbon nitrides under atmospheric circumstances [24]. Because of its low bandgap (2.7 eV), g-C_3N_4 can use both visible light and UV rays, unlike TiO_2. However, g-C_3N_4 has a number of limitations in its pure form, including quick recombination of photogenerated e^- and h^+ pairs, poor use of visible light, and limited electrical conductivity [25]. These limitations must be overcome in order to exhibit potent photocatalytic activity change g-C_3N_4's structure and characteristics. Numerous modification techniques have been adopted (Shown in Figure 9.2) to create the heterostructure design and enhance photocatalytic performance due to the special characteristics of semiconductors. The main objectives of g-C_3N_4-based photocatalysts are improved charge carrier migration rate, expanded active sites on the catalyst surface, and separation of the charge carriers created by light irradiations [26]. We can use g-C_3N_4 as a fundamental model system to further research the implicated photocatalytic

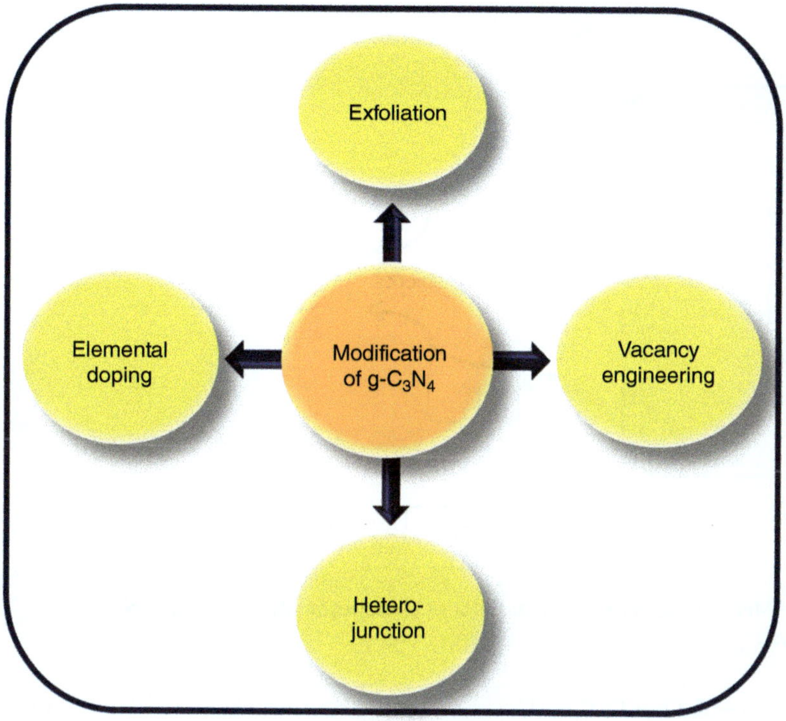

Figure 9.2 Different methods for modification of g-C_3N_4.

processes, in addition to producing a large-scale, high-efficiency photocatalyst based on g-C_3N_4.

9.3 General Mechanism of Photodegradation by Employing g-C_3N_4

On the photocatalyst surface, many oxidation–reduction reactions occur in response to light stimulation and induction, which define photocatalytic reactions [27]. The three steps for photodegradation associated with of g-C_3N_4 are as follows. (i) Enough photons are absorbed by g-C_3N_4 to create an e^- & h^+ pair by photo-generation; (ii) Successfully split up photogenerated electrons and holes move to the appropriate response sites; (iii) Free holes and electrons can either take part in the photodegradation process directly or they can first generate before indirectly destroying organic molecules by a variety of reactive radicals, as described in Eqs. (9.1)–(9.7). The general mechanism of photodegradation of antibiotics using g-C_3N_4 photocatalyst is shown in Figure 9.3.

$$g - C_3N_4 \xrightarrow{h\nu} e^-(CB) + h^+(VB) \tag{9.1}$$

$$O_2 + e^-(CB) \rightarrow O_2^{-\bullet} \tag{9.2}$$

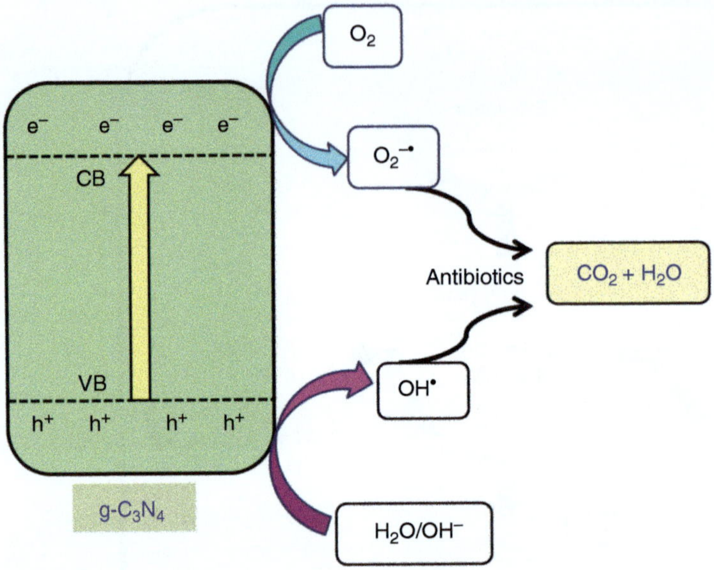

Figure 9.3 Mechanistic approach of antibiotics photodegradation using g-C_3N_4 photocatalyst.

$$h^+(VB) + H_2O \rightarrow OH^\bullet + H^+ \tag{9.3}$$

$$h^+(VB) + OH^- \rightarrow OH^\bullet \tag{9.4}$$

$$H_2O_2 \xrightarrow{hv} 2OH^\bullet \tag{9.5}$$

$$h^+(VB) + O_2^{-\bullet} \rightarrow {}^1O_2 \tag{9.6}$$

$$\left(h^+/O_2^{-\bullet}/OH^\bullet/1O_2\right) + \text{Antibiotics} \rightarrow CO_2 + H_2O \tag{9.7}$$

In particular, the CB potential of g-C_3N_4 (−1.3 eV/ NHE) exceeds the level of superoxide radical ($O_2^{-\bullet}$) formation ($O_2/O_2^{-\bullet}$, −0.33 eV/NHE), allowing photogenerated electrons to oxidize dissolved oxygen (O_2) with a large amount of oxidizing ability. Eq. (9.1) illustrates how e^- & h^+ pair may be generated while the light source falls on the catalyst; when the O_2 reacts with electron it forms $O_2^{-\bullet}$) (2). Next, hydroxyl radicals ($OH^\bullet$, Eqs. (9.3) and (9.4) are produced by the interaction with H_2O_2 and h^+. $OH^\bullet$ can be created when photogenerated holes combine with H_2O/OH^-. Pure g-C_3N_4 has a lower VB position than the potential for the formation of $OH^\bullet$ (+1.99 eV vs. NHE for $OH^\bullet/OH^-$), which reduces its oxidation capacity (+1.4 eV vs. NHE), due to a lack of $OH^\bullet$species additionally, the behavior of photogenerated carriers resonance energy transfer in the polymer g-C_3N_4 are related to the production of singlet oxygen (${}^1O_2$). Thus, the processes described in Eq. (9.6) can result in ${}^1O_2$. The breakdown of organic pollutants occurs in conjunction with h^+ and a number of the above generated reactive oxygen species ($O_2^{-\bullet}$, ${}^1O_2$, $OH^\bullet$, and h^+) (Eq. (9.7)).

9.3.1 Influencing Factors for Photocatalytic Degradation

To improve pure g-C_3N_4's efficiency, the parameters impacting its photocatalytic activity have been carefully researched. The important elements are discussed below in details.

(i) Surface morphology, size/thickness, and crystallinity:
   - Pollutant adsorption and photogenerated charge carrier combinations can be somewhat influenced by the crystallinity of g-C_3N_4. Additionally, improving crystallinity can decrease non-radiative losses, and facilitate exciton involvement in transferring energy.
   - Quantum confinement is exhibited by ultrathin or nanoscale photocatalytic materials, which widen the band gap and boost redox capacity. Furthermore, the material's domain size, which is similar to the length of the exciton diffusion (about nanometers), can be decreased by the nanostructure, increasing the likelihood of exciton dissociation.
   - Generating g-C_3N_4 with various morphologies results in the production of shorter carrier migration pathways and more active locations.

(ii) Band organization: The intrinsic bandgap (E_g) of semiconductor photocatalysts determines their light absorption range and redox capabilities. A low E_g value is advantageous for increasing the range of light absorption. For semiconductor photocatalysts to increase their redox capacity, the preferred VB and CB positions are more positive and negative, respectively. When constructing effective redox-capable photocatalysts that can use all of the light's energy, it is crucial to take the E_g value and the locations of the CB and VB into account.

(iii) Flaws: The photocatalytic process is greatly influenced by structural and impurity-induced flaws. The inclusion of suitable flaws can enhance e—h^+ separation and boost photoactivity. Excessive flaws, on the other hand, may act as recombination hubs for carriers, reduce the E_g value, broaden the photoresponse range, and promote exciton production.

(iv) Lighting, photocatalyst/substrate concentration, reaction temperature, pH level, coexisting ions, and surroundings are examples of reaction conditions. Light intensity and duration have a direct impact on the rates of photocatalytic reactions. Shielding effects and inadequate response sites must be avoided. In order to achieve this, the concentrations of the photocatalyst and reactants should also be carefully monitored. The stability of the catalyst, the types of reactants that are already present, and the production of reactive oxygen groups, such as $OH^{\bullet}$, $O_2^{-\bullet}$ are all reliant on the reaction solution's pH.

9.4 Antibiotics in Wastewater

Any class of drugs that can eliminate germs through certain interactions are known as "antibiotics." This class of medications primarily has an organic origin and is made to function effectively at very low doses. After intra-corporal delivery, they are completely removed from the body in a short amount of time. They are normally not biodegradable and, after a considerable amount of time, may be found in aquatic habitats. Due to their anthropogenic origins, they may be dangerous to organisms that live in these ecosystems [28]. Antibiotics are a sort of treatment where maintaining human health and reducing environmental pollution are incompatible. The survival of humans depends on these pharmaceutical categories, yet they also present serious environmental risks. They have established themselves as essential components of medical treatment for bacterial illnesses. However, the public's health continues to be seriously threatened by widespread abuse and misuse [29]. By considering the number of deaths and mortality rate in 1929, Alexander Fleming made the first antibiotic, the most significant discovery of the century. Despite this great feat, the evolution of resistance from some bacteria rapidly followed; nevertheless, during the past few decades, the global use of this family of drugs in both humans and cattle has expanded. To put this into context, in 2010, approximately 63 151 metric tons of antibiotics were used in the production of cattle alone, and by 2030, this amount is predicted to increase by nearly 67% [30]. Similarly, antibiotics are widely utilized in aquaculture (a globally expanding food industry). There is a global upsurge in antibiotics in man-made ecosystems, including wastewater treatment facilities and sewage systems, as a result of lax controls on the use of this class of medications. The bulk of freshwater, terrestrial, and marine habitats are ultimately produced [31]. Researchers and policymakers are also concerned about the prevalence of antibiotics in drinking water. Most wastewater treatment techniques now in use neither degrade nor have an impact on pharmaceutical wastes. Some of the present methods transform these components into hazardous intermediates, which are then discharged into receiving water bodies. Fluoroquinolones (FQs) are a significant class within these categories that are often utilized by people and animals [32]. This antibiotic has a wide spectrum and inhibits bacterial enzymes before killing them. The most widely prescribed antibiotic in the world, ciprofloxacin hydrochloride, a FQ derivative, has been found in various wastewater samples at nanogram levels (a limit that can cause antimicrobial resistance). In excessive doses, this drug may induce neurological issues, including seizures and tendon rupture. It is essential to remove pharmaceuticals before discharging them into drainage systems, bring them to a safe level because the majority of medicines are difficult to break down [33]. Now more than ever, a workable solution to reduce or eliminate these wastes is required.

9.5 Photocatalytic Antibiotic Degradation Using g-C_3N_4-based Composite

According to recent study, modified g-C_3N_4 enhances its photocatalytic activity, with remarkable qualities for catalytic applications. These modifications help to confine electrons and lessen the chance of e– and h^+ recombination, which alternatively would improve the photocatalytic system. Studies have also revealed that modified g-C_3N_4 makes use of solar radiation across a broad spectrum, whereas pure g-C_3N_4 could only absorb blue light. The study exploring how different antibiotics are degraded by g-C_3N_4-based photocatalysts are summarized in Table 9.1.

9.5.1 Antibiotic Degradation by Using Elemental Doped g-C_3N_4

According to studies, elemental doping of g-C_3N_4 increases its photocatalytic activity. Doping with metal or nonmetal atoms is commonly regarded as a successful and acceptable way of changing the electrical structure and band gap of g-C_3N_4 in order to improve its photocatalytic properties. These dopants lessen the possibility of electron–hole recombination and aid in electron trapping, which would otherwise disable the photocatalytic system. Doping also allowed for the use of a broader spectrum of solar energy, whereas pure g-C_3N_4 absorbed only blue light. An appropriate substrate for the distribution of fine metals is g-C_3N_4 with little agglomeration due to metal doping, and many amine groups decrease their size and boost stability, increasing suitability for attaching metal NPs [50]. The catalytically active sites of metal atoms boost the efficiency of g-C_3N_4's planned redox reactions. The investigations into the degradation of various antibiotics by element-doped g-C_3N_4-based photocatalysts are summarized in Table 9.1, wherein several research studies on the high photocatalytic efficiency modification of g-C_3N_4 have been reported.

Ou et al. (2020) described the production of porous Fe^{3+}-doped g-C_3N_4 with remarkable photocatalytic activity; it was then used as a photocatalytic system for the effective removal of sulfadiazine under visible light irradiation. A number of characterization results reveal that Fe^{3+} is efficiently doped into the g-C_3N_4 interlayer and is stabilized in g-C_3N_4 by the Fe-N coordination connection. Transient photocurrent results, diffuse reflectance spectroscopy (DRS), electrospray ionization (ESI), and scanning electron microscopy (SEM) data, as well as the findings showed that g-C_3N_4 doped with Fe^{3+} has a porous structure, a narrow bandgap, a reduced rate of photogenerated e– and h^+ recombination, and a wider spectrum of light irradiation to absorb. Because of the enhanced physical and chemical properties, the photocatalytic efficiency was significantly increased. Under white LED illumination for 90 minutes, the sulfadiazine is broken down using Fe^{3+}-doped g-C_3N_4 photocatalyst; sulfadiazine elimination was almost complete with 99.8% efficiency (in contrast to just 52.1% for bulk g-C_3N_4). The deterioration was attributed mostly to $^1O_2$ prior to the radiation, albeit a tiny amount of h^+, $OH^•$, and $O_2^{-•}$ also contributed to the degradation. The Fe^{3+}-doped g-C_3N_4 was discovered to be efficient for degrading a variety of antibiotic pollutants [34]. Viet et al. [35]investigated the synthesis of Ag-doped g-C_3N_4 capable of decomposing oxytetracycline (OTC). In actuality, the bandgap of g-C_3N_4 was reduced from 2.71 to 2.46 eV after Ag doping, which accelerated photoinduced e– and h^+ couples recombining to boost photocatalytic activity. After being exposed to visible light for two hours, the photocatalyst destroyed 98.7% of OTC at an inceptive concentration of 10 mg l^{-1} and a catalyst concentration of 0.2 mg l^{-1} (300 W xenon lamp). The formation of photoinduced e^- and h^+ couples was greatly boosted by the addition of Ag to g-C_3N_4, in addition to increasing the active site count and pore capacity. Because of the narrowing of the bandgap, the photocatalytic breakdown

Table 9.1 Studies on degradation of antibiotics by g-C_3N_4-based photocatalysts.

S. No.	Catalyst	Synthesis method	Antibiotic	Time (min)	Antibiotic initial conc. ($mg\,l^{-1}$)	Light source	Degradation efficiency (%)	References
1.	Fe^{3+}-g-C_3N_4	Thermal condensation	Sulfadiazine	90	2	White LED light (3 W)	99.8	[34]
2.	Ag-g-C_3N_4	Thermal condensation	Oxytetracycline	120	30	Xe lamp (300 W)	98.7	[35]
3.	Ag-g-C_3N_4	Thermal condensation	Tetracycline	120	20	Solar light	96.8	[36]
4.	Ba-g-C_3N_4	Thermal condensation	Tetracycline	120	20	Xe lamp (150 W)	91.9	[37]
5.	Ag-P@UCN	Co-pyrolysis	Sulfamethoxazole	30	5	Visible lamps	99.0	[38]
6.	Na-Cl-g-C_3N_4	Co-pyrolysis	Tetracycline	60	10	Xe lamp(350 W)	66.1	[39]
7.	I-K-g-C_3N_4	Co-pyrolysis	Sulfamethoxazole	45	10	Visible light lamps	99.0	[40]
8.	g-C_3N_4/MnO_2	Sonochemical synthesis	Tetracycline hydrochloride	135	20	LED irradiation (9 W, 220 V)	93.0	[41]
9.	NiO/WO_3@g-C_3N_4	Ultrasonication technique	Metronidazole	90	20	Sunlight	~100	[42]
10.	ZnTCPP/g-C_3N_4	Thermal polycondensation	Tetracycline	120	30	Visible light irradiation	80.3	[43]
11.	Polypyrrole sensitized zinc ferrite/g-C_3N_4	In situ polymerization method	Ciprofloxacin	120	20	Visible light irradiation	92.0(78.0)	[44]
12.	Carbon dots/g-C_3N_4	Hard templating method	Naproxen	25	5	Xe lamp (350 W)	98.6 (~63.1)	[45]
13.	BiOCl/g-C_3N_4	Hydrothermal	Tetracycline	60	20	Xe lamp (300 W)	97.1	[46]
14.	CTFNS/g-C_3N_4	Electrostatic self-assembly	Sulfamethazine	180	10	Xe lamp (500 W)	95.8	[47]
15.	$NaNbO_3$/g-C_3N_4	Hydrothermal-thermal polymerization	Ofloxacin	30	50	Simulated solar ($100\,mW\,cm^{-2}$)	99.5	[48]
16.	CeO_2/g-C_3N_4 (persulfate medium)	Thermal condensation	Norfloxacin	60	10	Xe lamp (150 W), $\lambda > 420$ nm)	88.6(63.8)	[49]

of OTC was significantly increased. It has been shown that OTC can be successfully degraded by OH•, and $O_2^{-\bullet}$ oxidation [35]. Minh Tri [36] developed an Ag-doped g-C_3N_4 catalyst and utilized it to remove hazardous antibiotic contaminants from hospital-contaminated water using sunlight. Tetracycline is completely broken down into smaller intermediate molecules; 96.8% of the tetracycline could be degraded by Ag-doped g-C_3N_4 in 120 minutes of solar light exposure. Not only did the Ag doping in g-C_3N_4 enhance optical absorbance, but it also boosted the generation and separation of photogenerated carriers. According to the findings of the mineralization and intermediate studies, TC was totally degraded into smaller intermediate products and, subsequently into CO_2 and H_2O [36]. Recently, it was demonstrated that the antibiotic tetracycline may be effectively broken down by barium-doped g-C_3N_4 (BCN), which is produced using a straightforward thermal polymerization procedure [37]. After 120 minutes of exposure to visible light, 2% barium-loaded BCN showed 91.94% TC degradation at pH 10. Barium NPs were added, which improved the photocatalytic activity of g-C_3N_4 by decreasing the band-gap energy of the material from 2.71 to 2.56 eV and increasing its surface area. Through a number of reactions, including ring cleavage, demethylation, and deamination, photogenerated holes are the primary oxidizing species that break down TC into simpler compounds with m/z values of 102.10 (hexan-2-ol) and 88.09 (Pentan-3-ol). Silver and phosphorus were co-doped on g-C_3N_4 (UCN) nanosheets made from urea by Nguyen et al. [38]. Under basic circumstances, having a bandgap of 2.51 eV, an Ag/P/UCN composite might degrade sulfamethoxazole (SMX) solution to 99% in 30 minutes. Uniformly distributed crystalline nano-Ag particles on the catalyst's surface increased light absorption capacity, served as an electron sink to gather photogenerated electrons, and lessened charge carrier recombination caused by light. The photocatalytic activity of Ag/P/UCN was shown to be durable and repeatable for up to six cycles [38]. It was also discovered that co-doping Na and Cl into the g-C_3N_4 (CN-NaCl) framework significantly increased g-C_3N_4 photocatalytic activity. Dicyanamide and sodium chloride were thermally condensed in a single process to produce CN-NaCl. Doping g-C_3N_4 with both Na and Cl lowers its bandgap from 2.76 to 2.67 eV, allowing photogenerated electrons and holes to quickly travel to the material's surface via sodium and chlorine ion channels and significantly separate from one another. After 60 min of exposure to visible light, the improved sample CN-NaCl-0.5 destroyed 66.6% of the tetracycline hydrochloride (TCH) [39]. In 45 min of visible light irradiation, using iodine and potassium co-doped g-C_3N_4, Paragas et al. [40] showed that sulfamethoxazole degrades to 99% of its original strength (IK- g-C_3N_4). The catalyst was created by co-pyrolyzing urea and potassium iodate and has shown recyclability up to three runs. Doping with iodine and potassium increased the rate of charge carrier transfer, leading to the fast production of excitons. These excitons mix with water molecules and oxygen species to form OH•, and $O_2^{-\bullet}$, respectively. The disintegration of sulfamethoxazole requires the presence of these radicals [40].

9.5.2 Antibiotic Degradation by Using g-C_3N_4-based Heterojunctions

By encouraging the separation of photo-induced electron-hole pairs and enabling charge carrier movements across the interface between heterostructures to prevent charge carrier recombination, the formation of heterostructures is a useful technique for enhancing the photocatalytic performance of g-C_3N_4 photocatalysts. Metal oxides, metal chalcogenides, and multicomponent oxides are just a few of the semiconductors that can form heterojunctions with g-C_3N_4 due to the material's unusual 2D layered structure and flexibility as a result of its polymeric nature. The efficacy and degradation of antibiotic substances in g-C_3N_4-based heterojunctions are currently being studied as shown in Table 1.

Yashas et al. [41] found the potential of GCN/MnO_2 nanocomposite for photocatalytic destruction of the harmful contaminant TCH. The sonochemical method used to create the GCN/MnO_2 adjusted their shape and imbued them with distinct photocatalytic and electrocatalytic capabilities. The enhanced photoelectrocatalytic activity of GCN/MnO_2 was indicated by its improved electro-catalytic response, enhanced absorption of visible light, suitable electrical structure of GCN and MnO_2, rapid electron mobility, and large active surface area [41]. A novel nanostructured NiO/WO_3@g-C_3N_4 was created using a straightforward and energy-efficient process, and it was then tested as a very effective reusable heterogeneous photocatalyst for the 30% H_2O_2 hydroxylation of benzene under visible light. A sustainable method has been used to break down metronidazole by a NiO/WO_3@g-C_3N_4 photocatalyst under solar light irradiation. Terephthalic acid was used as a probe molecule to look at the role of $OH^\bullet$, in metronidazole breakdown. Reusability of the photocatalyst was also carefully examined, and it was found that it could be used up to five times before showing any observable activity decline. The NiO/WO_3 have well immobilized on the surface support of g-C_3N_4, and it also absorbs light for C–H bond activation in benzene hydroxylation and metronidazole degradation [42]. g-C_3N_4 was successfully sensitized by a very efficient photocatalyst ZnTCPP, which is synthesized using a simple thermal polycondensation approach. The ZnTCPP molecules were successfully condensed on the g-C_3N_4 surface employing amide groups as bridging units, as shown by the FT-IR and XRD data. The best 10% ZnTCPP/ g-C_3N_4 composites demonstrated long-term reuse and elimination rates of 96% and 83%, respectively, for breaking down tetracycline (TC) under visible light. The superior visible light photocatalytic performance was primarily attributable to the highly efficient separation of electron–hole pairs and the improved solar light utilization, which was confirmed by photoluminescence (PL), UV–Vis DRS, electrochemical impedance spectra (EIS), and photocurrent responses. The creation of amide group bridging linking bonds, which help in the separation by acting as conduits, and the exchange of photoinduced charges between g-C_3N_4 and ZnTCPP. Furthermore, the sensitization of ZnTCPP is critical for improving visible light harvesting. Hydroxyl radicals ($OH^\bullet$) are the primary oxidative species in the ZnTCPP/g-C_3N_4 photodegradation system for TC elimination that is driven by visible light [43]. A hard-templating approach was used to successfully create a newly developed hybrid metal-free anchoring carbon dots to hollow carbon nitride nanospheres (HCNS/CDs) photocatalyst. Under visible light irradiation, the HCNS/CDs showed exceptional photocatalytic activity for the pharmaceuticals and personal care products (PPCPs') destruction. When compared to bulk g-C_3N_4 and HCNS, the HCNS/CDs displayed optimum photocatalytic activity for naproxen (NPX) degradation, which was 66.8 and 8.5 times greater, respectively. Surprisingly, 10 mg l^{-1} NPX was totally destroyed within 5 min of natural sun exposure. It was also discovered that $O_2^{-\bullet}$ played an important part in the photocatalytic process, which might result in the degradation of NPX [45]. The hydrothermal approach was used to successfully produce a 2D/2D BiOCl/g-C_3N_4 Z-scheme. BiOCl/g-C_3N_4-under artificial solar light, composite 3 exhibit the best photocatalytic efficiency for TC. Within 60 minutes, about 97.1% of the TC was degraded. In 180 minutes, TC achieved a TOC removal of 76.3%. The high photocatalytic performance of the BiOCl/g-C_3N_4-3 composite was also attributed to the fast charge transfer in the Z-scheme system, with h^+, and $O_2^{-\bullet}$ being the main reactive species. Potential applications for the 2D/2D BiOCl/g-C_3N_4-3 photocatalyst includes the treatment of antimicrobial wastewater and other areas [46]. The polymer-based 2D/2D heterojunction photocatalysts were produced using an electrostatic self-assembly method combining carbon nitride nanosheets (CNNS) with amine functionalization and covalent triazine-based frameworks nanosheets (CTFNS) with abundant carboxyl groups. The findings of the characterization demonstrate that CTFNS are widely spread over CNNS because CTFNS and CNNS have a high electrostatic attraction to one another. The enormous 2D nano-interface and the intensive contact between CNNS and CTFNS at the interface were credited with the extraordinary photocatalytic activity of CTFNS/CNNS composites. Transport of excited-state electrons would be greatly facilitated by the extensive interface contact area and close contact [47]. The high photoexcited electron–hole pair recombination rate of g-C_3N_4 limits its usage as a visible-light-driven photocatalyst. In order to boost the photocatalytic activity of g-C_3N_4, perovskite-type sodium niobate ($NaNbO_3$) nanorods (SNCN) were created by Zhang et al. [48] using a two-step hydrothermal and thermal polymerization method. Effective separation of charge carrier (e—h^+) pairs was made possible by transferring photoexcited electrons from 2D g-C_3N_4 to 1D $NaNbO_3$ consequently, reactive radical generation occurs [48]. Liu et al. [49] describes how to make a composite photocatalyst out of shuttlelike CeO_2 and g-C_3N_4 nanosheets and investigates the process of degrading Norfloxacin (NOR) utilizing this composite photocatalyst in conjunction with persulfate. Compared to pure CeO_2 and g-C_3N_4, the created composite photocatalyst has a greater NOR degradation efficiency. The development of strong oxidative SO^{4-} and $^1O_2$ in the presence of PS accelerated NOR degradation, which acted along with h^+, $OH^\bullet$, and $O_2^{-\bullet}$ to breakdown the NOR [49].

9.6 Conclusions and Perspectives

This chapter discusses the recent synthesis of a photocatalyst based on g-C_3N_4 for antibiotic breakdown. Despite standard metal-containing catalysts, g-C_3N_4 is a metal-free photocatalyst consisting of carbon and nitrogen. It is easily produced using readily accessible carbon- and nitrogen-rich precursors. Pyrolysis is the simplest way to make $g\text{-}C_3N_4$ out of all the physical, chemical, and mechanical processes. The generated $g\text{-}C_3N_4$ from this process is typically a bulk material, which has various drawbacks such as a small surface area, a high recombination rate, and a wide band gap, which restricts its use in a number of applications. As a result, the chapter described efforts to improve the photo-physical characteristics of $g\text{-}C_3N_4$ by structural and compositional tweaking. Textural features and pore properties may be tailored using hard and soft templates to increase surface area and offer a large number of active sites for adsorption and photocatalytic applications. Foreign element doping in the $g\text{-}C_3N_4$ matrix cleared the path for band engineering to minimize bandgap and increase visible light absorption. Although noble metal dopants provide benefits from surface plasmon resonance, their use is impractical due to their high cost. Creating heterojunction interfaces of $g\text{-}C_3N_4$ with other suitable semiconductors is a popular and sophisticated band engineering method. The $g\text{-}C_3N_4$ heterojunctions are fully explored in terms of band alignment and the number of components. We come to the conclusion that doping, nanostructure design, and the synthesis of hybrid nanocomposites can enhance $g\text{-}C_3N_4$'s efficiency. $g\text{-}C_3N_4$ is shown to be one of the best readily available materials for long-term environmental and energy solutions. Because bulk $g\text{-}C_3N_4$ has numerous restrictions, it must go through several modification procedures before it can be used. Future research may focus on these issues in an effort to develop an effective, environmentally friendly synthetic method for the synthesis of enhanced $g\text{-}C_3N_4$ that may be used for a range of applications. Although many ways have been employed to increase the efficiency of $g\text{-}C_3N_4$, it is mostly sensitive to blue-violet visible light and hence cannot absorb visible light efficiently. The following are some important factors that should be considered in future research:

- Controlling the chemical doping of $g\text{-}C_3N_4$ and achieving the desired nanostructure are demanding tasks.
- Some of the alteration procedures require the use of a template and corrosive chemicals, and hence are not considered green. The majority of $g\text{-}C_3N_4$-based composites contain valuable metals such as Ag, Cd, and Ti, which are both costly and highly harmful to the environment.
- In general, laboratory-prepared $g\text{-}C_3N_4$ has a high number of defects, a low water dispersion, and poor crystallinity.

Acknowledgments

The work greatly acknowledges the Department of Chemical Engineering of the National Institute of Technology, Raipur, for providing research opportunities.

References

1 Ganguly, P., Panneri, S., Hareesh, U.S. et al. (2019). *Recent Advances in Photocatalytic Detoxification of Water*. Elsevier Inc. https://doi.org/10.1016/B978-0-12-813926-4.00029-X.

2 Suyana, P., Ganguly, P., Nair, B.N. et al. (2021). Structural and compositional tuning in g-C_3N_4 based systems for photocatalytic antibiotic degradation. *Chem. Eng. J. Adv.* 8: 100148. https://doi.org/10.1016/j.ceja.2021.100148.

3 Kong, X., Liu, X., Zheng, Y. et al. (2021). Graphitic carbon nitride-based materials for photocatalytic antibacterial application. *Mater. Sci. Eng., R* 145: 100610. https://doi.org/10.1016/j.mser.2021.100610.

4 Siddiqui, M.R., AlOthman, Z.A., and Rahman, N. (2017). Analytical techniques in pharmaceutical analysis: a review. *Arabian J. Chem.* 10: S1409–S1421. https://doi.org/10.1016/j.arabjc.2013.04.016.

5 Kumar, A. and Pal, D. (2018). Antibiotic resistance and wastewater: correlation, impact and critical human health challenges. *J. Environ. Chem. Eng.* 6: 52–58. https://doi.org/10.1016/j.jece.2017.11.059.

6 John, A., Rajan, M.S., and Thomas, J. (2021). Carbon nitride-based photocatalysts for the mitigation of water pollution engendered by pharmaceutical compounds. *Environ. Sci. Pollut. Res.* 28: 24992–25013. https://doi.org/10.1007/s11356-021-13528-y.

7 Bai, X., Chen, W., Wang, B. et al. (2022). Photocatalytic degradation of some typical antibiotics: recent advances and future outlooks. *Int. J. Mol. Sci.* 23. https://doi.org/10.3390/ijms23158130.

8 Mishra, J., Pattanayak, D.S., Das, A.A. et al. (2019). Enhanced photocatalytic degradation of cyanide employing Fe-porphyrin sensitizer with hydroxyapatite palladium doped TiO_2 nano-composite system. *J. Mol. Liq.* 287: 110821. https://doi.org/10.1016/j.molliq.2019.04.098.

9 Pattanayak, D.S., Mishra, J., Nanda, J. et al. (2021). Photocatalytic degradation of cyanide using polyurethane foam immobilized Fe-TCPP-S-TiO_2-rGO nano composite. *J. Environ. Manage.* 297: 113312. https://doi.org/10.1016/j.jenvman.2021.113312.

10 Pattanayak, D.S., Mallick, N., Thakur, C., and Pal, D. (2020). Plant mediated green synthesis of silver nanoparticles for antimicrobial application: present status. *J. Indian Chem. Soc.* 97: 1108–1114.

11 Pattanayak, D.S., Pal, D., Thakur, C. et al. (2021). Bio-synthesis of iron nanoparticles for environmental remediation: status till date. *Mater. Today Proc.* 44: 3150–3155. https://doi.org/10.1016/j.matpr.2021.02.821.

12 Mohanty, L., Pattanayak, D.S., and Dash, S.K. (2021). An efficient ternary photocatalyst Ag/ZnO/g-C_3N_4 for degradation of RhB and MG under solar radiation. *J. Indian Chem. Soc.* 98: 100180. https://doi.org/10.1016/j.jics.2021.100180.

13 Mohanty, L., Pattanayak, D.S., Singhal, R. et al. (2022). Enhanced photocatalytic degradation of rhodamine B and malachite green employing $BiFeO_3$/g-C_3N_4 nanocomposites: an efficient visible-light photocatalyst. *Inorg. Chem. Commun.* 138: 109286. https://doi.org/10.1016/j.inoche.2022.109286.

14 Huang, R., Wu, J., Zhang, M. et al. (2021). Strategies to enhance photocatalytic activity of graphite carbon nitride-based photocatalysts. *Mater. Des.* 210: 110040. https://doi.org/10.1016/j.matdes.2021.110040.

15 Pattanayak, D.S., Pal, D., Mishra, J., and Thakur, C. (2022). Noble metal–free doped graphitic carbon nitride (g-C_3N_4) for efficient photodegradation of antibiotics: progress, limitations, and future directions. *Environ. Sci. Pollut. Res.* https://doi.org/10.1007/s11356-022-20170-9.

16 Behera, A., Babu, P., and Parida, K. (2021). Growth of macroporous TiO_2 on B-doped g-C_3N_4 nanosheets: A Z-scheme photocatalyst for H_2O_2 production and phenol oxidation under visible light. *Inorg. Chem. Front.* 8: 1489–1499. https://doi.org/10.1039/d0qi01327g.

17 Vaya, D., Kaushik, B., and Surolia, P.K. (2022). Recent advances in graphitic carbon nitride semiconductor: structure, synthesis and applications. *Mater. Sci. Semicond. Process.* 137: 106181. https://doi.org/10.1016/j.mssp.2021.106181.

18 Pattnaik, S.P., Behera, A., Acharya, R., and Parida, K. (2019). Green exfoliation of graphitic carbon nitride towards decolourization of congo-red under solar irradiation. *J. Environ. Chem. Eng.* https://doi.org/10.1016/j.jece.2019.103456.

19 Huang, D., Yan, X., Yan, M. et al. (2018). Graphitic carbon nitride-based heterojunction photoactive nanocomposites: applications and mechanism insight. *ACS Appl. Mater. Interfaces* 10: 21035–21055. https://doi.org/10.1021/acsami.8b03620.

20 Mishra, P., Behera, A., Kandi, D., and Parida, K. (2019). Facile construction of a novel $NiFe_2O_4$@P-doped g-C_3N_4 nanocomposite with enhanced visible-light-driven photocatalytic activity. *Nanoscale Adv.* 1: 1864–1879. https://doi.org/10.1039/c9na00018f.

21 Patnaik, S., Behera, A., and Parida, K. (2021). A review on g-C_3N_4/graphene nanocomposites: multifunctional roles of graphene in the nanohybrid photocatalyst toward photocatalytic applications. *Catal. Sci. Technol.* 11: 6018–6040. https://doi.org/10.1039/d1cy00784j.

22 Mohanty, L., Pattanayak, D.S., Pradhan, D., and Dash, S.K. (2022). Synthesis of novel p-n heterojunction g-C_3N_4/Bi_4 Ti_3O_{12} photocatalyst with improved solar-light-driven photocatalytic degradation of organic dyes. *Environ. Qual. Manage.* 1–15. https://doi.org/10.1002/tqem.21907.

23 Alaghmandfard, A. and Ghandi, K. (2022). A comprehensive review of graphitic carbon nitride (g-C_3N_4)–metal oxide-based nanocomposites: potential for photocatalysis and sensing. *Nanomaterials* 12. https://doi.org/10.3390/nano12020294.

24 Darkwah, W.K. and Oswald, K.A. (2019). Photocatalytic applications of heterostructure graphitic carbon nitride: pollutant degradation, hydrogen gas production (water splitting), and CO_2 reduction. *Nanoscale Res. Lett.* 14. https://doi.org/10.1186/s11671-019-3070-3.

25 Cao, S., Low, J., Yu, J., and Jaroniec, M. (2015). Polymeric photocatalysts based on graphitic carbon nitride. *Adv. Mater.* 27: 2150–2176. https://doi.org/10.1002/adma.201500033.

26 Lin, H., Wu, J., Zhou, F. et al. (2021). Graphitic carbon nitride-based photocatalysts in the applications of environmental catalysis. *J. Environ. Sci.* 124: 570–590. https://doi.org/10.1016/j.jes.2021.11.017.

27 Pattanayak, D.S., Pal, D., Thakur, C. et al. (2022). Catalytic potential of phyto-synthesized silver nanoparticles for the degradation of pollutants. In: *Sustainable Engineering, Energy, and the Environment*, 1e (ed. K.L. Wasewar and S.N. Rao), 465–481. Apple Academic Press https://doi.org/10.1201/9781003277484-36.

28 Kurt, A., Mert, B.K., Özengin, N. et al. (2017). Treatment of antibiotics in wastewater using advanced oxidation processes (AOPs). *Intech* 11: 13. https://doi.org/10.5772/67538.

29 Kraemer, S.A., Ramachandran, A., and Perron, G.G. (2019). Antibiotic pollution in the environment: from microbial ecology to public policy. *Microorganisms* 7: 1–24. https://doi.org/10.3390/microorganisms7060180.

30 Van Boeckel, T.P., Brower, C., Gilbert, M. et al. (2015). Global trends in antimicrobial use in food animals. *Proc. Natl. Acad. Sci. U. S. A.* 112: 5649–5654. https://doi.org/10.1073/pnas.1503141112.

31 Boy-Roura, M., Mas-Pla, J., Petrovic, M. et al. (2018). Towards the understanding of antibiotic occurrence and transport in groundwater: findings from the Baix Fluvià alluvial aquifer (NE Catalonia, Spain). *Sci. Total Environ.* 612: 1387–1406. https://doi.org/10.1016/j.scitotenv.2017.09.012.

32 Rodrigues-Silva, C., Porto, R.S., dos Santos, S.G. et al. (2019). Fluoroquinolones in hospital wastewater: analytical method, occurrence, treatment with ozone and residual antimicrobial activity evaluation. *J. Braz. Chem. Soc.* 30: 1447–1457. https://doi.org/10.21577/0103-5053.20190040.

33 Adeyemi, J.O., Ajiboye, T., and Onwudiwe, D.C. (2021). Mineralization of antibiotics in wastewater via photocatalysis. *Water Air Soil Pollut.* 232: 1–28. https://doi.org/10.1007/s11270-021-05167-3.

34 Ou, Q., Xu, S., Long, Y., and Zhang, X. (2020). Porous visible light-responsive Fe^{3+}-doped carbon nitride for efficient degradation of sulfadiazine. *Environ. Sci. Pollut. Res.* 27: 27849–27858. https://doi.org/10.1007/s11356-020-08749-6.

35 Viet, N.M., Trung, D.Q., Giang, B.L. et al. (2019). Noble metal -doped graphitic carbon nitride photocatalyst for enhancement photocatalytic decomposition of antibiotic pollutant in wastewater under visible light. *J. Water Process Eng.* 32: 100954. https://doi.org/10.1016/j.jwpe.2019.100954.

36 Minh Tri, N., Kim, J., Giang, B.L. et al. (2019). Ag-doped graphitic carbon nitride photocatalyst with remarkably enhanced photocatalytic activity towards antibiotic in hospital wastewater under solar light. *J. Ind. Eng. Chem.* 80: 597–605. https://doi.org/10.1016/j.jiec.2019.08.037.

37 Bui, T.S., Bansal, P., Lee, B.K. et al. (2020). Facile fabrication of novel Ba-doped g-C_3N_4 photocatalyst with remarkably enhanced photocatalytic activity towards tetracycline elimination under visible-light irradiation. *Appl. Surf. Sci.* 506: 144184. https://doi.org/10.1016/j.apsusc.2019.144184.

38 Nguyen, T.B., Huang, C.P., Doong, R. et al. (2020). Visible-light photodegradation of sulfamethoxazole (SMX) over Ag-P-codoped g-C_3N_4 (Ag-P@UCN) photocatalyst in water. *Chem. Eng. J.* 384: 123383. https://doi.org/10.1016/j.cej.2019.123383.

39 Wang, K.L., Li, Y., Sun, T. et al. (2019). Fabrication of Na, Cl co-doped graphitic carbon nitride with enhanced photocatalytic activity for degradation of dyes and antibiotics. *J. Mater. Sci. - Mater. Electron.* 30: 4446–4454. https://doi.org/10.1007/s10854-019-00733-2.

40 Paragas, L.K.B., de Luna, M.D.G., and Doong, R.A. (2018). Rapid removal of sulfamethoxazole from simulated water matrix by visible-light responsive iodine and potassium co-doped graphitic carbon nitride photocatalysts. *Chemosphere* 210: 1099–1107. https://doi.org/10.1016/j.chemosphere.2018.07.109.

41 Yashas, S.R., Shivaraju, H.P., Pema, G. et al. (2021). Sonochemical synthesis of graphitic carbon nitride-manganese oxide interfaces for enhanced photocatalytic degradation of tetracycline hydrochloride. *Environ. Sci. Pollut. Res.* 28: 4778–4789. https://doi.org/10.1007/s11356-020-10813-0.

42 Devi, M., Das, B., Barbhuiya, M.H. et al. (2019). Fabrication of nanostructured NiO/WO_3 with graphitic carbon nitride for visible light driven photocatalytic hydroxylation of benzene and metronidazole degradation. *New J. Chem.* 43: 14616–14624. https://doi.org/10.1039/c9nj02904d.

43 Ma, Z., Zeng, C., Hu, L. et al. (2019). A high-performance photocatalyst of ZnTCPP sensitized porous graphitic carbon nitride for antibiotic degradation under visible light irradiation. *Appl. Surf. Sci.* 484: 489–500. https://doi.org/10.1016/j.apsusc.2019.04.117.

44 Das, K.K., Patnaik, S., Mansingh, S. et al. (2020). Enhanced photocatalytic activities of polypyrrole sensitized zinc ferrite/graphitic carbon nitride n-n heterojunction towards ciprofloxacin degradation, hydrogen evolution and antibacterial studies. *J. Colloid Interface Sci.* 561: 551–567. https://doi.org/10.1016/j.jcis.2019.11.030.

45 Wu, Y., Wang, F., Jin, X. et al. (2020). Highly active metal-free carbon dots/g-C_3N_4 hollow porous nanospheres for solar-light-driven PPCPs remediation: mechanism insights, kinetics and effects of natural water matrices. *Water Res.* 172: 115492. https://doi.org/10.1016/j.watres.2020.115492.

46 Sun, Y., Qi, X., Li, R. et al. (2020). Hydrothermal synthesis of 2D/2D BiOCl/g-C_3N_4 Z-scheme: for TC degradation and antimicrobial activity evaluation. *Opt. Mater. (Amst.)* 108: 110170. https://doi.org/10.1016/j.optmat.2020.110170.

47 Cao, S., Zhang, Y., He, N. et al. (2020). Metal-free 2D/2D heterojunction of covalent triazine-based frameworks/graphitic carbon nitride with enhanced interfacial charge separation for highly efficient photocatalytic elimination of antibiotic pollutants. *J. Hazard Mater.* 391: 122204. https://doi.org/10.1016/j.jhazmat.2020.122204.

48 Zhang, D., Qi, J., Ji, H. et al. (2020). Photocatalytic degradation of ofloxacin by perovskite-type $NaNbO_3$ nanorods modified g-C_3N_4 heterojunction under simulated solar light: theoretical calculation, ofloxacin degradation pathways and toxicity evolution. *Chem. Eng. J.* 400. https://doi.org/10.1016/j.cej.2020.125918.

49 Liu, W., Zhou, J., and Yao, J. (2020). Shuttle-like CeO_2/g-C_3N_4 composite combined with persulfate for the enhanced photocatalytic degradation of norfloxacin under visible light. *Ecotoxicol. Environ. Saf.* 190: 110062. https://doi.org/10.1016/j.ecoenv.2019.110062.

50 Pattanayak, D.S., Pal, D., Mishra, J. et al. (2022). Doped graphitic carbon nitride (g-C_3N_4) catalysts for efficient photodegradation of tetracycline antibiotics in aquatic environments. *Environ. Sci. Pollut. Res.* https://doi.org/10.1007/s11356-022-19766-y.

10

Development of Noble Metal Supported Nanoparticles for Water Treatment

Hamyal Izhar[1], Fatima Izhar[1], Shoomaila Latif[2], Muhammad Imran[1], and Nazim Hussain[3]

[1] *Centre for Inorganic Chemistry, School of Chemistry, University of the Punjab, Lahore, 54000 Pakistan*
[2] *School of Physical Sciences, University of the Punjab, Lahore, 54000 Pakistan*
[3] *Centre for Applied Molecular Biology (CAMB), University of the Punjab, Lahore, 53700 Pakistan*

10.1 Noble Metal Nanoparticles

Because of their remarkable characteristics and variety of uses, noble metal nanoparticles are drawing huge interest from the scientific community. Unlike bulk analogs, nanoscale analogs have rare functional aspects, and researchers are discovering novel application possibilities in this regard. Tunability of characteristics by changing local environment, their composition, shape, or size demonstrate that these have remarkable skills. It is well established that many properties, such as electrical, optical, chemical, and others, can be precisely manipulated by applying the chemical composition of materials at the nanoscale [1]. Noble metal nanoparticles are much more stable and display less toxicity than other nanoscale systems. Hence, the use of these nanoparticles is greater than that of other nanosystems. Photocatalysts modified by noble metals have a high potential for and are highly efficient at removing several pollutants.

Having a variety of biomedical uses, including their usage in highly sensitive analytical tests [1, 2], drug and gene delivery [3–5], radiotherapy enhancement [6–9], and thermal ablation, nanoparticles and noble metal nanoparticles are adaptable agents. Moreover, for gene and drug delivery uses, noble metal nanoparticles have been suggested as nontoxic carriers. Furthermore, simultaneous therapy can be afforded by the nanoparticles-based systems. Nanotechnology-based systems could provide a less-invasive substitute, increasing the quality and expectancy of life of the patient, by making use of their extraordinary characteristics [10]. Among these, an attractive platform for cancer therapy in a wide range of clinical settings is represented by potential uses of noble metal nanoparticles [11].

Nanomaterials for Air and Water Purification, First Edition.
Edited by Deepu A. Gopakumar, Jyotishkumar Parameswaranpillai, Jinu Jacob George, and Midhun Dominic C.D.

10.2 Use of Noble Metal Nanoparticles for Purification of Water

Noble metals can be used in various forms to purify water and are discussed as follows:

10.2.1 Against Microorganisms

Due to their good photo-responsive performance, simple methods of preparation, and multiple bactericidal characteristics, noble metal nanoparticles, especially platinum (Pt), silver (Ag), and gold (Au), have been gaining great attention in antimicrobial field. Noble metals have stability and catalytic activity as well as feasible electronic structures [12]. Nanostructures, nanoparticles supported on active or inert materials, and ionic forms of noble metals are the most active against microorganisms.

10.2.2 Antioxidant, Antibacterial, Adsorbent, and Antibiofouling Applications

Noble metal nanoparticles are used as antioxidant, antibacterial, and antibiofouling materials. For example, silver nanoparticles were applied for Pb(II) and Cr(II) ion adsorption [13]. For antioxidant, antibacterial, and antibiofouling applications, Fe-Cu and Cu nanoparticles were investigated for water purification.

10.2.3 Drinking Water Disinfection

A comprehensive understanding of the mechanistic features of interaction between the species involved is needed by the use of nanoparticles in environmental remediation such as water purification. Nanotechnology is important when considering environmental remediation [14, 15]. Dehalogenation of various halocarbons is caused by noble metal nanoparticles, and their chemistry can be widened to pesticides of halocarbon [16]. In water purification applications, subsequently, noble metal (especially gold and silver) nanostructures have been utilized. Absorption spectra of these nanoparticles reveal that they are very sensitive to poisonous species present inside water, for example heavy metal ions [17, 18] (Pb, Cd, Hg, etc.) and pesticides [19, 20] (malathion, chlorpyrifos, endosulfan, etc.). Octahedral palladium nanoparticles have been reported to reduce bromate, a regulated cancer-causing oxyanion formed during disinfection of drinking water [21].

10.2.4 Decomposition of Pesticides

During wastewater treatment, noble metal nanoparticles are used as these can be detached from drinkable water and hence can be reused. For pesticide removal (chlorpyrifos, CP), citrate-stabilized silver nanoparticles have been investigated [22]. These noble metal nanoparticles have been stabilized using various chemical

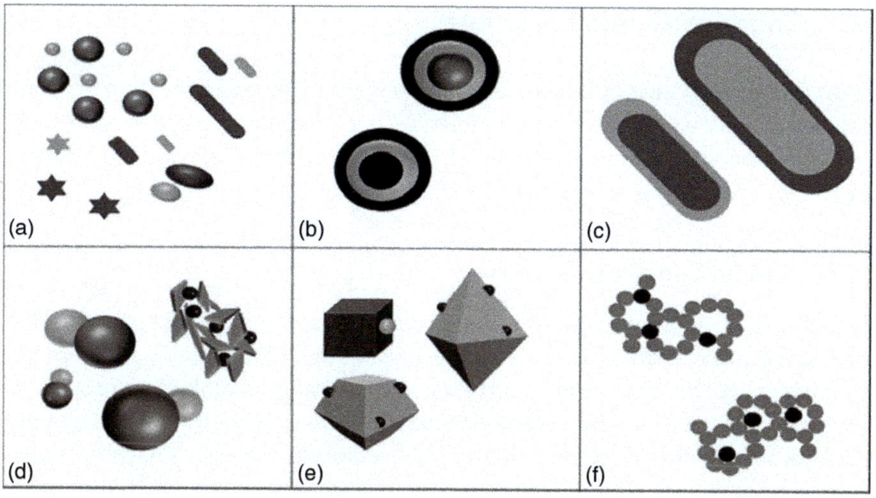

Figure 10.1 Morphology of composites of noble metals; (a) noble metal nanostructures that are unsupported (nanorice, nanostars, nanocluster, and nanorods); (b) nanoparticles having tertiary structures; (c) core–shelled nanorods; (d) supported noble metal nanoparticles; (e) faceted support noble metal nanoparticles; (f) noble metal nanoparticles absorbed inside aerogel network or in porosity [34]. Source: Reproduced from [34] with permission from Elsevier. License number 5404631281148.

compounds such as chitosan and polyethylene glycol [23]. With the help of inkjet printing silver ink, silver nanoparticles were also deposited on the surface of the membrane [24] (Figure 10.1).

10.2.5 "Green Chemistry" Approach

The usage of microorganisms for manufacturing noble metal nanoparticles has been a very famous trend among researchers as a "Green Chemistry" approach. For instance, silver-alginate biohydrogel was performed by reducing Ag cations by RhodococcusNCIM 2891 (actinobacteria) [25]. Generally, the first noble metal nanoparticles are firstly prepared, and after synthesis, they are deposited on the support [26], or vice versa. Other methods include:

- Sonochemical [27], thermal [28], radiolytic [29, 30], photochemical [31–33], and chemical [34–36] methods, where direct reduction of metal ions occurs on the support.
- Other procedures for noble metal nanoparticles deposited on the support are inkjet printing [24], dc-magnetron sputtering [37], electrophoresis, and sintering [38].
- The support as well as noble metal nanoparticles are manufactured during hydrolysis of metal alcoholates followed by calcination [39], laser pyrolysis [40], and anodization. For instance, silver, silver oxide, or titanium oxide is gained from one-step anodic oxidation of TiAg alloys [41] and the hydrothermal process [42].

10.3 Chemically Inert Support

Support that has no contribution to the primary mechanism of purification of water and does not react with pollutants is defined as "inert" support. For water purification, some examples of nanoparticles using chemically inert support are represented in Figure 10.1 and few are tabulated in Table 10.1.

10.3.1 Modification of Membranes

Nanoparticles of noble metals have also been utilized to modify the membranes to improve their performance. For example, polysulfone ultrafiltration membranes were impregnated with silver nanoparticles, and by doing so, biofouling resistance was noticed, triggered only by adsorbed noble metals [45]. It has also been reported that a cork filter with chitosan-Ag nanoparticles stopped the filter's bacterial clogging because of the antibacterial characteristics of Ag nanoparticles [46]. For better membrane performance, bimetallic nanoparticles were also reported. For example, in order to modify polyethersulfone (PES) membranes and chromium(VI) removal from water, multiwalled carbon nanotubes premodified with silver–iron nanoparticles have been utilized [47]. With Ag-Fe/MWCNT modified PES membranes, the elimination of Cr^{+6} ions increased from 9.3% to 94%, along with flux of water from 26.5 to 36.9 $L\,m^{-2}$. Similar findings have been reported after the modification of nanocellulose composites with nanoparticles of Pt and Ag [48].

Table 10.1 Examples of nanoparticles utilized for water purification by chemically inert support.

Noble metal nanoparticles (NMNPs)	Support	Application	References
NMNPs of Au and Ag	Alumina	Water treatment	[22]
NPs of noble metals: Cu, Ag, Rh, Ir, Ru, Pt, and Pd	Silica, alumina, and carbon	Nitrite and nitrate removal/water purification	[43]
Pd NPs	Silica	Water purification	[21]
Silver nanoparticles	Fe_3O_4 co-supported on fly ash and detachment of this composite through an outer magnet after treatment of water	Removal of Pb(II)/water treatment	[44]
Silver NPs formed on diatomite	Diatomite	Purification of water	[34]
Ag NPs	Ceramic tablet	Removal of bacteria	[34]

10.3.2 Use of Filters

It was suggested that filters could be effectively employed for water filtration. These filters have been utilized as supports for either Cu [49] or Ag nanoparticles [50].

10.4 Chemically Active Support

Deposition of noble metal nanoparticles can also be performed on active support, and these particles take part in the overall process. For example;

- On fly ash, Ag nanoparticles were deposited with zinc oxide [51], just like Ag-Fe_3O_4-fly ash composite [44], and pollutants were decomposed in this system.
- On the surface of mesoporous carbon merged with Nafion polymer under 254 nm UV irradiation and by using Ag nanoparticles, an improvement of 80% to 92% was observed in the photocatalytic breakdown of methyl orange.
- By using noble metals as an electron pool, charge carriers' recombination is retarded [31, 52–55].

10.5 Photocatalysis

Photocatalysis is one of the most attractive wastewater treatment methods, where pollutants can be degraded in the presence of light. From the valence band to the empty conduction band, excitation of an electron occurs, and a hole is created in the valence band [56]; whenever a semiconductor (composite material) possesses a specific band structure is irradiated and absorbs energy more than its bandgap. By transferring to some other material or particle surface, electrons and holes will be separated. Reactive oxygen species (ROS) are ultimately produced. Production of radicals like superoxide ($^{\cdot}O_2^-$), hydrogen peroxide (H_2O_2), and hydroxyl ($^{\cdot}OH$) occurs, and these are considered ROS. These species may react with all inorganic and organic compounds. In water technologies, the strongest oxidants are hydroxyl radicals, which can be safely manipulated. When oxygen is present in the system, it should be indicated that ROS are produced through both oxidation and reduction routes, that is, synthesis of $^{\cdot}O_2^-$ by oxygen through scavenging of electrons. The mechanism of semiconductor photocatalysis follows two steps:

- Manufacturing pairs of charge carriers; holes and electrons result from semiconductor excitation with the help of light energy or more than its band gap ($E \geq E_g$).
- To the surface of semiconductor, migration of charge carriers (bulk recombination), and reaction of adsorbed species with charge carriers (Figure 10.2).

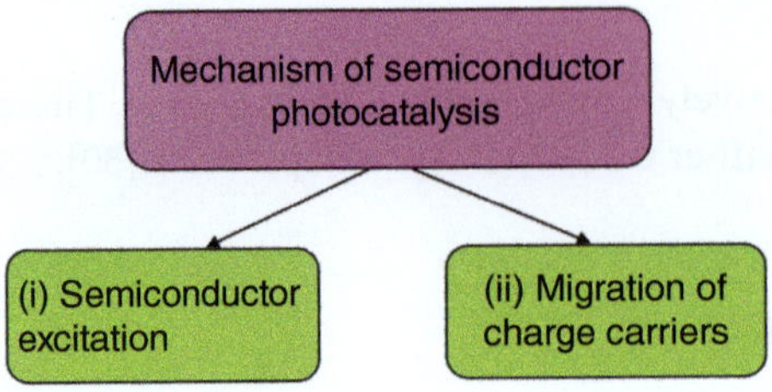

Figure 10.2 Different steps of the mechanism of semiconductor photocatalysis.

10.5.1 Titania as Photocatalyst and Improved Performance of Titania Photocatalysts

There are many photocatalysts used in the process of photocatalysis. Due to its high stability, activity, nontoxicity, and availability, titania is the most commonly applied and studied photocatalyst [57, 58].

There are a few limitations that are associated with titania as a photocatalyst, such as inactivity under visible light because of a large band gap and recombination of charge carriers. To improve these limitations, coupling with other semiconductors [59], surface modification [60–63], doping (with anions or cations [64–69]), and morphology arrangements [70, 71] have been performed on titania. In this regard, under UV irradiation, the photocatalytic activity of titania [53, 72] has been improved by coupling titania with noble metals in the form of metallic deposits [55, 73–76] and adsorbed complexes [77, 78]. As noble metals perform their function as electron sinks, enhanced photocatalytic activity was observed [73, 79, 80]. Noble metals working as an electron pool have also been evidenced by the production of various ROS [26, 81], a shortened lifetime of mobile electrons [82], an increase in photocurrent quantum yield, and a cathodic shift of flat band potential position [83] and decline in photoluminescence [84–88].

10.5.2 Plasmonic Photocatalysis

When conduction band electrons oscillate, their resonance with the oscillating electric field of the incident light is called surface plasmon resonance. Noble metals have characteristic absorption of visible light because of plasmon resonance, which could be used to activate titania [32, 89, 90]. Plasmonic characteristics of noble metals have been examined long ago in various fields such as optical data storage, medicine [91], surface-enhanced Raman spectroscopy (SERS) [91, 92], nanophotonics [93, 94], nano-lithography, chemical sensors [91], and biosensors [92]. However, their use in plasmonic photocatalysis is a decade old. There are three main mechanisms involved in plasmonic photocatalysis (Figure 10.3). The initial two mechanisms could be categorized as "plasmon-assisted photocatalysis" as the transfer of energy or charge occurs only under irradiation. However, the third mechanism can be categorized as "plasmon-assisted catalysis" since localized surface plasmon resonance only activates catalytic (thermal) reactions. A book chapter [95] and a few reviews [96–99] on plasmonic photocatalysis have been published, which could be consulted for deeper overview.

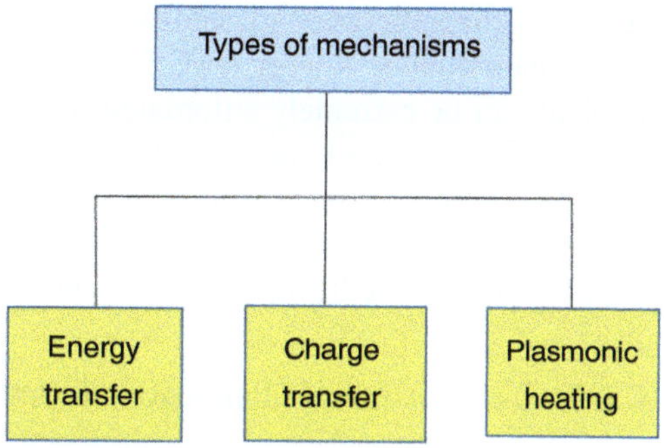

Figure 10.3 Three main mechanisms of plasmonic photocatalysis.

10.5.2.1 Improvement in the Activity of Plasmonic Photocatalysts

Although plasmonic photocatalysts are active for both microorganism inactivation and the oxidation or reduction of organic pollutants, there is still a need to improve their activity for possible commercial applications. This has attracted enormous research work on the activity, stability, and reusability of plasmonic photocatalysts. For a $BiO(OH)_{0.06}Br_{0.94}$, a remarkable example of an AgBr semiconductor modified with Ag nanoparticles has been recently reported with plenty of oxygen vacancies. It was observed that vis-response was triggered through both plasmonic excitations of silver nanoparticles with upcoming "hot" electron transfer to the conduction band of silver bromide, followed by one-electron reduction of adsorbed oxygen to $O_2^{\cdot-}$ and excitation of AgBr.

10.6 Removal of Pollutants

There are different ways for the removal of pollutants from wastewater.

10.6.1 Physical Procedures for Removal of Contaminants

10.6.1.1 Physicochemical Processes

The dangerous pollutants are separated from water using physicochemical methods. Some notable methods are coagulation, flocculation, filtration, adsorption, etc. These methods are often essential for efficient and fast elimination of pollutants, although no degradation of pollutants occurs in these methodologies. For instance, Fe and Ag are effective adsorbents for the exclusion of heavy metal ions chromium(VI), chromium(II), and lead(II) from aqueous solutions [47, 100]. This exclusion includes the reaction between the hydroxyl layer surrounding the spherical metallic core of iron or silver nanoparticles inside water and the metal ions [100].

10.6.1.2 Nanofiltration or Reverse Osmosis

For the treatment of produced water, the use of membrane filtration methods such as microfiltration, ultrafiltration, nanofiltration, and reverse osmosis provides many advantages. For example, for rejected components, the membrane is a positive barrier, so the difference in feed water quality will have a negligible influence on permeate quality. Energy costs for membrane equipment are often lower, and the plant can be extremely automated. For treatment of water, both ceramic and polymeric membranes have been utilized [101].

Nanofiltration, or reverse osmosis, is used for virus eradication from water because of its very small size. Nanofilters [46] and ultrafiltration membranes [45] with silver nanoparticles reportedly have antiviral characteristics.

10.6.2 Degradation of Chemical Pollutants

Contrast to the abovementioned physiochemical methods, pollutants' degradation occur in following chemical methods.

10.6.2.1 Catalytic Reduction

Horold et al. reported the usage of noble metal nanoparticles for water purification exhibiting catalytic reduction of nitrite and nitrate to nitrogen on the surface of supported noble metals (Cu, Ag, Rh, Ir, Ru, Pt, and Pd) and in the presence of hydrogen, [43]. In order to purify water, other catalytic reduction methods were also suggested. Few of them are:

(i) Reduction of 4-aminophenol during existence of sodium borohydride on silver nanoparticles settled in the polymer (Alginate biohydrogel) [25].
(ii) Utilizing 1 atmospheric hydrogen gas (1 atm H_2), at 25 °C and on supported and unsupported palladium nanoparticles, bromate reduction to bromide [21].
(iii) Decomposition of organophosphorothioate pesticide (chlorpyrifos, CP) to diethyl tiophosphate and 3,5,6-trichloro-2-pyridinol (TCP) on Au and Ag nanoparticles (supported and unsupported on alumina) [22].

10.6.2.2 Oxidative Decomposition

An increasing interest has been shown in the nanocomposites as effective environmental catalysts, with serious water pollution increasing. In environmental catalysis, noble metal-based nanocomposites have been widely studied by researchers. Serving as key functional parts, noble metals are usually joined with other nanomaterials for rationally designing nanocomposites. As an outcome of the removal of pollutants, increased catalytic characteristics are exhibited [102].

Various reports exhibit breakdowns of organic compounds using noble metal-modified titania photocatalysts via oxidative decomposition. For example, oxidative decomposition of methyl orange (MO), Rhodamine 6G (R6G) [103], Congo red [35], Plasmocorinth B, methylene blue (MB) [87, 104], antibiotics (trimethoprim [105], tetracycline hydrochloride [83]) and acids (salicylic acid [106], stearic acid [107], formic acid [40], and oxalic acid [108]).

10.6.2.3 Reductive Decomposition

Reductive decomposition of some pollutants, such as MB, nitrobenzene, and 2,2′-dipyridyl disulfide (RSSR), the enhancement of activity was also noticed by employing noble metal-based nanoparticles in different form [109].

10.7 Elimination of Microbiological Pollutants

The United Nations General Assembly (UNGA) stated clean and safe drinking water, a human right. Therefore, every possible attempt should be made to attain drinking water that is safe [110]. Pathogens may result in chronic (e.g. cancer and infectious hepatitis) and acute (e.g. watery bowel moments, i.e. diarrhea) diseases; hence, a sufficient grant of microbially safe water is the uppermost prime concern. With water polluted with feces of animals, generally, the greatest microbial risks are linked [110, 111]. Helminths, protozoa, viruses, and pathogenic bacteria are sources of feces. Therefore, their removal from water is necessary to make it safe for human consumption. Various methods have been employed for this purpose, including chlorine disinfection, UV irradiation, and ozonation, which are out of scope for this chapter. However, we will emphasize the direct antimicrobial action of noble metal nanoparticles in this regard.

10.7.1 Direct Antimicrobial Action Using Noble Metal Nanoparticles

10.7.1.1 Platinum

Platinum has historically well reported to inhibit bacterial cell division. Cisplatin an anticancer drug is well known in chemotherapy. Moreover, the microbiological toxicity of platinum nanoparticles has been monitored [112]. It was noticed that for both bacteria and fungi, platinum nanoparticles can become the cause of the discharge of cytosolic proteins and damage cell walls [113, 114]. Depending on its interaction with cellular components, toxic effects of platinum have been reported [115]. It has been reported that bactericidal characteristics are dependent on the size of platinum nanoparticles, and small Pt nanoparticles (less than 3 nm) exhibit bacteriotoxic characteristics [114].

10.7.1.2 Gold

Gold is the most stable element, and due to its high oxidizability, Au ions are toxic for almost all microorganisms. Bactericidal activity is rarely noticed for gold nanoparticles in either gram-positive or gram-negative bacteria [116, 117]. Compared to bactericidal activity; gold nanoparticles displayed fungicidal activity, which was dependent on size of the nanoparticles [115, 117]. An increase in interaction between binding sites of plasma membrane proteins and gold was observed because of enhancement in the surface area of Au nanoparticles. Resultantly, H^+-ATPase-mediated proton pumping was inhibited [118].

10.7.1.3 Copper

Since long ago, it has been established that, similar to silver, copper displays high antimicrobial activity. Copper has low toxicity and an attractive cost ($LD50 = 413$ mg/kg) and on the surface of bacteria, copper ions adsorb; where (i) the surface proteins' structures are denaturated [119], and (ii) in the bactericidal process, oxidative stress is induced by adsorbed copper ions [120].

It has also been reported that Cu_2O has higher activity in inactivation than silver and CuO [119]. Copper nanoparticles also have lower toxicity than Ag nanoparticles against some fungi species. Copper-graphene sponge was reported for drinking water purification as having a high efficiency of bacteria inactivation after filtration [120].

10.7.1.4 Silver

Silver is known to possess remarkable antimicrobial characteristics and has been utilized in several antimicrobial applications, such as, electric appliances, clothes, shoes, cosmetics, and even for the treatment of cutlery and teeth. Human health is not affected by the decreased Ag levels in water utilized for drinking (usually less than $5\,\mu g\,l^{-1}$). Furthermore, the immensely poisonous ionic Ag is not expected in the natural environment [113]. In contrast, when salts of Ag are utilized for maintenance of drinking water's bacteriological quality, greater levels of Ag (up to $0.1\,mg\,l^{-1}$) could be accepted without danger to health (The guideline limit of US-EPA is less than $0.1\,mg\,l^{-1}$) [121]. Generally, silver ions are adsorbed onto bacterial cell wall (negatively charged), followed by a breakdown and finally discharge of materials outside the cell, resulting in lysis of cell and ultimately death of cell [122, 123].

10.7.1.5 Other Noble Metals

Antimicrobial effects have also been reported for other nanoparticles, such as ruthenium and palladium.

For antimicrobial agents, ruthenium nanoparticles are used as acetylcholine (Ach)-modified structures and an oxide [124, 125]. Likewise, for gram-positive bacteria, Pd nanoparticles are more operative, and their size-dependent antimicrobial activity has been demonstrated them [126].

10.8 Conclusions

As outlined above, pure water is a need of every life, and its quality of access for humans is still a serious issue. For various key usages, the tunable characteristics of noble metal nanoparticles make them superior to other kinds of nanoparticles. For the disinfection of microorganisms and degradation of toxic pollutants, noble metal nanoparticles could be employed as an environmentally friendly and efficient means for water treatment. Due to the benefits and disadvantages of these metals, an appropriate examination of the characteristics of each metal is necessary to plan effective systems for water treatment. Moreover, mixtures of metals with other procedures appear to be pleasing as environmentally friendly, well-organized, and cost-effective treatment methods. There are many promising avenues in terms of material science and medical applications of noble metal nanoparticles, and constant struggles are required to bring these to reality.

References

1 Selvan, S.T., Tan, T.T.Y., Yi, D.K., and Jana, N.R. (2010). Functional and multifunctional nanoparticles for bioimaging and biosensing. *Langmuir* 26 (14): 11631–11641.

2 Baptista, P., Pereira, E., Eaton, P. et al. (2008). Gold nanoparticles for the development of clinical diagnosis methods. *Anal. Bioanal. Chem.* 391 (3): 943–950.

3 Yavuz, M.S., Cheng, Y., Chen, J. et al. (2009). Gold nanocages covered by smart polymers for controlled release with near-infrared light. *Nat. Mater.* 8 (12): 935–939.

4 Thomas, M. and Klibanov, A.M. (2003). Conjugation to gold nanoparticles enhances polyethylenimine's transfer of plasmid DNA into mammalian cells. *Proc. Natl. Acad. Sci. U.S.A.* 100 (16): 9138–9143.

5 Jones, M.R., Millstone, J.E., Giljohann, D.A. et al. (2009). Plasmonically controlled nucleic acid dehybridization with gold nanoprisms. *ChemPhysChem* 10 (9–10): 1461–1465.

6 Huang, X., Jain, P.K., El-Sayed, I.H., and El-Sayed, M.A. (2007). Gold nanoparticles: interesting optical properties and recent applications in cancer diagnostics and therapy. *Fut. Med.* https://doi.org/10.2217/17435889.2.5.681.

7 Hainfeld, J.F., Slatkin, D.N., and Smilowitz, H.M. (2004). The use of gold nanoparticles to enhance radiotherapy in mice. *Phys. Med. Biol.* 49 (18): N309.

8 Hainfeld, J.F., Dilmanian, F.A., Slatkin, D.N., and Smilowitz, H.M. (2008). Radiotherapy enhancement with gold nanoparticles. *J. Pharm. Pharmacol.* 60 (8): 977–985.

9 Hirsch, L.R., Stafford, R.J., Bankson, J.A. et al. (2003). Nanoshell-mediated near-infrared thermal therapy of tumors under magnetic resonance guidance. *Proc. Natl. Acad. Sci. U.S.A.* 100 (23): 13549–13554.

10 Cuenca, A.G., Jiang, H., Hochwald, S.N. et al. (2006). Emerging implications of nanotechnology on cancer diagnostics and therapeutics. *Cancer* 107 (3): 459–466.

11 Powell, A.C., Paciotti, G.F., and Libutti, S.K. (2010). Colloidal gold: a novel nanoparticle for targeted cancer therapeutics. In: *Cancer Nanotechnology*, Methods in Molecular Biology, vol. 624 (ed. S. Grobmyer and B. Moudgil), 375–384. Humana Press.

12 Ye, L., Cao, Z., Liu, X. et al. (2022). Noble metal-based nanomaterials as antibacterial agents. *J. Alloys Compd.* 904: 164091.

13 Ollis, D.F. (1985). Contaminant degradation in water. *Environmental Sci. Technol.* 19 (6): 480–484.

14 Warner, C.L., Addleman, R.S., Cinson, A.D. et al. (2010). High-performance, superparamagnetic, nanoparticle-based heavy metal sorbents for removal of contaminants from natural waters. *ChemSusChem* 3 (6): 749–757.

15 Campelo, J.M., Luna, D., Luque, R. et al. (2009). Sustainable preparation of supported metal nanoparticles and their applications in catalysis. *ChemSusChem* 2 (1): 18–45.

16 Nair, A.S. and Pradeep, T.J.C.S. (2003). Halocarbon mineralization and catalytic destruction by metal nanoparticles. *Curr. Sci.* 84 (12): 1560–1564.

17 Bootharaju, M.S. and Pradeep, T. (2010). Uptake of toxic metal ions from water by naked and monolayer protected silver nanoparticles: an X-ray photoelectron spectroscopic investigation. *J. Phys. Chem. C* 114 (18): 8328–8336.

18 Sumesh, E., Bootharaju, M.S., and Pradeep, T. (2011). A practical silver nanoparticle-based adsorbent for the removal of Hg^{2+} from water. *J. Hazard. Mater.* 189 (1-2): 450–457.

19 Nair, A.S., Tom, R.T., and Pradeep, T. (2003). Detection and extraction of endosulfan by metal nanoparticles. *J. Environ. Monit.* 5 (2): 363–365.

20 Nair, A.S. and Pradeep, T. (2007). Extraction of chlorpyrifos and malathion from water by metal nanoparticles. *J. Nanosci. Nanotechnol.* 7 (6): 1871–1877.

21 Wang, Y., Liu, J., Wang, P. et al. (2014). Palladium nanoparticles encapsulated in core–shell silica: a structured hydrogenation catalyst with enhanced activity for reduction of oxyanion water pollutants. *ACS Catal.* 4 (10): 3551–3559.

22 Bootharaju, M.S. and Pradeep, T.J.L. (2012). Understanding the degradation pathway of the pesticide, chlorpyrifos by noble metal nanoparticles. *Langmuir* 28 (5): 2671–2679.

23 Vimala, K., Mohan, Y.M., Sivudu, K.S. et al. (2010). Fabrication of porous chitosan films impregnated with silver nanoparticles: a facile approach for superior antibacterial application. *Colloids Surf., B* 76 (1): 248–258.

24 Lee, S.J., Heo, D.N., Heo, M. et al. (2017). Most simple preparation of an inkjet printing of silver nanoparticles on fibrous membrane for water purification: technological and commercial application. *J. Ind. Eng. Chem.* 46: 273–278.

25 Otari, S., Patil, R.M., Waghmare, S.R. et al. (2013). A novel microbial synthesis of catalytically active Ag–alginate biohydrogel and its antimicrobial activity. *Dalton Trans.* 42 (27): 9966–9975.

26 Priebe, J.B., Radnik, J., Lennox, A.J.J. et al. (2015). Solar hydrogen production by plasmonic Au–TiO_2 catalysts: impact of synthesis protocol and TiO_2 phase on charge transfer efficiency and H_2 evolution rates. *ACS Catal.* 5 (4): 2137–2148.

27 Cozzoli, P.D., Curri, M.L., Giannini, C., and Agostiano, A. (2006). Synthesis of TiO_2–Au composites by titania-nanorod-assisted generation of gold nanoparticles at aqueous/nonpolar interfaces. *Small* 2 (3): 413–421.

28 Dozzi, M.V., Prati, L., Canton, P., and Selli, E. (2009). Effects of gold nanoparticles deposition on the photocatalytic activity of titanium dioxide under visible light. *Phys. Chem. Chem. Phys.* 11 (33): 7171–7180.

29 Hai, Z., El Kolli, N., Uribe, D.B. et al. (2013). Modification of TiO_2 by bimetallic Au–Cu nanoparticles for wastewater treatment. *J. Mater. Chem. A* 1 (36): 10829–10835.

30 Nischk, M., Mazierski, P., Wei, Z. et al. (2016). Enhanced photocatalytic, electrochemical and photoelectrochemical properties of TiO_2 nanotubes arrays modified with Cu, AgCu and Bi nanoparticles obtained via radiolytic reduction. *Appl. Surf. Sci.* 387: 89–102.

31 Ohtani, B., Kakimoto, M., Nishimoto, S. et al. (1993). Photocatalytic reaction of neat alcohols by metal-loaded titanium (IV) oxide particles. *J. Photochem. Photobiol., A* 70 (3): 265–272.

32 Kowalska, E., Abe, R., and Ohtani, B. (2009). Visible light-induced photocatalytic reaction of gold-modified titanium (IV) oxide particles: action spectrum analysis. *Chem. Commun.* (2): 241–243.

33 Herrmann, J.M., Disdier, J., and Pichat, P. (1986). Photoassisted platinum deposition on TiO_2 powder using various platinum complexes. *J. Phys. Chem.* 90 (22): 6028–6034.

34 Kowalska, E., Endo, M., Wei, Z. et al. (2019). Noble metal nanoparticles for water purification. In: *Nanoscale Materials in Water Purification*, 553–579. Elsevier.

35 Pearson, A., Zheng, H., Kalantar-zadeh, K. et al. (2012). Decoration of TiO_2 nanotubes with metal nanoparticles using polyoxometalate as a UV-switchable reducing agent for enhanced visible and solar light photocatalysis. *Langmuir* 28 (40): 14470–14475.

36 Mendez-Medrano, M.G., Kowalska, E., Lehoux, A. et al. (2016). Surface modification of TiO_2 with Au nanoclusters for efficient water treatment and hydrogen generation under visible light. *J. Phys. Chem. C* 120 (43): 25010–25022.

37 Cojocaru, B., Neaţu, Ş., Sacaliuc-Pârvulescu, E. et al. (2011). Influence of gold particle size on the photocatalytic activity for acetone oxidation of Au/TiO_2 catalysts prepared by dc-magnetron sputtering. *Appl. Catal., B* 107 (1-2): 140–149.

38 Brennan, L.J., Purcell-Milton, F., Salmeron, A.S. et al. (2015). Hot plasmonic electrons for generation of enhanced photocurrent in gold-TiO_2 nanocomposites. *Nanoscale Res. Lett.* 10 (1): 1–12.

39 Tangale, N.P., Belhekar, A.A., Kale, K.B., and Awate, S.V. (2014). Enhanced mineralization of gaseous organic pollutant by photo-oxidation using Au-doped TiO_2/MCM-41. *Water Air Soil Pollut.* 225 (2): 1–13.

40 Bouhadoun, S., Guillard, C., Dapozze, F. et al. (2015). One step synthesis of N-doped and Au-loaded TiO_2 nanoparticles by laser pyrolysis: application in photocatalysis. *Appl. Catal., B* 174: 367–375.

41 Mazierski, P., Malankowska, A., Kobylański, M. et al. (2017). Photocatalytically active TiO_2/Ag_2O nanotube arrays interlaced with silver nanoparticles obtained from the one-step anodic oxidation of Ti–Ag alloys. *ACS Catal.* 7 (4): 2753–2764.

42 Awate, S.V., Belhekar, A.A., Bhagwat, S.V. et al. (2008). Effect of gold dispersion on the photocatalytic activity of mesoporous titania for the vapor-phase oxidation of acetone. *Int. J. Photoenergy* 2008: 789149.

43 Hörold, S., Vorlop, K.D., Tacke, T., and Sell, M. (1993). Development of catalysts for a selective nitrate and nitrite removal from drinking water. *Catal. Today* 17 (1-2): 21–30.

44 Joshi, M.K., Pant, H.R., Liao, N. et al. (2015). In-situ deposition of silver–iron oxide nanoparticles on the surface of fly ash for water purification. *J. Colloid Interface Sci.* 453: 159–168.

45 Zodrow, K., Brunet, L., Mahendra, S. et al. (2009). Polysulfone ultrafiltration membranes impregnated with silver nanoparticles show improved biofouling resistance and virus removal. *Water Res.* 43 (3): 715–723.

46 Kim, J.P., Kim, J.H., Kim, J. et al. (2016). A nanofilter composed of carbon nanotube-silver composites for virus removal and antibacterial activity improvement. *J. Environ. Sci.* 42: 275–283.

47 Masheane, M.L., Nthunya, L.N., Malinga, S.P. et al. (2017). Synthesis of Fe-Ag/f-MWCNT/PES nanostructured-hybrid membranes for removal of Cr (VI) from water. *Sep. Purif. Technol.* 184: 79–87.

48 Cruz-Tato, P., Ortiz-Quiles, E.O., Vega-Figueroa, K. et al. (2017). Metalized nanocellulose composites as a feasible material for membrane supports: design and applications for water treatment. *Environ. Sci. Technol.* 51 (8): 4585–4595.

49 Dankovich, T.A. and Smith, J.A. (2014). Incorporation of copper nanoparticles into paper for point-of-use water purification. *Water Res.* 63: 245–251.

50 Dankovich, T.A. (2014). Microwave-assisted incorporation of silver nanoparticles in paper for point-of-use water purification. *Environ. Sci.: Nano* 1 (4): 367–378.

51 Kim, H.J., Joshi, M.K., Pant, H.R. et al. (2015). One-pot hydrothermal synthesis of multifunctional Ag/ZnO/fly ash nanocomposite. *Colloids Surf., A* 469: 256–262.

52 Kraeutler, B. and Bard, A.J. (1978). Heterogeneous photocatalytic preparation of supported catalysts. Photodeposition of platinum on titanium dioxide powder and other substrates. *J. Am. Chem. Soc.* 100 (13): 4317–4318.

53 Nishimoto, S.-I., Ohtani, B., Kajiwara, H., and Kagiya, T. (1983). Photoinduced oxygen formation and silver-metal deposition in aqueous solutions of various silver salts by suspended titanium dioxide powder. *J. Chem. Soc., Faraday Trans. 1* 79 (11): 2685–2694.

54 Kowalska, E., Wei, Z.S., Karabiyik, B. et al. (2014). Development of plasmonic photocatalysts for environmental application. *Adv. Sci. Technol.* 93: 174–183.

55 Sclafani, A., Mozzanega, M.-N., and Herrmann, J.-M. (1997). Influence of silver deposits on the photocatalytic activity of titania. *J. Catal.* 168 (1): https://doi.org/10.1006/jcat.1997.1631.

56 Waqar, M., Imran, M., Adil, S.F. et al. (2019). Enhanced photoluminescence and photocatalytic efficiency of La-doped bismuth molybdate: its preparation and characterization. *Materials* 13 (1): 35.

57 Nakata, K. and Fujishima, A. (2012). TiO_2 photocatalysis: design and applications. *J. Photochem. Photobiol., C* 13 (3): 169–189.

58 Herrmann, J.-M., Guillard, C., and Pichat, P. (1993). Heterogeneous photocatalysis: an emerging technology for water treatment. *Catal. Today* 17 (1-2): 7–20.

59 Lalitha, K., Sadanandam, G., Kumari, V.D. et al. (2010). Highly stabilized and finely dispersed Cu_2O/TiO_2: a promising visible sensitive photocatalyst for continuous production of hydrogen from glycerol: water mixtures. *J. Phys. Chem. C* 114 (50): 22181–22189.

60 Kisch, H., Sakthivel, S., Janczarek, M. et al. (2007). A low-band gap, nitrogen-modified titania visible-light photocatalyst. *J. Phys. Chem. C* 111 (30): 11445–11449.

61 Macyk, W., Szaciłowski, K., Stochel, G. et al. (2010). Titanium (IV) complexes as direct TiO_2 photosensitizers. *Coord. Chem. Rev.* 254 (21-22): 2687–2701.

62 Qiu, X., Miyauchi, M., Sunada, K. et al. (2012). Hybrid Cu_xO/TiO_2 nanocomposites as risk-reduction materials in indoor environments. *ACS Nano* 6 (2): 1609–1618.

63 Janus, M., Tryba, B., Inagaki, M., and Morawski, A.W. (2004). New preparation of a carbon-TiO_2 photocatalyst by carbonization of n-hexane deposited on TiO_2. *Appl. Catal., B* 52 (1): 61–67.

64 Zaleska, A. (2008). Doped-TiO_2: a review. *Rec. Patent. Eng.* 2 (3): 157–164.

65 Asahi, R., Morikawa, T., Ohwaki, T. et al. (2001). Visible-light photocatalysis in nitrogen-doped titanium oxides. *Science* 293 (5528): 269–271.

66 Dozzi, M.V. and Selli, E. (2013). Doping TiO_2 with p-block elements: effects on photocatalytic activity. *J. Photochem. Photobiol., C* 14: 13–28.

67 Kuncewicz, J., Ząbek, P., Kruczała, K. et al. (2012). Photocatalysis involving a visible light-induced hole injection in a chromate (VI)–TiO_2 system. *J. Phys. Chem. C* 116 (41): 21762–21770.

68 Soria, J., Conesa, J.C., Augugliaro, V. et al. (1991). Dinitrogen photoreduction to ammonia over titanium dioxide powders doped with ferric ions. *J. Phys. Chem.* 95 (1): 274–282.

69 Ohno, T., Akiyoshi, M., Umebayashi, T. et al. (2004). Preparation of S-doped TiO_2 photocatalysts and their photocatalytic activities under visible light. *Appl. Catal., A* 265 (1): 115–121.

70 Wei, Z., Kowalska, E., Verrett, J. et al. (2015). Morphology-dependent photocatalytic activity of octahedral anatase particles prepared by ultrasonication–hydrothermal reaction of titanates. *Nanoscale* 7 (29): 12392–12404.

71 Janczarek, M., Kowalska, E., and Ohtani, B. (2016). Decahedral-shaped anatase titania photocatalyst particles: synthesis in a newly developed coaxial-flow gas-phase reactor. *Chem. Eng. J.* 289: 502–512.

72 Jaffrezic-Renault, N., Pichat, P., Foissy, A., and Merrier, R. (1986). Study of the effect of deposited platinum particles on the surface charge of titania aqueous suspensions by potentiometry, electrophoresis, and labeled-ion adsorption. *J. Phys. Chem.* 90 (12): 2733–2738.

73 Kowalska, E., Remita, H., Colbeau-Justin, C. et al. (2008). Modification of titanium dioxide with platinum ions and clusters: application in photocatalysis. *J. Phys. Chem. C* 112 (4): 1124–1131.

74 Bamwenda, G.R., Tsubota, S., Nakamura, T., and Haruta, M. (1995). Photoassisted hydrogen production from a water-ethanol solution: a comparison of activities of Au TiO_2 and Pt TiO_2. *J. Photochem. Photobiol., A* 89 (2): 177–189.

75 Ohtani, B., Kakimoto, M., Miyadzu, H. et al. (1988). Effect of surface-adsorbed 2-propanol on the photocatalytic reduction of silver and/or nitrate ions in acidic titania suspension. *J. Phys. Chem.* 92 (20): 5773–5777.

76 Hope, G.A. and Bard, A.J. (1983). Platinum/titanium dioxide (rutile) interface. Formation of ohmic and rectifying junctions. *J. Phys. Chem.* 87 (11): 1979–1984.

77 Zang, L., Lange, C., Abraham, I. et al. (1998). Amorphous microporous titania modified with platinum (IV) chloride a new type of hybrid photocatalyst for visible light detoxification. *J. Phys. Chem. B* 102 (52): 10765–10771.

78 Ohtani, B. and Nishimoto, S. (1993). Effect of surface adsorptions of aliphatic alcohols and silver ion on the photocatalytic activity of titania suspended in aqueous solutions. *J. Phys. Chem.* 97 (4): 920–926.

79 Pichat, P., JM, H., Disdier, J. et al. (1981). Photocatalytic hydrogen production from aliphatic alcohols over a bifunctional platinum on titanium dioxide catalyst.

80 Ohtani, B., Ogawa, Y., and Nishimoto, S. (1997). Photocatalytic activity of amorphous–anatase mixture of titanium (IV) oxide particles suspended in aqueous solutions. *J. Phys. Chem. B* 101 (19): 3746–3752.

81 Caretti, I., Keulemans, M., Verbruggen, S.W. et al. (2015). Light-induced processes in plasmonic gold/TiO_2 photocatalysts studied by electron paramagnetic resonance. *Top. Catal.* 58 (12): 776–782.

82 Carneiro, J.T., Savenije, T.J., and Mul, G. (2009). Experimental evidence for electron localization on Au upon photo-activation of Au/anatase catalysts. *Phys. Chem. Chem. Phys.* 11 (15): 2708–2714.

83 Smirnova, N., Vorobets, V., Linnik, O. et al. (2010). Photoelectrochemical and photocatalytic properties of mesoporous TiO_2 films modified with silver and gold nanoparticles. *Surf. Interface Anal.* 42 (6-7): 1205–1208.

84 Chen, W.-T., Chan, A., Al-Azri, Z.H.N. et al. (2015). Effect of TiO_2 polymorph and alcohol sacrificial agent on the activity of Au/TiO_2 photocatalysts for H_2 production in alcohol–water mixtures. *J. Catal.* 329: 499–513.

85 Dosado, A.G., Chen, W.T., Chan, A. et al. (2015). Novel Au/TiO_2 photocatalysts for hydrogen production in alcohol–water mixtures based on hydrogen titanate nanotube precursors. *J. Catal.* 330: 238–254.

86 Yu, J., Yue, L., Liu, S. et al. (2009). Hydrothermal preparation and photocatalytic activity of mesoporous Au–TiO_2 nanocomposite microspheres. *J. Colloid Interface Sci.* 334 (1): 58–64.

87 Li, F. and Li, X.Z. (2002). Photocatalytic properties of gold/gold ion-modified titanium dioxide for wastewater treatment. *Appl. Catal., A* 228 (1-2): 15–27.

88 Mkhalid, I.A. (2015). Visible light photocatalytic synthesis of aniline with an Au/$LaTiO_3$ nanocomposites. *J. Alloys Compd.* 631: 298–302.

89 Tian, Y. and Tatsuma, T. (2005). Mechanisms and applications of plasmon-induced charge separation at TiO_2 films loaded with gold nanoparticles. *J. Am. Chem. Soc.* 127 (20): 7632–7637.

90 Awazu, K., Fujimaki, M., Rockstuhl, C. et al. (2008). A plasmonic photocatalyst consisting of silver nanoparticles embedded in titanium dioxide. *J. Am. Chem. Soc.* 130 (5): 1676–1680.

91 Halas, N. (2005). Playing with plasmons: tuning the optical resonant properties of metallic nanoshells. *MRS Bull.* 30 (5): 362–367.

92 Haes, A.J., Haynes, C.L., McFarland, A.D. et al. (2005). Plasmonic materials for surface-enhanced sensing and spectroscopy. *MRS Bull.* 30 (5): 368–375.

93 Atwater, H.A., Maier, S., Polman, A. et al. (2005). The new "p–n junction": plasmonics enables photonic access to the nanoworld. *MRS Bull.* 30 (5): 385–389.

94 Okamoto, K., Niki, I., Shvartser, A. et al. (2007). Surface plasmon enhanced bright light emission from InGaN/GaN. *Phys. Status Solidi A* 204 (6): 2103–2107.

95 Louis, C. and Pluchery, O. (2017). *Gold Nanoparticles for Physics, Chemistry and Biology*. World Scientific.

96 Ueno, K. and Misawa, H. (2013). Surface plasmon-enhanced photochemical reactions. *J. Photochem. Photobiol., C* 15: 31–52.

97 Verbruggen, S.W. (2015). TiO_2 photocatalysis for the degradation of pollutants in gas phase: from morphological design to plasmonic enhancement. *J. Photochem. Photobiol., C* 24: 64–82.

98 Sarina, S., Waclawik, E.R., and Zhu, H. (2013). Photocatalysis on supported gold and silver nanoparticles under ultraviolet and visible light irradiation. *Green Chem.* 15 (7): 1814–1833.

99 Lou, Z., Wang, Z., Huang, B., and Dai, Y. (2014). Synthesis and activity of plasmonic photocatalysts. *ChemCatChem* 6 (9): 2456–2476.

100 Alqudami, A., Alhemiary, N.A., and Munassar, S. (2012). Removal of Pb (II) and Cd (II) ions from water by Fe and Ag nanoparticles prepared using electro-exploding wire technique. *Environ. Sci. Pollut. Res.* 19 (7): 2832–2841.

101 Zaidi, A., Simms, K., and Kok, S. (1992). The use of micro/ultrafiltration for the removal of oil and suspended solids from oilfield brines. *Water Sci. Technol.* 25 (10): 163–176.

102 Fu, Y., Yin, Z., Qin, L. et al. (2022). Recent progress of noble metals with tailored features in catalytic oxidation for organic pollutants degradation. *J. Hazard. Mater.* 422: 126950.

103 Yi, Z., Chen, J., Luo, J. et al. (2015). Surface-plasmon-enhanced band emission and enhanced photocatalytic activity of Au nanoparticles-decorated ZnO nanorods. *Plasmonics* 10 (6): 1373–1380.

104 Wang, X., Choi, J., Mitchell, D.R.G. et al. (2013). Enhanced photocatalytic activity: macroporous electrospun mats of mesoporous Au/TiO_2 nanofibers. *ChemCatChem* 5 (9): 2646–2654.

105 Oros-Ruiz, S., Zanella, R., Zanella, R., and Prado, B. (2013). Photocatalytic degradation of trimethoprim by metallic nanoparticles supported on TiO_2-P25. *J. Hazard. Mater.* 263: 28–35.

106 Zsolt, P., Andreea, R., Hidi, I.J. et al. (2013). Behavior of gold nanoparticles in a titania aerogel matrix: photocatalytic activity assessment and structure investigations. *Chin. J. Catal.* 34 (4): 734–740.

107 Kafizas, A., Kellici, S., Darr, J.A., and Parkin, I.P. (2009). Titanium dioxide and composite metal/metal oxide titania thin films on glass: a comparative study of photocatalytic activity. *J. Photochem. Photobiol., A* 204 (2-3): 183–190.

108 Iliev, V., Tomova, D., Bilyarska, L., and Tyuliev, G. (2007). Influence of the size of gold nanoparticles deposited on TiO_2 upon the photocatalytic destruction of oxalic acid. *J. Mol. Catal. A: Chem.* 263 (1-2): 32–38.

109 Costi, R., Saunders, A.E., Elmalem, E. et al. (2008). Visible light-induced charge retention and photocatalysis with hybrid CdSe–Au nanodumbbells. *Nano Lett.* 8 (2): 637–641.

110 World Health Organization (2011). *Guidelines for Drinking-Water Quality*, 4e. Geneva: WHO.

111 Davison, A., Howard, G., Stevens, M. et al. (2005). *Water Safety Plans: Managing Drinking-Water Quality from Catchment to Consumer*. World Health Organization.

112 Davis, C.P., Wagle, N., Anderson, M.D., and Warren, M.M. (1991). Bacterial and fungal killing by iontophoresis with long-lived electrodes. 35 (10): 2131–2134.

113 Ratte, H.T. (1999). Bioaccumulation and toxicity of silver compounds: a review. *Environ. Toxicol. Chem.* 18 (1): 89–108.

114 Asharani, P., Xinyi, N.G., Hande, M.P., and Valiyaveettil, S. (2010). DNA damage and p53-mediated growth arrest in human cells treated with platinum nanoparticles. *Nanomedicine* 5 (1): 51–64.

115 Ahmad, T., Wani, I.A., Lone, I.H. et al. (2013). Antifungal activity of gold nanoparticles prepared by solvothermal method. *Mater. Res. Bull.* 48 (1): 12–20.

116 Amin, R.M., Mohamed, M.B., Ramadan, M.A. et al. (2009). Rapid and sensitive microplate assay for screening the effect of silver and gold nanoparticles on bacteria. *Fut. Med.* 4 (6): 637–643.

117 Ovais, M., Raza, A., Naz, S. et al. (2017). Current state and prospects of the phytosynthesized colloidal gold nanoparticles and their applications in cancer theranostics. *Appl. Microbiol. Biotechnol.* 101 (9): 3551–3565.

118 Baram-Pinto, D., Shukla, S., Gedanken, A., and Sarid, R. (2010). Inhibition of HSV-1 attachment, entry, and cell-to-cell spread by functionalized multivalent gold nanoparticles. *Small* 6 (9): 1044–1050.

119 Sunada, K., Minoshima, M., and Hashimoto, K. (2012). Highly efficient antiviral and antibacterial activities of solid-state cuprous compounds. *J. Hazard. Mater.* 235: 265–270.

120 Deng, C.-H., Gong, J.L., Zeng, G.M. et al. (2017). Graphene sponge decorated with copper nanoparticles as a novel bactericidal filter for inactivation of *Escherichia coli*. *Chemosphere* 184: 347–357.

121 World Health Organization (2011). *Guidelines for Drinking-Water Quality*, vol. 216, 303–304. WHO.

122 Chwalibog, A., Sawosz, E., Hotowy, A. et al. (2010). Visualization of interaction between inorganic nanoparticles and bacteria or fungi. *Int. J. Nanomed.* 5: 1085.

123 Sambhy, V., MacBride, M.M., Peterson, B.R., and Sen, A. (2006). Silver bromide nanoparticle/polymer composites: dual action tunable antimicrobial materials. *J. Am. Chem. Soc.* 128 (30): 9798–9808.

124 Kannan, S. and Sundrarajan, M. (2015). Green synthesis of ruthenium oxide nanoparticles: characterization and its antibacterial activity. *Adv. Powder Technol.* 26 (6): 1505–1511.

125 Huang, X., Chen, G., Pan, J. et al. (2016). Effective PDT/PTT dual-modal phototherapeutic killing of pathogenic bacteria by using ruthenium nanoparticles. *J. Mater. Chem. B* 4 (37): 6258–6270.

126 Adams, C.P., Walker, K.A., Obare, S.O., and Docherty, K.M. (2014). Size-dependent antimicrobial effects of novel palladium nanoparticles. *PLoS One* 9 (1): e85981.

11

Immobilized Enzyme Systems for Wastewater Treatment

Ambreen Aisha[1], *Muhammad J. Iqbal*[2], *Razia Noreen*[3], *Naila Mukhtar*[4], *Asim Hussain*[5], *and Asim Rehman*[3]

[1] *Faisalabad Medical University, Department of Biochemistry, PMC colony Faisalabad, Faisalabad, 38000 Pakistan*
[2] *COMSATS University Islamabad, Department of Chemistry, Lahore Campus, Lahore, 54770 Pakistan*
[3] *Government College University, Department of Biochemistry, Faisalabad, 38000 Pakistan*
[4] *University of Swabi, 23320 Pakistan*
[5] *University of Agriculture, Department of Biochemistry, Faisalabad, 38000 Pakistan*

11.1 Immobilization Techniques

Enzyme immobilization, when performed properly, may deliver the potential benefits of enzyme delivery. Entrapment, adsorption, and cross-linking covalent chemical bonding are four of the most commonly utilized physical or chemical strategies for enzyme immobilization, as shown in Figure 11.1. The effectiveness of each of these approaches differs, according to Nisha et al. [1]. Physical immobilization methods such as ionic bonding, van der Waals forces, and physical constraint are used in both adsorption and entrapment. These approaches have the advantage of keeping the enzyme's catalytic activity because the enzyme's natural structure is not altered during the process [1].

A simple process, great efficiency, and low cost make adsorption one of the most widely used enzyme immobilization procedures nowadays [2, 3]. The immobilization technique involves immersing the support in an enzyme solution for a specific period of time. The enzyme and support, on the other hand, are easily disrupted due to the weak nonspecific forces that hold them together. Entrapment is a strategy for immobilizing enzymes that is used to prevent aggregates of enzyme, which decreases the enzyme catalytic activity. Enzymes are immobilized using approaches such as microencapsulation, metal–organic frameworks (MOF), or gel/fiber entrapment [4, 5]. Substrates flow through the networks to activate the reaction when enzymes are immobilized in this manner, while the enzyme stays crosslinked to the support. As a consequence, the enzyme's geometrical and functional stability improves, and leaching-induced enzymatic depletion is reduced to some degree. The enzyme configurations remain stable as a result of covalent

Nanomaterials for Air and Water Purification, First Edition.
Edited by Deepu A. Gopakumar, Jyotishkumar Parameswaranpillai, Jinu Jacob George, and Midhun Dominic C.D.

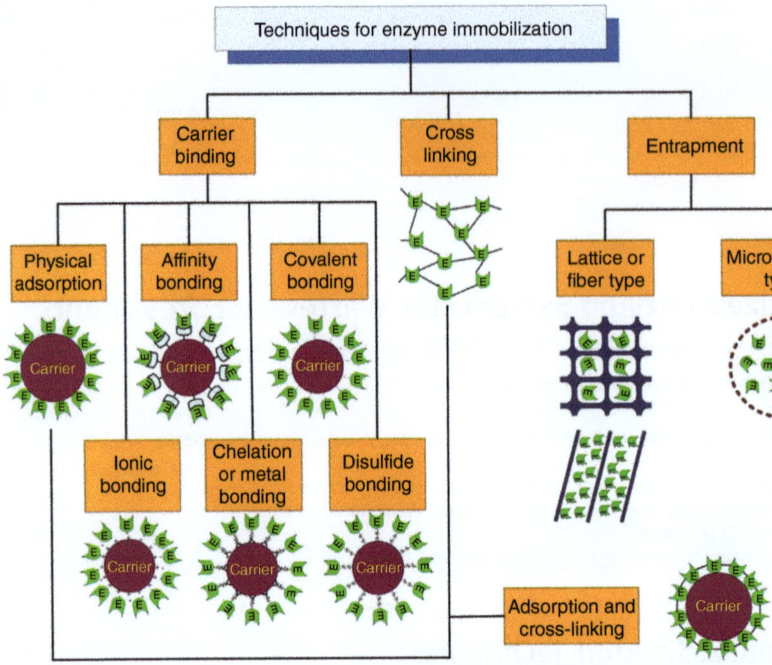

Figure 11.1 Comprehensive out look of enzyme immobilization techniques, with the major interactions involved.

bonds generated between the enzyme and the support. Consequently, the catalytic activity is very high. However, the presence of a high infusibility barrier that inhibits macromolecular substrates from flowing through the enzyme network is a disadvantage of entrapment. Covalent attachment and cross-linking enzymes, on the other hand, are chemical immobilization methods that rely on the information contained in covalent connections formed between enzymes and supports such as Glutaraldehyde. After immobilization, enzymes are modified in their conformation. This might lower the efficiency of enzyme, yet it strengthens enzyme stability by chemical cross linking [6].

11.2 Solid Matrix/Supports for Immobilization

The significance of solid support for the catalytic efficiency of immobilized enzyme technology is undisputed. The Extracellular matrix of the sorbents has an impact on adsorbed enzyme's functionality. For example, the support's physical and chemical structures, enzyme-carrier interactions, binding locations, and even the number of bonds may be influenced. The materials used for support should be readily accessible, affordable, and environmentally friendly [7]. The ideal potential properties of a support material are high catalytic efficiency, increased operational stability, insolubility, reduced toxicity, and resistance to severe environmental and microbial assault.

Organic auxiliary materials: Water-insoluble polysaccharides, such as collagen, biotin, dopamine, chitin, cellulose, and chitosan, are the most extensively utilized organic natural polymers for enzyme immobilization support materials. Chitosan, for example, is used as a support material because it is sustainable, nontoxic, inexpensive in cost, decomposable, and has strong enzymatic biocompatibility. Enzyme immobilization of SBP on chitosan beads showed improved enzymatic activity, tolerance to hydrolytic cleavage, and resilience to attack of microorganisms. Furthermore, Bilal et al. [8] found that immobilizing manganese peroxidase condensation on the support of chitosan nanoballs improved functional and storage strength and stability over the free enzyme. However, because of their hydrophilic properties and sparse affinities to bind enzymes, they have low mechanical stability. In contrast, a broad range of organic synthetic polymers, such as polystyrene, polyacrylate, polyacrylamide, polyamides, and vinyl, have been used as supports for immobilized enzymes [9]. Arslan [10] described, for example, the immobilization of horseradish peroxidase (HRP) onto polyethylene terephthalate (PET) fibers. PET fibers delivered renowned characteristics such as cost effectiveness, high surface areas, excellent mechanical toughness, improved temperature and storage stability, decreased sensitivity to microbiological attacks, and the capacity to be recycled. Miletic et al. [11] explained that synthetic polymers may allow for surface modification by polymerization of appropriate functional groups in order to get enhanced carrier properties. Wang et al. [12], exhibited HRP immobilization on polyacrylonitrile beads, which were treated with sodium chloride, hydrogen chloride, ethylenediamine, chitosan, and glutaraldehyde. When compared to free enzymes, these immobilized enzymes demonstrated momentous improvements in enzymatic characteristics and performance. According to Grigoras [13], material of inorganic supports have more persistence than to immobilized on organic supports, since they are more resistant to harsh working conditions. As a result, inorganic supports are preferable to organic supports in real-world applications. Clay, celite, kaolin, and silica are examples of inorganic natural minerals that are often employed as anchoring materials for enzyme immobilization. According to Sekuljica et al. [14], the adsorption of HRP onto kaolin proved to have excellent stability than organic matrixes. Owing to the superphysical stability of kaolin and its resilience to changes in factors chemically or physically changes like excessive pressure, temperature, and pH. Kim et al. [15] discovered that immobilizing HRP in fulvic acid-activated clays resulted in high enzymatic activity for the removal of phenol from wastewater. Furthermore, these inorganic natural supports are inexpensive, environmentally benign, and have great thermal and mechanical stability. However, the use of inorganic supports may produce abrasion in agitated vessels [16]. Inorganic synthetic minerals, whereas metals, metal oxides, and glass beads have been frequently employed as binding materials to immobilize enzymes. Materials with nanostructures, such as nanoparticles (NPs), nanocomposites, and carbon nanotubes, have recently received a lot of interest due to their practicality as support materials for enzyme immobilization. Nanomaterials, without a doubt, serve as sturdy and favorable binding supports due to their excellent properties, offering increased surface availability, better enzymatic efficiency, resistance to heat, and mechanical strengths [17]. Mohamed et al. [18] investigated and compared the enzyme activity of immobilized and pure HRP on Fe3O4 magnetic NPs. The investigations revealed that after immobilization enzymes produced greater enzymatic activity, vast temperature and pH range, and were more resistant to microbial assaults than free enzymes (Figure 11.2).

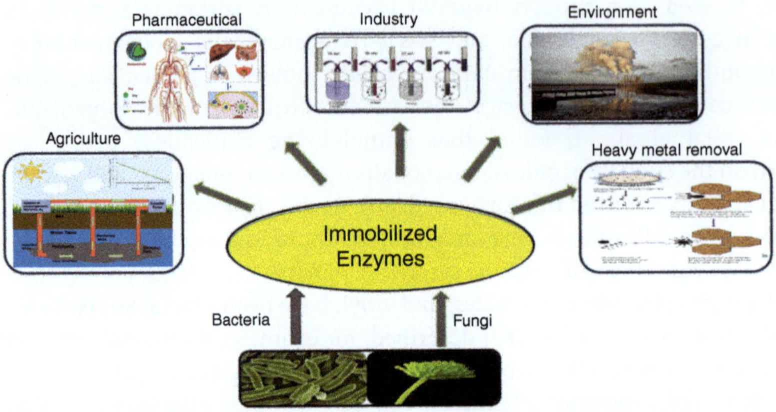

Figure 11.2 Immobilization of enzymes on organic or inorganic supports serves agricultural remediation, pharmaceutical applications, industry, environmental bioremediation, and detoxification of heavy metal pollutants.

11.3 Pollutants From Industry

Aromatic compounds, which include phenols and aromatic amines, are a significant class of pollutants that are rigorously controlled in many countries. They have been discovered in wastewater from a variety of sectors, including coal transformation, refining petroleum products, adhesives and polymers, timber preservation, metal plating, dyeing, and other chemical products, textile industries, mineral extraction and dressing, and papermaking [19]. Many industrial sites and sites of dumped waste are contaminated with crude oil (hydrocarbons). These locations pollute the streams and rivers and surface soils caused by gasoline, diesel, and fuel oil. Most of these NAPLs are hard to dispose of due to their volatility, high viscosity, and restricted mobility. [20].

Studies carried out to measure the pollutants from plastics and welding reveal organic industrial effluents like benzene, naphthalene, toluene, alkanes, and gaseous forms of methane, ethane and Butane (Figure 11.3) [21, 22]. There are already reports of fumes from laser etching and superheated steam welding of various polymers. It was observed that laser cutting produces a complex combination of organic molecules, including particle debris. All aromatics, including benzene, as well as PAHs, have been categorized as substances [23] (Figure 11.3).

Contaminants have severe consequences, if effluent fails to be scrubbed before release into rivers. Sludge and anaerobic circumstances may develop which predominantly contain organic substances. If they are thrown into water bodies like rivers, streams, or lakes, they may deplete naturally available oxygen sources and produce harmful chemicals that are detrimental to marine/fresh water life. Along with that, bacteria present in wastewater have the potential to cause infectious diseases. Inorganic chemical substances like detergents, phenols, agricultural pesticides, and fertilizers can be exceedingly toxic, carcinogenic, mutagenic, or teratogenic (Table 11.1).

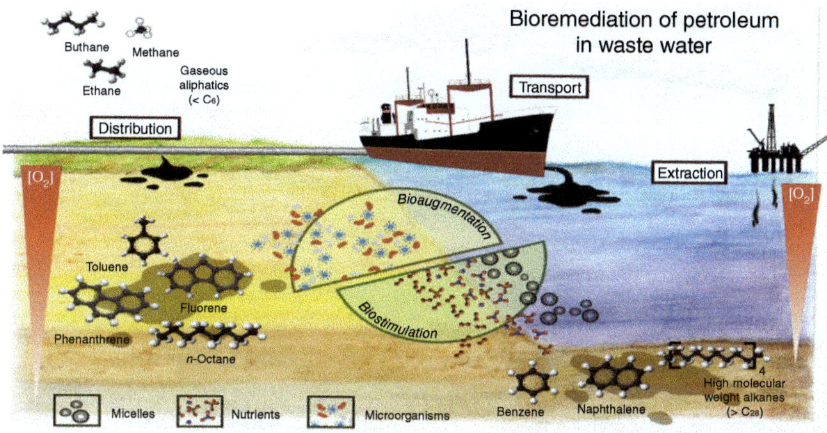

Figure 11.3 Removal of polycyclic hydrocarbons, aromatic hydrocarbons (benzene), alkanes derived from laser cutting and hot gas welding of plastics, etc., using detoxification processes of microbial bioaugmentation and biostimulation.

Table 11.1 Types of contaminants with their drastic outcomes.

Categories of contaminants	Sources and drastic outcomes
Solids in suspension	Sludge and anaerobic conditions may develop, if wastewater is not treated before being released to waterways.
Organic compounds that are biodegradable	Are mostly composed of protein-rich carbs and lipids. BOD and COD are routinely used to quantify them. They may deplete naturally occurring oxygen supplies and create toxic conditions that are harmful to marine life if they are dumped into river basins, streams, or lakes.
Microbes causing disease	Infectious illnesses may be spread by bacteria found in wastewater.
Pollutants of high priority	Chemical compounds, both organic and inorganic, may be extremely poisonous, carcinogenic, mutagenic, or teratogenic.
Organics that are refractory	Detergents, phenols, agricultural pesticides, and fertilizers are some of the substances that have a tendency to resist standard sewage treatment.

Source: [24].

11.4 Wastewater From Decolorization and Bleaching Operations

Bleaching processes produce muddy brown effluents containing hazardous and mutagenic chlorinated compounds that endanger the environment [25].

Laccases: Bleaching effluents may be treated sing peroxidase and laccases, among other methods (Figure 11.4). The Kraft method, which is widely employed

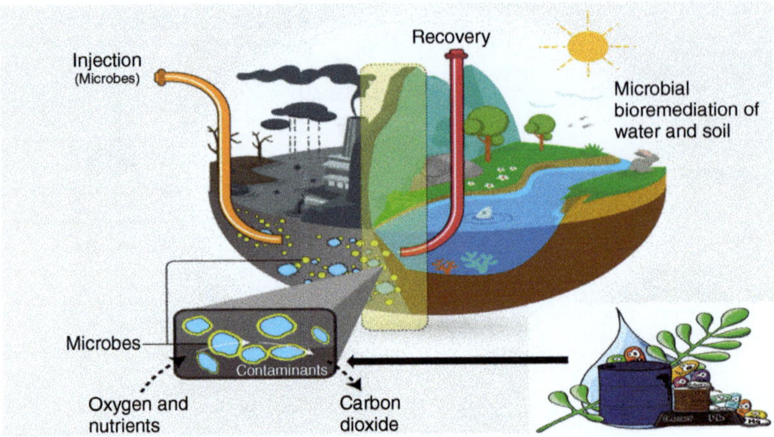

Figure 11.4 Microbial bioremediation of industrial effluents from paper, pulping, and decolorizing factories using immobilized laccases, oxidoreductases, and Keratinases.

in wood pulping, yields 5–8% (w w^{-1}) residual processed lignin from the textile fiber. The residual lignin is essential for the pulp's unique brown hue and is commercially cleaned using bleaching chemicals like as chlorine and chlorinated oxides [26].

Oxidoreductases and Keratinases, commercially accessible immobilized hydrolases, have the capacity to decolorize molasses effluent. Similarly, keratinase immobilized on commercial scale provided the maximum decolorized products (86.6–91.1%) of all commercial immobilized enzymes evaluated. Immobilized keratinase, rather than immobilized oxidoreductase, has the capacity to decolorize molasses effluent. Keratinase from Meiothermus taiwanensis WR-220 (KMT) bound on modified bagasse cellulose showed a decolorization product yield of 84.7–90.2%. For five days, it constantly eliminated 60.2–65.6% of the dyeing agents and 61.4–69.8% of the chemical oxygen demand (COD). Immobilized keratinase, rather than immobilized oxidoreductase, has the ability to decolorize treated municipal wastewater (TMWW) (Figure 11.4). Bagasse cellulose immobilized on altered KMT-wt cleared 84.7–90.2% of pigments [27].

11.5 The Involvement of Microbes in Detoxifying

Wide variety of microbial enzymes are employed for bioremidiation. The treatment aims to improve industrial effluents, tanning wastes, lignin biodegradation in the paper industry, and oxidative biotransformation of environmental contaminants (Table 11.2). Through oxidative coupling, certain microbes, including bacteria and fungi [28] and higher plants [29], detoxify harmful chemical molecules by using oxygenases (Figure 11.2).

Table 11.2 Microbial Enzymes and their effective detoxification reaction for industry.

Enzymes	Origin	Various reactions	Industry	Reference
Microbial cellulose	*Aspergillus niger* (fungal)	Removal of cellulose microfibrils	Paper and pulp industry	[30]
Microbial lipase	*Aspergillus niger* (fungal)	Applications in food, chemical, detergent manufacturing, cosmetic, and paper making industries	Hydrolysis, inter esterification, esterification, alcoholysis, and aminolysis	[31]
Microbial peroxidase	Plant and animal origin	The oxidation of phenolics and lignin at the expenditure of hydrogen peroxide (H_2O_2)	Tanning and lignin metabolism	[32]
Oxygenases	Bacterial enzymes	Dehalogenation reactions of alkyl halides derived from methane, ethanes, and ethylenes	Effluent treatment	[29, 33]
Monooxygenase		Denitrification, dehalogenation, ammonification, Desulfurization, hydroxylation, biodegradation, biotransformation, of various aromatic and aliphatic compounds	Industrial effluent	[34]
Dioxygenases i.e. Laccases	Plants, fungi, insects, and bacteria	Oxidative biotransformation of phenolics like (ortho and para diphenols, aminophenols, polyphenols), polyamines, aryl diamines and lignins	Environmental remediation	[35]

11.6 Pesticides, Including Herbicides, Insecticides, and Fungicides

Currently, crop protection pesticides, which comprise fungicides, insecticides, and herbicides, are sprayed on plants to control pests and weeds. In the near future, crop protection products are projected to continue to be extensively utilized across the globe. In addition to pesticide waste disposal and pesticide container and spray tank disinfection, pesticide industrial runoff contamination of surface and groundwater also contributes to environmental pollution effects [24]. It is estimated that three million tons of cyanide are utilized annually in different industrial pharmaceutical and estrogen removal operations across the globe. Among the novel research, Chagas revealed that chitosan beads cross-linked with glutaraldehyde showed high content of SBP immobilization for the clearance of phenolic compounds from waste water [24, 36].

11.7 Adsorption Treatment of Antibiotics From Effluents

Antibiotics include trimethoprim, fluoroquinolones, sulfonamides, and tetracyclines (TCs). The effectiveness of antibiotics such as TCs makes them one of the most widely prescribed medications for a broad variety of ailments [37]. There have been reports of antibiotics in wastewater influents and effluents, with around 50–90% of these being metabolites [38]. TCs were found in over 80% of industrial and home drainage systems, making them a prevalent antibacterial. Doxycycline (DOX) hydrochloride is a kind of tetracycline antibiotic [39]. Adsorption of DOX onto amorphous activated carbon produced from pumpkin seed coats by microwave-assisted pyrolysis. It has gained popularity over other TC (tetracyclin) antibiotics because of its broad antibacterial activity and ability to treat both gram-positive and gram-negative bacteria [40].

In early regimens of COVID-19 treatment, DOX was administered to reduce inflammation and infection [41]. Because of its complex structure and high solubility, DOX has significant residual toxicity in the aquatic environment. DOX pollutes bodies of water and enhances microorganism antibiotic resistance due to its high solubility in water [39]. A viable technique for eliminating DOX from wastewater is required. DOX has been removed from wastewater using a variety of techniques (physical, chemical, biological, and advanced oxidation). [42]. The tremendous demand for DOX, along with its imperfect metabolism and unpredictable atmospheric hazards, has produced a serious ecological concern that must be addressed. RHA was used in research to determine whether it may boost DOX antibiotic sorption effectively by binding to the antibiotic's active metabolite. Furthermore, the solidification process was used to conduct disposability research on exhausted RHA, and the leachate was exposed to a toxicity test to determine the DOX encapsulation ability. For experiment design and adsorption parameter optimization, RSM's central composite design was utilized. RHA was characterized using Fourier transform infrared spectroscopy (FTIR), X-ray diffraction (XRD), Brunauer-Emmett-Teller analysis (BET), scanning electron microscope (SEM) (energy dispersive X-ray [EDX]), and zeta potential analyzer.

Kaur et al. mentioned a feasible mechanism for amoxicillin trihydrate (AMT) deterioration. A Ti/RuO_2 anode was utilized to help in the mineralization of AMT. A 37.82% total organic carbon (TOC) removal indicated incomplete mineralization and the presence of AMT in degraded form. In a continuous reactor setup, Ti/RuO_2 electrodes were utilized to conduct electro-oxidation (EO) of waste water containing synthetic AMT as a model pollutant [43].

11.8 Removal of Antibiotics with Waste-Based Adsorbents

Numerous studies have focused on metals, antibiotics, or microplastics and their anaerobic, fungal, or phytoremediation in general, although few have investigated the influence of pharmaceutical industrial wastewater (PIWW) or the pharmacological waste in its entirety. The purpose of this research is to enlighten the qualitative elements, characteristics, and components of PIWW. The chapter considers not only the medications and their intermediary metabolites, but also specific polymers utilized in formulation and plastics used in packaging. Furthermore, it helps to rule out other generalized, costly, and inefficient treatments that improve just certain characteristics of water, such as COD, biochemical oxygen demand (BOD), and physicochemical parameters. It also explains why each approach was developed and its relevance. The extent of conventional primary, secondary, and tertiary treatment approaches, bioremediation techniques, and their restrictions have been reviewed and analyzed in order to widen their scope of use, future aspects, and usefulness. Furthermore, there is a vacuum since not all parts of the evaluation have been examined purely from the standpoint of the Pharmaceutical business [44].

Heteroatom-doped porous carbons are promising adsorbents [45]. The removal of antibiotics with waste-based adsorbents is highly efficient. Remarkable adsorption capacities (>1000 $mg\,g^{-1}$) were attained with some adsorbents. The synergic effect of adsorption and degradation can be achieved with photocatalysts. Porous carbons doped with complexes are promising adsorbents. Antibiotic removal using waste-based adsorbents is highly effective. With some adsorbents, remarkable capacities (more than 1000 $mg\,g^{-1}$) were achieved. It is possible to combine adsorption and degradation with photocatalysts [45].

Choline chloride was used as a hydrogen bond acceptor in the synthesis of deep eutectic solvents (DESs), while (hydrogen bond acceptor [HBA]) and glycolic acid acted as a hydrogen bond donors (HBDs) [39, 42, 43]. A RHA functionalization agent was used for the removal of ofloxacin (OFL) from a water body (streams/rivers). The binding was discovered using FTIR spectroscopy, Therapeutic Goods Administration (TGA), XRD, and field emission scanning electron microscope (FESEM) [46].

11.9 Effective Removal of Phenols and Dyes

Enzyme immobilization on magnetic nanocomposite has emerged as an attractive support material due to desired qualities such as convenience in separation, recovery efficiency, and the capacity to reuse the material. Magnetic nanocomposite is utilized for water reclamation in nanoencapsulation, nanotubes, and NPs as a support material [47, 48].

RHA is currently employed in diverse spectrum of industries, such as paints and pigments, the cement industry, insulators, rubber fillers, and other applications. Al-matrix composites supplemented with RHA are highly dependent on proper wetting by aluminum alloys. Because Al alloys barely wet RHA, the development of appropriate wetting methods is critical [49]. Rice husk physicochemical treatments resulted in the formation of SiC, SiO_2, elemental Si, activated carbon, magnesium silicide, and silicon nitride, among other materials [50].

11.10 Remediation Chemistry Using Biogenic Nanoparticles

To remove pollutants from the environment, "bioremediation" involves using living systems and plant extracts. Bioaugmentation, biostimulation, genetically modified organisms (GMOs), phytoremediation, and biomineralization are some of the treatment techniques used. Nanobioremediation is a more sophisticated technique that combines nanotechnology with bioremediation to speed up rehabilitation. To cure contaminants in water, biosynthesized nanomaterials, primarily in the form of NPs, are utilized in nanobioremediation procedures. Biosynthesis of NPs has recently emerged as a green technique as previously they were synthesized by physical or chemical means. Microbes such as bacteria, fungus, algae, and other forms of biomass are used to create NPs. Biogenesis of NPs is an environmentally benign and cost-effective alternative to standard technologies for large-scale NP synthesis. Aside from the benefits listed above, biosynthesis prevents the discharge of effluent toxins produced during the chemical synthesis of NPs. Biosynthesis is more favorable than other methods since it is ecologically benign and allows for size and characteristic modification. *Bacillus* subtilis and *Lactobacillus* subtilis employed for the synthesis of TiO_2 [51, 52].

Extracts of plants are used in biosynthesis to produce compounds that function as both reductants and caps, allowing for precise control over NP size. Filamentous fungi may also create NPs; Bansal et al. synthesized silica and titania NPs from fungal strains. Fungal strains may assist in generating silver NPs, since they are capable of producing massive concentrations in short period of time. Silver NPs are widely used in a variety of industries, ranging from electronics to medicine. Synthesis of nanoparticles from phytoconstituents is a popular method since the incubation time and other parameters that microorganisms need are not applicable [53, 54]. Anti-pollution measures include prevention and detection as well as rehabilitation. Bioremediation is a new restoration process that includes the breakdown of complicated models into individual bits using biologically synthesized materials.

Nanobioremediation utilizes various phylogenetically related NPs range including nanotubes, nanospheres, nanocomposites, nanofibers, and nanoclusters. In the treatment method, carbon nanotubes and zero-valent iron particles are widely utilized. Nanostructures with a dielectric center and metallic shell are being used to remove organic dyes from untreated wastewater. Ex situ or in situ reacting procedures are frequently used, with ex situ involving organic pollutant decomposition without toxic end products. While in situ, heavy metal techniques remove contaminants by absorption. Ex situ detoxification of organic contaminants is enhanced when biogenic pollutants are included in the treatment process. In an in-situ treatment method, pollutant content is reduced by introducing nanoscale zero-valent IONPs/other NPs into contaminated sites. Because NPs are colloidal, they may penetrate deeper into polluted areas. Gold and silver NPs are commonly employed in the decomposition of organic pigments; nevertheless, iron NPs have been successfully used in removal of pollutants [55]. Biofilm generation is a hazard to drinking groundwater because it causes erosion in the water. The application of nanocrystalline semiconductor thin films precludes the production of such biofilms. The application of nanocrystalline semiconductor thin films precludes the production of such biofilms.

11.10.1 Photocatalysis

Photocatalysis has shown to be a cost-effective and simple method in large-scale applications. TiO_2 with [56] a higher photocatalytic performance is utilized for environmental remediation. TiO_2 in various forms and sizes is used in photocatalysis. The photocatalyst is found in the core of the substrate, where microbial attachment occurs. The production of free radicals on the particle surface causes the bacterial cell walls to be broken during light irradiation. TiO_2NPs were biosynthesized from *Bacillus* subtilis, then slurry-coated onto a glass slide. Biofilm formation was then permitted. During irradiation, the hydroxy radical produced produces H_2O_2, which eliminates the bacteria in the biofilm. By increasing the active surface area, H_2O_2 generation may increase [57]. It has been discovered that NPs with metal, nonmetal, and co-doping may be employed to improve light responsiveness. Green synthesis Fe-doped NPs effectively exploit sunlight's in photocatalytic activity in the breakdown of organic contaminants.

Plant extract nanoparticles and iron nanoparticles derived from spinach leaves are used to address eutrophication, a major worldwide issue [56]. Iron NPs have been used effectively to remove BOD and COD from wastewater. Silver NPs are also made from Reishi mushroom extracts. This silver NP has strong DNA cleavage activity and has been demonstrated to be effective against a wide range of gram-positive and gram-negative bacteria. Gold NPs are also extracted from metal-resistant fungus and effectively employed in the treatment of rhodamine dye, which is carcinogenic and has been linked to developmental problems in humans. Fungal species were isolated, tested for gold tolerance, inoculated, and extracellular AuNPs were formed. Rhodamine B dye was reduced using biosynthesized NPs. Rhodamine B is first reduced by the additional electrons, which are subsequently dispersed, and charge density is built up on its surface as a result [58]. Dye adsorption is mediated by the molecules in biosynthesized NPs.

Microbe-derived NPs: Metal nanocatalysts derived from microorganisms have catalytic and antibacterial characteristics. Palladium nanocatalysts are made from palladium-tolerant microorganisms and are utilized in wastewater treatment. Samples were taken, and tolerant species were identified and cultivated. Palladium nanocatalysts with catalytic activity in the dehalogenation process were created. They bind to functional groups on cell envelopes and catalyze the reductive dechlorination of polychlorinated dibenzodioxins, allowing the chemical to be detoxified. Biogenic nanomagnetite (BnM) is a catalyst used to remediate organic water contaminants such as nitrobenzene and tetrachloroethylene. Geobacter sulfurreducens is used to produce BnM. A reduction product was identified when nitrobenzene was reduced. The inclusion of palladium as a catalyst boosted the reduction process much further. When palladium-BnM is treated with tetrachloroethylene, the pollutant reacts quickly and degrades [59].

11.10.2 Carbon Nanotube (CNT)-Based Nanotechnologies

Adsorption, membrane technology, and photocatalysis are examples of technologies based on carbon nanotubes (CNT) that have been integrated into standard wastewater treatment procedures [60, 61]. In the realm of nanostructured materials, carbon nanomaterials have emerged as viable support materials because of their distinctive morphological and structural characteristics [62]. With their superior operational stability, mechanical characteristics, chemical inertness, and biocompatibility to other nanomaterials, CNTs have captured the attention of several scientific groups [63, 64]. Ren et al. [65] reported immobilizing HRP on single-walled carbon nanotubes (SWCNT) significantly increased enzymatic efficiency. Nano-biocatalysts have been employed effectively in the foodstuff, diagnostic, biosensors, and pharmaceutical sectors, as well as in converting biomass setups, because of their unique properties.

11.10.3 Mn Oxide-Coated Carbon Nanotubes (MnO_2/CNTs)

As a pb (ii) absorber, Mn oxide-coated carbon nanotubes (MnO_2/CNTs) are employed. The loading of MnO_2 by Wang was confirmed by using XRD of amorphous phase Mn oxide-coated CNTs [66]. Metal ions are bound to maple sawdust [67] for adsorption of chromium(VI) from effluents.

11.10.4 Glutaraldehyde Activated Carbon Nanospheres (GCN)

HRP was effectively immobilized on carbon nanospheres modified by glutaraldehyde. Phenolic compounds in aqueous solution, particularly the chlorophenols, 4-methoxyphenol, and bisphenol A, displayed good removal efficiency, particularly when treated with activated carbon adsorbents, which effectively adsorbed and removed these compounds from the solution especially when activated carbon is used as an absorbent, it has maximal absorbance and removes effluents from the solution.

11.10.5 Surfactant Concentration and Fly Ash Nanoparticles (FANPs)

Surfactant concentration and fly ash nanoparticles (FANPs) are by-products of coal-burning power plants [20]. Polyaromatic hydrocarbons pollute groundwater, carrying gasoline, diesel, and fuel oil even to surface soil [68, 69]. On various industrial sites, disposal is a significant source of concern. The volatility, high viscosity, and poor mobility of several of these NAPLs make remediation of the polluted five sites problematic [44]. The use of foams may help ease these issues [70–72]. Due to instability of foams and the rapid decomposition of their liquid interfacial coatings, foams are difficult to use in applications. To obtain the necessary outcomes in contaminant bioremediation, foams need surface-active and stabilizing components to permit continuous formation and long-term foam stability [73, 74].

CO_2 as a component may be used to create such foams. CO_2 has poor viscosity and density on its own, and it is extremely mobile across cracks or permeable layers, causing deep channeling and poor displacement efficiency in heterogeneous 15 media. [75–77]. CO_2 foams stabilized by NPs, on the other hand, do not suffer from these limitations. NPs aggregate at fluid–fluid interfaces in foams and are emerging as a framework for foam stability [78]. Particles arrange themselves at surfaces to attain equilibrium and reduce surface energy, which reduces the coarsening of the contact [20]. CO_2 foams containing varying amounts of FANPs are used to cleanse petroleum in a microfluidic device. The percentage of NPs influences the effectiveness of CO_2 foam in removing NAPLs from the chip. Fly ash and surfactant blends reduce foam mobility and increase CO_2 sweep efficiency [79]. These mixes may also be used to make oil–water emulsions [80]. Foam stability and sweep efficiency are influenced by surfactant and NP concentrations, as well as by the kind of surfactant and NP used in them. NP-stabilized foams have been used in both oil recovery and oil-contaminated soil rehabilitation [81–85].

Guo and Aryana created CO_2 foams for increased oil recovery using SiO_2 and three surfactants: alpha-olefin sulfonate (AOS), lauramidopropyl betaine (LAPB), and sodium dodecyl sulfate (SDS) (EOR). Their findings show the effect of injection rate on oil recovery [86].

11.11 Conclusion

High catalytic effectiveness, stability, hydrophobicity, decreased toxicity, and environmental and microbiological attack resistance are desirable support material features. Organic auxiliary ingredients include collagen, biotin, dopamine, chitin, cellulose, and chitosan. Immobilization enhanced enzymatic activity, hydrolytic cleavage tolerance, and microbial resistance. Enzymes immobilized on organic or inorganic substrates are used to detoxify heavy metals. Bioaugmentation and biostimulation procedures may degrade PAH and NAPLs, polymers, metal coatings, dyeing, textile wastes, mines, medicines and dressings, and paper manufacturing. Biodegradation of paper pulp, including decolorizing effluents, employs immobilized Lipase, cellulose, POD, laccases, oxidoreductases, and Keratinases. Around 50–90% antibiotics, such as trimethoprim, fluoroquinolones, sulfonamides, and cyclines, are detected in sewerage. Which can be treated Materials such as rice husks and Ti/RuO_2 anodes were used to aid in the treatment of pharmaceutical effluents. It has been revealed that NPs from fungal sources, or metallic, nonmetallic, and codoping CNT, may promote light sensitivity. The green synthesis of Fe-doped NPs disintegrates organic pollutants via photocatalysis. HRP immobilized on glutaraldehyde-modified CNTs, Mn oxide-coated nanostructures for the removal of lead, chlorophenols, 4-methoxyphenol, and bisphenol A (phenols).

References

1 Nisha, S. et al. (2012). A review on methods, application and properties of immobilized enzyme. *Chem. Sci. Rev. Lett.* 1 (3): 148–155.

2 Alasonati, E. and Slaveykova, V.I.J.B.T. (2012). Effects of extraction methods on the composition and molar mass distributions of exopolymeric substances of the bacterium Sinorhizobium meliloti. *Bioresour. Technol.* 114: 603–609.

3 Anbazhagan, S. and Palani, S.J.U.S. (2018). Extraction of consortium of hydrolytic enzymes from waste activated sludge using ultrasonication and stirring with surfactants. *Ultrason. Sonochem.* 40: 874–880.

4 Koley, D. and Bard, A.J. (2010). Triton X-100 concentration effects on membrane permeability of a single HeLa cell by scanning electrochemical microscopy (SECM). *Proc. Natl. Acad. Sci.* 107 (39): 16783–16787.

5 Karn, S.K. and Kumar, A. (2019). Protease, lipase and amylase extraction and optimization from activated sludge of pulp and paper industry. *Indian J. Exp. Biol.* 57: 3.

6 Safi, C. et al. (2017). Energy consumption and water-soluble protein release by cell wall disruption of *Nannochloropsis gaditana*. *Bioresour. Technol.* 239: 204–210.

7 Rodrigues, R.C. et al. (2013). Modifying enzyme activity and selectivity by immobilization. *Chem. Soc. Rev.* 42 (15): 6290–6307.

8 Bilal, M. et al. (2016). Chitosan beads immobilized manganese peroxidase catalytic potential for detoxification and decolorization of textile effluent. *Int. J. Biol. Macromol.* 89: 181–189.

9 Brena, B. et al. (2013). Immobilization of enzymes: a literature survey. *Methods Mol. Biol.* 1051: 15–31.

10 Arslan, M. (2011). Immobilization horseradish peroxidase on amine-functionalized glycidyl methacrylate-g-poly (ethylene terephthalate) fibers for use in azo dye decolorization. *Polym. Bull.* 66 (7): 865–879.

11 Miletić, N., Nastasović, A., and Loos, K.J.B.T. (2012). Immobilization of biocatalysts for enzymatic polymerizations: possibilities, advantages, applications. *Bioresour. Technol.* 115: 126–135.

12 Wang, S. et al. (2016). Oxidative removal of phenol by HRP-immobilized beads and its environmental toxicology assessment. *Ecotoxicol. Environ. Saf.* 130: 234–239.

13 Grigoras, A.G. (2017). Catalase immobilization—a review. *Biochem. Eng. J.* 117: 1–20.

14 Šekuljica, N.Ž. et al. (2016). Immobilization of horseradish peroxidase onto kaolin. *Bioprocess Biosyst. Eng.* 39 (3): 461–472.

15 Kim, H.J. et al. (2012). Immobilization of horseradish peroxidase onto clay minerals using soil organic matter for phenol removal. *J. Mol. Catal. B Enzym.* 83: 8–15.

16 Longoria, A., Tinoco, R., and Torres, E. (2010). Enzyme technology of peroxidases: immobilization, chemical and genetic modification. In: *Biocatalysis Based on Heme Peroxidases: Peroxidases as Potential Industrial Biocatalysts*, 209–243. Berlin Heidelberg: Springer-Verlag https://doi.org/10.1007/978-3-642-12627-7_9.

17 Ding, S. et al. (2015). Increasing the activity of immobilized enzymes with nanoparticle conjugation. *Curr. Opin. Biotechnol.* 34: 242–250.

18 Mohamed, S.A. et al. (2017). Immobilization of horseradish peroxidase on Fe_3O_4 magnetic nanoparticles. *Electron. J. Biotechnol.* 27: 84–90.

19 Nicell, J.A., Al-Kassim, L., Bewtra, J.K., and Taylor, K.E. (1993 (Vol. 7, No. 1)). Wastewater treatment by enzyme catalysed polymerization and precipitation. In: *Biodeterioration Abstracts*, 1–8.

20 Wang, Z. et al. (2021). Optimum concentration of fly ash nanoparticles to stabilize CO_2 foams for aquifer and soil remediation. *J. Contam. Hydrol.* 242: 103853.

21 Yang, L. et al. (2019). Gas chromatography-Orbitrap mass spectrometry screening of organic chemicals in fly ash samples from industrial sources and implications for understanding the formation mechanisms of unintentional persistent organic pollutants. *Sci. Total Environ.* 664: 107–115.

22 Wang, T. et al. (2010). Air quality during the 2008 Beijing Olympics: secondary pollutants and regional impact. *Atmos. Chem. Phys.* 10 (16): 7603–7615.

23 Sims, J., Ellwood, P., and Taylor, H. (1993). Pollutants from laser cutting and hot gas welding of plastics. *Ann. Occup. Hyg.* 37 (6): 665–672.

24 Pandey, K. et al. (2017). Application of microbial enzymes in industrial waste water treatment. *Int. J. Curr. Microbiol. App. Sci.* 6 (8): 1243–1254.

25 Royer, G. et al. (1991). Continuous decolorization of bleached kraft effluents by Coriolus versicolor in the form of pellets. *J. Ind. Microbiol.* 7 (4): 269–277.

26 Kirk, T.K. and Yang, H.H. (1979). Partial delignification of unbleached kraft pulp with ligninolytic fungi. *Biotechnol. Lett.* 1 (9): 347–352.

27 Zhang, Z., Li, D., and Zhang, X. (2019). Enzymatic decolorization of melanoidins from molasses wastewater by immobilized keratinase. *Bioresour. Technol.* 280: 165–172.

28 Gianfreda, L., Xu, F., and Bollag, J.M. (1999). Laccases: a useful group of oxidoreductive enzymes. *Biorem. J.* 3 (1): 1–26.

29 Bollag, J.-M., Dec, J., and Krishnan, S.B. (1998). Use of plant material for the removal of pollutants by polymerization and binding to humic substances. Citeseer.

30 Sun, Y. and Cheng, J. (2002). Hydrolysis of lignocellulosic materials for ethanol production: a review. *Bioresour. Technol.* 83 (1): 1–11.

31 Melani, N.B., Tambourgi, E.B., and Silveira, E. (2020). Lipases: from production to applications. *Sep. Purif. Rev.* 49 (2): 143–158.

32 Singh, A.K. et al. (2021). Bioremediation of lignin derivatives and phenolics in wastewater with lignin modifying enzymes: status, opportunities and challenges. *Sci. Total Environ.* 777: 145988.

33 Fetzner, S. and Lingens, F. (1994). Bacterial dehalogenases: biochemistry, genetics, and biotechnological applications. *Microbiol. Rev.* 58 (4): 641–685.

34 Arora, P., Srivastava, A., and Singh, V. (2010). Application of monooxygenases in dehalogenation, desulphurization, denitrification and hydroxylation of aromatic compounds. *J. Bioremed. Biodegr.* 1: 112. OMICS Publishing Group J Bioremed Biodegrad ISSN.

35 Couto, S.R. and Herrera, J.L.T. (2006). Industrial and biotechnological applications of laccases: a review. *Biotechnol. Adv.* 24 (5): 500–513.

36 Chagas, P.M.B. et al. (2015). Immobilized soybean hull peroxidase for the oxidation of phenolic compounds in coffee processing wastewater. *Int. J. Biol. Macromol.* 81: 568–575.

37 Sayğılı, H. and Güzel, F. (2016). Effective removal of tetracycline from aqueous solution using activated carbon prepared from tomato (*Lycopersicon esculentum* Mill.) industrial processing waste. *Ecotoxicol. Environ. Saf.* 131: 22–29.

38 Li, W. et al. (2013). Occurrence, distribution and potential affecting factors of antibiotics in sewage sludge of wastewater treatment plants in China. *Sci. Total Environ.* 445: 306–313.

39 Kaur, G., Singh, N., and Rajor, A. (2021). Adsorption of doxycycline hydrochloride onto powdered activated carbon synthesized from pumpkin seed shell by microwave-assisted pyrolysis. *Environ. Technol. Innov.* 23: 101601.

40 Liu, S. et al. (2017). Facile synthesis of Cu (II) impregnated biochar with enhanced adsorption activity for the removal of doxycycline hydrochloride from water. *Sci. Total Environ.* 592: 546–553.

41 Sen Gupta, P.S. and Rana, M.K. (2020). Ivermectin, famotidine, and doxycycline: a suggested combinatorial therapeutic for the treatment of COVID-19. *ACS Pharmacol. Transl. Sci.* 3 (5): 1037–1038.

42 Kaur, G. et al. (2022). Removal of doxycycline hydrochloride from aqueous solution by rice husk ash using response surface methodology and disposability study. *Environ. Sci. Pollut. Res.* 1–15.

43 Kaur, R., Kushwaha, J.P., and Singh, N. (2019). Electro-oxidation of amoxicillin trihydrate in continuous reactor by Ti/RuO_2 anode. *Sci. Total Environ.* 677: 84–97.

44 Shah, A. and Shah, M. (2020). Characterisation and bioremediation of wastewater: a review exploring bioremediation as a sustainable technique for pharmaceutical wastewater. *Groundw. Sustain. Dev.* 11: 100383.

45 Juela, D.M. (2021). Promising adsorptive materials derived from agricultural and industrial wastes for antibiotic removal: a comprehensive review. *Sep. Purif. Technol.* 120286.

46 Kaur, G. et al. (2021). Deep eutectic solvent functionalized rice husk ash for effective adsorption of ofloxacin from aqueous environment. *J. Contam. Hydrol.* 242: 103847.

47 Cao, Y. et al. (2016). Magnetic AuNP@ Fe_3O_4 nanoparticles as reusable carriers for reversible enzyme immobilization. *Chem. Eng. J.* 286: 272–281.

48 Esmaeilnejad-Ahranjani, P. et al. (2015). Amine-functionalized magnetic nanocomposite particles for efficient immobilization of lipase: effects of functional molecule size on properties of the immobilized lipase. *RSC Adv.* 5 (42): 33313–33327.

49 Soltani, N. et al. (2018). Surface modification of rice-husk ash (RHA) by Si_3N_4 coating to promote its wetting by Al-Mg-Si alloys. *Mater. Chem. Phys.* 203: 223–234.

50 Bahrami, A. et al. (2017). Mechanical, thermal and electrical properties of monolayer and bilayer graded Al/SiC/rice husk ash (RHA) composite. *J. Alloys Compd.* 699: 308–322.

51 Jha, A.K. and Prasad, K. (2010). Ferroelectric $BaTiO_3$ nanoparticles: biosynthesis and characterization. *Colloids Surf., B* 75 (1): 330–334.

52 Kirthi, A.V. et al. (2011). Biosynthesis of titanium dioxide nanoparticles using bacterium *Bacillus* subtilis. *Mater. Lett.* 65 (17–18): 2745–2747.

53 Vijayalakshmi, U. et al. (2015). Green synthesis of silica nanoparticles and its corrosion resistance behavior on mild steel. *J. Indian Chem. Soc.* 92 (5): 675–678.

54 Bansal, V. et al. (2005). Fungus-mediated biosynthesis of silica and titania particles. *J. Mater. Chem.* 15 (26): 2583–2589.

55 Tratnyek, P.G. and Johnson, R.L. (2006). Nanotechnologies for environmental cleanup. *Nano Today* 1 (2): 44–48.

56 Turakhia, B., Turakhia, P., and Shah, S. (2018). Green synthesis of zero valent iron nanoparticles from *Spinacia oleracea* (spinach) and its application in waste water treatment. *J. Adv. Res. Appl. Sci* 5 (1): 46–51.

57 Dhandapani, P., Maruthamuthu, S., and Rajagopal, G. (2012). Bio-mediated synthesis of TiO_2 nanoparticles and its photocatalytic effect on aquatic biofilm. *J. Photochem. Photobiol. B Biol.* 110: 43–49.

58 Bastús, N.G. et al. (2014). Synthesis of highly monodisperse citrate-stabilized silver nanoparticles of up to 200 nm: kinetic control and catalytic properties. *Chem. Mater.* 26 (9): 2836–2846.

59 Banerjee, A. et al. (2021). Microbiological degradation of organic pollutants from industrial wastewater. In: *Nanobiotechnology for Green Environment* (ed. Routledge, Taylor & Francis Group), 83–114. CRC Press.

60 Hadavifar, M. et al. (2014). Adsorption of mercury ions from synthetic and real wastewater aqueous solution by functionalized multi-walled carbon nanotube with both amino and thiolated groups. *Chem. Eng. J.* 237: 217–228.

61 Czech, B. and Buda, W. (2015). Photocatalytic treatment of pharmaceutical wastewater using new multiwall-carbon nanotubes/TiO_2/SiO_2 nanocomposites. *Environ. Res.* 137: 176–184.

62 Khan, Z.U. et al. (2016). A review on composite papers of graphene oxide, carbon nanotube, polymer/GO, and polymer/CNT: processing strategies, properties, and relevance. *Polym.-Plast. Technol. Eng.* 55 (6): 559–581.

63 Georgakilas, V. et al. (2015). Broad family of carbon nanoallotropes: classification, chemistry, and applications of fullerenes, carbon dots, nanotubes, graphene, nanodiamonds, and combined superstructures. *Chem. Rev.* 115 (11): 4744–4822.

64 Mittal, G. et al. (2015). A review on carbon nanotubes and graphene as fillers in reinforced polymer nanocomposites. *J. Ind. Eng. Chem.* 21: 11–25.

65 Ren, L., Yan, D., and Zhong, W.J.C. (2012). Enhanced enzyme activity through electron transfer between single-walled carbon nanotubes and horseradish peroxidase. *Carbon* 50 (3): 1303–1310.

66 Wang, S.-G. et al. (2007). Removal of lead (II) from aqueous solution by adsorption onto manganese oxide-coated carbon nanotubes. *Sep. Purif. Technol.* 58 (1): 17–23.

67 Yu, L.J. et al. (2003). Adsorption of chromium from aqueous solutions by maple sawdust. *J. Hazard. Mater.* 100 (1–3): 53–63.

68 Ferguson, S.H. et al. (2004). A field trial of in situ chemical oxidation to remediate long-term diesel contaminated Antarctic soil. *Cold Reg. Sci. Technol.* 40 (1–2): 47–60.

69 Naidu, R. (2013). Recent advances in contaminated site remediation. *Water Air Soil Pollut.* 224 (12): 1–11.

70 Jeong, S.W. and Corapcioglu, M.Y. (2003). Physical model analysis of foam–TCE displacement in porous media. *AICHE J.* 49 (3): 782–788.

71 Longpré-Girard, M. et al. (2016). 2D sandbox experiments of surfactant foams for mobility control and enhanced LNAPL recovery in layered soils. *J. Contam. Hydrol.* 193: 63–73.

72 Lv, C., Chen, J., and Wang, X. (2017). Evaluation of surfactant performance in in situ foam flushing for remediation of dichlorodiphenyltrichloroethane-contaminated soil. *Int. J. Environ. Sci. Technol.* 14 (3): 631–638.

73 Forey, N. et al. (2020). Saponin foam for soil remediation: on the use of polymer or solid particles to enhance foam resistance against oil. *J. Contam. Hydrol.* 228: 103560.

74 Yekeen, N. et al. (2018). A comprehensive review of experimental studies of nanoparticles-stabilized foam for enhanced oil recovery. *J. Pet. Sci. Eng.* 164: 43–74.

75 Aryana, S.A. and Kovscek, A.R. (2012). Experiments and analysis of drainage displacement processes relevant to carbon dioxide injection. *Phys. Rev. E* 86 (6): 066310.

76 Cao, C. et al. (2020). A review of CO_2 storage in view of safety and cost-effectiveness. *Energies* 13 (3): 600.

77 Guo, F. and Aryana, S.A. (2018). Improved sweep efficiency due to foam flooding in a heterogeneous microfluidic device. *J. Pet. Sci. Eng.* 164: 155–163.

78 Garbin, V. et al. (2015). Interactions and stress relaxation in monolayers of soft nanoparticles at fluid-fluid interfaces. *Phys. Rev. Lett.* 114 (10): 108301.

79 Xing, D. et al. (2012). CO_2-soluble, nonionic, water-soluble surfactants that stabilize CO_2-in-brine foams. *SPE J.* 17 (04): 1172–1185.

80 Zhang, T. et al. (2010). Nanoparticle-stabilized emulsions for applications in enhanced oil recovery. In: *SPE Improved Oil Recovery Symposium*. OnePetro.

81 AlYousef, Z.A., Almobarky, M.A., and Schechter, D.S. (2018). The effect of nanoparticle aggregation on surfactant foam stability. *J. Colloid Interface Sci.* 511: 365–373.

82 Karakashev, S.I. et al. (2011). Formation and stability of foams stabilized by fine particles with similar size, contact angle and different shapes. *Colloids Surf., A* 382 (1–3): 132–138.

83 Li, S., Li, Z., and Wang, P. (2016). Experimental study of the stabilization of CO_2 foam by sodium dodecyl sulfate and hydrophobic nanoparticles. *Ind. Eng. Chem. Res.* 55 (5): 1243–1253.

84 Yekeen, N. et al. (2017). Bulk and bubble-scale experimental studies of influence of nanoparticles on foam stability. *Chin. J. Chem. Eng.* 25 (3): 347–357.

85 Karthick, A., Roy, B., and Chattopadhyay, P. (2019). A review on the application of chemical surfactant and surfactant foam for remediation of petroleum oil contaminated soil. *J. Environ. Manag.* 243: 187–205.

86 Guo, F. and Aryana, S. (2016). An experimental investigation of nanoparticle-stabilized CO_2 foam used in enhanced oil recovery. *Fuel* 186: 430–442.

12

Bioactive Nanoparticles for Water Purification

Madhushree Hegde[1], Akshatha Chandrashekar[1], Jineesh A. Gopi[2], Niranjana Prabhu[1], Deepu A. Gopakumar[3], Jinu Jacob George[3], Midhun Dominic C.D.[4], and Jyotishkumar Parameswaranpillai[2]

[1]*M.S. Ramaiah University of Applied Sciences, Faculty of Mathematical and Physical Sciences, Department of Chemistry, New BEL Road, MSR Nagar, Bengaluru 560054, Karnataka, India*
[2]*Alliance University, Division of Chemistry, Department of Science, Anekal - Chandapura Road, Bengaluru 562106, Karnataka, India*
[3]*Cochin University of Science and Technology, Department of Polymer Science and Rubber Technology, South Kalamassery, Ernakulam, Cochin 682022, Kerala, India*
[4]*Sacred Heart College (Autonomous), Department of Chemistry, Pandit Karuppan Road, Thevara 682013, Kochi, Kerala, India*

12.1 Introduction

The main purpose of water is its utilization for drinking, which is a fundamental need for sustainable living on earth. Major sources of water pollution are domestic sewage, improper waste management in urban areas, industrial pollution, mining activities, oil spills, agricultural runoff, etc. According to the United Nations, global freshwater use has increased by a factor of six over the last century [1, 2], and over two billion people live in countries that experience water stress [3]. The recent UN water report 2018 [4] revealed that two million tons of wastewater and other harmful effluents from various industries are drained into water bodies every day. In developing countries, about 90% of domestic sewage is discharged into local water bodies without treatment, affecting the water quality are creating serious human health risks. Conventional methods [5] for water remediation are shown in Figure 12.1. But these methods have limitations, such as the heavy usage of chemicals in the precipitation method, the high operational cost for ion exchange reactions, and the low absorption capacity of conventional absorbents [6].

There are requirements for innovative materials and technologies to meet the need for potable water. Nanotechnology could be a solution to these problems related to water treatment processes. The higher surface area-to-volume ratio and higher reactivity of nanomaterials make them suitable candidates for different environmental remediation processes [7]. Waterborne diseases are the major reasons for human morbidity and mortality worldwide, and prevention of these diseases is the Goal for 2030 set by the United Nations Sustainable Development [8].

Nanomaterials for Air and Water Purification, First Edition.
Edited by Deepu A. Gopakumar, Jyotishkumar Parameswaranpillai, Jinu Jacob George, and Midhun Dominic C.D.

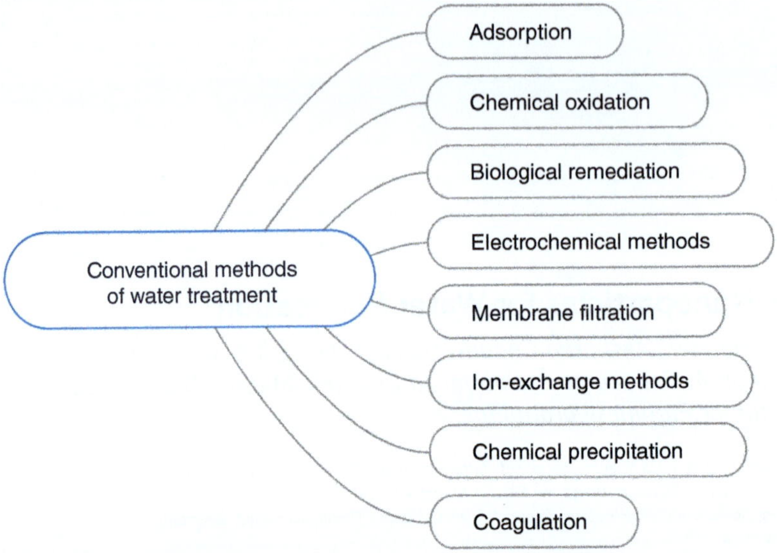

Figure 12.1 Conventional methods for water and wastewater treatment. Source: Data taken from Ouyang et al. [5].

Pathogens which are the major reasons behind waterborne diseases are *Shigella* spp, *Cryptosporidium parvum*, *Adenovirus 3*, *Leptospira*, *Naegleria fowleri*, *Escherichia coli*, *Schistosoma* spp., *Vibrio cholera*, and *Salmonella* [9]. Inactivating viruses and bacteria are the major approach behind the disinfection of water. Chlorine is used as the common oxidizing agent for the disinfection of drinking water, which can deactivate pathogens and give protection against the re-growth of various pathogens, including bacteria [10]. But the use of chlorine may cause some problems such as undesirable tastes, undesirable odor, as well as the toxic by-products during the oxidation of pollutants and pathogens [11]. Also, there are studies on the health issues associated with chlorine-resistant organisms such as *Cryptosporidium* spp. [12, 13]. Nowadays, nanomaterials are being explored as disinfectants that solve these issues. Among those nanoparticles (NPs), metal and metal oxide NPs show promise in terms of their properties such as antimicrobial and antifouling properties [14, 15]. Nanomaterials can be used as catalysts for the degradation of pollutants and absorbents for selective pollutants. They can also be incorporated into polymeric and inorganic membranes for better performance toward water treatment processes. But concerns exist regarding the ecotoxicity of nanomaterials if they leaked into the water body after the purification processes. Among different classes of nanomaterials, bionanomaterials are getting attention in water treatment processes due to their biodegradable nature and shorter lifetime [16]. Research and use of passive nanomaterials are not sufficient to tackle all challenges related to water purification. The fate, transport, and effects of the nanomaterials on living beings and ecosystems must be analyzed in detail, which is necessary to get regulatory acceptance that will permit the use of these technologies [1]. Characteristics of NPs that are important for water treatment are outlined by Ojha et al. [17], and the major characteristics among them are size, surface area, morphology, surface charge, surface hydrophobicity, electronic and optic properties, magnetic properties, and mechanical properties. The characterization techniques generally used to study these characteristics of NPs are various types of microscopy techniques, photon-correlation spectroscopy (PCS), dynamic light scattering (DLS), differential centrifugal sedimentation (DCS), nanoparticle tracking analysis (NTA), X-ray photoelectron spectroscopy (XPS), Fourier transform infrared spectroscopy (FTIR), Raman spectroscopy, Brunauere–Emmette–Teller (BET) technique, Zeta potential measurement, contact angle measurements, UV–visible diffuse reflectance spectroscopy, photoluminescence, and null ellipsometer [17–22]. Recent advancements in nanotechnology in environmental remediation show that waterborne pathogens could be destroyed, removed, or inactivated using nanobiosorbents, nanocatalysts, NPs, nanoparticle incorporated catalytic membranes, bioactive NPs, and nanobioreactors. In this chapter, the use of different types of bioactive NPs for water treatment is discussed. Also, the mechanism behind the effect of bioactive NPs on the pathogens present in the waterbodies and their antimicrobial and antibacterial properties are being discussed. Challenges and prospects are also outlined in this chapter.

12.2 Bioactive Nanoparticles

12.2.1 Bioactive Metallic Nanoparticles for Water Purification

Metal nanoparticles are toxic to pathogens through reactive oxygen species (ROS) generation by destroying nucleic acids, proteins, and cell membranes [23]. The antimicrobial activity of metallic NPs can be improved with the help of sunlight [24] and ionizing irradiation [25]. Silver, copper, and selenium NPs are widely studied for their antimicrobial properties in different applications. There are also concerns over the environmental and health aspects of metal nanoparticles. Recently, there have been studies on the preparation of metal NPs with biocompatible plant-derived secondary metabolites that have a less detrimental effect on the environment. Also, these metabolites show antibacterial effects. During the synthesis of metal NPs, extracts from different plants or algae with antimicrobial properties can be used to reduce metal ions [26].

12.2.1.1 Silver Nanoparticles for Water Treatment

Silver nanoparticles (Ag-NPs) are known for their broad spectrum of antimicrobial properties and their ability to tune their antimicrobial properties based on their size, morphology, surface chemistry, and the matrix in which are embedded. One of the initial studies that explained the antimicrobial effect of silver NPs was by Sondi and Salopek-Sondi on *E. coli*. They found that Ag-NPs can damage *E. coli* cells through the formation of "pits" in the cell wall of the bacteria. Liu et al. [27] and Shameli et al. [28] showed that the antibacterial properties of Ag-NPs change with their size of Ag-NPs. Wahab et al. [29] reviewed the antimicrobial properties of silver NPs in different fields of applications. Ag-NPs could act as bactericides against

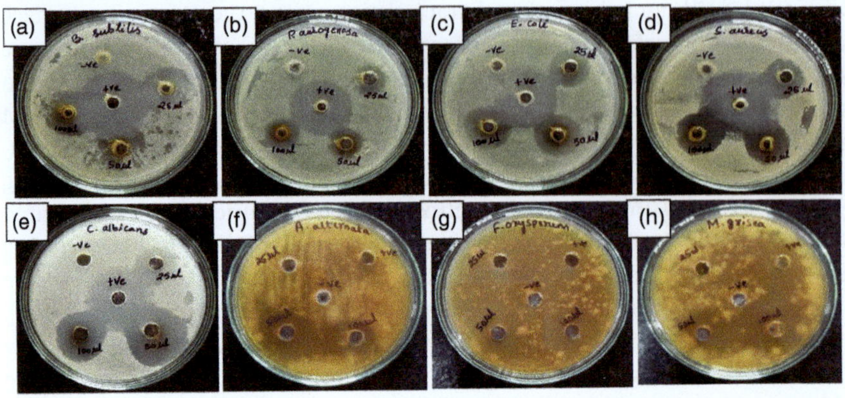

Figure 12.2 Antimicrobial activity of Ag-NPs against the test pathogens (a) B. *subtilis*, (b) *P. aeruginosa*, (c) *E. coli*, (d) *S. aureus*, (e) *C. albicans*, (f) *A. alternata*, (g) *F. oxysporum* and (h) *M. grisea*. Source: Bhat et al. [39]/with Permission from Elsevier/CC BY-NC-ND 4.0. (creative commons license: License No. 5420690976901).

different types of bacteria such as *E. coli*, *Pseudomonas aeruginosa*, *Bacillus cereus*, *Staphylococcus aureus*, and *Salmonella choleraesuis* bacteria. One of the major reasons behind the antibacterial properties of Ag-NPs is their ability to interact with the sulfur and phosphorus parts of bacteria. Silver NPs can also degrade organic substances, which is also an essential property required during water purification [30]. The nanocrystalline form of Ag-NPs possesses better efficacy in water purification through filtration techniques [31]. Gadkari et al. [32] reviewed the opportunities and challenges of silver NPs for water purification. Das et al. [33] reviewed the biosynthesis of Ag-NPs from plant extracts and sea weeds and their antibacterial characteristics. Devi and Joshi [34] screened 53 strains of various fungi for the synthesis of silver NPs, and these NPs showed antibacterial property against different bacteria such as *S. aureus*, *Streptococcus* pyogenes, and *Salmonella enterica*. Along with other antibacterial agents such as erythromycin, methicillin, chloramphenicol, and ciprofloxacin, Ag-NPs showed synergistic effects in antibacterial properties against *Klebsiella pneumoniae* and *Enterobacter aerogenes* [35]. Silver NPs also possess antifungal [36], antiviral [37], and anti-arthropod activities [38]. Bhat et al. synthesized Ag-NPs using *Ixora brachypoda* (DC) leaf extract and studied their antimicrobial activity using agar-well diffusion method. Figure 12.2 shows the antimicrobial activity of these Ag-NPs against various test pathogens and the Ag-NPs showed higher antibacterial activity against *Bacillus subtilis* and higher antifungal activity against *C. albicans*.

12.2.1.2 Copper Nanoparticles for Water Treatment

Copper nanoparticles (Cu-NPs) show good antimicrobial properties, which depend on factors such as temperature, humidity, dry or wet conditions, size and concentration of copper NPs, chemical environment, and method of preparation of copper nanoparticles [40]. Cu-NPs can be prepared mainly by chemical [41, 42] and biological [43–45], and these NPs have catalytic and antimicrobial properties that make them suitable for water treatment. Compared to silver NPs, Cu-NPs are cheaper options for water purification. The mechanism behind the antibacterial properties of Cu-NPs was studied by several researchers, and the mechanism is similar to the antibacterial mechanism of silver NPs [46–48]. Das et al. [49] studied the antimicrobial effect of linoleic acid-capped copper NPs and proposed that the inactivation of the enzymes in the bacteria and the generation of H_2O_2 are the reasons behind the antibacterial activity. Studies showed the interaction of copper NPs with –SH groups, amines, and carboxyl groups on the cell surface of bacteria [50, 51], followed by their penetration into the cells, destroying the DNA, RNA, protein, etc. in microorganisms [52, 53].

12.2.1.3 Selenium Nanoparticles in Water Treatment Processes

Selenium is an essential dietary trace element, and selenium nanoparticles (Se-NPs) can be prepared from physical, chemical, and biological methods, and they exhibit antimicrobial properties [54]. Biologically mediated Se-NPs possess better safety, eco-friendliness, and economic viability compared to Se-NPs prepared using physical and chemical techniques [55–57]. Alagesan and Venugopal [58] synthesized spherical Se-NPs with a diameter range of 45–90 nm by mixing *Withania somnifera* leaf extract and selenious acid solution. Along with the Se-NPs, secondary metabolites of *W. somnifera* were also formed. The NPs showed good antibacterial activity against *Bacillus* subtilis, *K. pneumoniae*, and *S. aureus*. Se-NPs attached polyamide membrane showed antibacterial effects with a bacteriostatic rate of 100% on *E. coli* and *S. aureus* [59]. Se-NPs mediated by *Ceropegia bulbosa Roxb* extract showed good antimicrobial, and photocatalytic activities and can be used in water purification techniques [60]. Table 12.1 lists a few recent studies on the antimicrobial properties of metal NPs for water treatment processes.

12.2.2 Bioactive Metal Oxides Nanoparticles for Water Purification

Metal oxides are widely studied for their various uses in wastewater treatment [75]. For example, metal oxides possess photocatalytic properties to degrade various water pollutants. Several modification techniques such as doping, surface functionalization, and combination with different metal oxides have been used to improve the photodegradation properties of metal oxides. Combination with electron scavenging agents such as metals and carbon nanomaterial also makes metal oxides as an efficient material to enhance the photocatalytic activity for the degradation of pollutants present in wastewater [76]. Recently, a significant number of studies were conducted to explore the antimicrobial properties of metal oxide NPs including the microbes that are resistant to antibiotics. Metal oxides of silver, titanium, aluminum, zinc, magnesium, nickel, chromium, manganese, copper, cerium, etc. have been studied for their antimicrobial properties. The factors affecting the effectiveness of metal oxides in controlling microbes, preventing biofilm formation, and the mechanisms of antimicrobial activities are listed in detail by Subhan et al. [77] and a few of those factors are shown in Figure 12.3.

Table 12.1 Recent studies on the antimicrobial properties of metal nanoparticles (NPs) for water treatment processes.

Sl No.	Nanoparticles	Method of synthesis of nanoparticles	Antimicrobial activity	References
1	Spherical Ag-NPs with size around 8 nm	Biosynthesis of Ag-NPs using *Moringa oleifera* flower extract.	Ag-NPs showed good antimicrobial properties against model bacteria such as *Klebsiella pneumoniae* and *Staphylococcus aureus* with inhibition zone of 17 and 29 mm, respectively.	[61]
2	Spherical Ag-NPs with the size range of 10–20 nm	Phytosynthesis of Ag-NPs using *Mangifera indica* flower extract as reducing agent.	Ag-NPs exhibited antibacterial property against *Staphylococcus* sp., *Klebsiella* sp., *Pantoea agglomerans*, and *Rahnella* sp. bacteria.	[62]
3	Spherical Ag-NPs with size range of 15–20 nm	Ag-NPs were prepared using flavonoids fraction of *Psidium guajava* L. leaves.	Ag-NPs exhibited inhibition against *Alcaligenes faecalis*, *Escherichia coli*, and *Aspergillus niger*.	[63]
4	Face-centered cubic crystalline structured Ag-NPs with average particle size of 220.8 ± 31.3 nm determined by DLS analysis	Ag-NPs were prepared using dried biomass of *Chlorella ellipsoidea* algae.	Disk diffusion Ag-NPs inhibited the growth of model bacteria such as *E. coli*, *K. pneumoniae* and *Pseudomonas Aeruginosa*, and *S. aureus*.	[64]
5	Ag-NPs with average particle size of 35 nm.	Ag-NPs synthesized using aqueous *Parkia speciosa* leaf extract in a green route.	Ag-NPs exhibited antimicrobial properties against *E. coli*, *S. aureus*, *P. aeruginosa*, and *B. subtilis*.	[65]
6	Ag-NPs dispersed carbonized bacterial cellulose (CBC) composite.	The composite prepared by the direct pyrolysis of bacterial cellulose by absorbing $Ag\text{-}NO_3$ under nitrogen atmosphere.	Composite containing 1.9% Ag exhibited antimicrobial property against *S. aureus* and *E. coli*.	[66]
7	Ag-NPs embedded copper oxide core–shell particles with size ranges from 200 to 700 nm.	Core–shell particles were synthesized by coating silver on CuO core.	Core–shell particles showed antibacterial activity against *S. aureus*.	[67]

8	Spherical Cu-NPs with average size of 183 nm	Cu-NPs were synthesized using *Cynodon dactylon* leaf extract.	Cu-NPs showed good antibacterial activity against *E. coli*, *B. subtilis*, *S. aureus*, and *K. pneumoniae*.	[68]
9	Spherical Cu-NPS with sizes in the range of 6.93–20.70 nm in diameter.	Cu-NPS were bio-synthesized by using extract of *C. sinensis*.	Cu-NPS could control the growth of *E. coli* and *S. aureus*.	[69]
10	Amorphous Cu-NPs with particle size in the range of 35.8–49.2 nm	Encapsulated Cu-NPs were synthesized using extract of *Moringa oleifera* leaves.	Cu-NPs good antibacterial activity against *E. coli*, *K. pneumoniae*, *S. aureus*, and *Enterococcus faecalis*.	[70]
11	Spherical Cu-NPs with size in the range of 2–10 nm.	Cu-NPs were synthesized using extract of *Celastrus paniculatus* Willd. leaf.	The nanoparticle showed antifungal property against *Fusarium oxysporum*.	[71]
12	Polydopamine decorated Cu-NPs in wood matrix.	The hydrophobic composite was prepared by self-polymerization of dopamine, followed by chemical deposition of Cu NPs along with fluorosilane modification.	The modified wood displayed good antibacterial activity against *E. coli* and *S. aureus*.	[72]
13	Polyvinylidene fluoride/Polyvinyl pyrrolidone/Ag-NPs electrospun membranes	The composite membranes were prepared in a single step which involves the formation of Ag NPs through reduction and stabilization using polyvinyl pyrrolidone.	Disk diffusion test showed that the membranes exhibit good antibacterial activity against *S. aureus*.	[73]
14	Spherical Se-NPs with diameter in the range of 45–90 nm.	The Se-NPs were prepared by the green synthesis which involves the mixing of extracts of *Withania somnifera* leaves selenious acid solution.	The Se-NPs exhibited good antibacterial activity against *B. subtilis*, *K. pneumoniae*, and *S. aureus*.	[58]
15	Spherical bio-Se NPs with an average size of 120 nm.	Se NPs were synthesized with the help of *Providencia* sp. DCX.	The NPs showed good antibacterial properties against *P. aeruginosa*, *E. coli*, V. *parahemo, lyticus*, *S. aureus*, *B. cereus* and *B. subtilis* in growth-inhibition studies and in cell survival images.	[74]

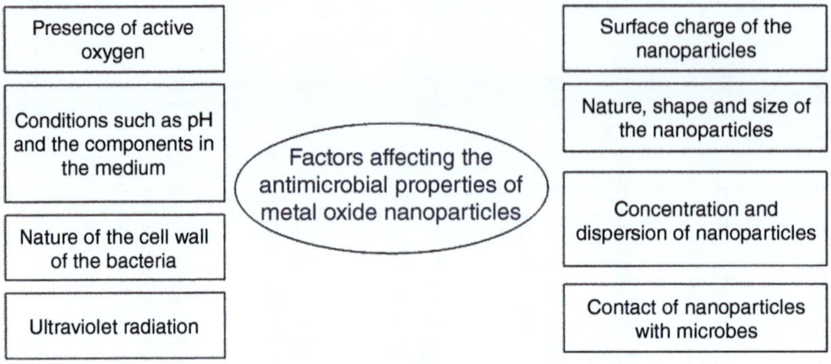

Figure 12.3 Factors affecting the antimicrobial properties of metal oxide nanoparticles.

Metal oxides exhibit good antimicrobial properties, and one of the major bactericidal mechanisms of metal and metal oxide nanocomposites is the generation of reactive oxygen species, which destroy the cell membrane of the bacteria and prevent the further growth of the cells. Metal oxide nanomaterials are also widely used in the fabrication of polymer nanocomposite membranes which can provide high activity against pathogens [78]. Aluminum oxide (Al_2O_3) NPs possess good antimicrobial properties and the combination of Al_2O_3 with other antimicrobial agents such as TiO_2 further enhances the antimicrobial properties of aluminum oxide NPs [79].

TiO_2 is widely studied for its antimicrobial properties against various water-borne pathogens such as Mycobacteria, Salmonella, Shigella, and protozoans [80–82]. These NPs can induce both biochemical and physical changes in the microbes, and these changes may lead to the inactivation or death of the microbes. The main mechanism behind the antimicrobial property of TiO_2 NPs is their photocatalytic activity and the resulting generation of free radicals [83]. This photocatalytic activity can inactivate viruses as well. The release of titanium ions from TiO_2 can also cause the death of microbes [84]. Zinc oxide (ZnO) nanoparticles also possess good antimicrobial properties against various pathogens in water. Weldegebrieal [85] reviewed the antibacterial and photocatalytic activity of ZnO NPs for wastewater treatment. In another review, Liu et al. [86] discussed key factors and mechanisms behind the photocatalytic and antibacterial properties of ZnO. The photocatalytic activity of ZnO can be improved by techniques such as doping with different atoms, depositing noble metal atoms on the surface, coupling with carbon-based nanomaterial, and creating heterojunctions [87]. ZnO can be used to improve various characteristics of polymeric materials including their antifouling properties [88]. Elmi et al. [89] studied the antibacterial property of zinc oxide nanoparticle in wastewater treatment and found that the antibacterial property of zinc oxide nanoparticle depends on the type of bacteria and the use of zinc oxide can be used as a complementary method with UV disinfection method for better efficiency. The mechanisms behind antimicrobial properties of ZnO are similar to those of TiO_2 and the generated free radicals can interact with pollutants and form other free radicals like hydrogen peroxide radicals [90]. Orou et al. [91] studied the antibacterial activities of ZnO nanorods and ZnO nanodisks and found that they can inhibit the growth of *B. subtilis* and *S. aureus* with minimal inhibitory concentration (MIC) of 128 and 256 μg ml^{-1}, respectively. The antibacterial action of zinc oxide NPs against *B. subtilis* bacteria and biofilm prevention was studied by Aswasthi et al. [92] and found that the presence of ZnO changes the surface morphology of cells, and this change is due to the oxidative stress generated by ZnO NPs. Table 12.2 lists recent studies on the antimicrobial properties of various metal oxides in water treatment processes.

12.2.3 Bioactive Bio-Based Nanoparticles for Water Purification

Chitosan is derived by partial deacetylation of chitin, and it has a structure consisting of alternating units of (1 → 4) linked *N*-acetyl glucosamine and glucosamine units. Chitosan is nontoxic, biodegradable, and has antimicrobial properties, and it can be used in various environmental remediation applications [105, 106]. Chitosan contains amino groups and primary and secondary hydroxyl groups, and chemical modification of the hydroxyl group is possible without changing its biophysical properties [107]. Divya and Jisha [108] listed the methods for the preparation of chitosan NPs, which are ionotropic gelation, microemulsion, polyelectrolyte complex, and reverse micellar techniques. The hydroxyl and amino functional group provide good adsorbing characteristics to chitosan. Chitosan NPs are studied for their adsorption capacity toward lead [109, 110], cadmium [111, 112], chromium [113], arsenic [114, 115], lanthanum ions [116], different organic dyes [117, 118], humic acids [119], etc., and found to be an effective material to get rid of trace metals as well as dyes from water. Chitosan-based hybrid materials and composites [120–122] and materials decorated with chitosan [123] also exhibit good adsorption toward pollutants present in water.

Chitosan shows antimicrobial properties against bacteria as well as fungi. Antimicrobial property of chitosan is based on the degree of acetylation, amount of chitosan, nature of pathogens, pH of the medium, etc. Chitosan showed antibacterial properties against various types of bacteria, such as *B. cereus*, *S. aureus*, *Lactobacillus plantarum*, *Lactobacillus brevis*, *Salmonella typhimurium*, *E. coli*, *P. aeruginosa* and *V. cholera* [124, 125]. The exact mechanism behind the antimicrobial properties is still unknown. There are several proposed mechanisms for the antimicrobial action of chitosan, and those mechanisms are reviewed by Rabea et al. [125]. One of the mechanisms is the leakage of intracellular constituents from the bacteria due to the interaction between positively charged chitosan molecules and negatively charged bacterial cell membranes [126, 127].

12.2.4 Carbon-Based Nanomaterials for Water Purification

Recently, extensive research was focused on the use of carbon-based nanomaterials for water remediation conditions [128].

12.2.4.1 Carbon Nanotubes (CNTs)

Carbon nanotubes (CNTs) are used for targeted adsorption of chemical and biological contaminants [129–132]. Functionalization of CNTs using

Table 12.2 Recent studies on antimicrobial properties of various metal oxides in water treatment processes.

Sl No.	Metal oxide nanoparticles	Synthesis and characteristics of nanoparticles	Antimicrobial activity	References
1	Aluminum oxide (Al_2O_3) nanoparticles	Spherical and hexagonal shaped Al_2O_3 NPs with size ranges between 50 and 100 nm were prepared from *Prunus × yedoensis* leaf extract.	Al_2O_3 NPs prepared at a pH of 7 displayed antibacterial characteristics against *S. aureus* and *E. coli*.	[93]
2	ZnO NPs	Zinc oxide NPs with 80–100 nm size ranges are synthesized using chemical reduction method.	Nanoparticle exhibited antibacterial property against *E. coli* and *S. aureus* in the ager well diffusion assay studies.	[94]
3	Rhamnolipid (RL) coated iron oxide (Fe_3O_4) NPs	Iron oxide NPs were prepared by a co-precipitation method and surface functionalized using RL. The RLs coated Fe_3O_4 NPs have an average size of 48 nm.	RL coated (Fe_3O_4) NPs possess antibacterial activity against *P. aeruginosa* and *S. aureus* and also prevent the biofilm formation.	[95]
4	Copper oxide (CuO) NPs	CuO NPs with the particle size in the range between 20 and 35 nm were prepared using *Cordia sebestena* flower aqueous extract.	Green synthesized CuO NPs exhibited antibacterial properties against *E. coli*, *S. aureus*, and *B. subtilis*. The antibacterial activity of CuO is due to their ability to interact with the bacteria through electrostatic interaction. This interaction leads to the oxidation of the cell of the bacteria, which results in the destruction of bacteria.	[96]
5	Iron oxide (α-Fe_2O_3) NPs	α-Fe_2O_3 NPs are synthesized using Sida cordifolia plant extract. Crystallite size the synthesized α-Fe_2O_3 is estimated around 20 nm.	Studies using ager well-diffusion method showed that NPs prevented the growth of *B. subtilis, S. aureus*, and *E. coli*.	[97]

6	FeO NPs	FeO NPs around 52.78 nm were prepared using *Ruellia tuberosa* (RT) leaf aqueous extract.	FeO NPs exhibited good antibacterial activity against *E. coli*, and *K. pneumoniae*.	[98]
7	Iron oxide NPs	Iron oxide NPs were prepared using *Agrewia optiva* and *Prunus persica* leaf extract.	Studies using agar well diffusion method showed that the iron oxide NPs possess antibacterial activity against various *S. bacteria* and various Coryne bacteria.	[99]
8	Titanium dioxide (TiO_2)	Modification using diethanolamine followed by doping with silver is done on titanium dioxide NPs.	Prepared nanoparticles showed good antibacterial characteristics due to the increase in the electron–hole separations at the surface of nano-TiO_2 by the silver ions and the release of strong oxidizing hydroxyl radicals.	[100]
9	Titanium dioxide	Spherically shaped TiO_2 NPs in the size range of 2–18 nm were synthesized using Monsonia burkeana (MB) plant extract.	TiO_2 NPs shows better antibacterial activity against *E. coli* compared to *S. aureus* due to the thicker cell wall of *S. aureus*.	[101]
10	Titanium dioxide	Spherical shaped TiO_2 NPs were prepared using titanium tetrabutoxide and extract of root of *Kniphofia foliosa*.	Disc diffusion study showed that biosynthesized TiO_2 NPs showed good antibacterial properties against the pathogens such as *S. aureus*, *E. coli*, *Klebsiella pneumonia*, and *Streptococcus pyogenes*.	[102]
11	Titanium dioxide/gold nanohybrid	Hydrothermal process followed by photo-reduction showed spiky TiO_2/Au nanohybrid particles.	Spiky TiO_2/Au NPs could inactivate the *S. aureus*.	[103]
12	α-Al_2O_3	Flakes-like α-Al_2O_3 particles with an average size of 20 nm were synthesized by low temperature solution-combustion method.	α-Al_2O_3 particles showed antibacterial activity against the test pathogens such as *Klebsiella aerogenes*, *E. coli*, *P.* desmolyticum, and *S. aureus* bacteria in agar-well diffusion method.	[104]

hexadecyltrimethylammonium bromide, triton X-100, and sodium dodecyl sulfate (SDS) helps disperse CNTs in aqueous solutions [133]. Wang et al. used superparamagnetic iron oxide nanoparticle-decorated CNTs to separate oil droplets from contaminated water [134]. Functionalized CNTs incorporating polymeric membranes have also been used for desalination [135–138]. CNT can be used to prepare electrochemical filters that can degrade various pollutants, such as tetracycline antibiotics present in water [139].

Carbon nanomaterials possess antibacterial properties, and these properties are reviewed by Saleemi et al. [140]. Liu et al. [141] reviewed the antimicrobial properties of CNTs, the mechanisms behind antimicrobial properties, the significance of CNTs in water disinfection and microbial control, safety concerns, etc. Various mechanisms have been proposed to explain the antimicrobial properties of carbon nanotubes. Cell membrane destruction via direct contact with CNTs and reducing the cell viability [142], impairment/destruction of cell membrane through oxidative stress followed by the leakage of intracellular matter [143], damage of cell membrane due to the piercing of the pathogen surface [144] are a few mechanisms that explain the antimicrobial properties of CNTs. Multiwalled CNTs (MWCNT) decorated with silver NPs show good antibacterial properties against *P. aeruginosa*, *S. aureus* [145], *E. coli*, *B. subtilis*, *C. albicans*, and *Aspergillus niger* [146]. Also, silver decorated CNTs can be used as coating materials on hollow fiber membranes for antibacterial activity in water purification [147]. Microwave irradiated 1-octadecanol functionalized CNTs exhibit antibacterial properties against *E. coli* from water [148]. Metal oxides such as zinc oxide, titanium dioxide, and ferric oxides can also be used as coating materials for CNTs for deactivating microorganisms in water [141, 149]. Brady-Estévez et al. [150, 151] fabricated a single walled carbon nanotubes (SWNT) filter with polyvinylidene fluoride (PVDF) microporous membrane to inactivate *E. coli K12* bacteria and *MS2 bacteriophage* virus from an aqueous matrix.

Ihsanulla et al. [152] synthesized silver doped-CNT membrane and studied its antifouling properties using *E. coli* and the studies showed that a 10% silver incorporated CNTs membrane exhibited high water permeate flux along with good antibacterial properties and this membrane could be used in continuous filtration system for the removal of various types of pollutants in the water. Yang et al. [153] grafted methyl triethylammonium chloride (MTAC) and poly(ethylene glycol) methyl ether methacrylate (EGMA) onto CNTs to prepare p-CNT/PES composite membranes using phase inversion method. The membrane showed good antifouling properties and antibacterial properties as well as good adsorption toward phenolic molecules. Vecitis et al. [154] prepared MWCNTs electrochemical microfilter for the removal and inactivation of viruses. At 2 and 3 V for 30 s, the filter could inactivate more than 75% of the sieved bacteria and more than 99.6% of the adsorbed viruses.

12.2.4.2 Graphene-Based Nanomaterials

The extraordinary properties of graphene and its derivatives make them good materials for various water treatment processes. They can be used for adsorption, filtration, and photodegradation processes in water treatment processes. Graphene, graphene oxides, and functionalized graphene were used for the adsorption of various pollutants in water such as safranin dye [155], crystal violet [156], methylene blue [157], and ibuprofen drug residue [158].

Graphene-based nanomaterials show antibacterial properties due to (i) presence of sharp edges of the one-atom-thick laminar structure and (ii) the oxygen-containing functional groups on their surface. Graphene-based nanomaterials also show synergistic effects along with other antibacterial agents. Graphene-based nanomaterials can be also incorporated into the polymeric membrane and improves the required properties for the water treatment of polymeric membranes [159]. Zeng et al. [160] summarized the mechanisms behind the microbial inactivation by graphene and its derivatives in water disinfection processes and microbial control. In another review, Zheng et al. [161] discussed antibacterial applications of graphene oxides and highlighted its biohazard effect in mammals. Lim et al. [162] prepared tannic acid functionalized GO and used it to fabricate a cross-linked membrane with hyperbranched polyethyleneimine (PEI). This membrane revealed good antibacterial activity against *E. coli* because of the intrinsic bactericidal properties of the tannic acid moieties present in the membrane. Similarly, PVDF-membrane with silver decorated GO prepared by phase inversion method exhibited enhanced hydrophilicity, antibacterial properties, and inhibited the formation of biofilm. The major reason behind the antibacterial property and antibiofilm formation is the slower release rate of Ag^+ in the membrane [163]. Graphene oxides have the potential to be used in the development of forward osmosis membranes due to their 2-D flake-like structure, and hydrophilicity along with its ability to provide good mechanical strength to the membrane and antibacterial effect [164].

A list of studies on the antimicrobial characteristics of graphene-based nanomaterials for water treatment is given in Table 12.3.

12.2.5 Hybrid Nanoparticles for Water Purification

Hybrid nanoparticles are widely studied for water treatment methods due to the synergistic effects they provide in terms of adsorption of pollutants, antimicrobial properties, and ease of separation after the water treatment [175]. For example, Tian et al. [176] prepared a nanocomposite by growing iron oxide nanoparticles and silver NPs on graphene oxide. The nanocomposite showed good antibacterial properties against test pathogens such as *E. coli* and *S. aureus*. This nanocomposite can be easily recyclable due to their magnetic properties induced by the iron oxide NPs. Similarly, Joghataeian et al. [177] prepared a water-dispersible magnetic carbon nanotube/core–shell nanohybrid, which contains grafted silver NPs. This hybrid showed antibacterial activity against two gram-negative bacteria (*E. coli* and *P. aeruginosa*) and two gram-positive bacteria (*B. subtilis* and *S. aureus*) in disk diffusion standard and MIC method. A nanohybrid based on dendritic poly(amidoamine) grafted functionalized MWCNTs and Ag NPs exhibited good antimicrobial activity in aqueous solutions [178]. Chang et.al [179] synthesized

Table 12.3 A list of recent studies on the use of graphene or their derivatives for water treatment.

Sl No.	Nanoparticles	Antimicrobial activity	Mechanism	References
1	Graphene oxide nanosheets	Graphene oxide nanosheets had shown higher antibacterial activity against *Staphylococcus aureus* compared *E. coli* due to thicker cell walls of *E. coli*.	The major mechanism behind the antibacterial property of this nanoparticle is due to the contact of the bacteria with the sharp edges of the NPs as well as the better charge transfer between the sharpened edge of the nanoparticle with the bacteria.	[165]
2	Graphite, graphene, and their other modified forms.	Among the nanomaterials studied, GO dispersion exhibits highest antibacterial activity against *E. coli*.	Cell membrane of bacteria is disrupted due to the contact between graphene nanosheets and cells. Also, oxidation stress due to the presence of NPs also led to antibacterial activity.	[166]
3	GO sheets	GO sheets with different lateral sizes are studied for their antibacterial properties against *E. coli* using colony counting test.	Results showed that the GO sheets with higher lateral sizes show higher antibacterial activity compared to the GO sheets with lower lateral sizes. This is due to the better covering of cells by the larger GO sheets, which leads to the reduction in the cell proliferation.	[167]
4	Nano assemblies of GO with Pluronic, which is a biocompatible block copolymer.	Nano assemblies were analyzed for their antibacterial activity against *E. coli*.	Combination of GO with pluronic improved the dispersion of GO. Also, pluronic acid also enhances the osmotic pressure generated on bacteria. These two factors enhance the antibacterial activity of GO considerably. Also, pluronic acid reduces the toxicity of GOPs toward human fibroblast cells.	[168]
5	Smart magnetic graphene was synthesized by microwave irradiation of GO and ferrocene precursors.	Antibacterial activity of smart magnetic graphene ($40\,\mu g\,l^{-1}$) against *E. coli* was tested using agar well diffusion method.	The smart magnetic graphene showed good adsorption towards various metal ions present in water. Also, it possesses good antibacterial activity due to various factors such as the piercing nature of 2D graphene into cells of bacteria, the surface roughness of smart magnetic graphene, and the antibacterial activity of iron oxide present on the surface of graphene.	[169]

6	Ag/Cu bimetallic NPs on the surface of graphene.	The antibacterial activity of the nanoparticle was analyzed by growth inhibition assay using *E. coli* cells.	The bimetallic Ag/CuNP–graphene hybrids exhibit good antibacterial activity against *E. coli*. This is due to the synergistic action of both metals that coexist on the surface.	[170]
7	Graphene–Cadmium sulfide (G–CdS) nanocomposite	G–CdS photocatalytic disinfection capacity was evaluated against *E. coli* and the result shows that G–CdS showed better inactivation efficiency compared to CdS NPs.	G–CdS induced higher oxidative stress compared to pure CdS to bacteria. This increased oxidative stress disrupts the antioxidant defense system of the cell, which leads to the death of the bacteria.	[171]
8	Guanidyl-functionalized graphene/polysulfone mixed matrix ultrafiltration membrane	The antibacterial property of membrane was studied by analyzing the growth of *E. coli* and *S. aureus* on the surface of membrane.	The membrane possessed good inhibitory effect toward the test bacteria. This may be due to factors such as the sharp edges of the RGO, the presence of guanidyl group which leads binding with phosphate group on the cell wall etc.	[172]
9	Copper oxide NPs doped in GO.	The antibacterial property of the hybrid material was tested against pathogenic *E. coli* and *S. typhimurium* using a standard plate count method.	The antibacterial property of nanocomposite is due to the inherent antibacterial property of Cu NP and interactions with the GO–CuO NP nanocomposite with the cell which leads to cellular defects.	[173]
10	Reduced graphene oxide–CuO nanocomposite film	The antibacterial properties were tested against *P. aeruginosa* using colony forming unit (CFU) counts.	The nanocomposite films displayed good antibacterial activity. This is due to the ability of nanocomposite films to transfer electrons from the bacterial cells. The nanorod can serve as a nano-knife which has the capability to pierce into the bacterial cells.	[174]

the magnetic graphene oxide–TiO_2 composites and these composites exhibited antibacterial activity against *E. coli* and could inactivate the bacteria within 30 min under solar irradiation. Song et al. [180] synthesized graphene oxide–silver nanocomposites and studied its antibacterial property using the test pathogens *E. coli* and *S. aureus*. SEM studies show that GO-Ag could destruct the wall membrane of *E. coli* easily compared to *S. aureus*. Also, oxidative stress generated by GO-Ag nanocomposites leads to better antibacterial activity. In another study, graphene manganese ferrite composite was prepared by a solvothermal process, and this composite showed good adsorption towards lead (Pb(II)) and cadmium (Cd(II)) ions and possess higher antimicrobial activity compared to graphene [181].

Hung et al. [182] prepared expanded graphite (EG) impregnated with different metal oxides using explosive combustion and blending methods. The antimicrobial activities of these composites were tested against gram-positive bacteria, gram-negative bacteria, and *B. anthracis* using various methods. Among the composites, EG/Ag_2O showed higher antibacterial activity compared to EG/CuO and EG/ZnO. This is due to the uniform distribution of Ag_2O particles in the EG matrix compared to other metal oxide NPs. Figure 12.4 shows the inhibition zone of the different metal oxide composites against various bacterial strains.

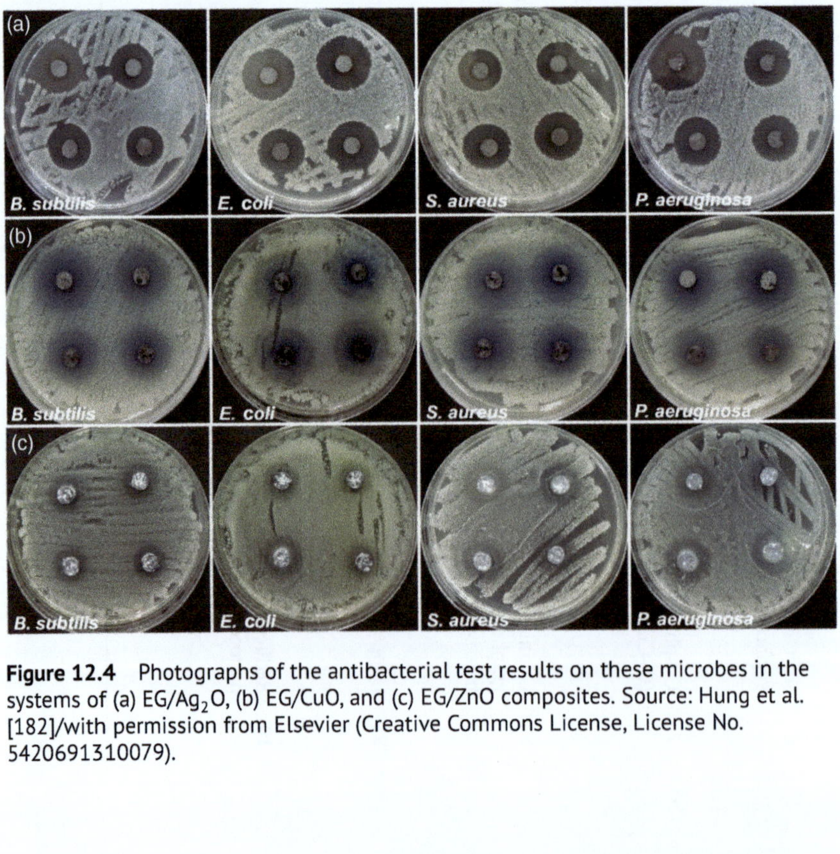

Figure 12.4 Photographs of the antibacterial test results on these microbes in the systems of (a) EG/Ag_2O, (b) EG/CuO, and (c) EG/ZnO composites. Source: Hung et al. [182]/with permission from Elsevier (Creative Commons License, License No. 5420691310079).

12.3 Concluding Remarks and Future Perspectives

Health issues due to pollution and increased awareness about the safety of drinking water have created a sudden growth in the field of research related to water remediation processes. There is good progress in the use of nanotechnology in various stages/processes of water treatment, such as adsorption and degradation of pollutants and killing or inactivating the pathogens present in the water. Bioactive nanomaterials have many advantages in the water remediation processes such as high efficiency in the treatment of water as well as large-scale production possible through a greener/biogenic route. Among bioactive nanomaterials, metal, metal oxides, bio-based particles as well as carbon-based nanomaterials are as suitable for various water treatment processes. Based on the nature of microbes, and the nature of bioactive NPs, the mechanism behind antimicrobial action will be different. By integrating various nanomaterials, multifunctional nanomaterials can be designed to achieve the characteristics required for water purification techniques. More studies need to be conducted to develop technologies to reduce the leaching of nanomaterials into drinking water during the treatment. More research on toxicology related to the use of bioactive nanomaterials in water treatment is needed to establish safe doses of these nanomaterials to reduce the health risk.

There are still areas to improve on in aspects related to the cost, safety, toxicity, production, commercialization, separation, and reusability of bioactive NPs. There is a need to develop digital monitoring techniques for the real-time evaluation of the efficiency of NPs in the water treatment processes. The treatment processes should be capable to minimize energy and chemical consumption as well as less waste generation. Nanotechnology-enabled water treatment processes are mainly used in certain situations such as point-of-use devices. To be used in large water treatment plants, still, innovation in bioactive nanoparticle-enabled technology is required. Also, toxicity studies are to be conducted extensively to cope with the regulations of the governing bodies. The use of biogenic methods to synthesize bioactive NPs and the use of bioderived materials as reducing, capping, and stabilizing agents for nanomaterials must be extensively investigated. Additionally, the green synthesis of nanomaterials from natural waste materials and utilizing these nanomaterials for water treatment could reduce the toxicity of waste materials. By addressing these points, bioactively engineered bioactive nanomaterials can possess great potential for providing solutions for the treatment of water.

References

1 Diallo, M.S., Duncan, J.S., Savage, N. et al. (2009). Nanotechnology solutions for improving water quality. In: *Nanotechnology Applications for Clean Water* (ed. N. Savage, J.S. Duncan, R.C. Sustich, et al.), 585–587. Elsevier.

2 United Nations Educational, Scientific and Cultural Organization (2021). UN World Water Development Report 2021: Valuing Water.

3 UNESCO World Water Assessment Programme (2018). The United Nations world water development report 2018: nature-based solutions for water.

4 International Initiative on Water Quality (IIWQ), The global water quality challenge & SDGs, UNESCO. https://en.unesco.org/waterquality-iiwq/wq-challenge (accessed 17 October 2022).

5 Ouyang, W., Chen, T., Shi, Y. et al. (2019). Physico-chemical processes. *Water Environ. Res.* 91 (10): 1350–1377.

6 Willner, M.R. and Vikesland, P.J. (2018). Nanomaterial enabled sensors for environmental contaminants. *J. Nanobiotechnol.* 16 (1): 1–16.

7 Thangavelu, L., Veeraragavan, G.R., Mallineni, S.K. et al. (2022). Role of nanoparticles in environmental remediation: an insight into heavy metal pollution from dentistry. *Bioinorg. Chem. Appl.* 2022: 1946724.

8 Griffiths, J.K. (2017). Waterborne diseases. In: *International Encyclopedia of Public Health* (ed. S.R. Quah), 388–401. Elsevier.

9 Pandey, P.K., Kass, P.H., Soupir, M.L. et al. (2014). Contamination of water resources by pathogenic bacteria. *AMB Express* 4 (1): 51.

10 Szewzyk, U., Szewzyk, R., Manz, W., and Schleifer, K.H. (2000). Microbiological safety of drinking water. *Annu. Rev. Microbiol.* 54: 81–127.

11 Mazhar, M.A., Khan, N.A., Ahmed, S. et al. (2020). Chlorination disinfection by-products in municipal drinking water – a review. *J. Clean. Prod.* 273: 123159.

12 Ferguson, C., De Roda Husman, A.M., Altavilla, N. et al. (2003). Fate and transport of surface water pathogens in watersheds. *Crit. Rev. Environ. Sci. Technol.* 33 (3): 299–361.

13 Kay, D., Crowther, J., Stapleton, C.M. et al. (2008). Faecal indicator organism concentrations and catchment export coefficients in the UK. *Water Res.* 42 (10–11): 2649–2661.

14 Ng, L.Y., Mohammad, A.W., Leo, C.P., and Hilal, N. (2013). Polymeric membranes incorporated with metal/metal oxide nanoparticles: a comprehensive review. *Desalination* 308: 15–33.

15 Theron, J., Walker, J.A., and Cloete, T.E. (2008). Nanotechnology and water treatment: applications and emerging opportunities. *Crit. Rev. Microbiol.* 34 (1): 43–69.

16 Mishra, R.K., Ha, S.K., Verma, K., and Tiwari, S.K. (2018). Recent progress in selected bio-nanomaterials and their engineering applications: an overview. *J. Sci. Adv. Mater. Devices* 3 (3): 263–288.

17 Ojha, A. (2020). Nanomaterials for removal of waterborne pathogens: opportunities and challenges. In: *Waterborne Pathogens: Detection and Control* (ed. M.N.V. Prasad and A. Grobelak), 385–432. Butterworth-Heinemann.

18 Joshi, R., Khandelwal, A., Shrivastava, M., and Singh, S.D. (2020). Characterization of nanomaterials using different techniques. In: *Soil Analysis: Recent Trends and Applications* (ed. A. Rakshit, S. Ghosh, S. Chakraborty, et al.), 187–198. Singapore: Springer.

19 Luo, P., Roca, A., Tiede, K. et al. (2018). Application of nanoparticle tracking analysis for characterising the fate of engineered nanoparticles in sediment-water systems. *J. Environ. Sci.* 64: 62–71.

20 Sahoo, S.K. and Hota, G. (2019). Amine-functionalized GO decorated with ZnO-$ZnFe_2O_4$ nanomaterials for remediation of Cr(VI) from water. *ACS Appl. Nano Mater.* 2 (2): 983–996.

21 Karunakaran, G., Suriyaprabha, R., Rajendran, V., and Kannan, N. (2015). Effect of contact angle, zeta potential and particles size on the in vitro studies of Al_2O_3 and SiO_2 nanoparticles. *IET Nanobiotechnol.* 9 (1): 27–34.

22 Issa, B., Obaidat, I.M., Albiss, B.A., and Haik, Y. (2013). Magnetic nanoparticles: surface effects and properties related to biomedicine applications. *Int. J. Mol. Sci.* 14 (11): 21266.

23 Mazur, P., Skiba-Kurek, I., Mrowiec, P. et al. (2020). Synergistic ROS-associated antimicrobial activity of silver nanoparticles and gentamicin against staphylococcus epidermidis. *Int. J. Nanomedicine* 15: 3551.

24 Allahverdiyev, A.M., Abamor, E.S., Bagirova, M. et al. (2011). Antileishmanial effect of silver nanoparticles and their enhanced antiparasitic activity under ultraviolet light. *Int. J. Nanomed.* 6: 2705.

25 van Phu, D., Lang, V.T.K., Lan, N.T.K. et al. (2010). Synthesis and antimicrobial effects of colloidal silver nanoparticles in chitosan by γ-irradiation. *J. Exp. Nanosci.* 5 (2): 169–179.

26 Amini, S.M. (2019). Preparation of antimicrobial metallic nanoparticles with bioactive compounds. *Mater. Sci. Eng. C* 103: 109809.

27 Liu, H.L., Dai, S.A., Fu, K.Y., and Hsu, S.H. (2010). Antibacterial properties of silver nanoparticles in three different sizes and their nanocomposites with a new waterborne polyurethane. *Int. J. Nanomed.* 5 (1): 1017.

28 Shameli, K., Ahmad, M.B., Jazayeri, S.D. et al. (2012). Investigation of antibacterial properties silver nanoparticles prepared via green method. *Chem. Cent. J.* 6 (1): 1–10.

29 Wahab, M.A., Li, L., Li, H., and Abdala, A. (2021). Silver nanoparticle-based nanocomposites for combating infectious pathogens: recent advances and future prospects. *Nano* 11 (3): 581.

30 Durán, N., Nakazato, G., and Seabra, A.B. (2016). Antimicrobial activity of biogenic silver nanoparticles, and silver chloride nanoparticles: an overview and comments. *Appl. Microbiol. Biotechnol.* 100 (15): 6555–6570.

31 Oyanedel-Craver, V.A. and Smith, J.A. (2008). Sustainable colloidal-silver-impregnated ceramic filter for point-of-use water treatment. *Environ. Sci. Technol.* 42 (3): 927–933.

32 Gadkari, R.R., Ali, S.W., Alagirusamy, R., and Das, A. (2018). Silver nanoparticles in water purification: opportunities and challenges. In: *Modern Age Environmental Problems and their Remediation* (ed. M. Oves, M. Zain Khan, M.I. Ismail, and I.), 229–237. Cham: Springer International Publishing.

33 Das, C.G.A., Kumar, V.G., Dhas, T.S. et al. (2020). Antibacterial activity of silver nanoparticles (biosynthesis): a short review on recent advances. *Biocatal. Agric. Biotechnol.* 27: 101593.

34 Devi, L.S. and Joshi, S.R. (2012). Antimicrobial and synergistic effects of silver nanoparticles synthesized using soil fungi of high altitudes of eastern himalaya. *Mycobiology* 40 (1): 27–34.

35 Bawaskar, M., Gaikwad, S., Ingle, A. et al. (2010). A new report on mycosynthesis of silver nanoparticles by fusarium culmorum. *Curr. Nanosci.* 6 (4): 376–380.

36 Gajbhiye, M., Kesharwani, J., Ingle, A. et al. (2009). Fungus-mediated synthesis of silver nanoparticles and their activity against pathogenic fungi in combination with fluconazole. *Nanomed. Nanotechnol. Biol. Med.* 5 (4): 382–386.

37 Galdiero, S., Falanga, A., Vitiello, M. et al. (2011). Silver nanoparticles as potential antiviral agents. *Molecules* 16 (10): 8894–8918.

38 Adhikari, U., Ghosh, A., and Chandra, G. (2013). Nano particles of herbal origin: a recent eco-friend trend in mosquito control. *Asian Pacific J. Trop. Dis.* 3 (2): 167.

39 Bhat, M., Chakraborty, B., Kumar, R.S. et al. (2021). Biogenic synthesis, characterization and antimicrobial activity of Ixora brachypoda (DC) leaf extract mediated silver nanoparticles. *J. King Saud Univ. Sci.* 33 (2): 101296.

40 Vincent, M., Hartemann, P., and Engels-Deutsch, M. (2016). Antimicrobial applications of copper. *Int. J. Hyg. Environ. Health* 219 (7): 585–591.

41 Chatterjee, A.K., Sarkar, R.K., Chattopadhyay, A.P. et al. (2012). A simple robust method for synthesis of metallic copper nanoparticles of high antibacterial potency against *E. coli*. *Nanotechnology* 23 (8): 085103.

42 Park, B.K., Jeong, S., Kim, D. et al. (2007). Synthesis and size control of monodisperse copper nanoparticles by polyol method. *J. Colloid Interface Sci.* 311 (2): 417–424.

43 Harne, S., Sharma, A., Dhaygude, M. et al. (2012). Novel route for rapid biosynthesis of copper nanoparticles using aqueous extract of *Calotropis procera* L. latex and their cytotoxicity on tumor cells. *Colloids Surf. B Biointerfaces* 95: 284–288.

44 Gautam, P.K., Singh, A., Misra, K. et al. (2019). Synthesis and applications of biogenic nanomaterials in drinking and wastewater treatment. *J. Environ. Manag.* 231: 734–748.

45 Hasan, S.S., Singh, S., Parikh, R.Y. et al. (2008). Bacterial synthesis of copper/copper oxide nanoparticles. *J. Nanosci. Nanotechnol.* 8 (6): 3191–3196.

46 Chatterjee, A.K., Chakraborty, R., and Basu, T. (2014). Mechanism of antibacterial activity of copper nanoparticles. *Nanotechnology* 25 (13): 135101.

47 Deryabin, D.G., Aleshina, E.S., Vasilchenko, A.S. et al. (2013). Investigation of copper nanoparticles antibacterial mechanisms tested by luminescent *Escherichia coli* strains. *Nanotechnol. Russ.* 8 (5–6): 402–408.

48 Ruparelia, J.P., Chatterjee, A.K., Duttagupta, S.P., and Mukherji, S. (2008). Strain specificity in antimicrobial activity of silver and copper nanoparticles. *Acta Biomater.* 4 (3): 707–716.

49 Das, R., Nath, S.S., and Bhattacharjee, R. (2011). Synthesis of linoleic acid capped copper nanoparticles and their fluorescence study. *J. Fluoresc.* 21 (3): 1165–1170.

50 Schrand, A.M., Rahman, M.F., Hussain, S.M. et al. (2010). Metal-based nanoparticles and their toxicity assessment. *Wiley Interdiscip. Rev. Nanomed. Nanobiotechnol.* 2 (5): 544–568.

51 Lee, H.J., Song, J.Y., and Kim, B.S. (2013). Biological synthesis of copper nanoparticles using Magnolia kobus leaf extract and their antibacterial activity. *J. Chem. Technol. Biotechnol.* 88 (11): 1971–1977.

52 Tee, J.K., Ong, C.N., Bay, B.H. et al. (2016). Oxidative stress by inorganic nanoparticles. *Wiley Interdiscip. Rev. Nanomed. Nanobiotechnol.* 8 (3): 414–438.

53 Salah, I., Parkin, I.P., and Allan, E. (2021). Copper as an antimicrobial agent: recent advances. *RSC Adv.* 11 (30): 18179–18186.

54 Huang, T., Holden, J.A., Heath, D.E. et al. (2019). Engineering highly effective antimicrobial selenium nanoparticles through control of particle size. *Nanoscale* 11 (31): 14937–14951.

55 Wadhwani, S.A., Shedbalkar, U.U., Singh, R., and Chopade, B.A. (2016, 2016). Biogenic selenium nanoparticles: current status and future prospects. *Appl. Microbiol. Biotechnol.* 100 (6): 2555–2566.

56 Zhang, W., Zhang, J., Ding, D. et al. (2017). Synthesis and antioxidant properties of *Lycium barbarum* polysaccharides capped selenium nanoparticles using tea extract. *Artif. Cells Nanomed. Biotechnol.* 46 (7): 1463–1470.

57 Alam, H., Khatoon, N., Raza, M. et al. (2019). Synthesis and characterization of nano selenium using plant biomolecules and their potential applications. *Bionanoscience* 9 (1): 96–104.

58 Alagesan, V. and Venugopal, S. (2019). Green synthesis of selenium nanoparticle using leaves extract of *Withania somnifera* and its biological applications and photocatalytic activities. *Bionanoscience* 9 (1): 105–116.

59 Xu, X., Zhang, H., Gao, T. et al. (2021). Antibacterial thin film nanocomposite forward osmosis membranes produced by in-situ reduction of selenium nanoparticles. *Process. Saf. Environ. Prot.* 153: 403–412.

60 Cittrarasu, V., Kaliannan, D., Dharman, K. et al. (2021). Green synthesis of selenium nanoparticles mediated from *Ceropegia bulbosa* Roxb extract and its cytotoxicity, antimicrobial, mosquitocidal and photocatalytic activities. *Sci. Rep.* 11 (1): 1–15.

61 Bindhu, M.R., Umadevi, M., Esmail, G.A. et al. (2020). Green synthesis and characterization of silver nanoparticles from *Moringa oleifera* flower and assessment of antimicrobial and sensing properties. *J. Photochem. Photobiol. B Biol.* 205: 111836.

62 Ameen, F., Srinivasan, P., Selvankumar, T. et al. (2019). Phytosynthesis of silver nanoparticles using *Mangifera indica* flower extract as bioreductant and their broad-spectrum antibacterial activity. *Bioorg. Chem.* 88: 102970.

63 Wang, L., Lu, F., Liu, Y. et al. (2018). Photocatalytic degradation of organic dyes and antimicrobial activity of silver nanoparticles fast synthesized by flavonoids fraction of *Psidium guajava* L. leaves. *J. Mol. Liq.* 263: 187–192.

64 Borah, D., Das, N., Das, N. et al. (2020). Alga-mediated facile green synthesis of silver nanoparticles: photophysical, catalytic and antibacterial activity. *Appl. Organomet. Chem.* 34 (5): e5597.

65 Ravichandran, V., Vasanthi, S., Shalini, S. et al. (2019). Green synthesis, characterization, antibacterial, antioxidant and photocatalytic activity of Parkia speciosa leaves extract mediated silver nanoparticles. *Results Phys.* 15: 102565.

66 Ma, B., Chaudhary, J.P., Zhu, J. et al. (2021). Construction of silver nanoparticles anchored in carbonized bacterial cellulose with enhanced antibacterial properties. *Colloids Surf., A* 611: 125845.

67 Bouazizi, N., Vieillard, J., Thebault, P. et al. (2018). Silver nanoparticle embedded copper oxide as an efficient core–shell for the catalytic reduction of 4-nitrophenol and antibacterial activity improvement. *Dalton Trans.* 47 (27): 9143–9155.

68 Chompunut, L., Wanaporn, T., Anupong, W. et al. (2022). Synthesis of copper nanoparticles from the aqueous extract of *Cynodon dactylon* and evaluation of its antimicrobial and photocatalytic properties. *Food Chem. Toxicol.* 166: 113245.

69 Jahan, I., Erci, F., and Isildak, I. (2021). Facile microwave-mediated green synthesis of non-toxic copper nanoparticles using *Citrus sinensis* aqueous fruit extract and their antibacterial potentials. *J. Drug Deliv. Sci. Technol.* 61: 102172.

70 Das, P.E., Abu-Yousef, I.A., Majdalawieh, A.F. et al. (2020). Green synthesis of encapsulated copper nanoparticles using a hydroalcoholic extract of moringa oleifera leaves and assessment of their antioxidant and antimicrobial activities. *Molecules* 25 (3): 555.

71 Mali, S.C., Dhaka, A., Githala, C.K., and Trivedi, R. (2020). Green synthesis of copper nanoparticles using *Celastrus paniculatus* Willd. leaf extract and their photocatalytic and antifungal properties. *Biotechnol. Rep.* 27: e00518.

72 Duan, X., Liu, S., Huang, E. et al. (2020). Superhydrophobic and antibacterial wood enabled by polydopamine-assisted decoration of copper nanoparticles. *Colloids Surf., A* 602: 125145.

73 Irfan, M., Uddin, Z., Ahmad, F. et al. (2022). Ecofriendly development of electrospun antibacterial membranes loaded with silver nanoparticles. *J. Ind. Text.* 51 (2S): 2412S–2425S.

74 Zhang, H., Li, Z., Dai, C. et al. (2021). Antibacterial properties and mechanism of selenium nanoparticles synthesized by *Providencia* sp. DCX. *Environ. Res.* 194: 110630.

75 Khan, S.T., Ahmad, F., Shahadat, M. et al. (2020). Metal and metal oxide nanoparticles for water decontamination and purification. *Environ. Nanotechnol. Water Purif.* 151–186.

76 Upadhyay, R.K., Soin, N., and Roy, S.S. (2013). Role of graphene/metal oxide composites as photocatalysts, adsorbents and disinfectants in water treatment: a review. *RSC Adv.* 4 (8): 3823–3851.

77 Subhan, M.A. (2020). Antibacterial property of metal oxide-based nanomaterials. In: *Nanotoxicity Prevention and Antibacterial Applications of Nanomaterials* (ed. S. Rajendran, A. Mukherjee, T.A. Nguyen, et al.), 283–300. Elsevier.

78 Agrawal, A., Sharma, A., Awasthi, K.K., and Awasthi, A. (2021). Metal oxides nanocomposite membrane for biofouling mitigation in wastewater treatment. *Mater. Today Chem.* 21: 100532.

79 Parham, S., Wicaksono, D.H.B., Bagherbaigi, S. et al. (2016). Antimicrobial treatment of different metal oxide nanoparticles: a critical review. *J. Chin. Chem. Soc.* 63 (4): 385–393.

80 Hajipour, M.J., Fromm, K.M., Akbar Ashkarran, A. et al. (2012). Antibacterial properties of nanoparticles. *Trends Biotechnol.* 30 (10): 499–511.

81 Brugnera, M.F., Miyata, M., Zocolo, G.J. et al. (2013). A photoelectrocatalytic process that disinfects water contaminated with *Mycobacterium kansasii* and *Mycobacterium avium*. *Water Res.* 47 (17): 6596–6605.

82 Sökmen, M., Değerli, S., and Aslan, A. (2008). Photocatalytic disinfection of *Giardia intestinalis* and *Acanthamoeba castellani* cysts in water. *Exp. Parasitol.* 119 (1): 44–48.

83 Lawton, L.A., Robertson, P.K.J., Cornish, B.J.P.A., and Jaspars, M. (1999). Detoxification of microcystins (Cyanobacterial Hepatotoxins) using TiO_2 photocatalytic oxidation. *Environ. Sci. Technol.* 33 (5): 771–775.

84 Jafry, H.R., Liga, M.V., Li, Q., and Barron, A.R. (2011). Simple route to enhanced photocatalytic activity of P25 titanium dioxide nanoparticles by silica addition. *Environ. Sci. Technol.* 45 (4): 1563–1568.

85 Weldegebrieal, G.K. (2020). Synthesis method, antibacterial and photocatalytic activity of ZnO nanoparticles for azo dyes in wastewater treatment: a review. *Inorg. Chem. Commun.* 120: 108140.

86 Liu, J., Wang, Y., Ma, J. et al. (2019). A review on bidirectional analogies between the photocatalysis and antibacterial properties of ZnO. *J. Alloys Compd.* 783: 898–918.

87 Qi, K., Cheng, B., Yu, J., and Ho, W. (2017). Review on the improvement of the photocatalytic and antibacterial activities of ZnO. *J. Alloys Compd.* 727: 792–820.

88 Sheikh, M., Pazirofteh, M., Dehghani, M. et al. (2020). Application of ZnO nanostructures in ceramic and polymeric membranes for water and wastewater technologies: a review. *Chem. Eng. J.* 391: 123475.

89 Elmi, F., Alinezhad, H., Moulana, Z. et al. (2014). The use of antibacterial activity of ZnO nanoparticles in the treatment of municipal wastewater. *Water Sci. Technol.* 70 (5): 763–770.

90 Bhatia, S. and Verma, N. (2017). Photocatalytic activity of ZnO nanoparticles with optimization of defects. *Mater. Res. Bull.* 95: 468–476.

91 Orou, S.F.C., Hang, K.J., Thien, M.T. et al. (2018). Antibacterial activity by ZnO nanorods and ZnO nanodisks: a model used to illustrate "Nanotoxicity Threshold.". *J. Ind. Eng. Chem.* 62: 333–340.

92 Awasthi, A., Sharma, P., Jangir, L. et al. (2020). Dose dependent enhanced antibacterial effects and reduced biofilm activity against *Bacillus subtilis* in presence of ZnO nanoparticles. *Mater. Sci. Eng. C* 113: 111021.

93 Manikandan, V., Jayanthi, P., Priyadharsan, A. et al. (2019). Green synthesis of pH-responsive Al_2O_3 nanoparticles: application to rapid removal of nitrate ions with enhanced antibacterial activity. *J. Photochem. Photobiol. A Chem.* 371: 205–215.

94 Ragunathan, R., Velusamy, S., Nallasamy, J.L. et al. (2022). Synthesis and enhanced photocatalytic activity of Zinc oxide-based nanoparticles and its antibacterial activity. *J. Nanomater.* 2022: 3863184.

95 Khalid, H.F., Tehseen, B., Sarwar, Y. et al. (2019). Biosurfactant coated silver and iron oxide nanoparticles with enhanced anti-biofilm and anti-adhesive properties. *J. Hazard. Mater.* 364: 441–448.

96 Prakash, S., Elavarasan, N., Venkatesan, A. et al. (2018). Green synthesis of copper oxide nanoparticles and its effective applications in Biginelli reaction, BTB photodegradation and antibacterial activity. *Adv. Powder Technol.* 29 (12): 3315–3326.

97 Pallela, P.N.V.K., Ummey, S., Ruddaraju, L.K. et al. (2019). Antibacterial efficacy of green synthesized α-Fe_2O_3 nanoparticles using *Sida cordifolia* plant extract. *Heliyon* 5 (11): e02765.

98 Vasantharaj, S., Sathiyavimal, S., Senthilkumar, P. et al. (2019). Biosynthesis of iron oxide nanoparticles using leaf extract of *Ruellia tuberosa*: antimicrobial properties and their applications in photocatalytic degradation. *J. Photochem. Photobiol., B* 192: 74–82.

99 Mirza, A.U., Kareem, A., Nami, S.A.A. et al. (2018). Biogenic synthesis of iron oxide nanoparticles using *Agrewia optiva* and *Prunus persica* phyto species: characterization, antibacterial and antioxidant activity. *J. Photochem. Photobiol., B* 185: 262–274.

100 Li, J., Xie, B., Xia, K. et al. (2018). Enhanced antibacterial activity of silver doped titanium dioxide-chitosan composites under visible light. *Dent. Mater.* 11 (8): 1403.

101 Ngoepe, N.M., Mathipa, M.M., and Hintsho-Mbita, N.C. (2020). Biosynthesis of titanium dioxide nanoparticles for the photodegradation of dyes and removal of bacteria. *Optik* 224: 165728.

102 Bekele, E.T., Gonfa, B.A., Zelekew, O.A. et al. (2020). Synthesis of titanium oxide nanoparticles using root extract of *Kniphofia foliosa* as a template, characterization, and its application on drug resistance bacteria. *J. Nanomater.* 2020: 1–10.

103 Tang, Y., Sun, H., Shang, Y. et al. (2019). Spiky nanohybrids of titanium dioxide/gold nanoparticles for enhanced photocatalytic degradation and anti-bacterial property. *J. Colloid Interface Sci.* 535: 516–523.

104 Prashanth, P.A., Raveendra, R.S., Hari Krishna, R. et al. (2018). Synthesis, characterizations, antibacterial and photoluminescence studies of solution combustion-derived α-Al_2O_3 nanoparticles. *J. Asian Ceramic Soc.* 3 (3): 345–351.

105 Badawy, M.E.I. and Rabea, E.I. (2011). A biopolymer chitosan and its derivatives as promising antimicrobial agents against plant pathogens and their applications in crop protection. *Int. J. Carbohydr. Chem.* 2011: 1–29.

106 Confederat, L.G., Tuchilus, C.G., Dragan, M. et al. (2021). Preparation and antimicrobial activity of chitosan and its derivatives: a concise review. *Molecules* 26 (12): 3694.

107 Lopes, P.P., Tanabe, E.H., and Bertuol, D.A. (2020). Chitosan as biomaterial in drug delivery and tissue engineering. In: *Handbook of Chitin and Chitosan Volume 3: Chitin and Chitosan based Polymer Materials for Various* (ed. S. Gopi, S. Thomas, and A. Pius), 407–431. Elsevier.

108 Divya, K. and Jisha, M.S. (2017). Chitosan nanoparticles preparation and applications. *Environ. Chem. Lett.* 16 (1): 101–112.

109 Khanniri, E., Yousefi, M., Mortazavian, A.M. et al. (2021). Effective removal of lead (II) using chitosan and microbial adsorbents: response surface methodology (RSM). *Int. J. Biol. Macromol.* 178: 53–62.

110 Mishra, P.C., Islam, M., and Patel, R.K. (2013). Removal of lead (II) by chitosan from aqueous medium. *Sep. Sci. Technol.* 48 (8): 1234–1242.

111 Babakhani, A. and Sartaj, M. (2020). Removal of Cadmium (II) from aqueous solution using tripolyphosphate cross-linked chitosan. *J. Environ. Chem. Eng.* 8 (4): 103842.

112 Boamah, P.O., Huang, Y., Hua, M. et al. (2015). Removal of cadmium from aqueous solution using low molecular weight chitosan derivative. *Carbohydr. Polym.* 122: 255–264.

113 Bagheri, M., Younesi, H., Hajati, S., and Borghei, S.M. (2015). Application of chitosan-citric acid nanoparticles for removal of chromium (VI). *Int. J. Biol. Macromol.* 80: 431–444.

114 Gérente, C., Andrès, Y., McKay, G., and Le Cloirec, P. (2010). Removal of arsenic(V) onto chitosan: from sorption mechanism explanation to dynamic water treatment process. *Chem. Eng. J.* 158 (3): 593–598.

115 Boddu, V.M., Abburi, K., Talbott, J.L. et al. (2008). Removal of arsenic (III) and arsenic (V) from aqueous medium using chitosan-coated biosorbent. *Water Res.* 42 (3): 633–642.

116 Mahmoud, M.E., Ibrahim, G.A.A., and Abdelwahab, M.S. (2021). Manganese dioxide nanoparticles decorated with chitosan for effective removal of lead and lanthanum ions from water by microwave sorption technique. *Mater. Sci. Eng. B* 267: 115091.

117 Vakili, M., Rafatullah, M., Salamatinia, B. et al. (2014). Application of chitosan and its derivatives as adsorbents for dye removal from water and wastewater: a review. *Carbohydr. Polym.* 113: 115–130.

118 Szyguła, A., Guibal, E., Ruiz, M., and Sastre, A.M. (2008). The removal of sulphonated azo-dyes by coagulation with chitosan. *Colloids Surf., A* 330 (2–3): 219–226.

119 Wu, F.C., Tseng, R.L., and Juang, R.S. (2002). Adsorption of dyes and humic acid from water using chitosan-encapsulated activated carbon. *J. Chem. Technol. Biotechnol.* 77 (11): 1269–1279.

120 Zulfikar, M.A., Afrita, S., Wahyuningrum, D., and Ledyastuti, M. (2016). Preparation of Fe_3O_4-chitosan hybrid nano-particles used for humic acid adsorption. *Environ. Nanotechnol. Monit. Manag.* 6: 64–75.

121 Soares, S.F., Fernandes, T., Sacramento, M. et al. (2019). Magnetic quaternary chitosan hybrid nanoparticles for the efficient uptake of diclofenac from water. *Carbohydr. Polym.* 203: 35–44.

122 Gokila, S., Gomathi, T., Sudha, P.N., and Anil, S. (2017). Removal of the heavy metal ion chromiuim(VI) using Chitosan and Alginate nanocomposites. *Int. J. Biol. Macromol.* 104: 1459–1468.

123 Jalalian, N. and Nabavi, S.R. (2020). Electrosprayed chitosan nanoparticles decorated on polyamide6 electrospun nanofibers as membrane for acid fuchsin dye filtration from water. *Surf. Interfaces* 21: 100779.

124 Hosseinnejad, M. and Jafari, S.M. (2016). Evaluation of different factors affecting antimicrobial properties of chitosan. *Int. J. Biol. Macromol.* 85: 467–475.

125 Rabea, E.I., Badawy, M.E.T., Stevens, C.V. et al. (2003). Chitosan as antimicrobial agent: applications and mode of action. *Biomacromolecules* 4 (6): 1457–1465.

126 Ikinci, G., Şenel, S., Akincibay, H. et al. (2002). Effect of chitosan on a periodontal pathogen *Porphyromonas gingivalis*. *Int. J. Pharm.* 235 (1–2): 121–127.

127 Jung, B.-O., Kim, C.-H., Choi, K.-S. et al. (1999). Preparation of amphiphilic chitosan and their antimicrobial activities. *J. Appl. Polym. Sci.* 72 (13): 1713–1719.

128 Dubey, R., Dutta, D., Sarkar, A., and Chattopadhyay, P. (2021). Functionalized carbon nanotubes: synthesis, properties and applications in water purification, drug delivery, and material and biomedical sciences. *Nanoscale Adv.* 3 (20): 5722–5744.

129 Lee, K.M., Wong, C.P.P., Tan, T.L., and Lai, C.W. (2018). Functionalized carbon nanotubes for adsorptive removal of water pollutants. *Mater. Sci. Eng. B* 236–237: 61–69.

130 Barrejón, M., Syrgiannis, Z., Burian, M. et al. (2019). Cross-linked carbon nanotube adsorbents for water treatment: tuning the sorption capacity through chemical functionalization. *ACS Appl. Mater. Interfaces* 11 (13): 12920–12930.

131 Mousavi, S.A. and Janjani, H. (2018). Antibiotics adsorption from aqueous solutions using carbon nanotubes: a systematic review. *Toxin Rev.* 39 (2): 87–98.

132 Jame, S.A. and Zhou, Z. (2016). Electrochemical carbon nanotube filters for water and wastewater treatment. *Nanotechnol. Rev.* 5 (1): 41–50.

133 Bai, Y., Park, I.S., Lee, S.J. et al. (2011). Aqueous dispersion of surfactant-modified multiwalled carbon nanotubes and their application as an antibacterial agent. *Carbon* 49 (11): 3663–3671.

134 Wang, H., Lin, K.Y., Jing, B. et al. (2013). Removal of oil droplets from contaminated water using magnetic carbon nanotubes. *Water Res.* 47 (12): 4198–4205.

135 Hong, Y., Zhang, J., Zhu, C. et al. (2019). Water desalination through rim functionalized carbon nanotubes. *J. Mater. Chem. A* 7 (8): 3583–3591.

136 Chan, W.F., Chen, H.Y., Surapathi, A. et al. (2013). Zwitterion functionalized carbon nanotube/polyamide nanocomposite membranes for water desalination. *ACS Nano* 7 (6): 5308–5319.

137 Das, R., Ali, M.E., Hamid, S.B.A. et al. (2014). Carbon nanotube membranes for water purification: a bright future in water desalination. *Desalination* 336 (1): 97–109.

138 Farahani, M.H.D.A., Hua, D., and Chung, T.S. (2018). Cross-linked mixed matrix membranes (MMMs) consisting of amine-functionalized multi-walled carbon nanotubes and P84 polyimide for organic solvent nanofiltration (OSN) with enhanced flux. *J. Membr. Sci.* 548: 319–331.

139 Liu, Y., Liu, H., Zhou, Z. et al. (2015). Degradation of the common aqueous antibiotic tetracycline using a carbon nanotube electrochemical filter. *Environ. Sci. Technol.* 49 (13): 7974–7980.

140 Saleemi, M.A., Kong, Y.L., Yong, P.V.C., and Wong, E.H. (2022). An overview of antimicrobial properties of carbon nanotubes-based nanocomposites. *Adv. Pharm. Bull.* 12 (3): 449.

141 Liu, D., Mao, Y., and Ding, L. (2018). Carbon nanotubes as antimicrobial agents for water disinfection and pathogen control. *J. Water Health* 16 (2): 171–180.

142 Kang, S., Pinault, M., Pfefferle, L.D., and Elimelech, M. (2007). Single-walled carbon nanotubes exhibit strong antimicrobial activity. *Langmuir* 23 (17): 8670–8673.

143 Kang, S., Mauter, M.S., and Elimelech, M. (2008). Physicochemical determinants of multiwalled carbon nanotube bacterial cytotoxicity. *Environ. Sci. Technol.* 42 (19): 7528–7534.

144 Nagai, H. and Toyokuni, S. (2012). Differences and similarities between carbon nanotubes and asbestos fibers during mesothelial carcinogenesis: shedding light on fiber entry mechanism. *Cancer Sci.* 103 (8): 1378.

145 Atiyah, A.A., Haider, A.J., and Dhahi, R.M. (2019). Cytotoxicity properties of functionalised carbon nanotubes on pathogenic bacteria. *IET Nanobiotechnol.* 13 (6): 597–601.

146 Hamouda, H.I., Abdel-Ghafar, H.M., and Mahmoud, M.H.H. (2021). Multi-walled carbon nanotubes decorated with silver nanoparticles for antimicrobial applications. *J. Environ. Chem. Eng.* 9 (2): 105034.

147 Yoosefi Booshehri, A., Wang, R., and Xu, R. (2013). The effect of re-generable silver nanoparticles/multi-walled carbon nanotubes coating on the antibacterial performance of hollow fiber membrane. *Chem. Eng. J.* 230: 251–259.

148 Al-Hakami, S.M., Khalil, A.B., Laoui, T., and Atieh, M.A. (2013). Fast disinfection of *Escherichia coli* bacteria using carbon nanotubes interaction with microwave radiation. *Bioinorg. Chem. Appl.* 2013.

149 Ali, Q., Ahmed, W., Lal, S., and Sen, T. (2017). Novel multifunctional carbon nanotube containing silver and iron oxide nanoparticles for antimicrobial applications in water treatment. *Mater. Today Proc.* 4 (1): 57–64.

150 Brady-Estévez, A.S., Nguyen, T.H., Gutierrez, L., and Elimelech, M. (2010). Impact of solution chemistry on viral removal by a single-walled carbon nanotube filter. *Water Res.* 44 (13): 3773–3780.

151 Brady-Estévez, A.S., Kang, S., and Elimelech, M. (2008). A single-walled-carbon-nanotube filter for removal of viral and bacterial pathogens. *Small* 4 (4): 481–484.

152 Ihsanullah, Laoui, T., Al-Amer, A.M. et al. (2015). Novel anti-microbial membrane for desalination pretreatment: a silver nanoparticle-doped carbon nanotube membrane. *Desalination* 376: 82–93.

153 Yang, Y., Nie, C., Deng, Y. et al. (2016). Improved antifouling and antimicrobial efficiency of ultrafiltration membranes with functional carbon nanotubes. *RSC Adv.* 6 (91): 88265–88276.

154 Vecitis, C.D., Schnoor, M.H., Rahaman, M.S. et al. (2011). Electrochemical multiwalled carbon nanotube filter for viral and bacterial removal and inactivation. *Environ. Sci. Technol.* 45 (8): 3672–3679.

155 Banerjee, P., Sau, S., Das, P., and Mukhopadhayay, A. (2015). Optimization and modelling of synthetic azo dye wastewater treatment using graphene oxide nanoplatelets: characterization toxicity evaluation and optimization using artificial neural network. *Ecotoxicol. Environ. Saf.* 119: 47–57.

156 Goswami, S., Banerjee, P., Datta, S. et al. (2017). Graphene oxide nanoplatelets synthesized with carbonized agro-waste biomass as green precursor and its application for the treatment of dye rich wastewater. *Process. Saf. Environ. Prot.* 106: 163–172.

157 Wu, Z., Zhong, H., Yuan, X. et al. (2014). Adsorptive removal of methylene blue by rhamnolipid-functionalized graphene oxide from wastewater. *Water Res.* 67: 330–344.

158 Banerjee, P., Das, P., Zaman, A., and Das, P. (2016). Application of graphene oxide nanoplatelets for adsorption of Ibuprofen from aqueous solutions: evaluation of process kinetics and thermodynamics. *Process. Saf. Environ. Prot.* 101: 45–53.

159 Zhu, J., Wang, J., Hou, J. et al. (2017). Graphene-based antimicrobial polymeric membranes: a review. *J. Mater. Chem. A* 5 (15): 6776–6793.

160 Zeng, X., Wang, G., Liu, Y., and Zhang, X. (2017). Graphene-based antimicrobial nanomaterials: rational design and applications for water disinfection and microbial control. *Environ. Sci. Nano* 4 (12): 2248–2266.

161 Zheng, H., Ma, R., Gao, M. et al. (2018). Antibacterial applications of graphene oxides: structure-activity relationships, molecular initiating events and biosafety. *Sci. Bull.* 63 (2): 133–142.

162 Lim, M.Y., Choi, Y.S., Kim, J. et al. (2017). Cross-linked graphene oxide membrane having high ion selectivity and antibacterial activity prepared using tannic acid-functionalized graphene oxide and polyethyleneimine. *J. Membr. Sci.* 521: 1–9.

163 Li, J., Liu, X., Lu, J. et al. (2016). Anti-bacterial properties of ultrafiltration membrane modified by graphene oxide with nano-silver particles. *J. Colloid Interface Sci.* 484: 107–115.

164 Wu, W., Shi, Y., Liu, G. et al. (2020). Recent development of graphene oxide based forward osmosis membrane for water treatment: a critical review. *Desalination* 491: 114452.

165 Akhavan, O. and Ghaderi, E. (2010). Toxicity of graphene and graphene oxide nanowalls against bacteria. *ACS Nano* 4 (10): 5731–5736.

166 Liu, S., Zeng, T.H., Hofmann, M. et al. (2011). Antibacterial activity of graphite, graphite oxide, graphene oxide, and reduced graphene oxide: membrane and oxidative stress. *ACS Nano* 5 (9): 6971–6980.

167 Liu, S., Hu, M., Zeng, T.H. et al. (2012). Lateral dimension-dependent antibacterial activity of graphene oxide sheets. *Langmuir* 28 (33): 12364–12372.

168 Karahan, H.E., Wei, L., Goh, K. et al. (2016). Synergism of water shock and a biocompatible block copolymer potentiates the antibacterial activity of graphene oxide. *Small* 12 (7): 951–962.

169 Gollavelli, G., Chang, C.C., and Ling, Y.C. (2013). Facile synthesis of smart magnetic graphene for safe drinking water: heavy metal removal and disinfection control. *ACS Sustain. Chem. Eng.* 1 (5): 462–472.

170 Perdikaki, A., Galeou, A., Pilatos, G. et al. (2016). Ag and Cu monometallic and Ag/Cu bimetallic nanoparticle–graphene composites with enhanced antibacterial performance. *ACS Appl. Mater. Interfaces* 8 (41): 27498–27510.

171 Deng, C.H., Gong, J.L., Zeng, G.M. et al. (2016). Graphene–CdS nanocomposite inactivation performance toward *Escherichia coli* in the presence of humic acid under visible light irradiation. *Chem. Eng. J.* 284: 41–53.

172 Zhang, G., Zhou, M., Xu, Z. et al. (2019). Guanidyl-functionalized graphene/polysulfone mixed matrix ultrafiltration membrane with superior permselective, antifouling and antibacterial properties for water treatment. *J. Colloid Interface Sci.* 540: 295–305.

173 Rajapaksha, P., Cheeseman, S., Hombsch, S. et al. (2019). Antibacterial properties of graphene oxide–copper oxide nanoparticle nanocomposites. *ACS Appl. Bio Mater.* 2 (12): 5687–5696.

174 Alayande, A.B., Obaid, M., and Kim, I.S. (2020). Antimicrobial mechanism of reduced graphene oxide-copper oxide (rGO-CuO) nanocomposite films: the case of *Pseudomonas aeruginosa* PAO1. *Mater. Sci. Eng. C* 109: 110596.

175 Kusiak-Nejman, E. and Morawski, A.W. (2019). TiO_2/graphene-based nanocomposites for water treatment: a brief overview of charge carrier transfer, antimicrobial and photocatalytic performance. *Appl. Catal. B Environ.* 253: 179–186.

176 Tian, T., Shi, X., Cheng, L. et al. (2014). Graphene-based nanocomposite as an effective, multifunctional, and recyclable antibacterial agent. *ACS Appl. Mater. Interfaces* 6 (11): 8542–8548.

177 Joghataeian, M., Bahari, A., Qavami, A., and Raeisi, M.J. (2020). An antibacterial study of a new magnetic carbon nanotube/core-shell nanohybrids. *J. Environ. Chem. Eng.* 8 (5): 104150.

178 Yuan, W., Jiang, G., Che, J. et al. (2008). Deposition of silver nanoparticles on multiwalled carbon nanotubes grafted with hyperbranched poly(amidoamine) and their antimicrobial effects. *J. Phys. Chem. C* 112 (48): 18754–18759.

179 Chang, Y.N., Ou, X.M., Zeng, G.M. et al. (2015). Synthesis of magnetic graphene oxide–TiO_2 and their antibacterial properties under solar irradiation. *Appl. Surf. Sci.* 343: 1–10.

180 Song, B., Zhang, C., Zeng, G. et al. (2016). Antibacterial properties and mechanism of graphene oxide-silver nanocomposites as bactericidal agents for water disinfection. *Arch. Biochem. Biophys.* 604: 167–176.

181 Chella, S., Kollu, P., Komarala, E.V.P.R. et al. (2015). Solvothermal synthesis of $MnFe_2O_4$-graphene composite—Investigation of its adsorption and antimicrobial properties. *Appl. Surf. Sci.* 327: 27–36.

182 Hung, W.C., Wu, K.H., Lyu, D.Y. et al. (2017). Preparation and characterization of expanded graphite/metal oxides for antimicrobial application. *Mater. Sci. Eng. C* 75: 1019–1025.

13

Cellulose Nanofibers-Based Porous Materials, a Green Platform to Construct Advanced Sorbents for Oil/Spill Cleanup

Aiswarya Poulose[1], Ajith Mathew[1], Deepu A. Gopakumar[1], Daniel Pasquini[2], A. Mathiazhagan[3], and Jinu Jacob George[1]

[1] *Cochin University of Science and Technology, Department of Polymer Science and Rubber Technology, Kalamassery, Cochin 682022, Kerala, India*
[2] *Federal University of Uberlandia-UFU, Chemistry Institute, Campus Santa Monica-Bloco1DCP593, Uberlandia, Brazil*
[3] *Cochin University of Science and Technology, Department of Ship Technology, Kalamassery, Cochin 682022, Kerala, India*

13.1 Introduction

With the rapid development of industries and offshore extraction, oily water is becoming a very prevalent pollutant that can seriously harm the water environment and ecological systems. The regular oil spills and organic pollutants released during the extraction of crude oil pose a serious threat to marine environments [1]. Annual crude oil spills into the ocean are estimated to be in the range of 100 million gallons [2]. Traditional methods for cleaning up oil spills include in situ burning, gas flotation, centrifugation, and chemical precipitation. But the majority of these reactive strategies suffer from poor separation, low selectivity, and energy-intensive consumption, which restricts their widespread use in oil/water separation [3]. Another method for limiting the spread of oil slicks on water is physical/mechanical recovery, which includes oil in situ burning, skimmers, and boomers. Oil skimmers are expensive and inefficient; in situ burning results in a byproduct of pollution; and mechanical extraction based on booming and skimming involves a comparatively high energy consumption and time expense. In order to assure better oil-water separation and the recovery of oil from spills, considerable thought is being devoted to the development of novel techniques and technologies. Hence, research world focuses on the use of sorbents and insoluble compounds to remove oil via absorption, adsorption, or both [2]. It is therefore preferred to use sorbents to cleanup oil spills because of their hydrophobicity, oleophilicity, absorption rate and capacity, buoyancy, oil retention, durability, and reusability. The absorption capacity depends on many elements, such as (i) the properties of the substrate, (ii) nature of spilled oil, and (iii) the marine environment. The quality of sorbent materials is thought to be

Nanomaterials for Air and Water Purification, First Edition.
Edited by Deepu A. Gopakumar, Jyotishkumar Parameswaranpillai, Jinu Jacob George, and Midhun Dominic C.D.

the most crucial element affecting the amount of oil uptake as well as the retained oils in the porous holes of the absorbent among the three parameters stated above [4]. Numerous components, including biomass activated carbon, zeolite, chitosan (CS), graphene, plant-based clay minerals, and industrial wastes, are used in the adsorption process. These substances, which come in powder, films, hydrogels, and aerogels, can all be utilized as adsorbents to clean water. The high surface area, ease of scaling up, and high adsorption efficiency of aerogels are just a few of their benefits.

13.2 Aerogels as Material for Oil Recovery

Aerogels are a family of lightweight, highly interconnected, porous solid materials created by adding air to a gel instead of liquid. They exhibit promising potential as a perfect oil sorbent candidate to rapidly absorb a huge amount of oil and to float in water due to their large specific surface area, high porosity, and low density [5]. In recent years, there has been a lot of research on the use of aerogels for oil sorption. The recovery of the oil and the recyclability of the sorbent material are aerogels key advantages for treating oil spills [6]. There have been reports of a variety of recovery techniques, including mechanical squeezing, distillation, and extraction. The use of organic solvents to produce a mixture of oil and the solvent, which is then removed, makes extraction possibly the least environmentally friendly of the three methods. Typically, distillation is required to separate the solvent from the mixture and allow for the recycling of the solvent. Figure 13.1 shows the recovery methods for oil separation and the recovery of sorbents.

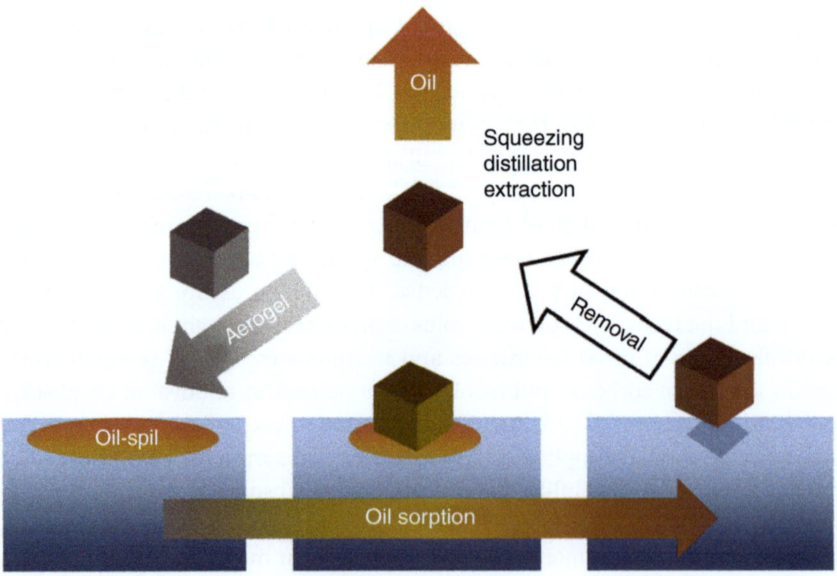

Figure 13.1 Oil recovery using aerogels.

The first commercially successful and frequently used aerogel is silica. Early in the 1930s, S. Kistler created the first silica aerogel, but its extensive use was constrained by the material's intrinsic rigidity and fragility as well as secondary pollutants. Currently used aerogel materials include carbon, metal–organic frameworks (MOFs), and polymers like polyaniline and polyvinylidene fluoride (PVDF). However, these materials have nonbiodegradable nature, and secondary contamination restricts their use. Due to their sustainability and environmental friendliness, green aerogel materials such as nanocellulose (NC), CS, and starch are recently getting greater attention. These polysaccharides are a viable substitute for silica and other nonbiodegradable polymers when creating aerogels for water purification due to their flexibility and good mechanical strength. In severe settings, these aerogels interact with water contaminants and adsorb/absorb pollutants by a variety of mechanisms, including electrostatic interactions, chelation, oleophilic, hydrophilic, and hydrophobic interactions. Due to the enormous concentrations required for large-scale oil spill cleanup, sorbent costs must undoubtedly be reasonable. When using sorbents extensively, disposal becomes a source of worry. Designing recyclable, affordable super-absorbents becomes difficult as a result [2].

13.3 Cellulose; a Biodegradable Polymer

The most prevalent, renewable, and nontoxic natural polymeric substance is cellulose, a linear polysaccharide of D-glucose residues linked by -(1-4)-glycosidic linkages (Figure 13.2). It possesses a number of commendable qualities, including biocompatibility, hydrophilicity, thermostability, and strong mechanical properties. Cellulose has demonstrated exceptional potential in drug loading, dye anionic adsorption, and metal adsorption. This biodegradable polymer is mostly found in nature as microfibrils in the cell walls of plants and trees, the tissues of algae, and the membrane of tunicate epidermal cells. Bacteria can also produce it in the form of nanofiber networks. Cellulosic materials use fibril aggregates, fibrils, nanocrystallites, and nanoscale disordered domains as examples of hierarchical structure design that extends from nanoscale to macroscopic dimensions. The complex multilevel structure of cellulose is composed of bundles or aggregates of ultrafine fibrils. A number of cellulose chains are present in the superfine fibril. Depending on the source of synthesis, each fibril is made up of repeated

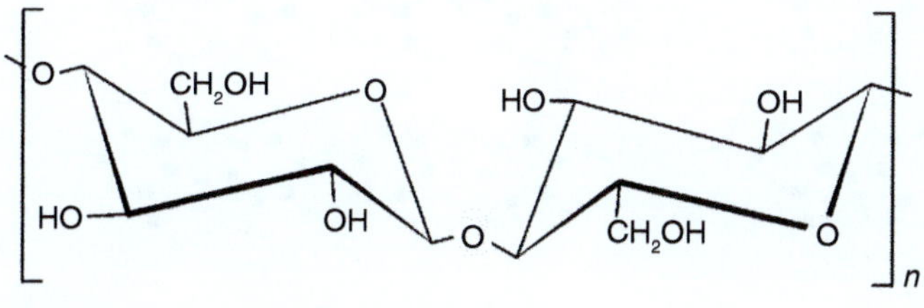

Figure 13.2 Structure of cellulose.

large ordered (crystalline) and small disordered (amorphous) domains with a cross-sectional dimension ranging from 2 to 20 nm [7].

The high density of hydroxyl groups that gives lignocellulosic substances their hydrophilic properties is one of their distinctive properties. Six distinct polymorphs of cellulose—I, II, IIII, IIIII, IV I, and IV II—are present, with the ability to change from one form to another. Though thought to have only one crystal structure, cellulose I, in its natural state, evidence supports the existence of two suballomorphs of cellulose I, referred to as Iα and Iβ. These two polymorphs exist in various ratios depending on the source of cellulose: Iα is common in celluloses made by bacteria and algae, and higher plant celluloses may contain both Iα and Iβ. Although cellulose chains adopt parallel topologies in both the Iα and Iβ structures, their hydrogen bonding patterns change, which suggests that the crystalline structure is different. (Figure 13.3) The sheets of cellulose(Iα and Iβ) are stacked "parallel up," and the hydroxylmethyl groups are positioned in a tg conformation so that their O_6 atom points towards the O_2 hydroxyl groups of the adjacent residue, resulting in a second inter-residue hydrogen bond [8].

The second-most researched allomorph, cellulose II, can be produced using one of two methods: (i) Chemical regeneration, which involves cellulose I being dissolved in a solvent and subsequently reprecipitated in water. (ii) The widely used method of "merçerization," named after its creator, John Mercer (1844), which entails swelling natural cellulose in concentrated sodium hydroxide solutions and producing cellulose II after the swelling agent has been removed. When cellulose I or II is subjected to different amines or ammonia (gas or liquefied), cellulose III is created after the swelling agent has been removed. Depending on whether the beginning form is I or

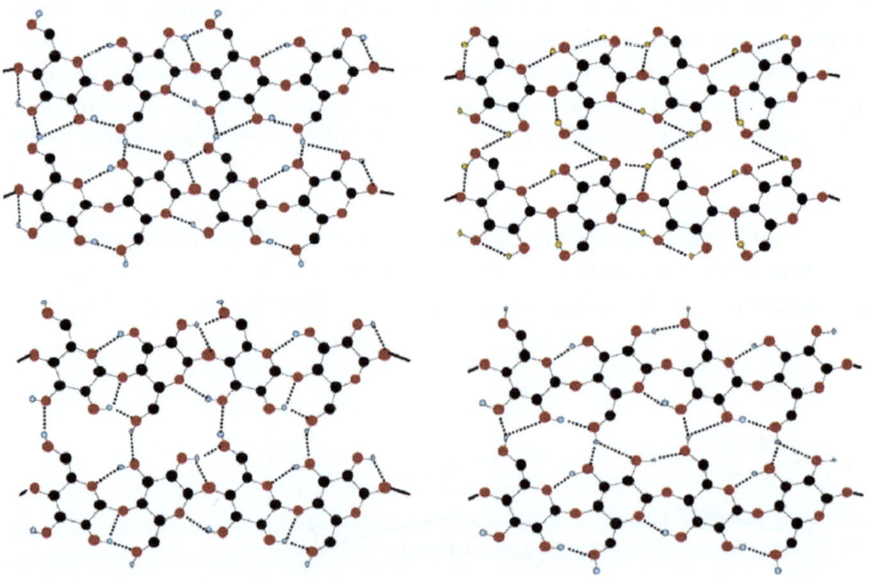

Figure 13.3 The two alternate hydrogen-bond networks in cellulose are shown in the top image; the dominant hydrogen-bond network in cellulose I is shown in the bottom image, with chains on the left at the origin of the unit cell and chains on the right at the center.

II, cellulose III will either take the shape of cellulose III_I or III_{II}. By heating cellulose III_I or III_{II} to 260 °C in glycerol, respectively, polymorphs IV_I and IV_{II} can be created. These two forms can return to their parent structures I or II, just like cellulose III. It has been demonstrated that cellulose IV is present in numerous plant primary cell walls in addition to the native cellulose I [9].

13.3.1 Sources of Cellulose

The most plentiful source of natural fibers is cellulose. Around 7.5×10^{10} tons of cellulose are extracted each year, according to estimates. The extraction of cellulose from numerous sources, including wood (wood, whether hard, or soft), seed, bast (flax and hemp), cane, leaf, straw (rice and wheat), fruit (coir), sisal, algae, fungi, bacteria, and minerals. Figure 13.4 shows the representation of various cellulose sources. The sources are listed in the order of conventional sources – which are regarded as primary sources – like wood and cotton. Based on its structural characteristics, wood can be divided into soft and hard woods [10]. Agriculture waste, which includes unprocessed wastes such as rice straw, banana rachis, and corncobs, is the next category. Waste from crops is now the second-largest source of cellulose. Another recently discovered source of cellulose is industrial waste, which is the processed waste generated by companies such as sugar cane bagasse, tomato, and garlic peels. Other specialized sources of cellulose include bacteria, algae, fungi, marine creatures (such as tunicates), and algae.

13.3.2 Cell Wall Composition

Plants have intricate composite structures that make up their cell walls. Interactions between cellulose and other elements of the cell wall (hemicellulose, lignin,

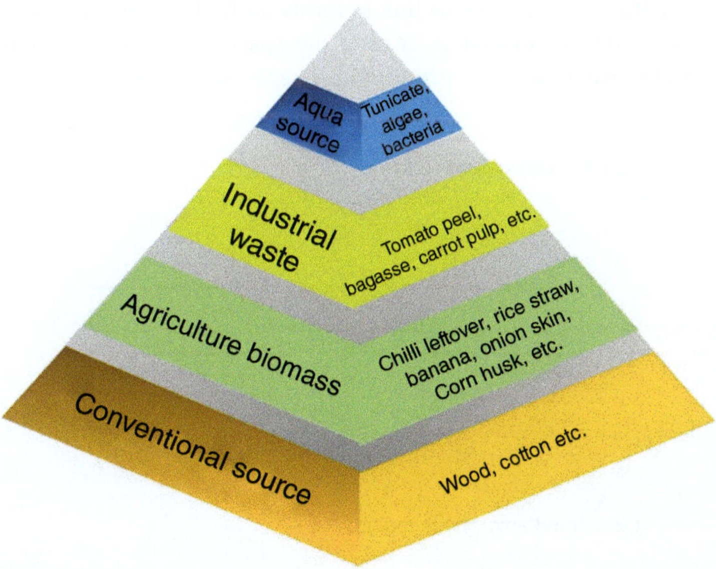

Figure 13.4 Representation of various sources of cellulose.

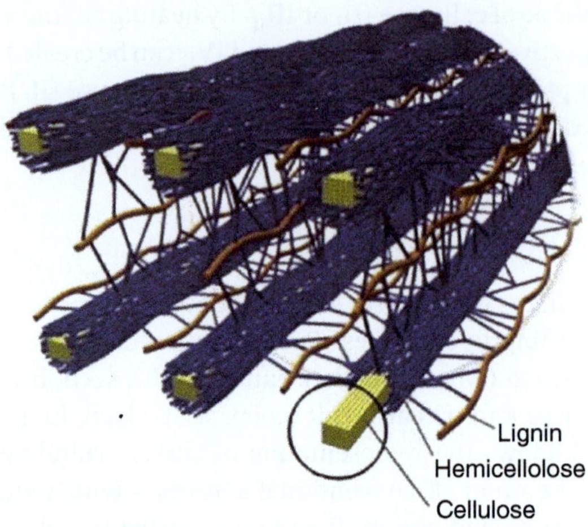

Figure 13.5 Cellulose surrounded by lignin and hemicellulose.

pectin, and water) happen due to hydrogen bonds, Van der Waals forces, electrostatic interactions, and hydrophobic interactions. Hydrogen bonds are broken and become disorganized in the amorphous area. Figure 13.5. depicts the crystalline and amorphous areas of cellulose. There are significant differences in hydrogen bonding and the orientation of cellulose in various sources.

The plant cell wall is made up of lignin, hemicellulose, and cellulose (Figure 13.6). Hemicellulose and pectin, which make up the matrix polysaccharides in plant cells, are produced in the Golgi apparatus and then transferred to the cell wall. Hemicellulose, a branching polymer of several sugar units, has a strong affinity for the cellulose found in cell walls. A complex polysaccharide called pectin promotes cell growth by creating gels that are swollen and allow microfibrils to go past one another or lock them after growth. Hemicellulose forms multiple hydrogen bonds with the cellulose microfibrils around it, sealing the space between them and preventing the fibrils from aggregating. Furthermore, glucuronic acid with carboxyl

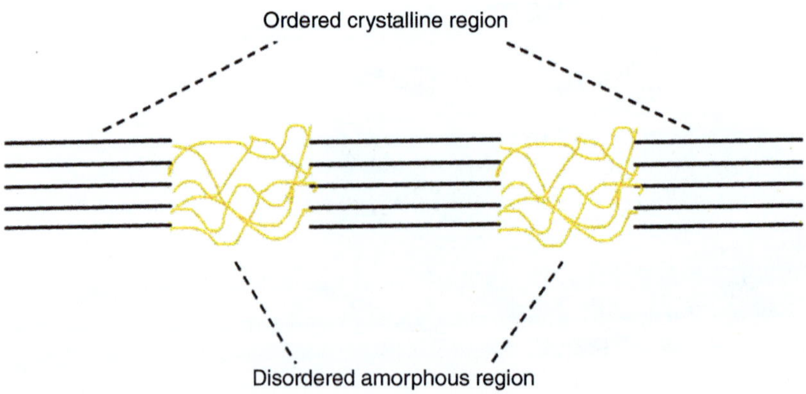

Figure 13.6 The amorphous and crystalline regions of cellulose.

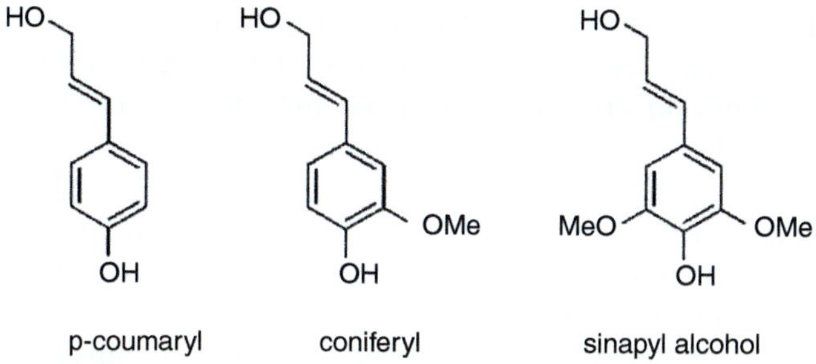

Figure 13.7 Units of lignin (14/3/23).

groups found in hemicelluloses allows for the delamination of fibers through the use of electrostatic repulsion forces [11, 12].

Lignin is a hydrophobic aromatic polymer that predominates in plant secondary cell walls and woody materials, and it helps to give these materials their mechanical and waterproofing qualities. Lignin engages in both electrostatic and hydrophobic interactions with cellulose and other components of the cell wall. Phenyl propane is a cross-linked heterogeneous and amorphous polymer that is used to make lignin. It is an optically inert, amorphous substance made up of three distinct monomers: coumaryl alcohol, coniferyl alcohol, and sinapyl alcohol, as shown in Figure 13.7 [13, 14].

13.3.3 Classification of Cellulose

NC refers to cellulose particles with at least one dimension between 1 and 100 nm. NC can be divided into two main categories based on the production conditions, which affect the dimensions, composition, and properties: (i) cellulose nanocrystals (CNC), also known as cellulose whiskers, and (ii) cellulose nanofibrils (CNF), also known as nanofibrillated cellulose (NFC), microfibrillated cellulose (MFC). NC also includes bacterial cellulose (BC) and electrospun cellulose nanofibers (ECNF). But whereas BC and ECNF are produced by a bottom-up process in which nanofibers are built up from low molecular weight sugars by bacteria or from dissolved cellulose using electrospinning, respectively, CNC and CNF are much more common because they are created by the disintegration of cellulose fibers into nanoscale particles [15].

13.3.3.1 Cellulose Nanocrystals

Cellulose nanocrystals, commonly known as nanowhiskers, are highly crystalline spherical structures with dimensions ranging from 2 to 25 nm and lengths of 100–750 nm depending on the source. Both crystalline and amorphous regions can be found in cellulose nanocrystals. The amorphous area is eliminated by acid hydrolysis, leaving the crystalline particles. Due to kinetic forces and the steric barrier, the amorphous parts of the cellulose chain are the first and most accessible to the acid for hydrolytic activity; while, the crystalline sections are less affected by the hydrolysis. Depending on the source, the crystallinity, size, and morphology will change. Sulfuric and hydrochloric acids can be used to perform acid hydrolysis [16, 17]. Sulfuric acid is frequently used as a hydrolyzing agent because, if nanocrystals are made using hydrochloric acid, due to their acidity, they have a constrained ability to disperse in solvents, and the suspension is unstable and prone to flocculation. Indeed, during hydrolysis, anionic sulphate ester groups ($-OSO_3^{2-}$) can be grafted onto the surface hydroxyl groups by sulfuric acid through an esterification reaction. The presence of these negatively charged sulfate ester groups induces the formation of a negative electrostatic layer covering the nanocrystals and promotes their dispersion in water. The high stability of sulfuric acid hydrolyzed CNC results in electrostatic repulsion between individual nanoparticles. However, it compromises the thermal stability of the nanoparticles. The basic chemical structure of cellulose contains three hydroxyl groups, which have the ability to initiate extensive intra-and intermolecular hydrogen bonding among the cellulose chains. This enables the crystalline packing of cellulose chains into highly compact systems. The crystalline structures of the resulting fiber are thought to have interchain hydrogen bonding that is inaccessible to water and the majority of conventional organic solvents. The chains are wider apart and loosely packed in the less ordered areas of nanofibrils, making them more accessible for interactions with other molecules like water. The composite's lower water uptake at equilibrium is caused by the densely packed crystalline zone of CNCs [16].

13.3.3.2 Bacterial Cellulose

BC, a unique type of natural cellulose, has recently attracted a lot of attention. Brown first described BC produced by certain bacteria, such as Acetobacter xylinum. According to various studies, bacteria produce cellulose to protect themselves from ultraviolet radiation, harsh chemical environments, and access to oxygen. Conversely, plant cell walls made of cellulose support the body against osmotic pressure. Being biocompatible and biodegradable, cellulose generated from either plants or microorganisms has the potential to be utilized in biomedical disciplines. Although BC has the same chain molecules connected by cellobiose as ordinary cellulose, it is free of contaminant molecules like lignin, hemicelluloses, and pectin that are often present in cellulose generated from plants [18].

BC has been used in a variety of fields, including paper repair, blood vessel regeneration, and wound dressing. Despite BC's special qualities, there are several restrictions on its use, such as its lack of antimicrobial capabilities, opportunistic transparency, and stress tolerance. BC composite, which consists of a matrix and reinforcement components, has been introduced to get around these restrictions. Fibers in BC have a porous natural configuration. It serves as a matrix for storing many particles made of various reinforcing materials. The anchoring and reinforcing materials give BC additional qualities that contribute to its biological nature and physiochemical characteristics [19].

13.3.3.3 Cellulose Nanofibers: Perfect Substrate for Fabrication of Cellulose Aerogels

Thin cellulose fibers, often known as cellulose nanofibers (CNFs), are extremely strong and stiff mechanically. Their micron-scale lengths contain both crystalline and amorphous parts, and their diameter is about 3 nm. They are extremely strong (nearly as strong as steel and Kevlar), have high crystallinity, hydrophilicity, a low coefficient of thermal expansion (1 107 K1), high stiffness (Young's modulus of 138 GPa), and an easily modifiable surface. Plant cellulose microfibers can be used to isolate CNF. It is frequently made by mechanically fibrillating cellulose, and the preparation technique and source of the cellulose determine the yield. To obtain CNF, cellulose biomass may be homogenized and microfluidized or ground to an ultrafine powder. Pretreating CNF with chemicals, enzymes, or even mechanical processes can sometimes be done to improve the materials' quality, use less energy, or accomplish other goals [20].

Using a Gaulin laboratory homogenizer, Turbak et al. and Herrick et al. discovered CNF for the first time in 1983. During fibrillation or delamination, the specific surface area and quantity of hydrogen bonds formed by the surface hydroxyl groups increase. CNFs are therefore more likely to form gels with higher viscosity. The main drawback of CNFs is that, once created, they have a propensity to gel, and their hydrophilicity prevents the dispersion of many hydrophobic polymers. The extraction procedure uses a lot of energy, which is significant. With the help of various surface modifications and pretreatments, all these challenges can be addressed. CNFs are commercially produced and employed in a variety of applications, including as composites, coatings, personal care goods, and construction, despite the disadvantages [21].

Because of their biodegradability, sustainability, and nanoscale structure, cellulose and other cellulosic materials have recently attracted a lot of attention for the production of aerogel. Duong Hai Minh received the 2016 Tech Connect Innovation Award for his groundbreaking work on a reusable cellulosic aerogel. He and his coworkers developed the first green cellulosic aerogels made from used paper in the world. Together with smart materials for biological applications, they are probably going to be exploited as coating materials for drug delivery. As a result, cellulosic aerogels have become extremely important in the field of research, not only because of their distinctive quintessential physical and chemical properties, true sustainability, and renewability, but also because of their regeneration technique and physical dissolution from non-derivatizing cellulose solvents, which are simple, inexpensive, hold great capabilities, and are also very easy to use. These specific features of cellulose aerogels find wide applications, including energy storage devices, sensors, pollutant treatment, thermal insulation materials, and so on [22].

Cellulosic aerogels are created after a series of meticulously carried-out stages under particular circumstances. The precursor material is first dispersed in a proper liquid to form a colloid and then left in order to enable gelation. The colloid solution forms the gel through a strong interaction with the added cross-linking agents. Numerous factors influence the gelation time of the colloidal solution, the precursor material, concentration of the colloid solution, the additives (if any), and the parameters of the solutions such as pH and temperature. After the gelation time, many precursor materials require some additional steps to strengthen the gel network. The drying of the gel network takes place after the completion of gelation time. The surface tension of the pores of the solid network is minimized

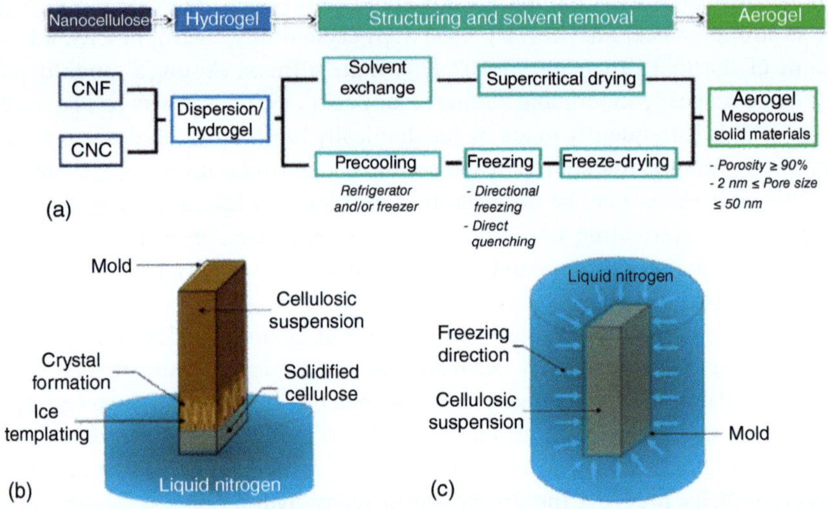

Figure 13.8 (a) Graphical representation showing the fabrication of aerogel using freeze drying and supercritical drying techniques. (b) Unidirectional freeze drying (c) rapid cooling in nitrogen bath.

with freeze-drying to form cryogel, thermal drying to form xerogel, or supercritical drying process to form aerogel [21, 23]. Figure 13.8 shows the graphical abstract for extraction of cellulose nanofiber aerogels by freeze drying and supercritical drying.

The supercritical drying technique, which is frequently used to turn gel into aerogel, converts liquid in a material into gas. Herein, the lack of surface tension and capillary stress in this technique allows for the maintenance of the nanostructured porous network. In comparison to the freeze-drying method, this method may generate a highly porous and more uniform pore structure while removing the pore liquid from the wet gel with the least amount of harm. Because the supercritical drying method requires more time, money, and solvent waste, it cannot be used to produce aerogels on a large scale for commercial use. In the freeze-drying method, the temperature is kept below the freezing point of the liquid samples to cause them to freeze. The frozen solvents are then removed using a sublimation process, and a porous interconnecting structure is produced as a result. Although not primarily in the mesoporous region, freeze-drying results in an open pore structure. Through the management of phase growth, the addition of dispersed phase liquids, the templating of structures, and the selection of the aerogel production technique, the morphology and, consequently, the structure–property relationships of aerogel may be regulated. Using unidirectional templating and slow cooling, a porous structure parallel to the freezing direction is provided. High precursor concentrations and high freezing rate cause microscopic crystals to form (for example, when immersed directly in liquid nitrogen), which creates tiny holes with large surface areas. This method has the advantage of being straightforward, affordable, and using an organic, harmless solvent [24]. Evaporative drying with water suffers pore collapse due to capillary forces; indeed, strong surface tension influences can result in

Figure 13.9 Extraction and fabrication of cellulose aerogels. Source: Jeff Dalton/Adobe Stock.

the partial collapse of the structure when the solvent inside the pores is ready to evaporate. In these cases, a solvent exchange is recommended before the evaporation stage to decrease the effect of capillary pressures on the collapse of pores.

The chemistry of aerogels is quite flexible, and it has been noted that a variety of variables, including the precursor materials, their ratios, preparation methods, and others, can change it. Pore size and surface area of the aerogels can also be tailored. Figure 13.9 shows the extraction and fabrication of cellulose aerogels form plant sources. Moreover, various functional groups have been added to the aerogels to give them the appropriate mechanical and physico-chemical properties. Depending on the intended use, hybrid aerogels can, in particular, have their characteristics changed and new capabilities introduced. On the other hand, new applications that were previously impractical would be made possible in the future if the characteristics of aerogels were improved [25].

13.4 Cellulose-Based Aerogels

A highly porous, nanostructured material with several benefits, cellulose-based aerogel exhibits the typical characteristics of aerogels. In fact, numerous research studies on them are carried out each year. Its distinctive qualities, such as its high mechanical strength, high degree of polymerization, high purity, and high crystallinity, make it a potential material for creating aerogels with flexible and pressure-resistant capabilities. Moreover, employing cellulose as the precursor material for the creation of aerogels has numerous advantageous characteristics, including the cellulose chain's functional groups, which are thought to act as a cross-linking agent for many other materials. Another benefit of the material is its potential for regeneration or reuse. In addition, chemically altering cellulose increases its mechanical strength, and producing cellulose aerogels with desired structural properties is not too difficult [25, 26].

Raw natural sorbents have a similar or occasionally lesser density to conventional sorbents due to their moderate oil sorption capability (i.e. 3–50 times their own weight), high biodegradability, and sorbents. Other benefits include their potential for recycling, better oil recovery, and comparatively simple dumping procedure than other varieties of traditional sorbents. Notwithstanding the foregoing benefits, most conventional natural sorbents also have a number of disadvantages, such as poor buoyancy and oil sorption selectivity, which are related to the higher water absorbance of the aerogels. Poor water repellency lessens the capacity of their microscopic pores to absorb oil. The characteristics of aerogel-type materials are combined with additional benefits of naturally occurring cellulose, such as large number of sources, renewability, biodegradability, and sites of surface modification [5].

13.4.1 Cellulose-Based Aerogels for Oil Absorption

Cellulose aerogels or sponges are next generation of sorbent materials, succeeding silica and synthetic polymer-based ones. As a result, after the necessary changes, cellulose aerogels appear to be excellent natural oil sorbents [5, 27]. The oil absorption properties of cellulose aerogels are influenced by a number of factors, including the capillary effect, Van der Waals forces, the hydrophobic interaction between the oils and absorbents, and morphological aerogel parameters (such as surface wettability, total pore volume, and pore structure). Aerogel-type sorbents with a similar pore size distribution and three-dimensional (3D) structure would benefit from the oil density's contribution to the saturated absorption capacity. It is still unknown how amphiphilic CNFs and constructed CNF aerogels are. Contrary to hydrophobic carbon-based aerogels and hydrophilic inorganic oxide aerogels, the amphiphilicity of CNF aerogels would allow the uptake of both polar and nonpolar liquids. Furthermore, the many surface hydroxyls enable tuning of the amphiphilicity toward increased hydrophobicity [28].

The aerogels' improved oil/water selectivity would also benefit from higher hydrophobicity and oleophilicity. Cellulose aerogels exhibit amphiphilicity (i.e. poor oil/water selective adsorption) due to the abundance of hydroxyl groups. As a result, they need to be modified to change the skeleton's natural hydrophilicity to hydrophobicity and oleophilicity. By adding roughness (i.e. micro- or nanoscale asperities) and low surface-energy materials, one can make surfaces hydrophobic. Typically, the water contact angle (CA) is used to measure the surface wettability. The water fondness of the surface of materials is divided into hydrophilic, hydrophobic, and superhydrophobic categories based on the CA values. Surfaces are referred to as hydrophilic if the CA is less than 90°; hydrophobic if the CA is between 90° and 150°. The surface is superhydrophobic if CA is greater than 150°. Additionally, a lower oil viscosity makes it easier for the oil to penetrate the porous network of aerogels, increasing their ability to absorb oil. In contrast, highly porous aerogels typically have more because they have more internal free volume available for oil sorption [28, 29]. In the case of hydrophobized cellulose aerogels, the hydrophobic interactions between modified cellulose and oils (or organic solvents) as well as the capillary effect of the pores are the main causes of the oil sorption. Thus, the majority of their adsorption process falls under physisorption.

13.4.2 Mechanism of Oil Absorption

Both absorption and adsorption are frequently used terms for sorption. The assimilation of a substance from one state into another is known as absorption (e.g. liquid being absorbed by a solid or gases being absorbed by water). Adsorption is the process by which ions and molecules physically bind to the surface of another molecule. An adsorbent is a porous substance with capillaries and pores that has less than 50% surplus liquid coating its surface [30]. Sorbate molecules gather on the surface of the sorbent during adsorption. In the case of oil adsorption, the oil molecules must first diffuse onto the sorbent surface, after which they must become trapped inside the sorbent due to capillary processes, before agglomerating into oil droplets in the sorbent's pores. Oil mostly comprises of nonpolar hydrocarbons, and Van der Waals forces and other physical noncovalent interactions are thought to be responsible for its adsorption. For fibrous sorbent, however, only adsorption and capillary action processes occur [31].

The wettability of any surface, which is directly related to the adsorption of oils, is determined by the surface's chemistry and shape. To determine wettability, one uses the water CA. A water droplet on a flat surface with a CA of less than 90° is said to be hydrophilic, one with a CA of between 90° and 150° to be hydrophobic, and one with a CA of more than 150° is said to be superhydrophobic, as shown in Figure 13.10. Due to the lack of competition between water and oil molecules, the hydrophobicity of the sorbent limits water sorption and hence increases oil sorption capacity. Superhydrophobicity can be introduced into a material that would otherwise not be superhydrophobic by raising the surface roughness and lowering the surface energy. Superhydrophobicity is a property of many natural materials, such as lotus leaves [32].

13.4.3 Hydrophobic Modification of Cellulose Aerogel

Chemical vapor deposition (CVD), atom layer deposition (ALD), cold plasma treatment, sol–gel, esterification, and fluorination are now the most widely utilized hydrophobization techniques for cellulose-based aerogels. TiO_2, SiO_2, alkoxysilanes, chlorosilanes, alkyl ketene dimer, (tridecafluoro-1,1,2,2-tetrahydrooctyl), stearoyl chloride, silanes, palmitoyl chloride, etc., are the examples of some hydrophobization agents used.

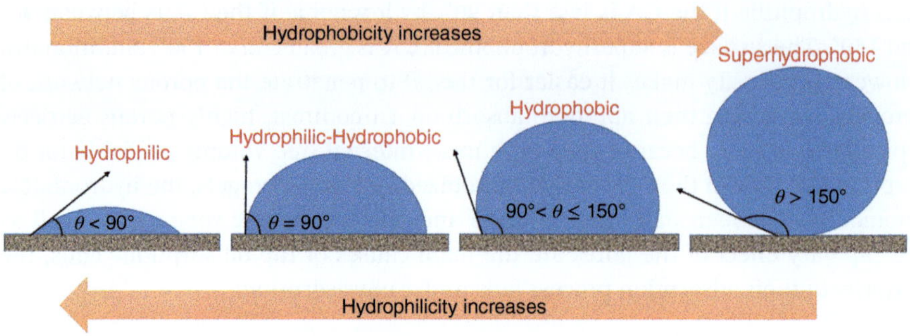

Figure 13.10 Wettability of water droplets on flat surface of a material.

13.4.3.1 Chemical-Vapor Deposition Method

CVD is a synthesis technique where the substrate is exposed to one or more volatile precursors in the presence of a microwave-induced plasma (a partly ionized gas containing free electrons), which react or decompose on the surface to generate a deposit. On the surface of the substrate, the plasma speeds up the reaction. The preparation of flexible, ultralight, and hydrophobic cellulose aerogels is made simple and efficient by using a process called thermal CVD in conjunction with chemical cross-linking of cellulose solution, lyophilization, and hydrophobic modification with methyltrichlorosilane (MTCS). The finding was that the cellulose aerogels treated with MTCS not only had a highly porous structure but also displayed long-lasting hydrophobicity. Particularly, the hydrophobic cellulose aerogels had outstanding adsorption characteristics as well as good recyclability, and they could quickly and effectively collect oils and organic solvents from water both on the surface and at the bottom [33]. Nguyen et.al found a formula for producing biodegradable cellulose aerogels from paper waste is both affordable and scalable. The green aerogel is macroporous and has a very low thermal conductivity and density of 0.029–0.032 $Wm^{-1} K^{-1}$ and 0.04 $g cm^{-3}$, respectively. It has a high absorption capacity and can hold 18 to 20 times its weight in fluids. Squeezing the aerogel can recover up to 99.8% of the liquid. It is possible to tailor the construction to absorb either polar (water) or non-polar liquids (oil). Methyltrimethoxysilane coating the aerogel increases its hydrophobicity without reducing its absorbency. The aerogel's mechanical flexibility and strength enable a variety of uses [34]. An inexpensive oil-absorbing material has been synthesized by Yang Mi et al. After straightforward silylation using the CVD technique, the aerogel acquired superhydrophobicity. The aerogel was modified to preferentially absorb oils from water with an unusually high absorption capacity of 80–197 times its own weight, outperforming all previously reported hydrophobic cellulose aerogels. Also, because of its great compressibility, the aerogel was able to mechanically squeeze out 85% of the oil, compared to only 55% for other compressible aerogels. These anisotropic cellulose/graphene aerogels have a low production cost and are readily available as raw materials (cellulose and graphite) for oil absorption studies [35]. For the first time, a simple and affordable approach for creating biocompatible cellulose aerogels using recycled paper waste and Kymene cross-linker has been devised by Feng et al. The recycled cellulose aerogels provide very stable super-hydrophobicity for more than five months and outstanding oil absorption capabilities of up to 95 g g^{-1} with the 0.25 wt% cellulose aerogel after being coated with methyltrimethoxysilane (MTMS) by CVD. Comprehensive measurements are also made of the effects of various cellulose fiber concentrations, varied compositions of cellulose and Kymene cross-linker, ambient temperatures, and environmental pH levels on the behavior of cellulose aerogels in terms of absorbing oil. According to the experimental findings, cellulose aerogels produce their greatest capacity for absorption at 50 °C [36]. Cheng et al. extracted aerogels from cotton, and to provide the created aerogels with a hydrophobic surface, silanization with MTMS is easily accomplished using CVD. The cotton aerogel exhibits the maximum oil absorption capacity over 100 g g^{-1} at an initial concentration of 0.25 wt%. Moreover, the cotton/cellulose aerogels have strong organic pollutant absorption capabilities. Using a pseudo-first-order model, the absorption kinetics of aerogels with various cotton concentrations was also studied. (Figure 13.11) Cotton/cellulose aerogels are shown to be promising materials for oil absorption and environmental pollution treatment by both equilibrium absorption and absorption kinetics [37].

13.4.3.2 Cold Plasma Treatment Method

Over the past 30 years, cold plasma technologies have been frequently used in the production of semiconductors, magnetic media, specialty glasses, metal coatings,

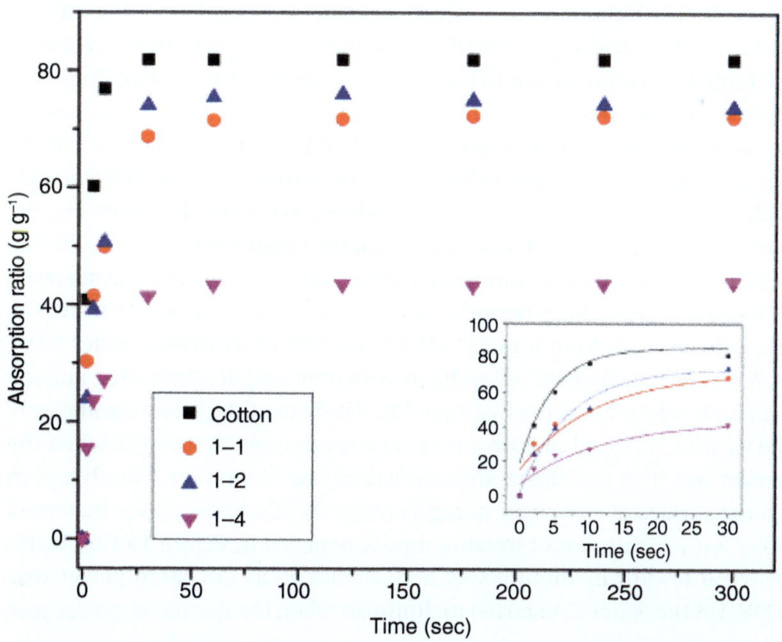

Figure 13.11 Absorption kinetic studies of aerogels with different cellulose concentrations. Source: Adapted from Cheng et al. [37].

and other materials. The effectiveness of these methods depends on their capacity to alter a material's surface properties by physically or chemically altering its outermost layers (1 m), without altering its bulk properties. Electrical discharge in low-pressure gases results in low-temperature plasmas. They are made up of a variety of highly reactive species, including electrons, photons, ions, radicals, and excited molecules. Device parameters, including chamber shape, gas flow rate, frequency, and power applied, in addition to the gas used, determine their chemical composition and physical characteristics. With this technique, polymer surfaces can have their chemistry and morphology altered to a depth of several tens of microns while retaining nearly all of their bulk attributes. As a result, radically different surface roughness, hydrophilicity, or chemistry can be obtained [38]. In order to improve reusability, a novel type of cellulose-based aerogel was created by Li et al. Based on the Schiff base reaction, the amino groups of chitosan, a type of inexpensive, renewable, environmentally friendly, and biodegradable polysaccharide similar to cellulose, were reacted with the aldehyde groups of oxidized cellulose. By using oxidation, cross-linking, freeze drying, and cold plasma modification, the desired cellulose aerogel was created. High oil/water selectivity, high absorption capacity, and excellent reusability were all features of the novel aerogel [39]. Li et al. also focused on the fabrication of an environmentally benign, economically viable adsorbent material that, in contrast to conventional adsorbents, can absorb up to 15 times their dry weight in oil while still having a high adsorbent capacity. Trimethylchlorosilane (TMCS) was used as the plasma in a cold plasma approach to modify the surface of cellulose aerogels [40]. By using the immersion approach, the NaOH/thiourea/H_2O solvent system, and regulating the rate of tetraethyl orthosilicate's hydrolysis-fasculation, cellulose-SiO_2 composite hydrogel was created tetraethyl-orthosilicate (TEOS). Using freeze-drying technology and the cold plasma modification, hydrophobic cellulose aerogels were fabricated by Shi and his coworkers. With a maximum water CA of 132°, the hydrophobic modification with CCl_4 as plasma successfully changed the surface of composite aerogels from hydrophilic to hydrophobic. After being modified to become hydrophobic, the composite aerogels still performed well in terms of heat insulation.

The change in water CA on the composite aerogel surface following plasma modification for a varied modification period is depicted in Figure 13.12a. The water CA of composite aerogels rose from 0 to 83° after 5 minutes of treatment under 50 W. The water CA was greater than 90° after the treatment period of 10 minutes, and the composite aerogels' surface became hydrophobic. The water CA did not significantly rise after the treatment period of 20 minutes. The maximum CA was 132 when the discharge power was 50 W and the modification time was 25 minutes. The change in water CA on the surface of composite aerogels when the discharge power increased from 0 to 70 W with 25 minutes of treating time is depicted in Figure 13.12b. Water CA and the trend toward hydrophobicity both increased as discharge power rose from 0 to 50 W. Yet the water CA started to diminish when the discharge power rose to 70 W [41].

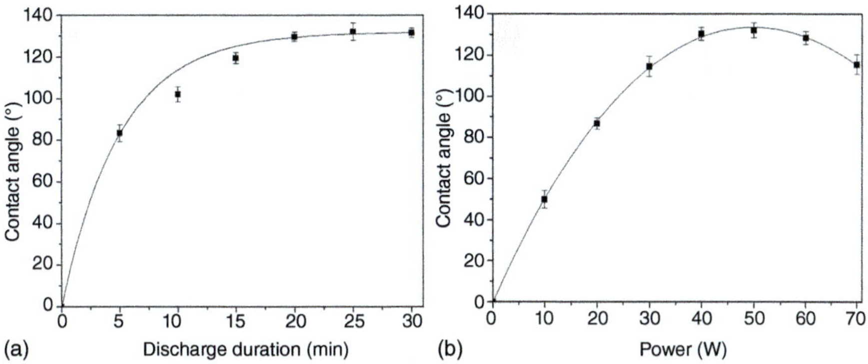

Figure 13.12 Effect of discharge duration and discharge power on water contact angle. Source: Shi et al. [41]/with permission of Elsevier.

13.4.3.3 Sol–Gel Process Method

A well-known technique for a variety of materials, including inorganic membranes, monolithic glasses and ceramics, thin films, and ultra-fine powders, is the sol–gel process. Even the synthesis of 1D nanomaterials uses it today. This method's benefit, which has helped it become so well-liked, is that it can be carried out at room temperature with the potential for simple chemical doping. The hydrolysis of a precursor-containing solution yields suspended colloidal particles, which is the basis of this procedure. A gel-like substance is created during the condensation process. Recycled jute bags containing 60–65% cellulose were utilized to create cellulose sponge using the sol–gel process, followed by freeze drying, to create an affordable alternative for dealing with oil spills. TEOS/Hexadecyltrimethoxy silane (TEOS/HDTMS) was applied as a dip coating on the sponge surface to make it extremely hydrophobic. Figure 13.13a,b, respectively, shows the oil sorption characteristics of super hydrophobic cellulose sponge at various contact times with diesel and motor oil both in static and dynamic settings. From the data, it is clear that aerogel has shown good adsorption capacity for both motor oil and diesel. For a dynamic system, it was 35.55 times greater for diesel oil compared to motor oil, which was 31.37 times greater. It was marginally lower for static systems, at 28.94 and 27.41 times for motor oil and diesel oil, respectively. Diesel oil was shown to have a higher level of sorption than motor oil. This may be due to the lower viscosity of diesel oil. In comparison to high-viscosity oil, the molecules of low-viscosity oil travel through the pore channels more quickly, leading to rapid swelling. Physisorption plays a crucial role in the sponge's ability to remove oil from an oil–water mixture. Oil and organic solvents are drawn into the sponge's interior by the driving force when it is in contact with them. The hydrophobic interactions between the oil and the sponge surface and the capillary effect of the 3D network of linked pores are the main causes of this driving force [42].

In order to create thermally and mechanically stable nanofiber aerogels, a simple method for creating hybrid sol–gel electrospun silica-cellulose diacetate

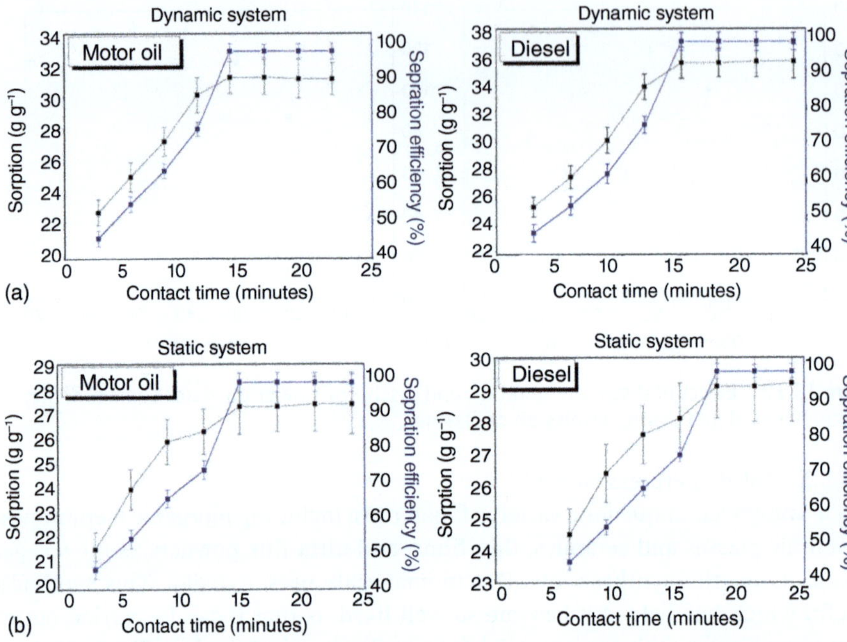

Figure 13.13 (a) Oil absorption characteristics of dynamic systems at different contact times (b) Oil absorption characteristics of static system at different contact times.

(CDA)-based nanofibers is provided by Pirzada et al. They mixed silica precursor mixture with the CDA solution instead of electrospinning silica nanoparticles with CDA or mixing silica fibers with CDA afterward, and allowed the system to go through sol–gel processing during electrospinning. Enhanced hydrophobicity and high oleophilicity are produced when CDA nanofibers and silica network are cross-linked, creating a thermally and physically resistant hybrid nanofiber aerogel (NFA). Applications that need for selective oil and water separation can explore the use of these materials' superhydrophobic and oleophilic properties [43].

13.4.3.4 Modification by Surface Roughness and Esterification

The hydrophobic behavior of solids is greatly influenced by their surface shape and chemical composition. For practical applications, it's crucial to adjust the surface to the necessary level of roughness in order to produce hydrophobic surfaces that are consistent and repeatable. It has been demonstrated that the reduction in the contact area between the liquid drop and the hydrophobic surface caused by an increase in surface roughness in the form of predictable peaks and troughs lowers the apparent surface free energy [44]. By lowering its surface energy, cellulose esterification is another method for producing hydrophobic cellulose surfaces. A high degree of substitution can be produced by esterifying cellulose with fatty acid derivatives that contain more reactive moieties, like acyl chloride. Esterification with stearic acid chloride (SAC) was used as a superabsorbent to create a highly porous superhydrophobic (CA higher than 159°) NFC/polyvinyl alcohol

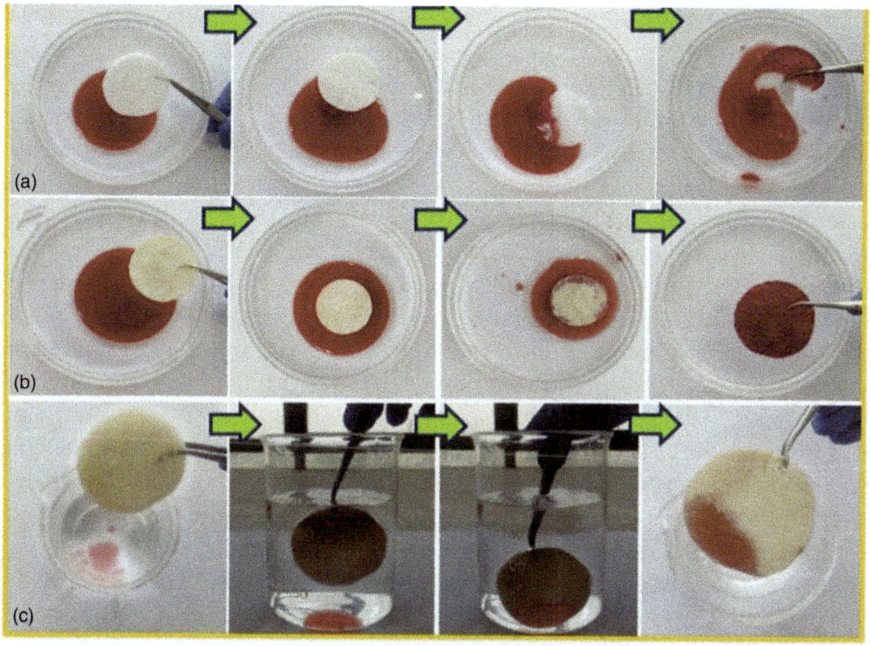

Figure 13.14 (a) Oil spillage cleanup; (a and b) from surface water, (diesel engine oil/water mixture) using (a) NFC/PVA aerogel, (b) NFC/PVA-SAC aerogel (c) from underwater system (chloroform/water mixture) using NFC/PVA-SAC aerogel. Source: Chhajed et al. [45]/with permission from Elsevier.

(NFC/PVA) aerogel for cleaning up oil and chemical spills in water. Due to their amphiphilic nature, unmodified aerogel NFC quickly absorbs both water and diesel engine oil. Due to the material's low mechanical strength and water absorption after absorption, unaltered aerogel readily disintegrates and settles. It was not able to selectively absorb oil using NFC/PVA aerogel that had not been changed. The modified aerogel, in contrast (Figure 13.14), preferentially absorbs only diesel engine oil and repels water. Chloroform was utilized to show the effectiveness of the separation in the undersea system. Chloroform stains with diesel engine oil for better visualization since its density is greater than that of water, causing it to sink to the bottom. The modified aerogel could also readily absorb chloroform from the water, as demonstrated in Figure 13.14c, where chloroform is likewise completely separated [45]. For the creation of the cellulose acetate-based silicon dioxide/titanium dioxide (CA-based SiO_2/TiO_2) hybrid microsphere composite aerogel film, a simple phase inversion method was used. The aerogel demonstrated efficient separation for surfactant-stabilized water-in-oil emulsions with nanoscale droplets. Instead of lyophilization, a water vapor-induced phase inversion was used to enhance the interconnectivity of the pores. By means of a simple hydrolysis/condensation procedure, the produced SiO_2 and TiO_2 microspheres provide micro-nanoscale hierarchical structures on the film surface. As a consequence, the films can separate surfactant-stabilized water-in-oil emulsions (with a droplet size of about 345 ± 39 nm) with high flux and high separation efficiency only at ordinary pressure. Low humidity could cause phase separation, which could result in large pore sizes. Moreover, the creation of SiO_2 and TiO_2 nanoparticles during the sol–gel reaction contributed to the film's rough surface. The average pore width of the aerogel films ranged from 451 to 878 nm, and they possessed large specific surface areas [46].

Esterified cellulose aerogels had been synthesized from oat hulls by esterification with citric acid using two different methods, such as ultrasonication and a reactive extrusion-assisted process. Both groups of samples that had undergone reactive extrusion and ultrasonication operations were found to have undergone modifications. In comparison to extruded sample (cellulose controlled sample without any reagent [ECW]) (control extruded sample, 7.75 $g\,g^{-1}$) and unmodified cellulose (9.38 $g\,g^{-1}$), all changed samples had water absorption capacity (WAC) values that varied from 1.63 to 1.71 ($g\,g^{-1}$). The effect of the torque action of the extruder can influence specific fibers' characteristics, which could explain why the ECW sample had a lower WAC value than unmodified cellulose. When compared to cellulose and ECW samples, oil absorption capacity rose in CA treated samples by a factor of two or three [47]. Using parenchyma cellulose from corn stalk pith (CSP) and naturally occurring tartaric acid (TA) as the grafting/cross-linking agent by esterification in *N,N*-dimethylacetamide/lithium chloride (DMAc/LiCl) system, an environmentally friendly absorbent material was created. When the dry absorbent comes into contact with water, it expands as a result of H_2O molecules diffusing into the super water absorbents (SWA) matrix. Typically, a key component of the diffusion process is the movement of water in the dynamical space that has formed between the macromolecular chains. The separation distance between polymeric chains also increases as a result of the absorbent swelling, which is linked to a larger-scale segmental motion. The inherent features of the polymer, solution types, and ambient temperature all affect how water absorbent diffuses and swells [48]. Tripadhi et al. proposed an alternative technique in which the gel is made in acetone, but the solvent is switched to water before freeze drying. By using this method, high pore volumes (>99%) and lightest cellulose acetate aerogels (4.3 $mg\,ml^{-1}$) are created. The 4 wv^{-1}% cellulose acetate (4%) aerogel also exhibits good mechanical qualities, with a maximum compression strain of 92% without failure and a compressive stress of 350 kPa, which is higher than that previously reported for cellulosic aerogels. It also has a density of 24.3 $mg\,ml^{-1}$. In an oil-in-water media, the modified aerogels were tested once more under shear conditions. For easier notice, the used kerosene was colored red. The images (Figure 13.15) demonstrate that a clear solution may be made in under a minute since the modified aerogel has virtually completely absorbed the colored oil. Even in stirred fluid, the modified aerogel maintains both oil and its mechanical integrity, demonstrating its applicability for cleaning up oil spills. A stirred oil-in-water system was created to further replicate high viscosity oil spills in a turbulent environment (typical of marine environments). The oil utilized for this system was used motor oil (spent car oil with a viscosity of about 170 cP) [49].

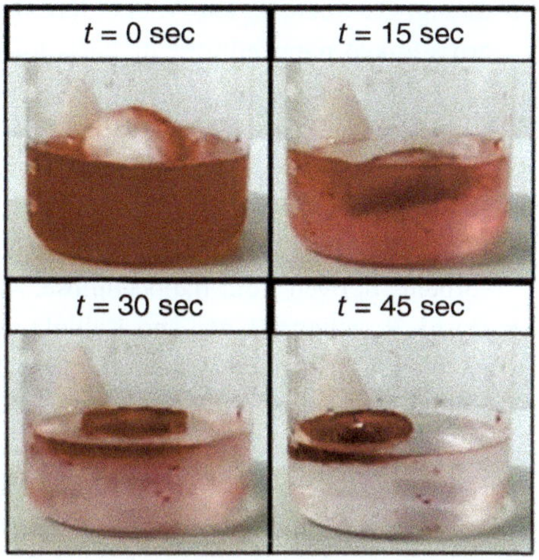

Figure 13.15 The reusability of the aerogel was tested for ten cycles of sorption and desorption [49].

13.5 Conclusion

The discharge effect of generated water has developed into a significant environmental issue as a result of the rising volume of wastewater produced by the petroleum sector. Due to the intricacy of the contaminants, a single treatment procedure is typically insufficient to eliminate all of them. Wastewater treatment for the petroleum sector can be done with cellulose-based aerogels since they are effective, affordable, and pollution-free. Cellulose-based materials, which are capable of efficiently removing pollutants, particularly in the cleaning of crude oil spills, the flocculation of solid suspended matter during drilling or oil recovery in the upstream oil industry, the adsorption of heavy metals or chemicals, and the separation of oil/water by cellulosic membrane in the downstream water treatment industry, among other applications. Furthermore, cellulose-based aerogels offer a wide range of good qualities, including abundant sources, low energy consumption, economic viability, environmental protection, and renewables, and they have a very bright future in the petroleum sector. Future research will mostly concentrate on improving already-existing technologies and treating and processing produced water physically, chemically, and/or biologically in order to meet reuse and discharge constraints.

References

1 Shi, G., Qian, Y., Tan, F. et al. (2019). Controllable synthesis of pomelo peel-based aerogel and its application in adsorption of oil/organic pollutants. *R. Soc. Open Sci.* 6: 181823.

2 Hammouda, S.B., Chen, Z., An, C., and Lee, K. (2021). Recent advances in developing cellulosic sorbent materials for oil spill cleanup: a state-of-the-art review. *J. Clean. Prod.* 311: 127630.

3 Shang, Q., Chen, J., Hu, Y. et al. (2021). Facile fabrication of superhydrophobic cross-linked nanocellulose aerogels for oil–water separation. *Polymers (Basel).* 13: 1–14.

4 Chau, M.Q., Truong, T.T., Hoang, A.T., and Le, T.H. (2021). Oil spill cleanup by raw cellulose-based absorbents: a green and sustainable approach. *Energy Sources Part A* 00: 1–14.

5 Liu, H., Geng, B., Chen, Y., and Wang, H. (2017). Review on the aerogel-type oil sorbents derived from nanocellulose. *ACS Sustain. Chem. Eng.* 5: 49–66.

6 Doshi, B., Sillanpää, M., and Kalliola, S. (2018). A review of bio-based materials for oil spill treatment. *Water Res.* 135: 262–277.

7 Seddiqi, H., Oliaei, E., Honarker, H. et al. (2021). Cellulose and its derivatives: towards biomedical applications. *Cellulose* 8 (Springer Netherlands: 1893–1931.

8 Habibi, Y., Lucia, L.A., and Rojas, O.J. (2010). Cellulose nanocrystals: chemistry, self-assembly, and applications. *Chem. Rev.* 110: 3479–3500.

9 Azizi Samir, M.A.S., Alloin, F., and Dufresne, A. (2005). Review of recent research into cellulosic whiskers, their properties and their application in nanocomposite field. *Biomacromolecules* 6: 612–626.

10 Faruk, O., Bledzki, A.K., Fink, H.P., and Sain, M. (2012). Biocomposites reinforced with natural fibers: 2000–2010. *Prog. Polym. Sci.* 37: 1552–1596.

11 Chaker, A., Alila, S., Mutjé, P. et al. (2013). Key role of the hemicellulose content and the cell morphology on the nanofibrillation effectiveness of cellulose pulps. *Cellulose* 20: 2863–2875.

12 Rajinipriya, M., Nagalakshmaiah, M., Robert, M., and Elkoun, S. (2018). Importance of agricultural and industrial waste in the field of nanocellulose and recent industrial developments of wood based nanocellulose: a review. *ACS Sustain. Chem. Eng.* 6: 2807–2828.

13 Zhang, J., Chen, Y., and Brook, M.A. (2014). Reductive degradation of lignin and model compounds by hydrosilanes. *ACS Sustain. Chem. Eng.* 2: 1983–1991.

14 Doherty, W.O.S., Mousavioun, P., and Fellows, C.M. (2011). Value-adding to cellulosic ethanol: lignin polymers. *Ind. Crop. Prod.* 33: 259–276.

15 Nechyporchuk, O., Belgacem, M.N., and Bras, J. (2016). Production of cellulose nanofibrils: a review of recent advances. *Ind. Crop. Prod.* 93: 2–25.

16 Ng, H.M., Sin, L., Tee, T. et al. (2015). Extraction of cellulose nanocrystals from plant sources for application as reinforcing agent in polymers. *Composites, Part B* 75: 176–200.

17 Mariano, M., El Kissi, N., and Dufresne, A. (2014). Cellulose nanocrystals and related nanocomposites: review of some properties and challenges. *J. Polym. Sci., Part B* 52: 791–806.

18 Huang, Y., Zhu, C., Yang, J. et al. (2014). Recent advances in bacterial cellulose. *Cellulose* 21: 1–30.

19 Esa, F., Tasirin, S.M., and Rahman, N.A. (2014). Overview of bacterial cellulose production and application. *Agric. Agric. Sci. Procedia* 2: 113–119.

20 Guan, Q.F., Yang, H., Han, Z. et al. (2021). Sustainable cellulose-nanofiber-based hydrogels. *ACS Nano* 15: 7889–7898.

21 Abdul Khalil, H.P.S., Adnan, A.S., Yahya, E.B. et al. (2020). A review on plant cellulose nanofibre-based aerogels for biomedical applications. *Polymers (Basel)* 12: 1759.

22 Zaman, A., Huang, F., Jiang, M. et al. (2020). Preparation, properties, and applications of natural cellulosic aerogels: a review. *Energy Built Environ.* 1: 60–76.

23 Gupta, P., Singh, B., Agrawal, A.K., and Maji, P.K. (2018). Low density and high strength nanofibrillated cellulose aerogel for thermal insulation application. *Mater. Des.* 158: 224–236.

24 Ahankari, S., Paliwal, P., Subhedar, A., and Kargarzadeh, H. (2021). Recent developments in nanocellulose-based aerogels in thermal applications: a review. *ACS Nano* 15: 3849–3874.

25 Yahya, E.B., Alzalouk, M.M., Alfallous, K.A., and Abogmaza, A.F. Antibacterial cellulose-based aerogels for wound healing application : a review. *Biomed. Res. Ther.* 7: 4032–4040.

26 Yang, J., Xia, Y., Xu, P., and Chen, B. (2018). Super-elastic and highly hydrophobic/superoleophilic sodium alginate/cellulose aerogel for oil/water separation. *Cellulose* 25: 3533–3544.

27 Pinnow, M., Fink, H.P., Fanter, C., and Kunze, J. (2008). Characterization of highly porous materials from cellulose carbamate. *Macromol. Symp.* 262: 129–139.

28 Jiang, F. and Hsieh, Y.L. (2014). Amphiphilic superabsorbent cellulose nanofibril aerogels. *J. Mater. Chem. A* 2: 6337–6342.

29 Wang, Y., Yadav, S., Heinlein, T. et al. (2014). Ultra-light nanocomposite aerogels of bacterial cellulose and reduced graphene oxide for specific absorption and separation of organic liquids. *RSC Adv.* 4: 21553–21558.

30 Praba Karana, C., Rengasamy, R.S., and Das, D. (2011). Oil spill cleanup by structured fibre assembly. *Indian J. Fibre Text. Res.* 36: 190–200.

31 Wahi, R., Chuah, L.A., Choong, T.S.Y. et al. (2013). Oil removal from aqueous state by natural fibrous sorbent: an overview Department of Chemistry, Universiti Malaysia Sarawak , 94300 Kota Samarahan, Sarawak,. *Sep. Purif. Technol.* 113: 51–63. https://doi.org/10.1016/j.seppur.2013.04.015.

32 Gupta, S. and Tai, N. (2016). Carbon materials as oil sorbents: a review on the synthesis and performance. *J. Mater. Chem. A* 1550–1565. https://doi.org/10.1039/c5ta08321d.

33 Liao, Q., Su, X., Zhu, W. et al. (2016). RSC advances flexible and durable cellulose aerogels for highly e ff ective oil/water separation. *RSC Adv.* 6: 63773–63781.

34 Nguyen, S.T., Feng, J., Ng, S. et al. (2014). Colloids and surfaces a: physicochemical and engineering aspects advanced thermal insulation and absorption properties of recycled cellulose aerogels. *Colloids Surf. ,A* 445: 128–134.

35 Mi, H.-Y., Jing, X., Politowicz, A.L. et al. (2018). https://doi.org/10.1016/j.carbon.2018.02.033). Highly compressible ultra-light anisotropic cellulose/graphene aerogel fabricated by bidirectional freeze drying for selective oil absorption. *Carbon* 132: 199–209.

36 Feng, J., Nguyen, S.T., Fan, Z., and Duong, H.M. (2015). Advanced fabrication and oil absorption properties of super-hydrophobic recycled cellulose aerogels. *Chem. Eng. J.* 270: 168–175.

37 Cheng, H., Gu, B., Pennefather, M.P. et al. (2017). https://doi.org/10.1016/j.matdes.2017.05.082). Cotton aerogels and cotton-cellulose aerogels from environmental waste for oil spillage cleanup. *Mater. Des.* 130: 452–458.

38 Hasırcı, V. and Hasirci, N. (2005). Control of polymeric biomaterial surfaces 2. *Surf. Interfaces Biomater.* https://doi.org/10.1533/9781845690809.1.29, 29–59.

39 Li, A.Z., Shao, L., Hu, W. et al. (2018). https://doi.org/10.1016/j.carbpol.2018.03.027). Excellent reusable chitosan/cellulose aerogel as an oil and organic solvent absorbent. *Carbohydr. Polym.* 191: 183–190.

40 Lin, R., Li, A., Zheng, T. et al. (2015). Hydrophobic and flexible cellulose aerogel as an efficient, green and reusable oil sorbent. *RSC Adv.* 5: 82027–82033.

41 Shi, J., Lu, L., Guo, W. et al. (2013). Heat insulation performance, mechanics and hydrophobic modification of cellulose-SiO_2 composite aerogels. *Carbohydr. Polym.* 98: 282–289.

42 Ahuja, D., Dhiman, S., Rattan, G. et al. (2021). Superhydrophobic modification of cellulose sponge fabricated from discarded jute bags for oil water separation. *J. Environ. Chem. Eng.* 9: 105063.

43 Pirzada, T., Ashrafi, Z., Xie, W., and Khan, S.A. (2020). Cellulose silica hybrid nanofiber aerogels: from sol–gel electrospun nanofibers to multifunctional aerogels. *Adv. Funct. Mater.* 30: 1907359.

44 Mahadik, D.B., Venkateswara, A., Parale, V. et al. (2011). Effect of surface composition and roughness on the apparent surface free energy of silica aerogel materials. *Appl. Phys. Lett.* 99.

45 Chhajed, M., Yadav, C., Agrawal, A.K., and Maji, P.K. (2019). Esterified superhydrophobic nanofibrillated cellulose based aerogel for oil spill treatment. *Carbohydr. Polym.* 226: 10.

46 Yang, X., Yadav, C., Agrawal, A. et al. (2018). Cellulose acetate-based SiO_2/TiO_2 hybrid microsphere composite aerogel films for water-in-oil emulsion separation. *Appl. Surf. Sci.* 435: 609–616.

47 Gil Giraldo, G.A., Mantovan, J., Marim, B.M. et al. (2021). Surface modification of cellulose from oat hull with citric acid using ultrasonication and reactive extrusion assisted processes. *Polysaccharides* 2: 218–233.

48 Tan, T., Zhou, J., Gao, X. et al. (2021). Industrial crops & products synthesis, characterization and water-absorption behavior of tartaric acid-modified cellulose gel fromcorn stalk pith. *Ind. Crop. Prod.* 169: 113641.

49 Tripathi, A., Parsons, G.N., Rojas, O.J., and Khan, S.A. (2017). Featherlight, mechanically robust cellulose ester aerogels for environmental remediation. *ACS Omega* 2: 4297–4305.

14

Novel Carbon-Based Nanomaterials (CNMs) for Air Purification

Neelakandan M. Santhosh

Jožef Stefan Institute, Department of Gaseous Electronics (F6), Jamova Cesta 39, Ljubljana, 1000 Slovenia

14.1 Introduction

Air pollution is one of the most significant environmental concerns nowadays, affecting most countries worldwide. Considering the seriousness of the issue and its major threat to global health, all parts of society should focus on resolving this issue by identifying it as a global health priority in sustainable international development. Therefore, the research on the data based on air quality and related health impacts is evolving rapidly. According to the World Health Organization (WHO), approximately 6.7 million premature deaths are reported annually associated with the combined effects of ambient and household air pollution. Also, household air pollution causes approximately 3.2 million deaths per year, according to the statistics for 2020, including over 237 000 deaths of children under the age of 5 [1]. The severe environmental and economic impact caused by air pollution must be resolved as early as possible to ensure the quality of people's lives and environmental sustainability.

Air pollution is caused by the changes occurring in the natural atmospheric conditions due to the excess addition of any biological, physical, or chemical substances expelled from biogenic, geogenic, or anthropogenic sources. Such air pollutants can be in particulate or gaseous forms [2]. Amid the variety of air pollutants, the microscale particles are severely affecting human and other living organism life and health care. Such microscale particles or a mixture of small particles, including chemical, organic, or inorganic substances, are known as particulate matter (PM) [3, 4]. According to the aerodynamic diameter, the PMs are divided into different groups: ultrafine $PM_{0.3}$(diameter <0.3 μm), fine $PM_{2.5}$ (diameter <2.5 μm), and PM_{10} (diameter <10 μm) [5, 6]. Exposure to any of these PMs in the long-term could cause damage to human health, especially lung and cardiovascular systems. The gaseous form refers to different chemical molecules like ozone (O_3), sulfur dioxide (SO_2), and carbon monoxide (CO) [7]. Besides, the environmental issues raised by PMs, haze, and sand storms also exponentially increase and lead to irreversible damage to the ecosystem.

Nanomaterials for Air and Water Purification, First Edition.
Edited by Deepu A. Gopakumar, Jyotishkumar Parameswaranpillai, Jinu Jacob George, and Midhun Dominic C.D.

Several factors contribute to deteriorating air at an alarming rate, including mineral powders, coal, and carbon powders (produced by industrial emissions), automobile exhausts, chemical vapors, and forest fire smoke [8]. Besides this, naturally occurring pollen, microorganisms, and aerosols with viruses also cause air pollution. Apart from this, human activities also largely cause air pollution, mainly from the combustion of fossil fuels. Fossil fuels are one of the largest energy-supplying systems for human needs. Continued exploration and combustion have negative consequences for the ecosystem. Production of various volatile organic compounds (VOC), gas wastes from industries, and unsophisticated degradation of home/industry wastes largely affect air quality and severely damage environmental systems. Therefore, it is important to identify the issues caused by various particulates and develop methods to separate them from the air to improve air quality.

Different strategies have been explored for the remediation and control of air pollution. However, the performance of most of the strategies was unsatisfactory and suffered serious drawbacks in terms of high cost, time consumption, and the generation of a secondary pollutant. An overview of different air pollutant controlling methods and their performances is outlined in Table 14.1.

Considering these features, air filtration is an important aspect of PM removal techniques because it effectively filters the PM matter and controls air pollution. In this method, the employed filter membranes are the key to filtering the differently sized PMs. In order to replace such unsatisfactory-performing techniques, alternate methods have been researched and resulted in the application of nanotechnology and engineering principles [18]. The emergence of multifunctional nanomaterials and advancements in nanotechnologies significantly contributed to the progress in air filtration technology. Multifunctional properties such as porous structures, antibacterial activities, high-temperature stability, excellent mechanical properties, tunable super-hydrophobic/ hydrophilic properties, and high chemical stability of novel membrane materials have been attracting increasing research attention. The excellent physical and chemical properties of novel membrane materials and their easy tailoring behavior have given a wide range of opportunities to be used efficiently for filtration applications.

Nanotechnology and nanoengineering are the fastest-growing scientific areas, often called the next Industrial Revolution" [19]. Due to its ability to hold both basic research and technology development, nanotechnology is considered one of the most feasible and reliable techniques to reduce the environmental impacts of air pollution. Techniques that are used to effectively remove PMs and other gaseous contaminants with the assistance of nanotechnology are known as nanobioremediation [20]. Such approaches offer economical and eco-friendly solutions for controlling air pollution and improving sustainable development by effectively removing environmental pollutants from the air through nanoparticles. Even though conventional bioremediation using plants, enzymes, and microbes demonstrated good efficiency, improving their performance with nanoparticle assistance is currently gaining research attention. Several reports describe that the nanoparticles or nanostructures can be used as nanoadsorbents, nanocatalysts, nanofilters or nanostructured membranes, and nanosensors [21, 22].

Table 14.1 Overview of the different air pollutant controlling techniques for removing PMs, VOCs and other components.

Pollutant	Purification method	Efficiency	Advantages	Drawbacks	Reference
PM	Fabric filters (bag house)	>99% for $PM_{2.5}$	High efficiency toward dust and PM; low cost	Temperature sensitivity; Filter media regeneration	[8, 9]
	Wet Scrubbers	~50% for $PM_{2.5}$	Simultaneous removal of gas and PMs	Corrosion, high energy consumption	[8, 10]
	Electrostatic precipitators	Poor for $PM_{2.5}$	Low energy consumption and cost	Temperature and humidity sensitive	[8, 11]
	Cyclones/centrifugal forces	Poor for $PM_{2.5}$	Low cost; Simple process	Low efficiency	[8, 10]
VOC	Thermal oxidation	>99% efficiency	High efficiency and suitable for large-scale applications	High energy consumption, undesirable by-products	[12]
	Catalytic oxidation	>99% efficiency	Energy efficient, lower formation of by-products	High cost for catalysts	[12]
	Adsorption	>90% efficiency	Low cost, useful for large scale	High regeneration cost, secondary pollution	[13, 14]
Nitrogen oxides	Three-way catalyst	>95% efficiency	High efficiency	High cost and not suitable for diesel systems	[15]
	Selective catalytic reduction	80–90% efficiency	High conversion rate and suitable for high dust conditions	Possible ammonia leaks and expensive catalysts	[12]
	Selective non-catalytic reduction	30–75% efficiency	Catalyst-free and low cost	Possible ammonia leak, narrow temperature window	[16]

Source: Adapted and modified Cao et al. [17].

14.2 Nanomaterials for Air Purification

Size-dependent properties and a high aspect ratio and surface area of nanomaterials make them favorable for detecting highly sensitive contaminants and separating them from a mixture with high precision. In such a way, nanomaterials could contribute to lowering the use of critical raw materials, minimizing waste production, and controlling the number of toxic substances, which could be useful for environmental protection. Therefore, nanomaterials-based research for controlling and improving air quality considers that a nanoscale adsorbent can significantly lower air pollution and improve air quality, and such adsorbents are known as nanoadsorbents. CNMs (nanotubes and graphene) and metal oxides (zinc oxide) are the main candidates used as adsorbents. High surface area, larger pore diameter, and pore volume help these structures increase the reactive sites in the backbone, which could help bind with the PM or any other contaminants. Nanocatalysts are other types of nanomaterials that are widely used in the remediation of air [21, 22]. In these types, nanomaterials with semiconducting features for photocatalytic remediation are explored by exposing the nanomaterials to a light source with a similar energy to their band gap. Such a process results in the formation of electron–hole pair and participates in the remediation of the air as a function of the size of the catalyst material applied. Nanocatalyst systems are known for selectively transforming chemical functionalities with ultra-high product yield while simultaneously offering recovery of catalyst materials. Titanium dioxide, graphene/titanium composites, zinc oxides, and manganese-based composites are reported as highly efficient nanocatalysts for air purification. Besides these, nanostructured membranes with different ranges of pore sizes are widely used for the selective permeation and separation of PMs and other contaminants. Various studies reported optimizing and fabricating nanostructured membranes with different pore dimensions and activity centers for selectively separating diverse gas contaminants [23, 24]. Among all the membrane-type materials for separating the gas contaminants, carbon nanomaterials (CNMs) are the front-runner candidates because of their multi-functionality, dimensionality, morphological features, and structural properties.

14.2.1 CNMs: Structures and Properties

Carbon is one of the most abundant materials in the earth's crust and is present in the form of large blocks of carbonate rocks and coal. Carbon is known for existing in both crystalline and amorphous forms and in various forms of allotropes. Owing to the excellent chemical, physical, electrical, and optical properties of carbon and its allotropes, carbon is considered one of the most promising materials for various applications, including electrical, electronic, sensing, and energy storage. Over the last four decades, research interest in carbon and its nanostructures has emerged because of their advanced properties at the nanoscale. The discovery of fullerene stimulated the research on nanocarbon-based research and its applications. Following this, carbon nanotubes (CNTs) were discovered in multiwalled and single-walled forms. More recently, a two-dimensional (2D), single-layered form of carbon atoms was discovered, known as graphene. Following this, novel carbon structures with sp^2 and sp^3 hybridizations emerged in the form of diamond, few-layer graphene, defected graphene, vertical graphene, and graphitic nanoribbons. Depending on the dimension (0D, 1D, 2D, or 3D) and orientation (horizontal or vertical), the structural and chemical features of the nanocarbons are varied, which can be tailored using additional processing to make them suitable for

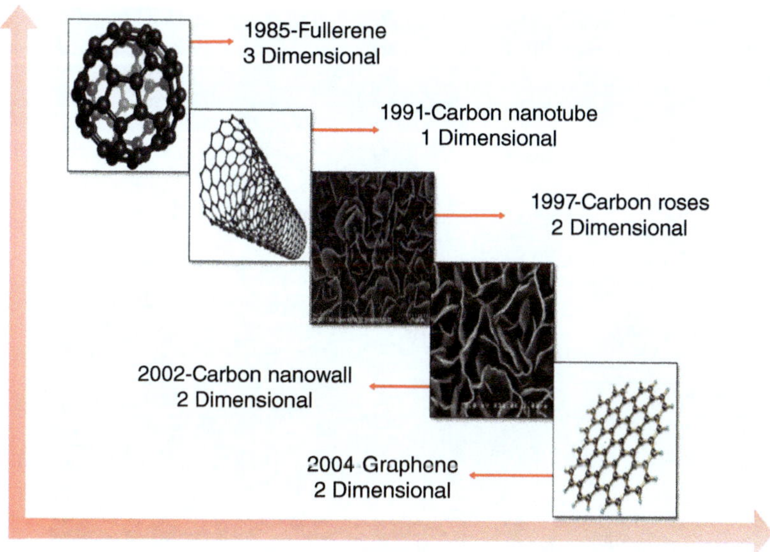

Figure 14.1 Milestones in carbon nanostructure research. Source: Mukhopadhyay [9]/Taylor and Francis group.

advanced applications. A time-dependent discovery of different morphologies of carbon-based nanostructures is presented in Figure 14.1.

The first discovered variety of CNM, fullerene, known by the chemical formula C60, has attracted interesting research attention due to its unique thermal and chemical characteristics, which allow acting as an effective electron acceptor. It is known that the surface area of fullerene is affected by the surrounding area. The unique adsorptive capacity of fullerene for organic, organic metallic, hydrocarbons, and polycyclic aromatic compounds is influenced by the dispersive interaction forces. Therefore, such features make them a promising candidate for filtration membranes. Another form of carbon is when the carbon nanosheet rolled in a 1D pattern turns into nanotubes. Depending on the number of rolled carbon sheets, nanotubes are classified into single-walled and multi-walled varieties. The structural and electrical properties of nanotubes are highly influenced by the atomic arrangement of carbon atoms and the diameter and length of the nanotubes.

The 2D form of carbon is one of the most promising nanostructures in the carbon family, called graphene. Graphene is a 2D, one carbon atom-thick layer arranged in a hexagonal array. Graphene possesses a highly crystalline feature in which the carbon atoms are bonded with sp^2 hybridization. Graphene is considered the building unit for all the other dimensional allotropes of carbon [29]. The large surface area and easy functionalisability of the graphene structures make them interesting for membrane applications for gas separation and permeation. Besides the horizontal orientation, vertically oriented 1D and 2D carbon nanostructures were also gaining interest after the discovery of vertical graphene. Vertically oriented graphene or carbon nanowalls are 1–10 stacks of graphene sheets aligned vertically on a substrate surface. Owing to their unique orientation, exposed sharp edges, non-stacking

Figure 14.2 SEM micrographs of different CNMs: (a) nanoparticles [25], (b) nanotubes [26], (c) graphitic carbon [27], and (d) nanowalls [28].

morphology, and large aspect ratio, they have potential in a wide range of applications, including gas separation and filtration. In a similar way, vertical CNTs are also reported, in which the densely packed nanotubes could act as an effective membrane for separation applications. The surface morphology of carbon nanoparticles, vertical graphene, CNTs, and graphitic carbon is presented in Figure 14.2.

The conventional methods used to synthesize the CNMs are the following: (i) chemical vapor deposition: assure high-quality synthesis with low-batch yield; typically CNMs are produced at a temperature range between 700 and 1500 °C. These techniques are employed in the presence of carbon-containing precursor gases with or without the assistance of catalyst nanoparticles; (ii) laser ablation: production of CNMs by the use of continuous waves or pulsed lasers and ablating a carbon source at a higher temperature. During this method, the carbon source is evaporated by the laser, and the vapor will initiate the growth of CNMs on a catalyst-coated substrate; (iii) arc-discharge methods: a carbon source is evaporated by a neutral-gas-containing plasma discharged by a high current. The current is passed between parallel electrodes, one of which is a carbon source; (iv) plasma deposition: plasma is a gaseous discharge generated by applying a strong electromagnetic field, accelerating electrons to collide with the neutral gas. These collisions lead to the dissociation, ionization, and excitation of the gaseous species, producing more electrons, ions, photons, radicals, and excited background gas. During CNM production, a carbon-containing precursor gas is used for the ignition of the plasma, and the plasma discharge parameters vary to obtain high-quality CNMs with different morphology and functionality; (v) wet-chemical methods: using other chemical reagents and reactions, carbon resources are reduced to different CNMs by delamination and growth. Apart from these techniques, chemical and mechanical exfoliations are widely used in graphene production. Also, chemical or thermal reduction of graphene oxide is widely used to synthesize graphene-based CNMs. Considering the large surface area and high adsorption capacities, CNM-based air filters are becoming a promising research field for successful air purification.

14.3 CNMs: Applications for Air Purification

Due to the high adsorptive capacity and large surface area of CNMs, CNM-based materials are ideal for improving air remediation, especially for removing hazardous air pollutants such as PMs, VOCs, and nitrogen oxides. The significant properties that make carbon nanostructures a promising material for air purification are:

1. Large surface area: All CNMs have a very high surface area to volume ratio, providing the ability to capture a large concentration of pollutants per unit mass. This makes them highly effective for removing pollutants from the air.
2. Adsorption capacity: The high adsorption capacity of CNMs helps to adsorb a wide range of pollutants, including gases, VOCs, and PM.
3. Excellent chemical stability: Chemically and structurally stable CNMs have high degradation capacities, even in harsh environments, and are resistant to sudden chemical reactions, attracting them for long-term use as air purifiers.
4. Reusability: CNMs can be reused several times for air purification, allowing a cost-effective option for long-term use.
5. Low energy consumption: The low energy consumption of CNMs for air purification is suitable for the environmentally friendly candidate for air purifiers.

Though the properties of CNMS are suitable for air filtration, implementing them for air filtration requires a pre-treatment of the materials in some cases. This pretreatment allows the changes in the physical, chemical, and structural modifications of CNMs to create a passage in the structure for the pollutants. One of the most common and direct pretreatment methods is the activation of CNMs by exposing them to high temperatures in the presence of a gas. Such heat treatment creates structural defects and a high porosity in the material with a large surface area. Such improvements in surface area and porosity will effectively remove pollutants selectively from the air. Several reports demonstrated that activated carbon could be used to remove NO by oxidizing nitrogen with activated carbon [30–32]. CNMs are also pretreated via different physical and mechanical modifications, such as grinding or milling. Continuous grinding or milling enables changes in the structure and surface of the materials and creates a large number of structural defects in the materials. These newly created defects could also act as active sites on the surface of the CNMs, leading to improved adsorption capacity and selectivity for specific pollutants with improved efficiency in purification. Besides, surface functionalization and heteroatom doping of CNMs are also reported as efficient treatment processes to improve the surface properties of the materials for enhanced efficiency. Surface functionalization of the CNM surface is a simple tool for incorporating different functional entities, such as reactive radicals, metallic nanoparticles, and pendant molecules, into the graphene lattice and modifying their properties for air filtration. Surface functionalization can be performed in two ways: (i) In-situ treatment, modification, and functionalization of CNMs during growth by mixing the functional group containing source into the carbon precursor mixture; or (ii) through postprocessing treatment, where the CNMs are treated with the functional resources after the synthesis process. The effects of the in situ treatment during growth allow the incorporation of functional entities into the bulk of the material. In contrast, the postprocessing techniques allow the controlled engineering of CNMs with targeted species. Thermal treatment, chemical treatment, and plasma-assisted treatments are the most widely used surface engineering techniques to tailor the surface properties of the CNMs. Among these different methods, the plasma-enabled treatment process has the advantage of tailoring the CNM surfaces with controlled functionalities, considering the ability of plasma to create defects and functionalize the surface simultaneously. Based on their structure, morphology, dimension, and functionality, several CNM-based membranes are reported for efficient air purification and separation.

The available conventional techniques to filter the pollutants and purify the air either use mechanical or electrostatic features. One of the most common membranes is a fibrous media filter (FMF), which is permeable material with sparsely distributed fibers upon which particles are deposited by the different filtration mechanisms. Structurally, FMF is an open, 3D structure consisting of multiple layers of randomly oriented fibers, which are responsible for separating pollutants from the air. A combination of interception, diffusion, and inertial impaction is the mechanism of filtration using such membranes. Since the separation is influenced by the inertia force, the separation occurs based on the mass of the materials. The diffusion mechanism applies when the small particles interact with the fibrous structures. The high-efficiency particulate absorbing (HEPA) filter is the well-known class of FMF and gives sufficient separation currently; the efficiency of such filters against aerosols and small-sized particulates is an issue in the long run. Moreover, the higher operating costs of this technology and the problems with secondary pollution from the pollutants deposited on the filters make it difficult to use in the long run.

Another class of conventional filters are adsorbent media filters (AMF), which are mainly used for separating pollutant gases and VOCs. Activated carbon is one of the most promising materials for AMF compared to alumina, zeolite, and other materials. However, AMF suffers from issues with selectivity against various gases and restricts long-term use and commercialization. Also, the high cost and fast saturation rate limit their potential for advanced applications. Besides this, ionizers have been used as a filter medium to attach the pollutant particles in electronic separation techniques. Ionizers are low-cost operations and more suitable for large areas compared to FMFs and AMFs. However, the slow reaction time of ionizers is a limitation that needs to be improved. In this scenario, filter membranes based on CNMs become attractive and have enormous potential to be used in the current scenario. Based on this, several carbon-based air filters have emerged. On the basis of a number of layers, crystalline and functional properties, CNMs can be synthesized in different forms and enables the sensing and purification of harmful air pollutants, including VOCs, CO, O_3, SO_2, nitrogen oxides, heavy metals, and PMs [33, 34] The major mechanisms of air purification are adsorption, membrane filtration, sensing, and catalytic activities.

Adsorption is the most frequently used purification technique considering the ability of engineered and property-tailored CNMs to exhibit affinity toward different pollutants, gases, and molecules. The use of adsorption techniques is common in the field of environmental remediation. These techniques can be applied in various forms, such as gas adsorption, liquid adsorption, and adsorption of both organic and inorganic pollutants. Essentially, adsorption is a process by which an adsorbate, present in a gaseous or liquid bulk, adheres to the surface of adsorbent material. Unlike absorption, in which a substance diffuses into another substance, the adsorbate only adheres to the surface of the adsorbent material. This process occurs naturally in the environment and is widely used in various industrial processes. In the adsorption process, each contaminant has its own unique adsorption capacity due to the differences in bonding interactions between their molecules. For example, in the case of nanotubes, organic compounds are primarily adsorbed onto CNTs through various bonding interactions, such as electrostatic, hydrophobic, and hydrogen bonding interactions. These interactions can also influence the electronic and transport properties of the nanotubes. In the case of heavy metal adsorption, CNTs can attract metal ions through electrostatic attraction, surface precipitation, physical adsorption, and complexation with functional groups. Apart from that, the geometrical features of CNMs have also influenced the mechanisms, which are influenced by carrier localization and charge distribution [35].

Membrane filtration is another type of gas separation in which the CNM structures are finely tuned for separating specific gas from a mixture of gases. Designing a CNM-based membrane is a risky task considering the multiple complex steps involved in the designing process. In general, CNM membrane designing involves six steps starting from the precursor selection, polymeric matrix preparation, pretreatment, carbonization, posttreatment, and module construction. In these steps, carbonization is the most crucial, which defines the structural properties, functional entities and morphological features of CNM membranes [36]. The precursor materials that have been used for the carbonization process are thermosetting resin, graphite, coal, bio-based materials and biowastes, and the carbonization is carried under an inert atmosphere, vacuum or nitrogen-containing environment. Recently, various synthetic precursors, including polyimide and derivatives, phenol-based resin, poly furfuryl alcohol (PFA), polyvinylidene chloride–acrylate terpolymer (PVDC–AC), formaldehyde gels, biowastes, cellulose, and others, have been reported as precursors for membrane fabrications [27, 37–41]. The advantages of membranes for filtration processes are their high surface area, low cost of production, easy cleaning procedure, low maintenance cost, and high efficiency. Gas separation through membranes relies on several factors, including the differences in how the components of a gas mixture interact with the membrane, the critical pore size and distribution of the membrane, and the mean free path length

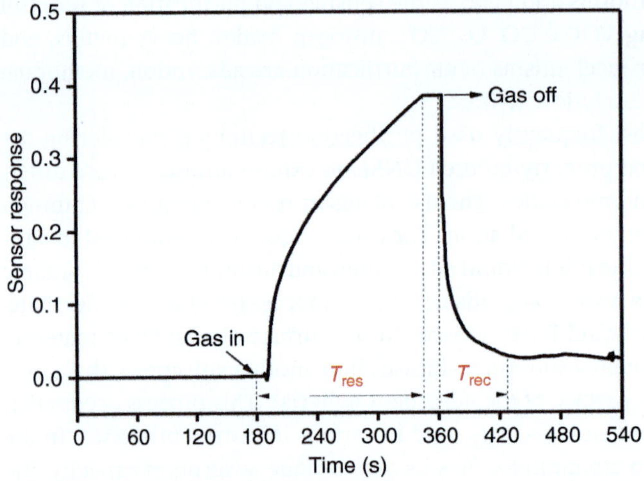

Figure 14.3 Sensor response of a carbon nanotube-based sensor against ethanol at room temperature. Source: Reused from Santhosh et al. [44].

of gas molecules. Various models have been suggested to explain gas transport through carbon membranes, with molecular sieve (activated diffusion), surface diffusion (selective adsorption), Knudsen diffusion, and viscous flow (Poiseuille flow) being the most commonly accepted mechanisms [42]. The influence of such a mechanism on gas transport is determined by the physical and chemical properties of the membrane, the gases being separated, and the operating conditions of the permeation system, such as temperature and pressure [42, 43].

The other class of applications that have been used in air purification industry by employing carbon materials is gas sensing. This allows us to understand the quality of the air and immediately sense the presence of pollutant gases if they are detected. The properties of CNMs have been considered to fabricate gas sensors are based on different transduction principles (changes in their electrical or optical properties, mass-sensitive, etc.). Typically, the gas sensors work on the principle of change in resistance in which the resistance of the sensor either increases or decreases when exposed to a pollutant gas. This could be due to the increase in resistance caused by the change in contact resistance between individual layers of CNMs by the adsorption of the pollutant or external gas molecule, which increases the macroscopic resistance of the sensor. Figure 14.3 presents the sensor response of carbon nanotube-based sensor sensing against ethanol vapor.

Based on these types of principles, CNMs have been widely used for the removal of various pollutants from the environment. Various forms of CNMs and their targeted pollutant gases reported in the literature are provided in Table 14.2.

Table 14.2 Different varieties of CNMs and their targeted pollutant with the indicated mechanism.

CNMs	Targeted pollutant	Mechanism	References
Carbon nanotubes deposited on quartz filters	VOCs	Sensing and purification carried out by π-π interaction	[45]
Single-wall and multiwall carbon nanotubes	NO*x*	NO, and O_2 pass through the nanotubes; NO gets absorbed on the surface after gets oxidized to NO_2	[46]
Modified nanotubes using 3 aminopropyl tri ethoxysilane (APTS)	CO_2	Amine functionalized nanotube surfaces provide sufficient active sites for CO_2 adsorption	[47]
Fullerene B_{40}	CO_2	Strong chemisorption of fullerene structures improves the adsorption capacity.	[48]
Fullerene-like boron nitride nanocage	N_2O	Adsorbing and decomposition of N_2O at the surface	[49]
Graphene oxide composites	CO_2, NH_3, SO_2, H_2S, and N_2	Highly functionalize graphene surface enhances the adsorption	[50, 51]
Nanotube/NaClO	Isopropyl vapor	Physical adsorption by Van der Waals forces and chemical adsorption onto adsorbent surface functional groups	[52]
Heteroatom doped nanotubes (silicon and boron)	CO and CH_3OH	Tailored electronic properties of doped nanotubes enhance the adsorption properties	[53]
Nanotube on polyethylene terephthalate (PET)	C_2H_5OH	High surface area and electronic properties enhance the chemisorption of the gases	[54]
Plasma-tailored carbon nanotubes	Ethanol	Plasma surface engineering improves the electronic properties and creates more functional groups for the adsorption	[44]
Reduced graphene/polyaniline composite	CO_2	Nitrogen doping improves the bonding between carbon and CO_2 surfaces	[55]
Graphene/Mn_3O_4	CO_2	Due to the physical adsorption enhanced by the formation of intermolecular electrostatic interactions.	[56]
Carbon nanotubes	Formaldehyde	Chemisorption plays a critical role in the removal of formaldehyde	[34]
Carbon nanofibers	Formaldehyde	Nitrogen doping creates more active sites and improves the adsorption efficiency	[57]

(Continued)

Table 14.2 (Continued)

CNMs	Targeted pollutant	Mechanism	References
Biochar	Magnesium, copper, and nickel	Chemisorption, along with diffusion, play a major role in the adsorption	[58]
Nitrogen-doped biochar	Copper and cadmium	Nitrogen doping enhances the adsorption capacity and the adsorption attained by both complexation and cation π bonding	[59]
Carbon aerogel	Phenol; P-nitrophenol; resorcinol; tetracycline; oxytetracycline; doxycycline; sulfamethazine	Intraparticle diffusion and pore diffusion were the critical factors in the adsorption rate	[60]
Carbon aerogel/$Na_2Ti_3O_7$ composite	Rubidium and caesium	Ion exchange between Rb^+/Cs^+ and Na^+ in the interlayer of $Na_2Ti_3O_7$ and the complexation of Rb^+/Cs^+ with the –OH functional groups are the key to high metal removal	[61]
Hierarchical carbon-nanotube/quartz-fibre	Aerosol	Microfibers with high stability in a hierarchical structure and CNTs with a high specific surface area contribute to the high filtration efficiency	[62]

14.4 Future Perspectives

Despite their potential, there are still challenges associated with the use of CNMs for air filtration. These include issues related to structural and physical properties, cost-efficiency, scalability, and toxicity. Therefore, the major challenges that currently need to be addressed are:

1. Tailoring the intrinsic properties of CNMs since the intrinsic CNMs are inert toward environmental contaminants.
2. Designing advanced functional CNM hybrids with metal derivatives or polymer composites with precise functional termination. Hybrid structures could accumulate and separate the particulate with different sizes.
3. Identify suitable CNMs and fine-tune their properties to improve their selectivity and sensitivity towards air pollutants to sufficiently detect and remove them.
4. Fast, facile, and low-cost protocol for designing advanced CNMs in an environmentally benign approach.

14.5 Conclusion

Carbon nanomaterials, including fullerenes, CNTs, and graphene, have shown promising potential for selective air filtration applications. Unique properties of CNMs, such as various morphologies, high surface area, high aspect ratio, and high mechanical strength, make them emerging materials for air purification applications. Advanced synthesis and surface engineering methods, such as conventional chemical vapor deposition and plasma-enabled designing techniques, can be used to design CNMs with desired structures, morphologies, and properties. Though the CNMs have shown great potential toward air filtration, challenges associated with the use of CNMs for air filtration need to be addressed, and future research can help address these challenges and improve the effectiveness of air filtration technologies by establishing a low-cost CNM design technique for fabricating advanced filtration membranes with improved selectivity and sensitivity. Such techniques will open up a future paradigm for a cleaner environment and sustainable development.

References

1 World Health Organization (WHO) Household air pollution (n.d.). https://www.who.int/news-room/fact-sheets/detail/household-air-pollution-and-health (accessed 19 January 2023).

2 Spinazzè, A., Campagnolo, D., Cattaneo, A. et al. (2020). Indoor gaseous air pollutants determinants in office buildings—the OFFICAIR project. *Indoor Air* 30: 76–87. https://doi.org/10.1111/INA.12609.

3 Zhang, R., Liu, C., Zhou, G. et al. (2018). Morphology and property investigation of primary particulate matter particles from different sources. *Nano Res.* 11: 3182–3192. https://doi.org/10.1007/S12274-017-1724-Y/METRICS.

4 Jeong, S., Cho, H., Han, S. et al. (2017). High efficiency, transparent, reusable, and active $PM_{2.5}$ filters by hierarchical Ag nanowire percolation network. *Nano Lett.* 17: 4339–4346. https://doi.org/10.1021/ACS.NANOLETT.7B01404/ASSET/IMAGES/LARGE/NL-2017-01404N_0005.JPEG.

5 Zhao, X., Wang, S., Yin, X. et al. (2016). Slip-effect functional air filter for efficient purification of $PM_{2.5}$. *Sci. Rep.* 6 (1): 1–11. https://doi.org/10.1038/srep35472.

6 Harrison, R.M. and Yin, J. (2000). Particulate matter in the atmosphere: which particle properties are important for its effects on health? *Sci. Total Environ.* 249: 85–101. https://doi.org/10.1016/S0048-9697(99)00513-6.

7 Zhao, Q., Li, X., Zhou, Q. et al. (2020). Nanomaterials developed for removing air pollutants. *Adv. Nanomater. Pollut. Sensing Environ. Catal.* 203–247. https://doi.org/10.1016/B978-0-12-814796-2.00006-X.

8 Pui, D.Y.H., Chen, S.C., and Zuo, Z. (2014). $PM_{2.5}$ in China: measurements, sources, visibility and health effects, and mitigation. *Particuology* 13: 1–26. https://doi.org/10.1016/J.PARTIC.2013.11.001.

9 Mukhopadhyay, A. (2009). Pulse-jet filtration: an effective way to control industrial pollution part I: theory, selection and design of pulse-jet filter. *Micromachines* 9 (11): 565. https://doi.org/10.3390/mi9110565.

10 Darcovich, K., Jonasson, K.A., and Capes, C.E. (1997). Developments in the control of fine particulate air emissions. *Adv. Powder Technol.* 8: 179–215. https://doi.org/10.1016/S0921-8831(08)60463-X.

11 Mizuno, A. (2000). Electrostatic precipitation. *IEEE Trans. Dielectr. Electr. Insul.* 7: 615–624. https://doi.org/10.1109/94.879357.

12 Kamal, M.S., Razzak, S.A., and Hossain, M.M. (2016). Catalytic oxidation of volatile organic compounds (VOCs) – a review. *Atmos Environ.* 140: 117–134. https://doi.org/10.1016/J.ATMOSENV.2016.05.031.

13 Liu, Y., Shao, M., Fu, L. et al. (2008). Source profiles of volatile organic compounds (VOCs) measured in China: Part I. *Atmos Environ.* 42: 6247–6260. https://doi.org/10.1016/J.ATMOSENV.2008.01.070.

14 Yi, F.Y., Lin, X.D., Chen, S.X., and Wei, X.Q. (2009). Adsorption of VOC on modified activated carbon fiber. *J. Porous Mater.* 16: 521–526. https://doi.org/10.1007/S10934-008-9228-5/FIGURES/8.

15 Granger, P. and Parvulescu, V.I. (2011). Catalytic NO_x abatement systems for mobile sources: from three-way to lean burn after-treatment technologies. *Chem. Rev.* 111: 3155–3207. https://doi.org/10.1021/CR100168G.

16 Perring, A.E., Pusede, S.E., and Cohen, R.C. (2013). An observational perspective on the atmospheric impacts of alkyl and multifunctional nitrates on ozone and secondary organic aerosol. *Chem. Rev.* 113: 5848–5870. https://doi.org/10.1021/CR300520X.

17 Cao, J.J., Huang, Y., and Zhang, Q. (2021). Ambient air purification by nanotechnologies: from theory to application. *Catalysts* 11: 1276. https://doi.org/10.3390/CATAL11111276.

18 Madima, N., Mishra, S.B., Inamuddin, I., and Mishra, A.K. (2020). Carbon-based nanomaterials for remediation of organic and inorganic pollutants from wastewater. a review. *Environ. Chem. Lett.* 18: 1169–1191. https://doi.org/10.1007/S10311-020-01001-0.

19 Roco, M.C. (2005). The emergence and policy implications of converging new technologies integrated from the nanoscale. *J. Nanopart. Res.* 7: 129–143. https://doi.org/10.1007/S11051-005-3733-0.

20 Gabal, E., Chatterjee, S., Ahmed, F.K., and Abd-Elsalam, K.A. (2020). Carbon nanomaterial applications in air pollution remediation. *Carbon Nanomater. Agri-Food Environ. Appl.* 133–153. https://doi.org/10.1016/B978-0-12-819786-8.00007-4.

21 Tai, X.H., Lai, C.W., Juan, J.C., and Lee, K.M. (2020). Nanocatalyst-based catalytic oxidation processes. *Nanomater. Air Rem.* 133–150. https://doi.org/10.1016/B978-0-12-818821-7.00007-5.

22 Theerthagiri, J., Duraimurugan, K., Kim, H.S., and Madhavan, J. (2019). Graphitic carbon nitride-based nanostructured materials for photocatalytic applications. *Photocatal. Funct. Mater. Environ. Rem.* 291–307. https://doi.org/10.1002/9781119529941.CH10.

23 Samantara, A.K., Ratha, S., and Raj, S. (2019). Functionalized graphene nanocomposites in air filtration applications, functionalized graphene nanocomposites and their derivatives: synthesis. *Process. Appl.* 65–89. https://doi.org/10.1016/B978-0-12-814548-7.00004-0.

24 Rubina, A., Rubinová, O., and Blasinski, P. (2019). Application of nanofilters for ventilation. *Inter. Rev. Appl. Sci. Eng.* 10: 51–56. https://doi.org/10.1556/1848.2019.0008.

25 Kovacevic, E., Strunskus, T., Santhosh, N.M. et al. (2021). Thermal stability studies of plasma deposited hydrogenated carbon nitride nanostructures. *Carbon N Y* 184: 82–90. https://doi.org/10.1016/J.CARBON.2021.08.008.

26 Santhosh, N.M., Upadhyay, K.K., Filipič, G. et al. (2023). Widening the limit of capacitance at high frequency for AC line-filtering applications using aqueous carbon-based supercapacitors. *Carbon N Y* 203: 686–694. https://doi.org/10.1016/J.CARBON.2022.12.026.

27 Santhosh, N.M., Vasudevan, A., Jurov, A. et al. (2020). Oriented carbon nanostructures from plasma reformed resorcinol-formaldehyde polymer gels for gas sensor applications. *Nanomaterials* 10: 1704. https://doi.org/10.3390/NANO10091704.

28 Santhosh, N.M. et al. (2020). N-graphene nanowalls via plasma nitrogen incorporation and substitution: the experimental evidence. *Nanomicro. Let.t* 12 (1): 53. https://doi.org/10.1007/s40820-020-0395-5.

29 Geim, A.K. and Novoselov, K.S. (2007). The rise of graphene. *Nat. Mater.* 6 (3): 183–191. https://doi.org/10.1038/nmat1849.

30 Sousa, J.P.S., Pereira, M.F.R., and Figueiredo, J.L. (2013). Modified activated carbon as catalyst for NO oxidation. *Fuel Process. Technol.* 106: 727–733. https://doi.org/10.1016/J.FUPROC.2012.10.008.

31 Ahmed, S.N., Baldwin, R., Derbyshire, F. et al. (1993). Catalytic reduction of nitric oxide over activated carbons. *Fuel* 72: 287–292. https://doi.org/10.1016/0016-2361(93)90044-3.

32 Rubel, A.M., Stewar, M.L., and Stencel, J.M. (1995). Activated carbon for control of nitrogen oxide emissions. *J. Mater Res.* 10: 562–567. https://doi.org/10.1557/JMR.1995.0562.

33 Romero-Guzmán, L., Reyes-Gutiérrez, L.R., Romero-Guzmán, E.T. et al. (2017). Carbon nanotube filters for removal of air pollutants from mobile sources. *J. Miner. Mater. Character. Eng.* 6: 105–118. https://doi.org/10.4236/JMMCE.2018.61009.

34 Gulati, S., Lingam B, H.N., Kumar, S. et al. (2022). Improving the air quality with functionalized carbon nanotubes: sensing and remediation applications in the real world. *Chemosphere* 299: 134468. https://doi.org/10.1016/J.CHEMOSPHERE.2022.134468.

35 Chung, J.H., Hasyimah, N., and Hussein, N. (2021). View of application of carbon nanotubes (CNTs) for remediation of emerging pollutants - a review. *Trop. Aquat. Soil Pollut.* https://doi.org/10.53623/tasp.v2i1.27.

36 Saufi, S.M. and Ismail, A.F. (2004). Fabrication of carbon membranes for gas separation—a review. *Carbon NY* 42: 241–259. https://doi.org/10.1016/J.CARBON.2003.10.022.

37 Seo, D.H., Rider, A.E., Kumar, S. et al. (2013). Vertical graphene gas- and bio-sensors via catalyst-free, reactive plasma reforming of natural honey. *Carbon N Y* 60: 221–228. https://doi.org/10.1016/j.carbon.2013.04.015.

38 Seo, D.H., Pineda, S., Yick, S. et al. (2015). Plasma-enabled sustainable elemental lifecycles: honeycomb-derived graphenes for next-generation biosensors and supercapacitors. *Green Chem.* 17 (4): 2164–2171. https://doi.org/10.1039/c4gc02135e.

39 Shusen, W., Meiyun, Z., and Zhizhong, W. (1996). Asymmetric molecular sieve carbon membranes. *J. Membr. Sci.*, Elsevier. https://www.sciencedirect.com/science/article/pii/0376738895002057 (accessed 7 April 2023).

40 Hatori, H., Kobayashi, T., Hanzawa, Y. et al. (2000). Mesoporous carbon membranes from polyimide blended with poly (ethylene glycol). *J. Appl. Polym. Sci.* 79 (5): 836–841. https://doi.org/10.1002/1097-4628(20010131)79:5<836::AID-APP80>3.0.CO;2-1.

41 M. Shiflett, Foley, H.C. (n.d.). Reproducible production of nanoporous carbon membranes, Pascal-Francis. Inist. Fr. (New York, undefined NY), undefined 2001. https://pascal-francis.inist.fr/vibad/index.php?action=getRecordDetail&idt=1042191 (accessed 7 April 2023).

42 Salleh, W. and Ismail, A.F. (2015). Carbon membranes for gas separation processes: Recent progress and future perspective. *J. Membr. Sci. Res* 1 (2015): 2–15. http://www.msrjournal.com/&url=http:/www.msrjournal.com/article_12301.html (accessed 7 April 2023).

43 Julbe, A., Farrusseng, D., and Guizard, C. (n.d.). Porous ceramic membranes for catalytic reactors—overview and new ideas. *J. Membr. Sci.* https://www.sciencedirect.com/science/article/pii/S0376738800003756 (accessed April 7 2023).

44 Santhosh, N.M., Vasudevan, A., Jurov, A. et al. (2020). Improving sensing properties of entangled carbon nanotube-based gas sensors by atmospheric plasma surface treatment. *Microelectron Eng.* 232: 111403. https://doi.org/10.1016/J.MEE.2020.111403.

45 Amade, R., Hussain, S., Ocaña, I.R., and Bertran, E. (2014). Growth and functionalisation of carbon nanotubes on quartz filter for environmental applications. *J. Environ. Eng. Ecol. Sci.* 3: 2. https://doi.org/10.7243/2050-1323-3-2.

46 De Zhang, W. and Zhang, W.H. (2009). Carbon nanotubes as active components for gas sensors. *J. Sens.* 2009: https://doi.org/10.1155/2009/160698.

47 Su, F., Lu, C., Cnen, W. et al. (2009). Capture of CO_2 from flue gas via multi-walled carbon nanotubes. *Sci. Total Environ.* 407: 3017–3023. https://doi.org/10.1016/J.SCITOTENV.2009.01.007.

48 Dong, H., Lin, B., Gilmore, K. et al. (2015). B40 fullerene: an efficient material for CO_2 capture, storage and separation. *Curr. Appl. Phys.* 15: 1084–1089. https://doi.org/10.1016/J.CAP.2015.06.008.

49 Esrafili, M.D. (2017). N_2O reduction over a fullerene-like boron nitride nanocage: a DFT study. *Phys. Lett. A* 381: 2085–2091. https://doi.org/10.1016/J.PHYSLETA.2017.04.009.

50 Wu, S., He, Q., Tan, C. et al. (2013). Graphene-based electrochemical sensors. *Small* 9: 1160–1172. https://doi.org/10.1002/SMLL.201202896.

51 Seredych, M. and Bandosz, T.J. (2012). Manganese oxide and graphite oxide/MnO_2 composites as reactive adsorbents of ammonia at ambient conditions. *Microporous Mesoporous Mater.* 150: 55–63. https://doi.org/10.1016/J.MICROMESO.2011.09.010.

52 Hsu, S. and Lu, C. (2007). Modification of single-walled carbon nanotubes for enhancing isopropyl alcohol vapor adsorption from air streams. *Sep. Sci. Technol.* 42: 2751–2766. https://doi.org/10.1080/01496390701515060.

53 Azam, M.A., Alias, F.M., Tack, L.W. et al. (2017). Electronic properties and gas adsorption behaviour of pristine, silicon-, and boron-doped (8, 0) single-walled carbon nanotube: a first principles study. *J. Mol. Graph Model* 75: 85–93. https://doi.org/10.1016/J.JMGM.2017.05.003.

54 Alshammari, A.S., Alenezi, M.R., Lai, K.T., and Silva, S.R.P. (2017). Inkjet printing of polymer functionalised CNT gas sensor with enhanced sensing properties. *Mater Lett.* 189: 299–302. https://doi.org/10.1016/J.MATLET.2016.11.033.

55 Christian Kemp, K., Chandra, V., Saleh, M., and Kim, K.S. (2013). Reversible CO_2 adsorption by an activated nitrogen doped graphene/polyaniline material. *Nanotechnology* 24: 235703. https://doi.org/10.1088/0957-4484/24/23/235703.

56 Zhou, D., Liu, Q., Cheng, Q.Y. et al. (2012). Graphene-manganese oxide hybrid porous material and its application in carbon dioxide adsorption. *Chin. Sci. Bull.* 57: 3059–3064. https://doi.org/10.1007/S11434-012-5158-3.

57 Lee, K.J., Shiratori, N., Lee, G.H. et al. (2010). Activated carbon nanofiber produced from electrospun polyacrylonitrile nanofiber as a highly efficient formaldehyde adsorbent. *Carbon N Y* 48: 4248–4255. https://doi.org/10.1016/J.CARBON.2010.07.034.

58 Choudhary, M., Kumar, R., and Neogi, S. (2020). Activated biochar derived from Opuntia ficus-indica for the efficient adsorption of malachite green dye, Cu^{+2} and Ni^{+2} from water. *J. Hazard Mater.* 392: 122441. https://doi.org/10.1016/J.JHAZMAT.2020.122441.

59 Yu, W., Lian, F., Cui, G., and Liu, Z. (2018). N-doping effectively enhances the adsorption capacity of biochar for heavy metal ions from aqueous solution. *Chemosphere* 193: 8–16. https://doi.org/10.1016/J.CHEMOSPHERE.2017.10.134.

60 Li, K., Zhou, M., Liang, L. et al. (2019). Ultrahigh-surface-area activated carbon aerogels derived from glucose for high-performance organic pollutants adsorption. *J. Colloid Interface Sci.* 546: 333–343. https://doi.org/10.1016/J.JCIS.2019.03.076.

61 Liu, G., Mei, H., Tan, X. et al. (2018). Enhancement of Rb^+ and Cs^+ removal in 3D carbon aerogel-supported $Na_2Ti_3O_7$. *J. Mol. Liq.* 262: 476–483. https://doi.org/10.1016/J.MOLLIQ.2018.04.117.

62 Li, P., Wang, C., Li, Z. et al. (2014). Hierarchical carbon-nanotube/quartz-fiber films with gradient nanostructures for high efficiency and long service life air filters. *RSC Adv.* 4: 54115–54121. https://doi.org/10.1039/C4RA08746A.

15

Metal/Metal Oxide Nanoparticles Modified Polymeric Composite Membranes for Water Treatment

Akshatha Chandrashekar[1], Madhushree Hegde[1], Jineesh A. Gopi[2], Niranjana Prabhu[1], Deepu A. Gopakumar[3], Jinu Jacob George[3], Midhun Dominic C.D.[4], and Jyotishkumar Parameswaranpillai[2]

[1]*M.S. Ramaiah University of Applied Sciences, Faculty of Mathematical and Physical Sciences, Department of Chemistry, New BEL Road, MSR Nagar, Bengaluru 560054, Karnataka, India*
[2]*Alliance University, Division of Chemistry, Department of Science, Anekal - Chandapura Road, Bengaluru 562106, Karnataka, India*
[3]*Cochin University of Science and Technology, Department of Polymer Science and Rubber Technology, South Kalamassery, Ernakulam, Cochin 682022, Kerala, India*
[4]*Sacred Heart College (Autonomous), Department of Chemistry, Pandit Karuppan Road, Thevara 682013, Kochi, Kerala, India*

15.1 Introduction

Over the decades, the growth in the human population led to a sudden increase in the consumption of water and other natural resources. Along with various human activities, climate change has also deteriorated the water quality to a greater extent [1]. The release of a huge amount of organic and inorganic chemicals (pollutants) in an uncontrollable manner has contaminated the water resources and affected the balance of the aquatic ecosystem, which has deteriorated human health. Polluted water is a present-day challenge for researchers and scientists, as it has become a threat to the health of living beings as well as the environment. Poor human awareness of the importance of the preservation of water resources is one of the main reasons behind water pollution. Approximately 60% of the world's population has serious health problems due to the contamination of water through wastewater discharge. It has also led to severe water scarcity issues. Therefore, it is crucial to remove toxic contaminants from the wastewater runoff before discharge into our environment or consumption. Various techniques are used in industries to remove harmful contaminants [2, 3] from wastewater, and they are listed in Figure 15.1.

Among these techniques, membrane filtration seems to be the most suitable method to solve the present water purification challenges due to the advantages of this technique, which are shown in Figure 15.2.

The impurities can be removed based on the pore size of the membranes, and the driving force behind membrane filtration is pressure. The filtration in pressure-driven membranes is classified into micro, ultra, and nanofiltration methods and

Nanomaterials for Air and Water Purification, First Edition.
Edited by Deepu A. Gopakumar, Jyotishkumar Parameswaranpillai, Jinu Jacob George, and Midhun Dominic C.D.

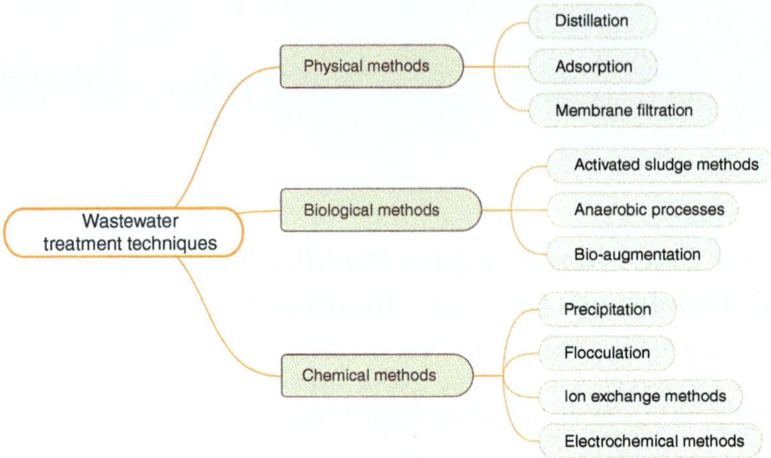

Figure 15.1 Different types of wastewater treatment techniques.

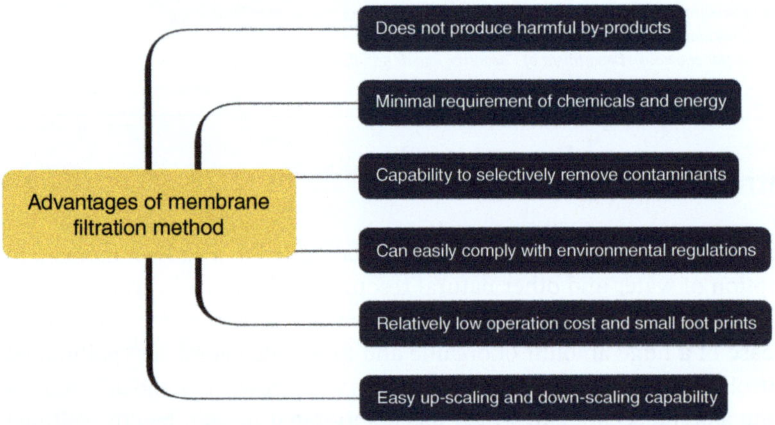

Figure 15.2 Advantages of membrane filtration techniques in wastewater treatment.

reverse and forward osmosis methods. Based on their composition, membranes for water treatment are categorized into polymeric membranes and inorganic membranes [4]. The advantages and disadvantages of polymeric membranes in water treatment processes [4] are listed in Figure 15.3.

Polymers that are commonly used for the preparation of filtration membranes are polysulfone (PSF), polyethersulfone (PES), polyacrylonitrile (PAN), and polyvinylidene fluoride (PVDF). The selection of these polymers is justified by their good chemical resistance, mechanical properties, and flexibility in processibility [5, 6]. The incorporation of certain nanomaterials can induce many functional properties, like antifouling properties, antimicrobial properties, photocatalytic activity, mechanical strength, chemical resistance, etc., in the polymeric membrane due to their unique inherent properties and large surface area. Among the nanomaterials, metal-based or metal oxide-based particles are widely explored

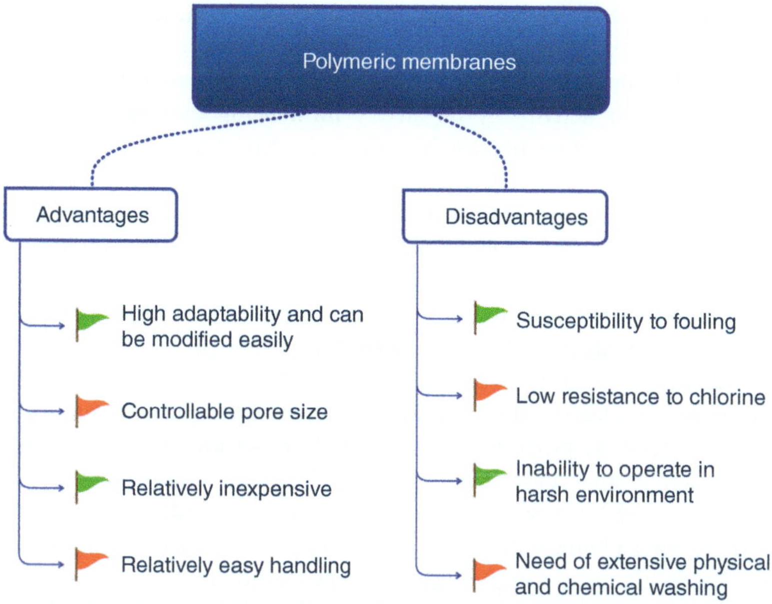

Figure 15.3 Advantages and disadvantages of polymeric membranes in water treatment processes.

in polymeric membranes due to their antimicrobial properties, their ability to reduce the absorption of foulants by the membrane as well as their ability to destruct the cell structure of microbes that causes fouling [7, 8]. This chapter gives an overview of the preparation and characteristics of metal and metal oxide incorporated polymeric membranes for water purification. The challenges related to polymeric–metal and polymeric–metal oxide membranes in water purification and the future requirement of research in this field are also discussed.

15.2 Metal and Metal Oxide Nanomaterials Incorporated Polymeric Membrane for Water Treatment

Nanomaterials have different physiochemical characteristics compared to their bulky form making them useful in various fields of science, particularly in water treatment. The modification of polymers with surface functionalized nanoparticles is found to be more efficient in the elimination of pollutants from water compared to the sole use of polymers or nanoparticles [9]. Traditionally, nanoparticles are introduced into its polymeric membrane to enhance the permeability, selectivity, mechanical strength, hydrophobicity, and antifouling properties. Nanoparticles can also induce functionalities in the polymeric membranes, such as adsorption of pollutants, photocatalytic capability to degrade the pollutants, and antimicrobial activity [10]. Among metals, silver has been widely studied for its usage in water treatment, such as the removal of heavy metals, halogenated organic compounds, and microorganisms even if the pollutants are in ultra-low concentrations [11]. For example, silver nanoparticles are used for the decomposition of both chemical and microbiological pollutants [12, 13]. Silver nanoparticles can interact with bacteria and reduce their activity by negatively influencing their permeability and respiration. This interaction breaks the cell of the bacteria and damages its DNA [4, 14].

Similarly, copper nanoparticles possess antibacterial and anti-fungicidal properties against different types of microorganisms [15]. This may be due to the generation of reactive oxygen species and superoxide anions, which lead to protein oxidation and DNA degradation in microorganisms [16]. Also, cupric ions released from Cu nanoparticles can react with phosphorus or sulfhydryl groups present in the DNA of microorganisms, which leads to DNA degradation [17].

Metal oxide nanoparticles can be multifunctional in water purification techniques due to their catalytic properties, antibacterial properties, ability to absorb hydrophilic hydroxyl groups, etc. There are several studies related to the use of titanium dioxide (TiO_2) nanoparticles in the polymeric membrane for water purification due to their photocatalytic properties and photo-induced hydrophilicity [18, 19]. The mechanism for photocatalytic properties and photo-induced hydrophilicity properties of TiO_2 are explained elsewhere [19–21]. ZnO is also a widely studied inorganic nanomaterial for water purification due to its outstanding antimicrobial and antifungal properties [22–24]. Aluminum oxide (Al_2O_3) nanoparticles can increase hydrophilicity and reduce biofouling in polymeric membranes [25, 26]. Oxides of iron such as Fe_2O_3 and Fe_3O_4 impart antibacterial and antifungal activities and hydrophilicity to the polymeric matrix. Also, oxides of iron can be used for the removal of dye present in the wastewater [27–29].

The interaction between polymer and nanoparticles may induce a positive influence in terms of antifouling properties, flux enhancement, retention of mechanical properties, and flux enhancement [30, 31]. To achieve greater performance, nanoparticles should be present in the pore structure and on the surface of the membrane. So, the conventional heterogeneous mixing of polymers and nanoparticles may not be suitable. Agglomeration of nanoparticles leads to impaired performance of the membrane with nonuniform selectivity and nonuniform antifouling properties across the membrane, therefore proper dispersion is required. Also, there should not be any detachment of nanoparticles from the membrane during the water treatment process, as the detachment of nanoparticles will contaminate the water and lead to serious concerns. Therefore, good interaction between nanoparticles and the polymeric membrane is essential to avoid the detachment of nanoparticles from the membrane [32].

15.2.1 Fabrication of Metal and Metal Oxide-Incorporated Polymeric Membrane

Different methods for the fabrication of polymeric membranes are discussed and compared by Liao et al. [33] and Bassyouni et al. [34]. Among these methods, phase inversion methods, layered interfacial polymerization, electrospinning, stretching, sintering, and etching methods are widely used for the assembly of polymeric membranes. The phase inversion method is one of the most widely explored methods to prepare metal or metal oxide nanoparticles incorporated into polymeric membranes. Among various types of phase inversion methods, nonsolvent-induced phase separation methods are mainly used to fabricate polymeric membranes with different morphologies. In phase separation methods, the affinity between solvent and non-solvent governs the creation of porous substructures.

Flat-sheet polymeric composite membranes were mostly prepared by *nonsolvent-induced phase separation method*. In this method, metallic nanoparticles were dispersed in the solvent first and then added to the polymer solution to form the cast solution. Later the cast solution was immersed in a nonsolvent to undergo coagulation or precipitation to produce a flat membrane [35, 36]. As the amount of nanoparticle increases, there is an enhancement in the hydrophilicity and hence selectivity of the membrane. But the addition of the nanoparticles should be optimized for the membrane, as the higher amount of nanoparticles may lead to problems such as the formation of macro-voids, poor mechanical properties, unflavored pore sizes, and pore distribution, agglomeration of nanoparticles in the membrane, and poor performance [34, 35, 37]. In the phase inversion process, nanoparticles are in the polymeric solution before casting, and this leads to the presence of nanoparticles not only on the surface of the membrane but also inside the polymeric matrix in the membrane.

Figure 15.4 shows the scanning electron micrographs (SEM) of polyether sulfone (PES) and nanosilver-incorporated HPEI/PES membranes, which were fabricated using the phase inversion method with different amounts of silver and energy-dispersive X-ray spectroscopy (EDS) spectrum of nanosilver incorporated HPEI/PES membranes. These SEM images indicate that the presence of silver nanoparticles and the amount of HPEI affect the morphology of the PES membrane. Scanning electron micrographs for the neat PES membrane showed a finger-like structure, while nanosilver-incorporated HPEI/PES membranes showed clear microvoids in their internal structure [38].

Recently, a significant number of works have been carried out on the selective modification of the surface of the polymeric membrane with nanoparticles. In these methods, nanoparticles are located on the polymeric membrane as the top layer, and this enhances the selectivity, performance, and span of the membrane. To enhance the interaction between the surface of the polymeric membrane and the nanoparticle, surface functionalization/modifications of nanoparticles are found to be highly efficient. Different types of methods to enhance the interaction between polymers and nanoparticles are compiled by García-Ivars et al. [32] and the major methods from this compilation are listed in Figure 15.5. Surface modification is beneficial to increase the dispersion of nanoparticles in the polymer matrix, which in turn, improves the mechanical properties of the membrane and its functional abilities.

Thin-film composite membranes can be prepared by the interfacial polymerization method [34, 39]. Morphology and the nature of the membranes can be altered through a selection of monomers, the nature of additives and their amounts, the type of solvent, and the nature of the substrate membrane [40]. To prepare

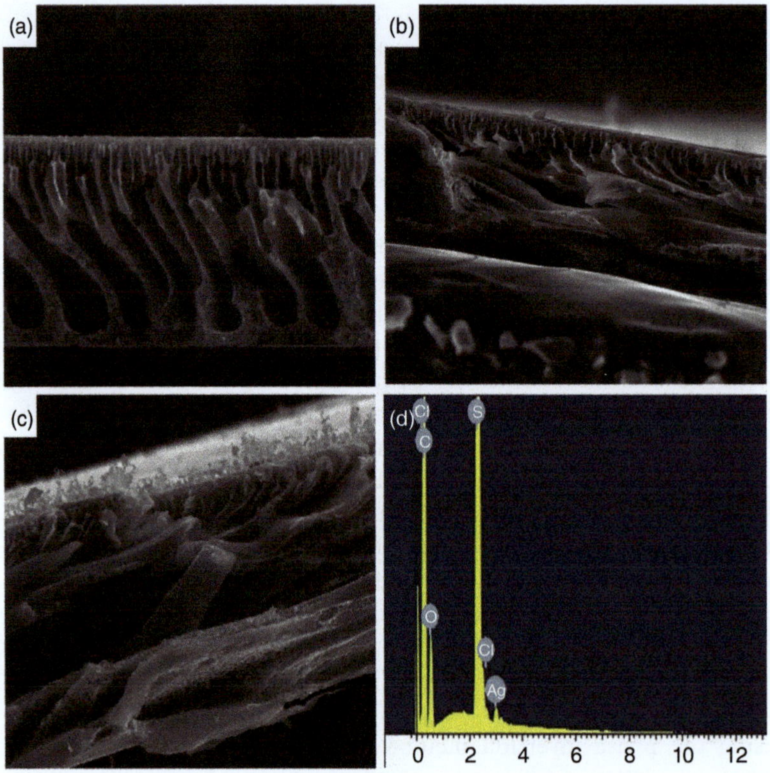

Figure 15.4 SEM images of PES (a), nAg@1% HPEI/PES, and nAg@3%HPEI/PES, and the EDS spectrum of nAg@3%HPEI/PES. Source: Okoro et al. [38]/with permission from Elsevier/CC BY-NC-ND 4.0 [under creative common license].

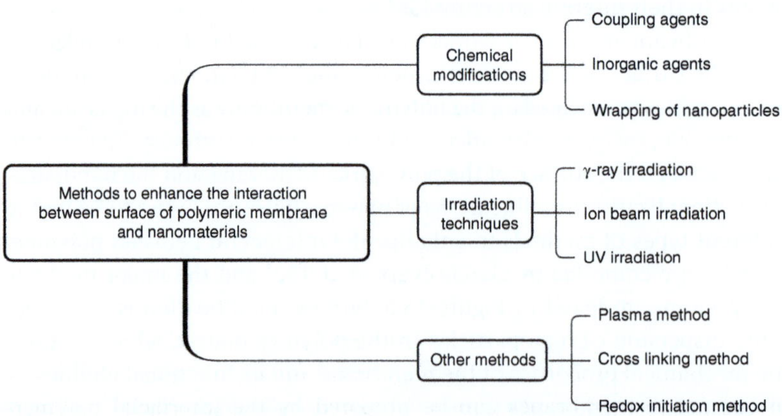

Figure 15.5 Methods to enhance the interaction between polymeric membranes and nanoparticles.

microporous membranes for microfiltration and ultrafiltration, the stretching method is widely used [34]. In this method, the polymer is melted and extruded to get a thin sheet, which, on strain, produces porosity. Cold stretching, as well as hot stretching, can be used to generate porosity. Polypropylene microporous membranes prepared by grafting TiO_2 nanoparticles were prepared by the stretching method to enhance the hydrophilicity of the polypropylene membrane [41]. Similar methods were adopted for making polymer membranes with metal or metal oxide nanoparticles embedded in the membrane [42–44]. Electrospinning can be used to prepare polymeric membranes for desalination and filtration processes. In this technique, a high voltage was applied between the grounded collector and the polymeric solution. When the electrode potential reaches above a particular value, it will overcome the droplet's surface tension and form a charged liquid jet and form fibers. By varying the applied electrode potential, flow rate, and viscosity of the polymer solution, a fibrous membrane with different morphologies can be obtained. Yin et al. [45] prepared palladium nanoparticles loaded MXene/polymer nanocomposites through the electrospinning method for wastewater treatment and the prepared membrane showed excellent catalytic properties. Similarly, Liu et al. [46] fabricated silver nanoparticle-containing poly(vinyl alcohol)/poly(acrylic acid) nanocomposites through electrospinning to use as a photocatalyst for dye degradation in wastewater treatment.

15.2.2 Antifouling Property of Metal and Metal Oxide-Incorporated Polymeric Membrane

During usage, contaminants such as colloids, particles, biomolecules, salts, etc., in the water will be collected and attached to the membrane, which leads to fouling. There are two types of fouling: external, and internal fouling. In external fouling, foulants will attach to the surface of the membrane and in internal fouling, foulants will be in the internal pores of the membrane. External fouling leads to a reduction in the water flux and delays the water treatment process. External fouling can be removed by cleaning if the foulants are weakly bound to the membrane. But cleaning will be difficult if the foulants are strongly attached to the membrane. Membranes for micro- and ultrafiltration generally face both external as well as internal fouling. But membranes for nanofiltration and reverse osmosis face external fouling due to their tight skin layers. Fouling will increase the operation time, cost, and energy required for the operation. Therefore, a membrane with antifouling properties is essential for the effective purification of water [47, 48]. One of the major reasons behind the fouling of the membrane is its hydrophobicity. The incorporation of nanoparticles is found to be one of the most efficient methods to increase the hydrophilicity of the membrane, which will improve the antifouling properties, photocatalytic, and photodegradation properties of the membrane. In other words, the increase in the nanoparticle concentration in the membrane imparts hydrophilicity so that the foulants will be repelled from the surface of the membrane and washed away during the filtration process. Jhaveri et al. [49] had written a comprehensive review on the improvement in antifouling properties of the membrane by the incorporation of different nanomaterials in the membrane. Table 15.1 lists a few selected

Table 15.1 Polymeric membrane with metal and metal oxide nanoparticle with antifouling properties.

Sl. No	Polymer	Nanoparticles(NPs)	Fabrication techniques	Antifouling characteristics	References
1	Polyvinylidene fluoride (PVDF)	TiO_2	Phase inversion method	Membrane with TiO_2 showed lower fouling in 1% bovine serum albumin (BSA) solution.	[19]
2	PES	Iron oxide (FeO)	Interfacial polymerization	The membrane showed antibiofouling properties in the studies using seawater and showed increased removal of hydrophobic contaminants.	[50]
3	Emulsion poly(vinyl chloride (EPVC)	TiO_2	Phase inversion method	Nano-TiO_2 embedded membranes exhibited higher fouling resistance compared to unfilled EPVC in BSA solution.	[51]
4	PVC	SiO_2	Nonsolvent-induced phase separation	Incorporation of SiO_2 NPs induced hydrophilicity, which leads to the repulsion of hydrophobic foulant protein and showed higher flux recovery.	[52]
5	PVDF	Silver NPs	Phase inversion method followed by coating silver ion solution and subsequent reduction	Incorporation of silver NPs enhanced antiorganic fouling performance of PVDF membrane.	[53]
6	PES	Silver NPs	Phase inversion method	1.5% AgNPs reduced the fouling nature of the membrane by 66%.	[54]
7	Polyethersulfone (PES)	Sulfated TiO_2 NPs	Membrane was fabricated using nonsolvent- induced phase inversion method by immersing precipitation.	Antifouling properties of membrane were studied using BSA solution and found that membranes with 2 wt% sulphated TiO_2 showed the lowest fouling rate due to its less surface roughness in this membrane.	[55]
8	PES	Titania NPs	Nonsolvent-induced phase separation	The incorporation of 0.5 wt% titania nanoparticles improved fouling resistance in the BSA solution.	[56]
9	PES	Cobalt NPs	Nonsolvent- induced phase inversion process (NIPS)	The PES-cobalt nanocomposite membranes showed good antibiofouling characteristics due to the enhanced hydrophilicity of the membrane nanocomposite	[57]
10	PES	Cobalt ferrite-copper oxide NPs	Mixed matrix membrane using a blending technique	The nanocomposite membrane showed higher surface hydrophilicity and antifouling properties compared to the PES membrane.	[58]

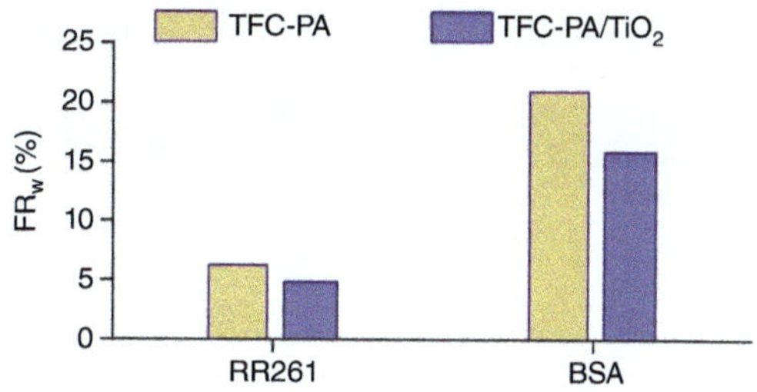

Figure 15.6 Irreversible fouling factors of uncoated and coated membranes. Source: Ngo et al. [59]/Elsevier/CC BY 4.0 [under creative commons license].

literatures related to antifouling properties of metal and metal oxide incorporated polymeric membranes.

Ngo et al. [59] studied the effect of UV-irradiated TiO_2 nanoparticles on the antifouling properties of polyamide thin film membranes. The irreversible fouling factors of uncoated and TiO_2 coated membranes are given in Figure 15.6. The experiment was conducted with high-fouling tendency compounds such as Reactive red dye (RR261) or proteins (BSA) as feed solutions. The results showed that the antifouling property of the thin film PA membrane (TFC-PA) was enhanced after coating with TiO_2 nanoparticles, and this is mainly due to the improved hydrophilicity of the membrane.

15.2.3 Antibacterial Properties of Metal and Metal Oxide-Incorporated Polymeric Membrane

Metal and metal oxide nanoparticles are known for their enhanced antimicrobial properties compared to conventional materials and they can be embedded into polymer matrices, which induce biocidal properties in the composites. One of the major reasons behind the antimicrobial properties of the metal or metal oxide nanoparticle-incorporated polymeric membrane is the presence of metal atoms as active agents or the metal ions released from the composites [60]. There are three stages in the formation of biofilm: (i) planktonic and initial interaction stages, (ii) development and maturation stages, and (iii) bacteria detachment stages. With the help of biocides, the growth of biofilm can be prevented. Metals or metal oxides are known for their antibacterial properties and are widely studied for their biocidal activities. The addition of metal or metal oxide nanoparticles disturbs the first stage in the biofilm formation, which is the initial interaction stage.

Copper, silver, titanium oxide, silver oxide (Ag_2O), silica oxide (SiO_2), cupric oxide (CuO), etc., show antibacterial characteristics [61]. The mechanism of antimicrobial activities of different nanomaterials are compiled by Dizaj et al. [62] and different kinds of mechanisms for the antimicrobial characteristics of metal and metal oxide nanoparticles are compiled by Jagadeeshan and Parsanathan [63]. Figure 15.7 depicts the various mechanisms of nanoparticles during the antibacterial activity. One of the major bactericidal mechanisms of metal and metal oxide nanocomposites is the generation of reactive oxygen species, which destroy the cell membrane of the bacteria, and prevents the further growth of the cells. The collapse of the cell membrane leads to the leakage of the internal cellular components [64].

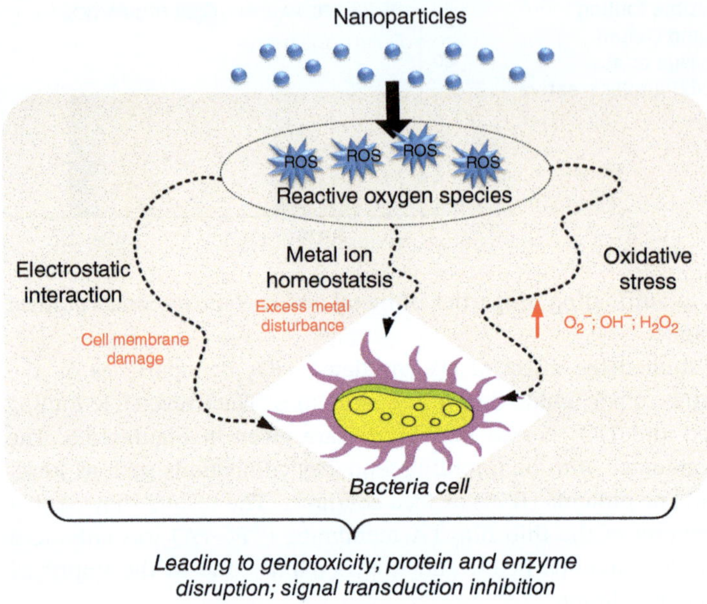

Figure 15.7 Mechanisms of antibacterial activity of the nanoparticles (inspired from [62] and [63]).

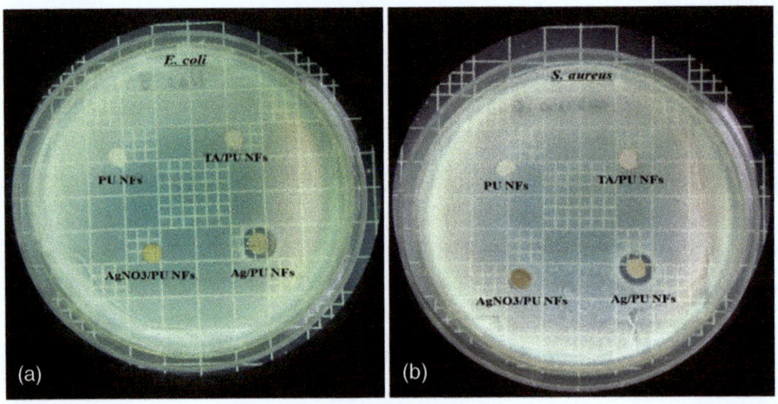

Figure 15.8 Antibacterial studies on different samples against *E.* coli (A) and *S. aureus* (B) using zone inhibition tests. Source: Pant et al. [65]/MDPI/CC BY 4.0 [Under a creative common license].

Pant et al. [65] studied the antibacterial property of silver nanoparticles incorporated polyurethane nanofiber (PU NF). Antibacterial properties of different samples such as pristine PU NFs, PU nanofibers with Tannic acid (TA), a PU NF mat with $AgNO_3$, and PU NF mats with Ag nanoparticles were studied using zone inhibition test against *Escherichia coli* (A) and *Staphylococcus aureus* (B) and shown in Figure 15.8. In this study, Ag/PU NFs showed excellent antibacterial properties against both bacteria with larger inhibition zone compared to other samples.

Table 15.2 lists a few recent studies on the antibacterial activity of polymeric membranes incorporated with metal or metal oxide nanoparticles.

15.2.4 Photocatalytic Properties of Metal and Metal Oxide-Incorporated Polymeric Membrane

Metal oxide nanoparticles such as ZnO, TiO_2, and iron oxide possess photocatalytic properties that can be used for the degradation of pollutants in wastewater [27]. The photocatalytic membrane can be prepared by incorporating these nanoparticles into a polymer matrix, which leads to synergistic characteristics such as the photodegradation of pollutants as well as the separation of pollutants in water. The performance of the photocatalytic membrane depends on the polymer matrix, nature of nanoparticles, nanoparticle content, dispersion of nanoparticles, and the morphology of the membrane. This polymeric photocatalytic membrane has the potential to be developed as a stand-alone filtration unit with upscaling capability and photocatalytic activity [75]. Transition metal oxides with controlled crystalline and structural features possess photocatalytic activities for the degradation of organic pollutants and possess good stability in water. The main mechanism of photocatalytic degradation of organic contaminants is due to the oxygen on the surface of the metal oxide. The absorption of a photon by the metal oxide activates the oxygen, which leads to the degradation of organic pollutants in the water into less toxic products. Doping with other atoms enhances photodegradation by altering the band gap of metal oxide semiconductors [76]. Romay et al. [77] discussed and stated that parameters such as the selection of the membrane materials, method of synthesis, targeted functionality, factors affecting stability, and photocatalytic property, influence the performance of photocatalytic polymeric membranes.

Various studies on the use of metal oxide-incorporated polymeric membrane as the photocatalytic agent for the degradation of pollutants are listed in Table 15.3.

15.2.5 Adsorption and Chemical Stability of Metal and Metal Oxide-Incorporated Polymeric Membrane

The adsorption method has gained significant interest in the water treatment process due to its low cost, ease of the process, and high efficiency of pollutant removal. The adsorption process can be either physisorption or chemisorption. Nanomaterials are widely explored for the development of polymer composites with better removal efficiency for water pollutants. It was found that the polymer matrix acts as an ideal platform and helps the adsorption of targeted water pollutants onto its surface with the help of nanoparticles embedded within it [34, 88, 89]. The targeted pollutants are adsorbed onto the surface of the polymeric nanocomposite membrane through surface adsorption, ion exchange processes, complexation, diffusion, etc. But their recyclability, reusability, and recovery are still challenging as nanoparticles may leach out after each water treatment cycle.

Table 15.4 lists a few studies related to the adsorption of pollutants in water with the help of metal and metal oxide incorporated polymeric membranes.

Table 15.2 Recent literature on the antibacterial activity of polymeric membranes incorporated with metal or metal oxide nanoparticles.

SL. No	Polymer	NP	Fabrication techniques	Antibacterial properties	Reference
1	PES	Cu	Membranes were fabricated using phase inversion method, in which copper NPs were dispersed in PES casting solution	The growth of bacterial colonies (*E. coli*) reduced with an increase in copper nanoparticles and the CuNPs/PES blended membranes showed enhanced antibacterial properties.	[66]
2	PES	Silver	Dip-coating method: Membrane was dipped in silver nitrate solution followed by reduction.	Halo test and shake flask tests using *E. coli NovaBlue* bacteria showed antibacterial property of the membrane with Ag NPs.	[67]
3	PSF and cellulose acetate (CA)	ZnO	Electrospinning technique	ZnO incorporated nanofibrous membrane showed excellent antibacterial activity against *E. coli*. The membrane with ZnO nanoparticles showed a bacterial growth inhibition zone diameter of 10 ± 0.6 mm.	[68]
4	PES	Silver (Ag)	Phase inversion technique with silver immobilization.	The immobilization of silver NPs on the PES membrane increased the antibacterial property against *E. coli* with complete elimination of the bacteria.	[69]
5	Poly(vinyl alcohol) (PVA) and konjac glucomannan (KGM)-based nanofiber membranes	ZnO	Electrospinning and thermal cross-linking.	Incorporation of ZnO in PVA/KGM displayed better antibacterial properties against *E. coli* and *Bacillus* subtilis. 1.0 wt% and 2.0 wt% ZnO incorporated PVA/KGM membrane showed excellent antibacterial activity for both *E. coli* and *B*. subtilis.	[70]
6	PVDF	Cu	Electroless copper deposition on PDA-coated PVDF	The Cu nanoparticle deposited PVDF membrane showed high antibacterial activity against *E. coli*.	[71]
7	Polyacrylonitrile (PAN)	Cu and Zn	NPs were added to the PAN membrane by the spark ablation method.	The incorporation of Zn NPs exhibited good antibacterial property against *S. aureus* and limited antibacterial activity against *E. coli*. The incorporation of copper nanoparticles exhibited showed limited antibacterial potential against both *S. aureus* and *E. coli* bacteria.	[72]
8	PAN	Ag, CuO, and ZnO	Electrospinning technique	Their antibacterial activity using disk diffusion tests and growth kinetic models was studied against *E. coli* and *S. aureus*. Studies showed that the addition of nanoparticles led to the inactivation of bacteria.	[73]
9	Polyether Sulfone (PES) and cellulose acetate	Copper oxide (CuO)	The solution casting method was used to prepare copper oxide from mixed matrices of PES and cellulose acetate.	The incorporation of copper oxide NPs showed more than 75% inhibition of the model gram-negative bacteria, *E. coli*.	[74]

Table 15.3 Studies on the photocatalytic activities of the metal/metal oxide incorporated polymeric membrane for the degradation of various pollutants in water.

Sl No.	Polymer	NP	Preparation of membrane	Photocatalytic activity	Reference
1	PVDF	TiO_2	Phase inversion method	PVDF/TiO_2 membrane showed photocatalytic nature toward the degradation of the reactive black (RB5) dye.	[19]
2	PES	TiO_2@Ni	Phase inversion method with magnetic field assistance	In the presence of UV light, the membrane composites with TiO_2 attached to Ni showed photocatalytic activity towards the degradation of Methylene blue (MB).	[78]
3	PVA/poly(acrylic acid) (PAA)	Silver nanoparticles modified by graphene oxide	Electrospinning technique	The prepared silver NP incorporated into PVA/PAA membrane showed catalytic degradation activity toward MB, and the activity was retained even after eight cycles of degradation.	[46]
4	Polyvinyl acetate (PVAc)	TiO_2	Air jet spinning technique	The photodegradation studies of TiO_2 membrane exhibited good efficiency in the degradation of methylene blue dye	[79]
5	Polyethylene terephthalate (PET)	CuO	CuO NPs incorporated PET nanofibers were fabricated by electrospinning followed by, the dipping method, and reduction.	PET nanofibers/CuO NPs exhibited excellent photocatalytic activity for the degradation of MB dye from water.	[80]
6	Cellulose acetate	ZnO	Cellulose acetate polymeric membranes were prepared by the phase inversion method.	Cellulose acetate impregnated with ZnO NPs had shown faster degradation of MB dye under sunlight compared to degradation under UV light.	[81]
7	Cotton linter cellulose membranes	ZnO	Dissolution-regeneration process.	In the photocatalytic studies, 1.6 wt% ZnO incorporated into cellulose membrane removed 87.6% of MB dye in 100 min under UV light. The membrane was reusable even after six cycles of usage.	[82]

(*Continued*)

Table 15.3 (Continued)

Sl No.	Polymer	NP	Preparation of membrane	Photocatalytic activity	Reference
8	PES	Sulfonated graphene oxide /ZnO	Phase inversion method	PES-sulphonated graphene oxide-ZnO membranes exhibited higher photocatalytic efficiency for crystal violet and ciprofloxacin.	[83]
9	PVDF	TiO_2 and graphene oxide (GO)	PVDF membrane was coated with TiO_2 and GO after activation of the membrane using argon plasma and after grafting with acrylic acid.	GO enhanced the photocatalytic efficiency of TiO_2 for the degradation of phenol/m-nitrophenol.	[84]
10	PAN-ethanolamine (PAN-ETA)	Ag-AgBr NPs	PAN-ETA membrane was coated using Ag-AgBr nanoparticles with the help of chitosan (CS)-TiO_2 layer using vacuum filtration and in situ partial reduction.	The membrane exhibited photocatalytic activity for the degradation of MB and rhodamine B solutions up to 90.75% and 96.81%, respectively, under visible light irradiation for 30 min.	[85]
11	PVDF	SnO_2-GO hybrid NPs	Delay phase conversion method	SnO_2-GO/PVDF membranes exhibited excellent photodegradation of bovine serum albumin (BSA) compared to neat PVDF membranes. In addition to that, the incorporation of NPs enhanced hydrophilicity.	[86]
12	Polyamide-6 (PA-6)	TiO_2	Electrospinning	The nanofiber membrane composite showed excellent photocatalytic activity for the degradation of MB dye. A degradation of up to 84% and 99% of MB after 2 and 6 h of UV irradiation, respectively.	[87]

Table 15.4 Adsorption studies using metal and metal oxide-incorporated polymeric membranes.

Sl. No	Polymer	NP	Fabrication method	Adsorption behavior	References
1	PES	PANI/Fe_3O_4	Phase inversion method	0.1 wt% modified Fe_3O_4 into PES membrane showed higher absorption toward Cu(II) ions present in water. The membrane nanocomposite has reusability after desorption using EDTA as a regenerator.	[90]
2	PES	Metformine-modified silica-coated Fe_3O_4	Phase inversion method and	0.1 wt % metformin-modified silica-coated Fe_3O_4 NPs displayed excellent copper removal property (about 92%). The membrane showed acceptable reusability for several cycles and desorption was possible with the help of EDTA.	[91]
3	Cellulose acetate	ZnO	Impregnation method	ZnO NPs could enhance the adsorption characteristics of cellulose acetate membrane toward the adsorption of metolachlor and acetochlor. This membrane could adsorb different herbicides present in water.	[92]
4	PES	γ-alumina	Phase inversion method	Alumina-incorporated PES membranes showed copper ion removal efficiency. The prepared membrane showed sufficient adsorption efficiency even after four cycles of filtration.	[93]
5	PVA	SiO_2	Wet-phase inversion method.	With the help of an electrodialysis system, the adsorption efficiency of SiO_2-incorporated PVA membrane towards Zinc ions has been studied and the results showed that incorporation of TiO_2 enhanced the adsorption efficiency of the PVA membrane.	[94]
6	PSF	Nano-TiO_2 and graphene oxide (GO)	Phase inversion method	PSF membrane with TiO_2/GO fillers showed removal of ammonia and organic content in water by 9.27% and 60.98%.	[95]
7	PVDF	Chitosan-coated iron oxide nanomaterials	The glass vacuum filter assembly technique was used to make the PVDF membrane incorporated with chitosan-coated iron oxide	PVDF membranes incorporated with chitosan-coated iron oxide displayed a removal capacity of 74.6 $mg\,g^{-1}$ for Evans blue (EB). Under acidic conditions, the adsorption capacity got enhanced to about 434.78 $mg\,g^{-1}$.	[96]
8	PVA/PVDF	Al_2O_3, SiO_2, and CuO	Phase inversion method	The nanocomposite membranes exhibited excellent removal efficiency towards Malachite green (MG), Methylene blue (MB), Congo red (CR), and Safranin O (SO) from an aqueous solution.	[97]

Compared to polymeric membranes, inorganic membranes possess more resistance to strong acids, alkalis, and oxidants such as chlorine. Chlorine is a widely used disinfectant in water, and it can react with electron-rich functional groups present in the polymeric membranes. This makes polymeric membrane vulnerable to use as membranes in the water treatment process. So, it is important to increase the chemical stability of polymeric membranes [4]. Nanoparticles can play huge roles in improving the chemical stability of the polymeric membrane. Metal and metal oxide nanoparticles can also enhance the mechanical properties of the polymeric membrane [98]. Fouling on the membrane is the major reason behind the flux decline in the wastewater treatment process. The incorporation of metal and metal oxide nanoparticles can also solve this flux decline issue to a greater extent [99].

15.3 Prospects of Metal and Metal Oxide-Incorporated Polymeric Membranes

The selection of suitable nanomaterials is based on their performance, availability, practicality, stability, agglomeration, and cost. In the last two decades, significant progress has been made in the field of polymer membranes and their modification using nanoparticles. But still, more research is required to solve the limitations of these polymeric membranes and understand the nanoscale phenomena that determine the efficiency of the polymer nanocomposite membrane. The effect of two or more nanoparticles or multielement nanostructures in the polymer membrane nanocomposite in terms of the stability of the membrane as well as the synergistic effects in terms of the performance properties of the membrane need to be explored in detail. Proper dispersion of nanomaterials, as well as the exposure of nanomaterials to water, is still a challenge, and modifications in fabrication techniques must be explored to achieve these characteristics. Chemical resistance is an important characteristic of a polymeric membrane, especially for long-term usage. Yet, extensive research is required to explore the usage of metal and metal oxide nanoparticles for the enhancement of the chemical stability of the membrane such as its resistance to chlorine. Large-scale manufacturing of polymeric membranes is an area that is being developed for industrial applications. Also, the release of nanoparticles from the polymeric membrane during long usage needs to be investigated in detail as this impacts health and the environment. Future research should focus on the development of highly stable, effectively functional, and scalable nanocomposite membranes for various applications in the field of water treatment.

15.4 Conclusions

Various types of polymers, such as PES, PA, PI, PVDF, and cellulose acetate, are used for producing polymeric membranes in water treatment. To improve the efficiency of the polymeric membrane, the incorporation of nanoparticles such as metal nanoparticles and metal oxide nanoparticles has been explored in the last two decades. Properties of the polymeric membrane, such as hydrophilicity, antifouling properties, antimicrobial properties, photocatalytic activity, self-cleaning ability, and adsorption capacity, can be enhanced by the addition of metal and metal oxide nanoparticles. Along with these properties, other characteristics of the polymer membrane, such as mechanical strength, chemical resistance, and thermal resistance, for the membrane could also be enhanced by the incorporation of these nanomaterials. Metallic nanoparticles such as Ag and Cu, and metallic oxide nanoparticles ZnO, Fe_3O_4, TiO_2, and ZrO_2 are used to increase these properties of polymeric membranes. The properties of polymeric membrane nanocomposite are based on the characteristics of polymers and nanomaterials, the dispersion of nanomaterials, amount of nanomaterials, the morphology of nanomaterials and polymeric membrane, and the interaction between polymers and nanomaterials.

References

1 Montanari, A. and Koutsoyiannis, D. (2012). A blueprint for process-based modeling of uncertain hydrological systems. *Water Resour. Res.* 48: W09555. https://doi.org/10.1029/2011WR011412.

2 Singh, N.B. and Susan, A.B.H. (2018). Polymer nanocomposites for water treatments. In: *Polymer-based Nanocomposites for Energy and Environmental Applications* (ed. M. Jawaid and M. Mansoob), 569–595. Woodhead Publishing https://doi.org/10.1016/B978-0-08-102262-7.00021-0.

3 Rajasulochana, P. and Preethy, V. (2016). Comparison on efficiency of various techniques in treatment of waste and sewage water – a comprehensive review. *Resour. Technol.* 2: 175–184. https://doi.org/10.1016/J.REFFIT.2016.09.004.

4 Wen, Y., Yuan, J., Ma, X. et al. (2019). Polymeric nanocomposite membranes for water treatment: a review. *Environ. Chem. Lett.* 17: 1539–1551. https://doi.org/10.1007/s10311-019-00895-9.

5 Xu, X., Yang, Y., Liu, T., and Chu, B. (2022). Cost-effective polymer-based membranes for drinking water purification. *Giant* 10: 100099. https://doi.org/10.1016/J.GIANT.2022.100099.

6 Werber, J.R., Osuji, C.O., and Elimelech, M. (2016). Materials for next-generation desalination and water purification membranes. *Nat. Rev. Mater.* 1: 1–15. https://doi.org/10.1038/natrevmats.2016.18.

7 Zhang, R., Liu, Y., He, M. et al. (2016). Antifouling membranes for sustainable water purification: strategies and mechanisms. *Chem. Soc. Rev.* 45: 5888–5924. https://doi.org/10.1039/C5CS00579E.

8 María Arsuaga, J., Sotto, A., del Rosario, G. et al. (2013). Influence of the type, size, and distribution of metal oxide particles on the properties of nanocomposite ultrafiltration membranes. *J. Membr. Sci.* 428: 131–141. https://doi.org/10.1016/J.MEMSCI.2012.11.008.

9 Makvandi, P., Iftekhar, S., Pizzetti, F. et al. (2021). Functionalization of polymers and nanomaterials for water treatment, food packaging, textile and biomedical

applications: a review. *Environ. Chem. Lett.* 19: 583–611. https://doi.org/10.1007/s10311-020-01089-4.

10 Yin, J. and Deng, B. (2015). Polymer-matrix nanocomposite membranes for water treatment. *J. Membr. Sci.* 479: 256–275. https://doi.org/10.1016/J.MEMSCI.2014.11.019.

11 Pradeep, T. and Anshup (2009). Noble metal nanoparticles for water purification: a critical review. *Thin Solid Films* 517: 6441–6478. https://doi.org/10.1016/J.TSF.2009.03.195.

12 Kowalska, E., Endo, M., Wei, Z. et al. (2019). Noble metal nanoparticles for water purification. In: *Nanoscale Materials in Water Purification* (ed. S. Thomas, D. Pasquini, S.-Y. Leu, and D.A. Gopakumar), 553–579. Elsevier https://doi.org/10.1016/B978-0-12-813926-4.00027-6.

13 Sana, S.S., Haldhar, R., Parameswaranpillai, J. et al. (2022). Silver nanoparticles-based composite for dye removal: a comprehensive review. *Cleaner Mater.* 100161. https://doi.org/10.1016/j.clema.2022.100161.

14 Ursino, C., Castro-Muñoz, R., Drioli, E. et al. (2018). Progress of nanocomposite membranes for water treatment. *Membranes* 8: 18. https://doi.org/10.3390/MEMBRANES8020018.

15 Muthulakshmi, L., Varada Rajalu, A., Kaliaraj, G.S. et al. (2019). Preparation of cellulose/copper nanoparticles bionanocomposite films using a bioflocculant polymer as reducing agent for antibacterial and anticorrosion applications. *Compos. Part B Eng.* 175: 107177. https://doi.org/10.1016/J.COMPOSITESB.2019.107177.

16 Tamayo, L., Azócar, M., Kogan, M. et al. (2016). Copper-polymer nanocomposites: an excellent and cost-effective biocide for use on antibacterial surfaces. *Mater. Sci. Eng. C* 69: 1391–1409. https://doi.org/10.1016/J.MSEC.2016.08.041.

17 Ruparclia, J.P., Chatterjee, A.K., Duttagupta, S.P., and Mukherji, S. (2008). Strain specificity in antimicrobial activity of silver and copper nanoparticles. *Acta Biomater.* 4: 707–716. https://doi.org/10.1016/J.ACTBIO.2007.11.006.

18 Bouziane Errahmani, K., Benhabiles, O., Bellebia, S. et al. (2021). Photocatalytic nanocomposite polymer-TiO_2 membranes for pollutant removal from wastewater. *Catalysts* 11 (3): 402. https://doi.org/10.3390/catal11030402.

19 Tetteh, E.K., Rathilal, S., Asante-Sackey, D., and Chollom, M.N. (2021). Prospects of synthesized magnetic TiO_2-based membranes for wastewater treatment: a review. *Dent. Mater.* 14 (13): 3524. https://doi.org/10.3390/ma14133524.

20 Rahimpour, A., Madaeni, S.S., Taheri, A.H., and Mansourpanah, Y. (2008). Coupling TiO_2 nanoparticles with UV irradiation for modification of polyethersulfone ultrafiltration membranes. *J. Membr. Sci.* 313: 158–169. https://doi.org/10.1016/J.MEMSCI.2007.12.075.

21 Miyauchi, M., Nakajima, A., Watanabe, T., and Hashimoto, K. (2002). Photocatalysis and photoinduced hydrophilicity of various metal oxide thin films. *Chem. Mater.* 14: 2812–2816. https://doi.org/10.1021/CM020076P.

22 Madaeni, S.S., Ghaemi, N., and Rajabi, H. (2015). Advances in polymeric membranes for water treatment. In: *Advances in Membrane Technologies for Water*

Treatment (ed. A. Basile, A. Cassano, and N.K. Rastogi), 3–41. Wood Head publishing https://doi.org/10.1016/B978-1-78242-121-4.00001-0.

23 Leo, C.P., Cathie Lee, W.P., Ahmad, A.L., and Mohammad, A.W. (2012). Polysulfone membranes blended with ZnO nanoparticles for reducing fouling by oleic acid. *Sep. Purif. Technol.* 89: 51–56. https://doi.org/10.1016/J.SEPPUR.2012.01.002.

24 Shen, L., Bian, X., Lu, X. et al. (2012). Preparation and characterization of ZnO/polyethersulfone (PES) hybrid membranes. *Desalination* 293: 21–29. https://doi.org/10.1016/J.DESAL.2012.02.019.

25 Saleh, T.A. and Gupta, V.K. (2012). Synthesis and characterization of alumina nano-particles polyamide membrane with enhanced flux rejection performance. *Sep. Purif. Technol.* 89: 245–251. https://doi.org/10.1016/J.SEPPUR.2012.01.039.

26 Homayoonfal, M., Mehrnia, M.R., Rahmani, S., and Mohades Mojtahedi, Y. (2015). Fabrication of alumina/polysulfone nanocomposite membranes with biofouling mitigation approach in membrane bioreactors. *J. Ind. Eng. Chem.* 22: 357–367. https://doi.org/10.1016/J.JIEC.2014.07.031.

27 Naseem, T. and Durrani, T. (2021). The role of some important metal oxide nanoparticles for wastewater and antibacterial applications: a review. *Environ. Chem. Ecotoxicol.* 3: 59–75. https://doi.org/10.1016/j.enceco.2020.12.001.

28 Leonel, A.G., Mansur, A.A., and Mansur, H.S. (2021). Advanced functional nanostructures based on magnetic iron oxide nanomaterials for water remediation: a review. *Water Res.* 190: 116693. https://doi.org/10.1016/j.watres.2020.116693.

29 Parveen, S., Wani, A.H., Shah, M.A. et al. (2018). Preparation, characterization and antifungal activity of iron oxide nanoparticles. *Microb. Pathog.* 115: 287–292. https://doi.org/10.1016/j.micpath.2017.12.068.

30 Van Der Bruggen, B. (2009). Chemical modification of polyethersulfone nanofiltration membranes: a review. *J. Appl. Polym. Sci.* 114: 630–642. https://doi.org/10.1002/APP.30578.

31 Venna, S.R., Spore, A., Tian, Z. et al. (2017). Polyphosphazene polymer development for mixed matrix membranes using SIFSIX-Cu-2i as performance enhancement filler particles. *J. Membr. Sci.* 535: 103–112. https://doi.org/10.1016/J.MEMSCI.2017.04.033.

32 García-Ivars, J., Corbatón-Báguena, M.J., and Iborra-Clar, M.I. (2019). Development of mixed matrix membranes: incorporation of metal nanoparticles in polymeric membranes. In: *Nanoscale Materials in Water Purification* (ed. S. Thomas, D. Pasquini, S.-Y. Leu, and D.A. Gopakumar), 153–178. Elsevier https://doi.org/10.1016/B978-0-12-813926-4.00011-2.

33 Liao, Y., Loh, C.H., Tian, M. et al. (2018). Progress in electrospun polymeric nanofibrous membranes for water treatment: fabrication, modification and applications. *Prog. Polym. Sci.* 77: 69–94. https://doi.org/10.1016/J.PROGPOLYMSCI.2017.10.003.

34 Bassyouni, M., Abdel-Aziz, M.H., Zoromba, M.S. et al. (2019). A review of polymeric nanocomposite membranes for water purification. *J. Ind. Eng. Chem.* 73: 19–46. https://doi.org/10.1016/J.JIEC.2019.01.045.

35 Garcia-Ivars, J., Iborra-Clar, M.I., Alcaina-Miranda, M.I., and Van der Bruggen, B. (2015). Comparison between hydrophilic and hydrophobic metal nanoparticles on the phase separation phenomena during formation of asymmetric polyethersulphone membranes. *J. Membr. Sci.* 493: 709–722. https://doi.org/10.1016/J.MEMSCI.2015.07.009.

36 Dong, X., Lu, D., Harris, T.A., and Escobar, I.C. (2021). Polymers and solvents used in membrane fabrication: a review focusing on sustainable membrane development. *Membranes* 11 (5): 309.

37 Li, J.F., Xu, Z.L., Yang, H. et al. (2009). Effect of TiO_2 nanoparticles on the surface morphology and performance of microporous PES membrane. *Appl. Surf. Sci.* 255: 4725–4732. https://doi.org/10.1016/J.APSUSC.2008.07.139.

38 Okoro, H.K., Ndlwana, L., Ikhile, M.I. et al. (2021). Hyperbranched polyethylenimine-modified polyethersulfone (HPEI/PES) and nAg@HPEI/PES membranes with enhanced ultrafiltration, antibacterial, and antifouling properties. *Heliyon* 7: e07961. https://doi.org/10.1016/J.HELIYON.2021.E07961.

39 Purkait, M.K., Sinha, M.K., Mondal, P., and Singh, R. (2018). Introduction to membranes. *Interface. Sci. Technol.* 25: 1–37. https://doi.org/10.1016/B978-0-12-813961-5.00001-2.

40 Gohil, J.M. and Choudhury, R.R. (2019). Introduction to nanostructured and nano-enhanced polymeric membranes: preparation, function, and application for water purification. In: *Nanoscale Materials in Water Purification* (ed. S. Thomas, D. Pasquini, S.-Y. Leu, and D.A. Gopakumar), 25–57. Elsevier https://doi.org/10.1016/B978-0-12-813926-4.00038-0.

41 Saffar, A., Carreau, P.J., Kamal, M.R., and Ajji, A. (2014). Hydrophilic modification of polypropylene microporous membranes by grafting TiO_2 nanoparticles with acrylic acid groups on the surface. *Polymer* 55: 6069–6075. https://doi.org/10.1016/J.POLYMER.2014.09.069.

42 Rajlaxmi, Gupta, N., Behere, R.P. et al. (2021). Polymer nanocomposite membranes and their application for flow catalysis and photocatalytic degradation of organic pollutants. *Mater. Today Chem.* 22: 100600. https://doi.org/10.1016/j.mtchem.2021.100600.

43 Asad, A., Sameoto, D., and Sadrzadeh, M. (2020). Overview of membrane technology. In: *Nanocomposite Membranes for Water and Gas Separation* (ed. M. Sadrzadeh and T. Mohammadi), 1–28. Elseveir https://doi.org/10.1016/B978-0-12-816710-6.00001-8.

44 Jain, H. and Garg, M.C. (2021). Fabrication of polymeric nanocomposite forward osmosis membranes for water desalination—a review. *Environ. Technol. Innov.* 23: 101561. https://doi.org/10.1016/J.ETI.2021.101561.

45 Yin, J., Zhan, F., Jiao, T. et al. (2020). Highly efficient catalytic performances of nitro compounds via hierarchical PdNPs-loaded MXene/polymer nanocomposites synthesized through electrospinning strategy for wastewater treatment. *Chin. Chem. Lett.* 31: 992–995. https://doi.org/10.1016/J.CCLET.2019.08.047.

46 Liu, Y., Hou, C., Jiao, T. et al. (2018). Self-assembled AgNP-containing nanocomposites constructed by electrospinning as efficient dye photocatalyst materials for wastewater treatment. *Nano* 8: 35. https://doi.org/10.3390/NANO8010035.

47 Shahkaramipour, N., Tran, T.N., Ramanan, S., and Lin, H. (2017). Membranes with surface-enhanced antifouling properties for water purification. *Membranes* 7: 13. https://doi.org/10.3390/MEMBRANES7010013.

48 Geise, G.M., Lee, H.S., Miller, D.J. et al. (2010). Water purification by membranes: the role of polymer science. *J. Polym. Sci., Part B* 48: 1685–1718. https://doi.org/10.1002/POLB.22037.

49 Jhaveri, J.H. and Murthy, Z.V.P. (2016). A comprehensive review on anti-fouling nanocomposite membranes for pressure driven membrane separation processes. *Desalination* 379: 137–154. https://doi.org/10.1016/J.DESAL.2015.11.009.

50 Lakhotiya, S.R., Mukhopadhyay, M., and Kumari, P. (2019). Iron oxide (FeO) nanoparticles embedded thin-film nanocomposite nanofiltration (NF) membrane for water treatment. *Sep. Purif. Technol.* 211: 98–107. https://doi.org/10.1016/j.seppur.2018.09.034.

51 Rabiee, H., Farahani, M.H.D.A., and Vatanpour, V. (2014). Preparation and characterization of emulsion poly(vinyl chloride) (EPVC)/TiO_2 nanocomposite ultrafiltration membrane. *J. Membr. Sci.* 472: 185–193. https://doi.org/10.1016/J.MEMSCI.2014.08.051.

52 Yu, Z., Liu, X., Zhao, F. et al. (2015). Fabrication of a low-cost nano-SiO_2/PVC composite ultrafiltration membrane and its antifouling performance. *J. Appl. Polym. Sci.* 132: 41267. https://doi.org/10.1002/APP.41267.

53 Li, J.H., Shao, X.S., Zhou, Q. et al. (2013). The double effects of silver nanoparticles on the PVDF membrane: surface hydrophilicity and antifouling performance. *Appl. Surf. Sci.* 265: 663–670. https://doi.org/10.1016/J.APSUSC.2012.11.072.

54 Rana, S., Nazar, U., Ali, J. et al. (2017). Improved antifouling potential of polyether sulfone polymeric membrane containing silver nanoparticles: self-cleaning membranes. *Environ. Technol.* 39: 1413–1421. https://doi.org/10.1080/09593330.2017.1330900.

55 Guo, J. and Kim, J. (2017). Modifications of polyethersulfone membrane by doping sulfated-TiO_2 nanoparticles for improving anti-fouling property in wastewater treatment. *RSC Adv.* 2017: 33822–33828. https://doi.org/10.1039/C7RA06406C.

56 Low, Z.X., Wang, Z., Leong, S. et al. (2015). Enhancement of the antifouling properties and filtration performance of poly(ethersulfone) ultrafiltration membranes by incorporation of nanoporous titania nanoparticles. *Ind. Eng. Chem. Res.* 54: 11188–11198. https://doi.org/10.1021/acs.iecr.5b03147.

57 Gzara, L., Ahmad Rehan, Z., Khan, S.B. et al. (2016). Preparation and characterization of PES-cobalt nanocomposite membranes with enhanced anti-fouling properties and performances. *J. Taiwan Inst. Chem. Eng.* 65: 405–419. https://doi.org/10.1016/J.JTICE.2016.04.012.

58 Zareei, F. and Hosseini, S.M. (2019). A new type of polyethersulfone based composite nanofiltration membrane decorated by cobalt ferrite-copper oxide nanoparticles with enhanced performance and antifouling property. *Sep. Purif. Technol.* 226: 48–58. https://doi.org/10.1016/J.SEPPUR.2019.05.077.

59 Ngo, T.H.A., Nguyen, D.T., Do, K.D. et al. (2016). Surface modification of polyamide thin film composite membrane by coating of titanium dioxide nanoparticles. *J. Sci. Adv. Mater. Devices* 1: 468–475. https://doi.org/10.1016/J.JSAMD.2016.10.002.

60 Palza, H. (2015). Antimicrobial polymers with metal nanoparticles. *Int. J. Mol. Sci.* 16: 2099–2116. https://doi.org/10.3390/IJMS16012099.

61 Zunita, M., Makertihartha, I.G.B.N., Saputra, F.A. et al. (2018). Metal oxide based antibacterial membrane. *IOP Conf. Ser. Mater. Sci. Eng.* 395: 012021. https://doi.org/10.1088/1757-899X/395/1/012021.

62 Dizaj, S.M., Lotfipour, F., Barzegar-Jalali, M. et al. (2014). Antimicrobial activity of the metals and metal oxide nanoparticles. *Mater. Sci. Eng. C* 44: 278–284. https://doi.org/10.1016/J.MSEC.2014.08.031.

63 Jagadeeshan, S. and Parsanathan, R. (2019). Nano-metal oxides for antibacterial activity. In: *Advanced Nanostructured Materials for Environmental Remediation* (ed. M. Noushad, S. Rajendran, and F. Gracia), 59–90. Springer Cham https://doi.org/10.1007/978-3-030-04477-0_3.

64 Pachaiappan, R., Rajendran, S., Show, P.L. et al. (2021). Metal/metal oxide nanocomposites for bactericidal effect: a review. *Chemosphere* 272: 128607. https://doi.org/10.1016/J.CHEMOSPHERE.2020.128607.

65 Pant, B., Park, M., and Park, S.J. (2019). One-step synthesis of silver nanoparticles embedded polyurethane nano-fiber/net structured membrane as an effective antibacterial medium. *Polymers* 11: 1185. https://doi.org/10.3390/POLYM11071185.

66 Ozay, Y., Dizge, N., Gulsen, H.E. et al. (2016). Investigation of electroactive and antibacterial properties of polyethersulfone membranes blended with copper nanoparticles. *CleanRooms* 44: 930–937. https://doi.org/10.1002/CLEN.201500953.

67 Sawada, I., Fachrul, R., Ito, T. et al. (2012). Development of a hydrophilic polymer membrane containing silver nanoparticles with both organic antifouling and antibacterial properties. *J. Membr. Sci.* 387–388: 1–6. https://doi.org/10.1016/J.MEMSCI.2011.06.020.

68 Mousa, H.M., Hamdy, M., Yassin, M.A. et al. (2022). Characterization of nanofiber composite membrane for high water flux and antibacterial properties. *Colloids Surf., A* 651: 129655. https://doi.org/10.1016/J.COLSURFA.2022.129655.

69 Wai, K.P., Koo, C.H., Pang, Y.L. et al. (2020). In situ immobilization of silver on polydopamine-coated composite membrane for enhanced antibacterial properties. *J. Water Process. Eng.* 33: 100989. https://doi.org/10.1016/J.JWPE.2019.100989.

70 Lv, D., Wang, R., Tang, G. et al. (2019). Ecofriendly electrospun membranes loaded with visible-light-responding nanoparticles for multifunctional usages: highly efficient air filtration, dye scavenging, and bactericidal activity. *ACS Appl. Mater. Interfaces* 11: 12880–12889. https://doi.org/10.1021/acsami.9b01508.

71 Li, R., Wu, Y., Shen, L. et al. (2018). A novel strategy to develop antifouling and antibacterial conductive Cu/polydopamine/polyvinylidene fluoride membranes for water treatment. *J. Colloid Interface Sci.* 531: 493–501. https://doi.org/10.1016/J.JCIS.2018.07.090.

72 Abedalwafa, M.A., Zhang, H., Mei, Q. et al. (2020). A novel method to construct antimicrobial surface by decorating polyacrylonitrile nanofibrous membrane with nanoparticles. *J. Ind. Text.* 51: 587–598. https://doi.org/10.1177/1528083720934198.

73 Shalaby, T., Hamad, H., Ibrahim, E. et al. (2018). Electrospun nanofibers hybrid composites membranes for highly efficient antibacterial activity. *Ecotoxicol. Environ. Saf.* 162: 354–364. https://doi.org/10.1016/J.ECOENV.2018.07.016.

74 Gul, S., Khan, S.A., Rehan, Z.A. et al. (2019). Antibacterial CuO-PES-CA nancomposite membranes supported Cu0 nanoparticles for water permeability and reduction of organic pollutants. *J. Mater. Sci. Mater. Electron.* 30: 10835–10847. https://doi.org/10.1007/s10854-019-01427-5.

75 Kuvarega, A.T. and Mamba, B.B. (2016). Photocatalytic membranes for efficient water treatment. In: *Semiconductor Photocatalysis – Materials, Mechanisms and Applications* (ed. C. Cao). Intech Open https://doi.org/10.5772/62584.

76 Gusain, R., Gupta, K., Joshi, P., and Khatri, O.P. (2019). Adsorptive removal and photocatalytic degradation of organic pollutants using metal oxides and their composites: a comprehensive review. *Adv. Colloid Interf. Sci.* 272: 102009. https://doi.org/10.1016/J.CIS.2019.102009.

77 Romay, M., Diban, N., Rivero, M.J. et al. (2020). Critical issues and guidelines to improve the performance of photocatalytic polymeric membranes. *Catalysts* 10: 570. https://doi.org/10.3390/CATAL10050570.

78 Sun, T., Liu, Y., Shen, L. et al. (2020). Magnetic field assisted arrangement of photocatalytic TiO_2 particles on membrane surface to enhance membrane antifouling performance for water treatment. *J. Colloid Interface Sci.* 570: 273–285. https://doi.org/10.1016/J.JCIS.2020.03.008.

79 Abdal-Hay, A., Hamdy Makhlouf, A.S., and Khalil, K.A. (2015). Novel, facile, single-step technique of polymer/TiO_2 nanofiber composites membrane for photodegradation of methylene blue. *ACS Appl. Mater. Interfaces* 7: 13329–13341. https://doi.org/10.1021/acsami.5b01418.

80 Yasin, S.A., Abbas, J.A., Saeed, I.A., and Ahmed, I.H. (2020). The application of green synthesis of metal oxide nanoparticles embedded in polyethylene terephthalate nanofibers in the study of the photocatalytic degradation of methylene blue. *Polym. Bull.* 77: 3473–3484. https://doi.org/10.1007/s00289-019-02919-4.

81 Abu-Dalo, M.A., Al-Rosan, S.A., Albiss, B.A. et al. (2021). Photocatalytic degradation of methylene blue using polymeric membranes based on cellulose acetate impregnated with ZnO nanostructures. *Polymers* 13: 3451. https://doi.org/10.3390/POLYM13193451.

82 Azmi, A., Lau, K.S., Chin, S.X. et al. (2022). Regenerated cellulose membrane incorporating photocatalytic zinc oxide as a bifunctional membrane for decoloration of methylene blue. *Polym. Adv. Technol.* 33: 843–850. https://doi.org/10.1002/PAT.5560.

83 Boopathy, G., Gangasalam, A., and Mahalingam, A. (2020). Photocatalytic removal of organic pollutants and self-cleaning performance of PES membrane incorporated sulfonated graphene oxide/ZnO nanocomposite. *J. Chem. Technol. Biotechnol.* 95: 3012–3023. https://doi.org/10.1002/JCTB.6462.

84 Tran, M.L., Fu, C.C., Wei, T.H. et al. (2020). Surface coating of titania and graphene oxide onto plasma-activated polymer membranes as efficient photocatalysts for organics removal from water. *J. Water Process Eng.* 37: 101488. https://doi.org/10.1016/J.JWPE.2020.101488.

85 Wang, Y., Dai, L., Qu, K. et al. (2021). Novel Ag-AgBr decorated composite membrane for dye rejection and photodegradation under visible light. *Front. Chem. Sci. Eng.* 15: 892–901. https://doi.org/10.1007/S11705-020-2011-0.

86 Chen, Z., Chen, G.E., Xie, H.Y. et al. (2021). Photocatalytic antifouling properties of novel PVDF membranes improved by incorporation of SnO_2-GO nanocomposite for water treatment. *Sep. Purif. Technol.* 259: 118184. https://doi.org/10.1016/J.SEPPUR.2020.118184.

87 Daels, N., Radoicic, M., Radetic, M. et al. (2014). Functionalisation of electrospun polymer nanofibre membranes with TiO_2 nanoparticles in view of dissolved organic matter photodegradation. *Sep. Purif. Technol.* 133: 282–290. https://doi.org/10.1016/J.SEPPUR.2014.06.040.

88 Radoor, S., Karayil, J., Jayakumar, A. et al. (2021). An efficient removal of malachite green dye from aqueous environment using ZSM-5 zeolite/polyvinyl alcohol/carboxymethyl cellulose/sodium alginate bio composite. *J. Polym. Environ.* 29: 2126–2139. https://doi.org/10.1007/s10924-020-02024-y.

89 Radoor, S., Karayil, J., Parameswaranpillai, J., and Siengchin, S. (2020). Removal of anionic dye Congo red from aqueous environment using polyvinyl alcohol/sodium alginate/ZSM-5 zeolite membrane. *Sci. Rep.* 10: 15452. https://doi.org/10.1038/s41598-020-72398-5.

90 Daraei, P., Madaeni, S.S., Ghaemi, N. et al. (2012). Novel polyethersulfone nanocomposite membrane prepared by PANI/Fe_3O_4 nanoparticles with enhanced performance for Cu(II) removal from water. *J. Membr. Sci.* 415–416: 250–259. https://doi.org/10.1016/J.MEMSCI.2012.05.007.

91 Ghaemi, N., Madaeni, S.S., Daraei, P. et al. (2015). Polyethersulfone membrane enhanced with iron oxide nanoparticles for copper removal from water: application of new functionalized Fe_3O_4 nanoparticles. *Chem. Eng. J.* 263: 101–112. https://doi.org/10.1016/J.CEJ.2014.10.103.

92 Shad, S., Bashir, N., Belinga-Desaunay Nault, M.F., and Lynch, I. (2021). Incorporation of biogenic zinc nanoparticles into a polymeric membrane: impact on the capture of organic herbicides. *Clean Eng. Technol.* 5: 100339. https://doi.org/10.1016/J.CLET.2021.100339.

93 Ghaemi, N. (2016). A new approach to copper ion removal from water by polymeric nanocomposite membrane embedded with γ-alumina nanoparticles. *Appl. Surf. Sci.* 364: 221–228. https://doi.org/10.1016/J.APSUSC.2015.12.109.

94 Căprărescu, S., Modrogan, C., Purcar, V. et al. (2021). Study of polyvinyl alcohol-SiO_2 nanoparticles polymeric membrane in wastewater treatment containing zinc ions. *Polymers* 13: 1875. https://doi.org/10.3390/POLYM13111875.

95 Kusworo, T.D., Susanto, H., Aryanti, N. et al. (2021). Preparation and characterization of photocatalytic PSf-TiO_2/GO nanohybrid membrane for the degradation of organic contaminants in natural rubber wastewater. *J. Environ. Chem. Eng.* 9: 105066. https://doi.org/10.1016/J.JECE.2021.105066.

96 Shin, J.H., Yang, J.E., Park, J.E. et al. (2022). Rapid and efficient removal of anionic dye in water using a chitosan-coated iron oxide-immobilized polyvinylidene fluoride membrane. *ACS Omega* 7: 8759–8766. https://doi.org/10.1021/acsomega.1c06991.

97 Isawi, H. (2019). Evaluating the performance of different nano-enhanced ultrafiltration membranes for the removal of organic pollutants from wastewater. *J. Water. Process Eng.* 31: 100833. https://doi.org/10.1016/J.JWPE.2019.100833.

98 Soroko, I. and Livingston, A. (2009). Impact of TiO_2 nanoparticles on morphology and performance of crosslinked polyimide organic solvent nanofiltration (OSN) membranes. *J. Membr. Sci.* 343: 189–198. https://doi.org/10.1016/J.MEMSCI.2009.07.026.

99 Yan, L., Li, Y.S., Xiang, C.B., and Xianda, S. (2006). Effect of nano-sized Al_2O_3-particle addition on PVDF ultrafiltration membrane performance. *J. Membr. Sci.* 276: 162–167. https://doi.org/10.1016/J.MEMSCI.2005.09.044.

16

Nanocellulose-Based Ultralight Porous Material for Various Environmental Applications

Ajith Mathew[1], *Aiswarya Poulose*[1], *Deepu A. Gopakumar*[1], *Daniel Pasquini*[2], *Yves Grohens*[3], *and Jinu Jacob George*[1]

[1]*Cochin University of Science and Technology, Department of Polymer Science and Rubber Technology, Kalamassery, Cochin 682022, Kerala, India*
[2]*Chemistry Institute, Federal University of Uberlandia-UFU, Uberlandia, Campus Santa Monica-Bloco1DCP593, Brazil*
[3]*Universite de Bretagne Sud, IRDL-FRE CNRS 3744, Lorient, Lorient 56100, France*

16.1 Introduction

Worldwide environmental challenges include local water shortages and domestic air pollution [1]. So, in this chapter, we discuss an overview of aerogel fabricated with cellulose nanofiber (CNF) for the successful eradication of contaminants from air and water systems. The global problem of water scarcity continues to grow due to climate change and population growth. Population growth in the world increases the rate (>4600 km^3 $year^{-1}$) of water usage worldwide. It could reach 6000 km^3 $year^{-1}$ due to the increase in the worldwide population by 2050 [2]. Presently, of the total fresh water content, the agricultural sector uses 69%, Business domains 19%, and corporations 12% [3]. Each year, the total amount of water used by these industries will rise by 1% [4]. In addition to rising water use, increased industrial, agricultural, and home waste pollution rates degrade water quality and harm aquatic and terrestrial ecosystems. Due to the unsanitary environment, inadequate systems for purifying and distributing water threaten public health and contribute to the spread of diseases like cholera, diarrhea, dysentery, hepatitis A, typhoid, and polio [5]. Significant environmental problems have recently been brought on by water contamination from various organic pollutants, such as regular oil spills, pharmaceutical residues, and other synthetic chemical leaks [6, 7]. Pharmaceutical residues have emerged as a major global concern due to their low biodegradability, ease of bioaccumulation, high persistence, and possible negative effects on ecosystems and human health [8]. These problems worsen each year due to the buildup of these pollutants in the environment [6]. Traditional methods that rely on petroleum-based, nonbiodegradable synthetic polymers could add another source of environmental pollution [9]. Various techniques like precipitation, ion exchange, flocculation, membrane filtration, electrolysis, distillation, and evaporation are

Nanomaterials for Air and Water Purification, First Edition.
Edited by Deepu A. Gopakumar, Jyotishkumar Parameswaranpillai, Jinu Jacob George, and Midhun Dominic C.D.

commonly used for water purification. When compared to membrane methods, the main disadvantages of distillation, evaporation, and electrolysis processes are their expensive costs and slow pace of freshwater outflow. Considering their simple nature and potency, membrane-based purification techniques are widely employed currently [5]. Reverse osmosis is the basic principle behind membrane-based filtration techniques, which are used for 70% of the desalination process globally. However, the challenges with reverse osmosis still include the need for membrane fouling and high working pressures [10]. In addition to diseases brought on by water contamination air pollution, one of the significant contributors to respiratory diseases and various illnesses. Worldwide, air pollution claims the lives of almost seven million people each year. Total of 95% of population on the planet reside in regions where air quality is below WHO criteria [11]. Industry, automotive exhaust, domestic energy usage for cooking and heating, biowaste combustion, and electricity generation are only a few of the factors that contribute to ambient air pollution. According to estimates, 4.2 million people die each year as a result of heart disease, acute, lung cancer, chronic respiratory diseases, and stroke due to these outdoor pollutants. In addition, indoor air allergens such as smoke from cooking fires responsible for 3.7 million fatalities annually. Many forms of pollutants, including particulate matter CO, (PM), CH_4, CO_2, polyaromatic hydrocarbons (PAHs) and volatile organic compounds (VOCs; phenol, HCHO, etc.) are created when faeces, fossil fuels, and timber are burned on inefficient or open furnaces. These PMs are dangerous pollutants [12–14]. Moreover, the increase in the concentration of atmospheric CO_2 has garnered substantial focus alongside PM around the world because it contributes to nearly 75% of all greenhouse gas emissions and is a key cause of global warming [15]. According to a recent study, the atmospheric CO_2 concentration increased from 270 to 400 ppm following the industrial revolution [16]. As a result, research has focused on various techniques and equipment for CO_2 or PM capture. Traditional air filters are made from fossil compounds, including polyimide, polyethylene, and polyacrylonitrile. The utilization of these materials as air filters is hindered by several obstacles, primarily their size-based mechanism that relies on diffusion, sieving, and interception. Furthermore, their expensive cost, deficiency of active functional groups, and inability to biodegrade present additional complications, making them less feasible for this purpose [17, 18]. The world's increasing demands have drawn researchers globally for the creation of materials with high-performance and intelligent. Therefore, to achieve elements regarding sustainable development, it is essential to increase the production of products with value-added with less environmental problems [19]. As better adsorbents than thin membranes, aerogels are becoming more and more important in the field of water and air filtration. They are highly porous in nature (~99%) [20], lightweight solid skeletons with nanostructures that have a large area on the surface and low density [21]. Their extensive porosity and surface area provide them with the ability to adsorb or absorb pollutants that exceed their own weight by a significant margin [22, 23]. S.S Kistler developed the first silica-based aerogel in early 1930s. However, silica aerogels' wider applicability was constrained by their intrinsic stiffness and fragility as well as secondary pollutants. Aerogels are created using both inorganic and organic components in addition to silica. Currently used aerogel materials include metal–organic frameworks (MOFs), carbon, Poly-vinylidene fluoride (PVDF) and polymers like polyaniline (PA). However, these materials' non-biodegradable nature and secondary contamination restrict their use [5]. Natural bio-resource exploration has vast potential to decrease our reliance on petroleum products since it is driven by the overall trend of sustainable economic development [24]. Cellulose, soybean protein, silk fibroin, chitin, and keratin are the some of the biopolymers, which can be used as an alternative material for the fabrication of air filters due to their biodegradability, ease of availability, straight forward to process, and recyclable [17]. Cellulose is one of the most prevalent biopolymers among all other biopolymers. It has been used in a wide range of applications, such as production of automobile parts, removing pesticides and heavy metals from water using biocomposites, desalination of water, air purification, and fuel separation in the automotive industry. It has also been used for packaging, in ballistics as a transparent membrane for protection, and in the biomedical field [11].

16.2 Morphology and Structure of Cellulose

Cellulose, the most abundant natural polymer, forms the structural framework of plant cells and various microorganisms, including fungi, bacteria, and algae. It is composed of β-D-glucopyranose units connected by β-(1–4) glycosidic bonds, with a degree of polymerization ranging from several hundreds to over ten thousand. The general formula representing cellulose is $(C_6H_{10}O_5)n$. These glucose units are organized into microfibrils, which are held together by intermolecular and intramolecular hydrogen bonds, as well as intermolecular Van der Waals forces. The presence of both nonreducing and reducing ends contributes to the stability and structural integrity of cellulose [25, 26]. The interactions mentioned above, including the noncovalent intermolecular forces, play a crucial role in maintaining the stability and firm structure of cellulose. They are particularly important for plants and certain marine organisms [27]. The physical properties of cellulose are determined by the three hydroxyl (OH) groups present on each anhydroglucose unit at positions C2, C3, and C6. The reactivity order of these –OH functional groups has been reported as OH–C6 = OH–C2 > OH–C3, based on etherification reactions. The –OH group at the C6 position exhibits a higher reactivity, approximately ten times greater than that of the C2 or C3 positions. Due to the abundance of –OH groups, cellulose is highly polar [28]. The –OH groups also contribute to the compactness of the crystalline structure and facilitate the formation of an extensive hydrogen bonding network. This network imparts insolubility to cellulose in most aqueous and organic solvents. In order to dissolve cellulose for homogeneous reactions, the hydrogen bonding network must be disrupted [29, 30]. Nanocellulose (NC) refers to cellulose with dimensional scales ranging from 1 to 100 nm. It can be isolated from natural cellulose fibrils and is categorized into three main types based on their sources: cellulose nanocrystals (CNCs), cellulose nanofibrils (CNFs), and bacterial nanocellulose (BNC). NC retains the intrinsic properties of cellulose, including biodegradability,

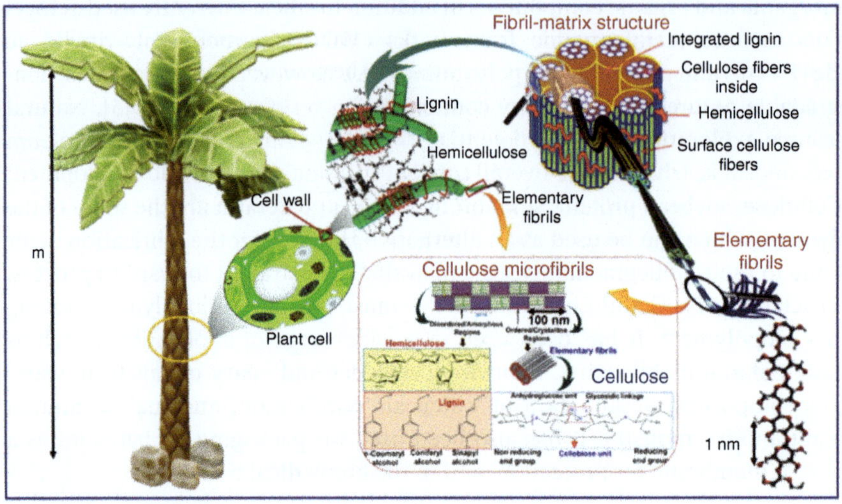

Figure 16.1 The hierarchical structure of wood illustrates the organization of cellulose microfibrils within plant cell walls and the chemical composition of cellulose, hemicellulose, and lignin. Source: Yadav et al. [28] Reproduced with the permission from Elsevier.

renewability, hydrophilicity, ease of chemical modification, high tensile strength, high elastic modulus (130–150 GPa), high specific surface area (SSA) (up to several hundreds of $m^2\,g^{-1}$), and low density ($1.6\,g\,cm^{-1}$) [30, 31] (Figure 16.1).

16.3 Types of Nanocellulose

16.3.1 Cellulose Nanofiber (CNF)

CNF, also referred to as microfibrillated cellulose, nanofibrils, or nano-fibrillated cellulose (NFC), is a composite structure composed of elementary fibrils or microfibrils that contain both ordered (crystalline) and disordered (amorphous) domains. In plant fibers, these microfibrils exist as bundles of entangled and flexible cellulose chains. The term "microfibrils" typically describes fibrous cellulose structures that are 2–10 nm thick and tens of microns long, while cellulose nanofibers usually consist of 10–50 microfibrils with diameters ranging from 20 to 60 nm. They exhibit a diameter less than 100 nm and comprise both amorphous and crystalline regions [32, 33]. However, CNF exhibits a couple of inherent limitations arising from its physical properties. First, the abundance of –OH groups within CNF leads to strong intermolecular and intramolecular hydrogen interactions, resulting in the formation of a gel-like structure during production. Second, CNF displays high hydrophilicity, which hinders its application in various fields such as paper coating (due to increased water repellence) and composites (due to the tendency to form agglomerates in petrochemical polymers) [34]. To overcome these drawbacks, chemical surface modifications are employed to reduce the number of hydroxyl interactions and enhance compatibility with different matrices. Such modifications aim to improve the performance of CNF in various industrial applications. CNF offers distinct advantages, including the ability to be produced on an industrial scale with different functional groups and through various processes. Mechanical treatments such as grinding/refining, microfluidization, high-pressure homogenization and ultrasonication are commonly utilized for CNF preparation. Additionally, nonconventional methods like steam explosion, cryocrushing, aqueous counter extrusion and collision have been explored [32, 33].

16.3.2 Cellulose Nanocrystals (CNC)

CNC, also known as cellulose nanowhiskers (CNW), nanocrystalline cellulose (NCC), or cellulose crystallites, is obtained through a common technique called acid hydrolysis [35]. In this process, cellulose is treated with sulfuric acid (H_2SO_4) under controlled temperature and time conditions. The acid hydrolysis selectively removes the amorphous regions within the microfibrils, causing the remaining crystalline sections to aggregate [36]. Other acids such as hydrochloric acid (HCl), phosphoric acid (H_3PO_4), and hydrobromic acid (HBr) have also been utilized for CNC extraction. Typically, CNC exhibits relatively lower aspect ratios compared to CNFs, with diameters ranging from 3 to 5 nm [37] and lengths extending from 100 to 500 nm. The crystal structure of CNCs resembles whiskers, featuring tapered ends likely attributed to the acid hydrolysis process. The concept of producing CNCs through acid hydrolysis was first introduced by Nickerson and Habrle in 1947 [38]. Concentrated acids such as HCl or H2SO4 (64 wt.% sulfuric acid) effectively hydrolyze the amorphous regions while preserving the crystalline regions, referred to as "cellulose micelles." CNC structures are characterized by their rigidity, elongation, and high crystalline content, ranging from 54 to 88% [28]. These crystals exhibit limited flexibility due to the removal of amorphous material, resulting in a high percentage of Iβ crystal structure (68–94%). The diameters of CNCs are nonuniform due to the random cleaving of cellulose chains during hydrolysis [28, 39]. The degree of crystallinity varies among different cellulose sources, impacting the size and morphology of the liberated crystals. During sulfuric acid hydrolysis, CNCs with higher colloidal stability and a highly repulsive nature are generated, characterized by their negative charge [40].

16.3.3 Bacterial Nanocellulose (BNC)

BNC is a type of cellulose produced by certain aerobic, nonpathogenic bacteria through a biotechnological process. It exhibits both cellulose properties and nanomaterial characteristics. BNC is constructed from ribbon-like fibers, which are bundles composed of elemental fibrils. The diameter of these fibers ranges from 20 to 100 nm, while the elemental fibrils measure 2 to 20 nm in diameter [41, 42]. The production of BNC involves two key processes: fermentation and purification. It begins by cultivating bacteria in an aqueous culture medium containing glucose, phosphate, carbon, and nitrogen. The growth conditions, such as the nutrition

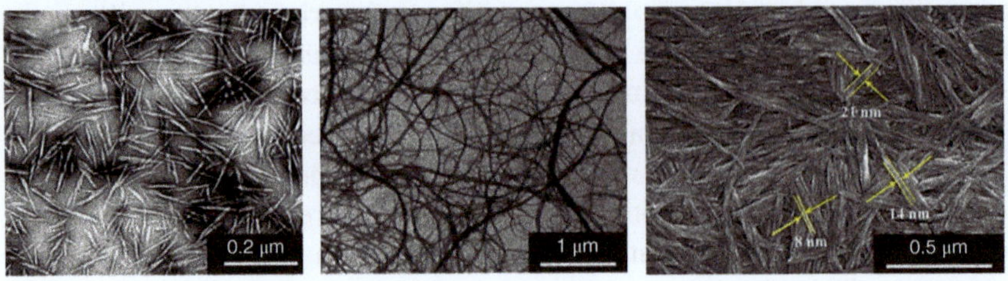

Figure 16.2 SEM image of CNC, CNF, and BNC. Source: Paul et al. [5]/Reproduced with permission from Elsevier.

source, oxygen ratio, bacterial strain type, incubation period, and bioreactor cultivation conditions, need to be carefully adjusted to control the form and quality of BNC tubes [43, 44]. During production, the bacteria synthesize β-1,4-glucan chains from various carbon sources. These chains extend outward through pores on the cell membrane called terminal complexes (TCs). The β-1,4-glucan chains assemble in the culture medium, forming protofibrils that subsequently crystallize into micro- and macro-fibrils, bundles, and eventually develop into ribbon-like structures. A single ribbon comprises approximately 1000 individual β-1,4-glucan chains [45]. BNC offers several advantages, including significantly higher crystallinity (84–89%), excellent chemical purity without impurities like hemicellulose, lignin, or pectin, high strength compared to other biomaterials, high water absorption capacity, high biodegradability, good moldability, and smaller fiber cross-section, contributing to its high porosity. The applications of BNC primarily revolve around biomedical research. It finds use in artificial skin, antimicrobial materials for wound dressings and wound healing, scaffolds for tissue regeneration (e.g. bone and cartilage), artificial blood vessels, cell therapy, drug delivery systems, and artificial membranes. Additionally, BNC is employed in various fields such as water treatment membranes, papermaking, power generation, textiles, food and food packaging, acoustics, and cosmetics [46] (Figure 16.2).

16.3.4 Aerogels

Aerogel is a unique type of material characterized by its three-dimensional (3D) porous structure, where the liquid components are replaced by gas during the drying process while preserving the integrity of the gel network. The creation of aerogel was first achieved by Kistler in 1931 [47, 48]. There are three primary categories of lightweight porous materials: foam, sponge, and aerogel. Foam is formed by introducing gases (typically air) into a liquid, solid, or hydrogel, resulting in a dispersed gas structure. The pore diameter in foam is generally larger than 50 nm, and the pore structure can be classified as open-hole, closed-hole, or reticulated hole. The term "sponge" refers to a low-density porous elastic polymer material with an open-cell structure, where gas is dispersed within a solid matrix. The macropores present in the sponge, including channels and holes, allow the free passage of water and other molecules [49]. Traditionally, the name "aerogel" has been used for materials with a nanoscale pore structure that are created through supercritical drying. On the other hand, porous materials produced through evaporation and freeze drying processes are known as xerogel and cryogel, respectively [50]. Because of superior adsorption capabilities, aerogels are replacing thin membranes for the purification of air and water. They are lightweight nanostructured solid skeletons that are very porous (99%), have a low density, are extremely close to the air, and have a large area on the surface [5]. Huge porosity and high surface area make the aerogels have the advantage of adsorption and absorption of contaminants many times their weight. Membranes and aerogels made up of inorganic minerals (graphene, carbon nanotubes (CNT)), commodity polymers (HDPE, PET, and LDPE) and synthetic polymers (e.g. polyurethane foam and polypropylene) are highly hazardous that are not biodegradable and release harmful elements when ignited, which results in major environmental problems. However, most of these materials have several limitations such as low degradation rate, low absorption capacity, generation of secondary pollutants, inadequate buoyancy, poor reusability [49, 51]. It is very important to consider a polymer's capacity to degrade after a particular period of use because the modification it undergoes could cause buildup and pollution. NC-based aerogels possess several advantageous properties, such as good biocompatibility, easy degradation, and green reproducibility. They combine the benefits of inorganic and polymer aerogels, including low density, high porosity, and SSA, making them appealing for diverse functional applications such as thermal insulation, energy storage, and efficient adsorption. These aerogels can also overcome the limitations of traditional aerogel materials such as silica, carbon, and tin dioxide, which can be brittle and require complex equipment for synthesis. Researchers have recognized NC-based aerogels as an environmentally friendly and cost-effective alternative [30]. Additionally, the Young's modulus of CNF aerogels is generally higher compared to CNC aerogels due to their higher aspect ratio and smaller lateral dimension, which promotes the formation of a robust entangled network [52].

16.4 Preparation of Nanocellulose-Based Aerogels

16.4.1 Preparation of Hydrogels

Aerogels are frequently employed in the filtration of water and gases because of their excellent adsorption and absorption capacities. The sol–gel approach, self-assembly mode, 3D technology, and emulsion process are the most frequently used techniques for creating aerogels. The sol–gel approach, one of the common methods used for the fabrication of aerogel. This method is simple and can create monolithic hydrogel. There are three processes in the sol–gel method: gel formation, aging, and drying [5]. Intramolecular and intermolecular hydrogen bond formation help for the formation of better entanglement and self-aggregation of nanocrystals or nanofibrils because of the presence of abundant hydroxyl groups on NC surface. So, the fabrication of hydrogels via NC is quite simple. Accordingly, the requirement for the crosslinking agent is not necessary for the process [30]. The degree of crosslinking is influenced by the concentration of NC. Greater concentration will enhance the stiffness of the

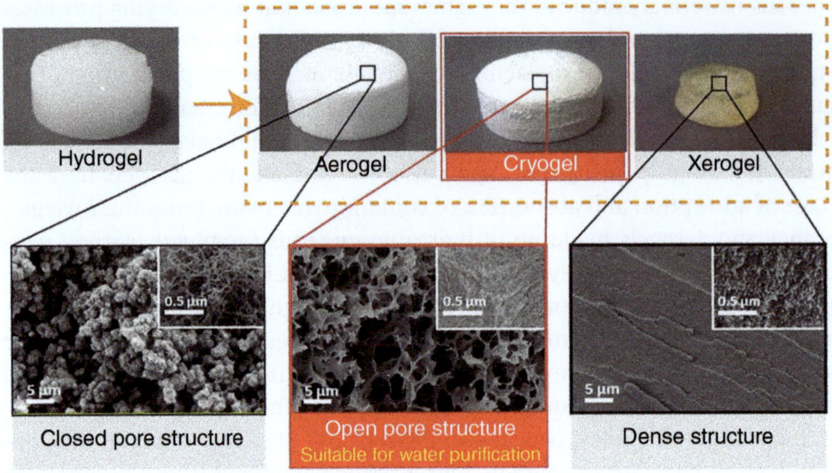

Figure 16.3 Physical image and SEM images of cellulose hydrogel, aerogel, cryogel, and xerogel forms. Source: Paul et al. [5]/Reproduced with permission from Elsevier.

aerogel structures, which compromise the surface area and porosity [5]. Negatively charged functional groups like carboxymethyl, carboxyl, or sulfonic groups can be incorporated onto the surface of NC. This modification facilitates the formation of highly stable and homogeneous aqueous NC dispersion by leveraging electrostatic repulsions between the negatively charged NC particles. The presence of these functional groups promotes the generation of a 3D network through the combination of hydrogen bonding and entanglement of the long NC fibrils. This network structure acts as a backbone, contributing to improved strength and modulus of the resulting gel-like material. Research has shown the effectiveness of such functionalization techniques in enhancing the properties of NC-based materials (Figure 16.3).

The incorporation of modified NC with negatively charged functional groups helps maintain the structural integrity of the gel during the drying process, preventing significant shrinkage. This is particularly beneficial for the fabrication of aerogels with low solid NC content. However, the dispersion of these modified NC in organic solvents with low polarity, such as tert-butyl alcohol (TBA) or ethanol commonly used in aerogel drying, may be less effective. The gelation behavior of NC can be classified into two categories: chemical crosslinking and physical crosslinking [30]. Chemical crosslinking involves the formation of irreversible covalent bonds between cellulose chains using specific crosslinking agents (multifunctional monomers) like citric acid or glutaraldehyde [53]. On the other hand, physical crosslinking relies on weak interactions such as Van der Waals forces, hydrogen bonds, electrostatic interactions, and ionic interactions to create reversible connections between polymer chains [54]. Inorganic salts such as $CaSO_4$ and $CaCl_2$ can also be introduced to the NC dispersion to form reversible crosslinks. For instance, carboxylic CNFs can chelate with multivalent metal ions (e.g. Zn^{2+}, Cu^{2+}, and Co^{2+}) via ionic gelation to form a hydrogel [55, 56]. The chemical method provides better control over the porosity and SSA of the gel compared to the physical method. However, it is generally more expensive and time-consuming.

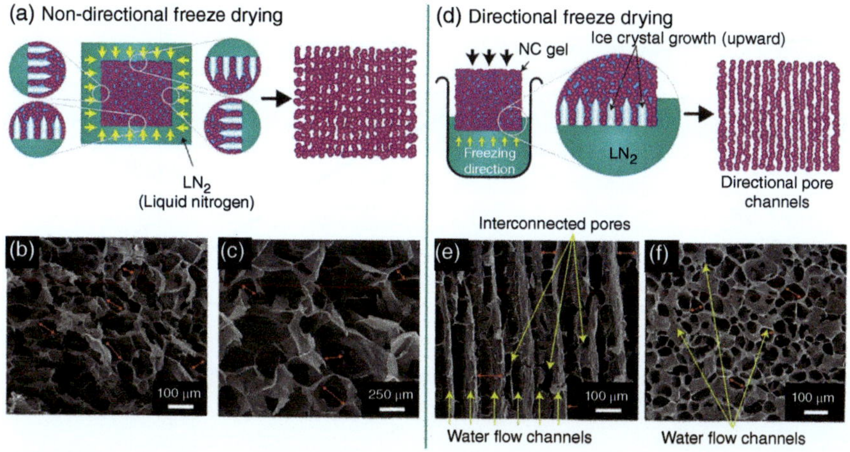

Figure 16.4 Directional freezing and ice crystal growth in hydrogel. (a) Nondirectional freezing of NC gel. (b–c) Cross-sectional and vertical sectional SEM images of nondirectional freeze-dried aerogel. (d) Refrigeration (directional) freezing of gel. (e–f) Cross-sectional and vertical sectional SEM image of directional freeze-dried aerogel, respectively. Source: Paul et al. [5]/Reproduced with permission from Elsevier.

Research has investigated various approaches to gelation in NC systems, offering valuable insights into tailoring the properties and performance of NC-based gels [30] (Figure 16.4).

16.5 Drying Process

The solvent has to be removed from the gel in order to obtain the aerogel. There are three types of drying process freeze-drying (FD), supercritical drying (SCD) or ambient pressure drying (APD). The aerogels that are prepared can be classified as either aerogel (SCD), xerogel (APD), or cryogel FD depending on the density pore structure that is retained after the complete elimination of solvents. The parameters of the aerogel such as surface area, shape, pore structure, and pore size are all influenced by the drying process [5] (Figure 16.5).

16.5.1 Ambient Pressure Drying (APD)

The APD technique is a simple and cost-effective method used in the production of aerogels. The choice of solvent plays a crucial role in this technique. An ideal solvent for APD should have a low surface tension to minimize internal capillary pressure. Initially, the solvent is introduced to the hydrogel through the solvent exchange method, and then it is dried under ambient pressure conditions [57]. During the evaporation of the liquid from the gel pores, capillary stresses are generated at the liquid-vapor interface, leading to shrinkage and reduced porosity in the resulting aerogel. Therefore, the selection of an appropriate solvent is critical in the APD technique to mitigate these effects [58]. In a study by Li et al. they employed a pretreatment process for CNF gel by freezing and exchanging water with 2-propanol,

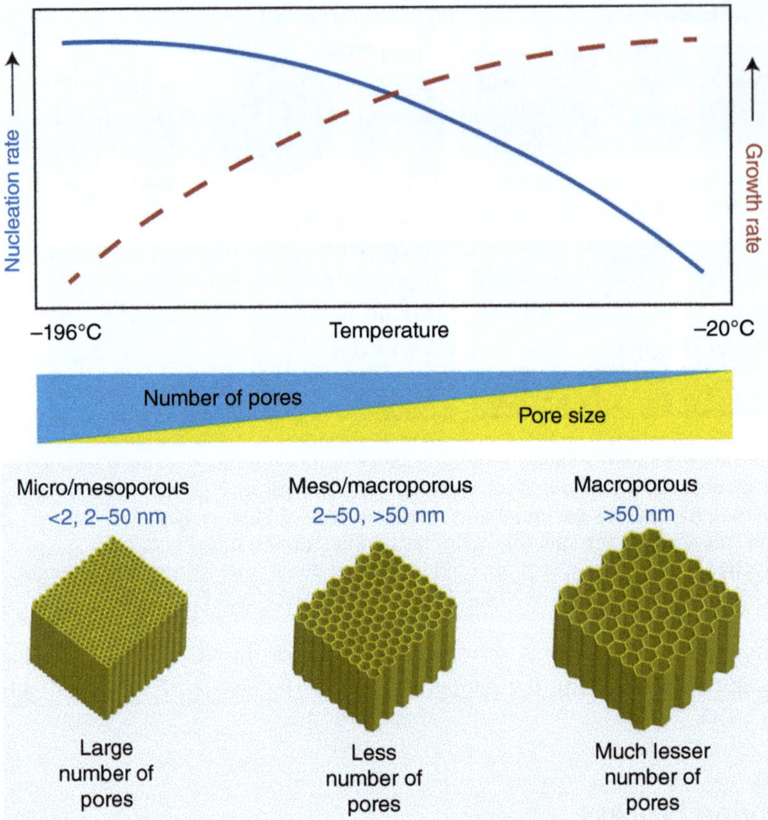

Figure 16.5 Relationship between aerogel freezing temperature and pore size. Source: Paul et al. [5] Reproduced with the permission from Elsevier.

which has a significantly lower surface tension. Subsequently, the gel was dried under ambient pressure conditions, resulting in an aerogel formation without collapse or noticeable shrinkage [59]. Research has emphasized the importance of solvent properties and their impact on the drying process of aerogels, highlighting the significance of solvent selection in achieving desired aerogel characteristics.

16.5.2 Freeze Drying (FD)

FD, also known as lyophilization, is a commonly employed technique for the production of aerogels. It involves freezing the solvent within the gel by utilizing refrigerators, dry ice, or liquid nitrogen at temperatures ranging from −20 to −196 °C [60]. The frozen solvent, typically water, is then directly sublimated under low pressure and temperature conditions, preserving the porous structure of the gel. This process avoids the formation of capillary pressure during drying and maintains the integrity of the skeletal structure [61]. FD offers several advantages such as high efficiency, environmental friendliness, and cost-effectiveness, making it a popular choice for NC aerogel production [30]. The pore structure and surface area of the resulting aerogel can be controlled by manipulating the rate and direction of ice crystallization during the freezing process. Slow freezing leads to the formation of larger ice crystals and macro-porous structures, while a higher freezing rate promotes a more desirable mesoporous pore size distribution. There are two types of FD techniques: directional and nondirectional. Directional FD is preferred as it produces a unidirectional (anisotropic) pore structure, which enhances water flux and allows control over ice growth direction. Nondirectional FD, on the other hand, results in pores oriented in different directions, which hinders water flux. However, it should be noted that FD can cause shrinkage and structural breakage of the cryogel during crystal growth [5]. Research has focused on optimizing FD parameters to tailor the pore structure and properties of NC aerogels, offering opportunities for their applications in various fields.

16.5.3 Supercritical Drying (SCD)

To overcome the drawbacks associated with FD and APD techniques, the SCD method can be employed [5]. In SCD, the solvent within the NC gel, primarily water, is replaced with supercritical fluids that possess unique solubility, diffusivity, and lower surface tension properties. By adjusting temperature and pressure to reach the critical point of the solvent, the gel is subjected to slow pressure release and cooling. This process allows the solvent to undergo supercritical transformations from liquid to gas phases, resulting in the formation of a structurally robust aerogel [30]. One of the advantages of SCD is the absence of apparent surface tension due to the presence of the solvent. As a result, the wet gel can be converted into an aerogel while preserving its skeleton structure. At the critical point, there is no clear distinction between the liquid and vapor phases; both coexist, and the molecules are free to move. This unique characteristic of the supercritical state helps avoid capillary forces that can cause structural damage [62]. Supercritical fluids such as carbon dioxide (CO_2) and ethanol are commonly used for SCD of NC gels. CO_2 is favored due to its critical temperature, which is slightly above room temperature at 31.3 °C, making it more stable and safer to handle. Supercritical CO_2 drying exhibits higher efficiency in maintaining the structure of NC gels compared to FD [63]. Therefore, it is much safer and used more frequently. Supercritical CO_2 drying is more efficient than FD at maintaining the NC gel structure. However, it should be noted that the application of supercritical drying is currently limited and mostly employed for high-quality products, primarily due to the complex process, longer production cycle, high temperature and pressure requirements, time consumption, and higher solvent replacement costs associated with this method [5, 30].

16.6 Application of NC-Based Light Weight Porous Material

16.6.1 NC-Based Aerogels: Advancements in Water Decontamination

NC aerogels possess a distinctive 3D network structure with a combination of stiffness and flexibility, along with micro/mesoporous porosity, interconnected pore

Figure 16.6 (a) The water contact angle before and after 30 days of exposure. (b,c) Water (dyed with methyl blue) and oil (dyed with Sudan III) were dropped on the surface of KNA. (d) Removal of oil (dyed with Sudan III) from the surface of the water with KNA. Source: Zhang et al. [64]/Reproduced with permission from Elsevier.

structure, high surface area, and customizable surface properties. These exceptional characteristics make NC aerogels well-suited as sorbents for the removal of various contaminants from water, employing both adsorption and absorption processes [5]. The removal of major toxic pollutants such as dyes, metal ions, oils, and organic solvents will be discussed in detail, highlighting the mechanisms involved, including electrostatic interactions, hydrophilic/hydrophobic interactions, and chelation (Figure 16.6).

16.6.2 Dye Removal

As the textile sector experiences unprecedented growth worldwide, accompanied by an increasing global population, the use of dyes to color textiles has become integral. However, this dyeing process requires a significant amount of water, leading to environmental challenges. Alarmingly, approximately 50 000 tons of dyes are discharged annually into water bodies without any prior treatment [65]. These water-soluble organic dyes, which are commonly employed in industries such as textiles, paints, leather, and paper, pose a significant concern as they are toxic, chemically stable, and difficult to degrade [66, 67]. The accumulation of organic dyes in water bodies not only obstructs sunlight penetration, hindering the photosynthesis of aquatic plants, but also presents serious health risks to both humans and animals due to their mutagenic and carcinogenic properties. Moreover, the release of untreated dyes into the environment can have detrimental effects on ecosystems, leading to the destruction of aquatic life [68]. The organic dyes can accumulate in both humans and animals and result in major health issues since they are mutagenic or carcinogenic. NC aerogels offer a promising solution for the removal of organic dyes from water. The porosity, surface area, surface functional groups, presence of functional nanofillers, pH of water, pollutant concentration, contact time, and hydrophilicity are key factors influencing the adsorption efficiency of NC aerogels in dye removal [5]. Recent research by Beh et al. demonstrated the exceptional removal capacity of CNF aerogels derived from sago pith waste for methylene blue (MB), attributed to their high surface area resulting from the small diameter of CNF fibers (ranging from 15 to 30 nm) [69]. By harnessing the unique properties of NC aerogels, it is possible to address the challenge of organic dye contamination in water, mitigating the environmental impact and safeguarding human health. The development and utilization of NC aerogels in the textile industry hold significant potential for promoting sustainable and environmentally friendly dyeing processes.

16.6.3 Removal of Heavy Metals

The presence of heavy metals in water poses significant health risks, as these compounds can accumulate in biological tissues, resist natural degradation, and cause adverse effects over an extended period of time. Heavy metal ions, such as copper, lead, mercury, cadmium, chromium, cobalt, arsenic, and cesium, are among the primary culprits. While trace amounts of certain metal ions are necessary for metabolic activity and preventing illnesses, excessive concentrations can lead to severe health problems, including muscular and neurological degeneration [70, 71]. Lead ions (Pb (II)), commonly found in wastewater from sources like batteries and printing, are particularly challenging to decompose. Similarly, copper ions (Cu (II)), widely utilized in appliances, circuit conduction, and insecticides, can induce toxic symptoms like hemolysis, diarrhea, and even renal failure. To address the removal of heavy metal ions, the properties of aerogels must be carefully tailored to optimize adsorption performance. The pore structure of aerogels determines their mechanical strength and facilitates effective water diffusion [72]. Additionally, crosslinking between NC and the incorporation of nanofillers enhance wet strength and introduce additional surface functional groups, thereby enhancing the aerogel's absorption capacity. The rate of pollutant diffusion is influenced by the solvent's pH [5]. Recent studies have demonstrated the effectiveness of NC-based aerogels for heavy metal ion removal. Rong et al. developed a CNF/CS/montmorillonite aerogel (CCMA) using a directional FD method, showcasing its adsorption capabilities for Cu(II), Pb(II), and Cd(II) [73]. Another research by Mo et al. resulted in a 3D honeycomb-structured aerogel utilizing tempo-oxidized CNF (TO-CNF), graphene oxide (GO), and trimethylolpropane-tris-(2-methyl-1-aziridine) propionate (TMPTAP) through unidirectional freeze casting. This aerogel exhibited lengthy interconnected channels that facilitated rapid diffusion of metal ions, enabling efficient adsorption of Pb(II), Cu(II), Zn(II), Cd(II), and Mn(II) ions [74].

16.6.4 Oil/Water Separation

The Earth has experienced several oil spills in its oceans and coastal waterways during the past 50 years. Petroleum products that float and appear as thin layers above the water's surface can cause long-term adverse effects on marine life and can take months to clean up [75]. To deal with the dangers of oil spills and their harmful impacts on the planet, multiple treatment approaches are commonly used. The three categories of most frequently utilized techniques are physical, chemical, and biological techniques. The two physical techniques that are frequently used to gather oil are skimmers and booms; however, their capacity to remove oil is limited. The chemical process uses in situ burning, dispersion, and solidification, all of which are expensive, challenging, and harmful to the environment. The biological approach is a time-consuming process that uses microorganisms to break down the hydrocarbons in the oil, but the pH, temperature, organic species, oxygen content, and other factors all affect their effectiveness [76]. For the treatment of oily waste water, the absorption method is considered an alternative technique for the effective removal of oil from water. The advantages of this method are its high capacity for oil removal, simple operation, and low consumption of energy [5]. For the recovery of oil spills, a variety of absorbents have been used, including zeolites, silica aerogels, and polystyrene fibers. The practical usage of these materials is constrained by a number of limitations, including weak sorption capability, poor buoyancy, low oil selectivity, and ecological unfriendliness. Cellulose is nontoxic, inexpensive, and biodegradable polymer. Raw celluloses are amphiphilic and have a tendency to absorb water and oil in an unselective manner. A lot of focus has been placed on turning amphiphilic cellulose into very hydrophobic materials by changing its surface and morphology to increase their selectivity to oil and to make them good candidates for oil absorption. Low-surface-energy hydrophobic groups are introduced to cellulose aerogel matrix through physical or chemical processes. Some of the researchers achieved hydrophobic surface by oleophilic TiO_2 coating, silica nanoparticles coating, graphene oxides, silanization with methyl trimethoxy silane or treatment with trimethyl-chlorosilane via the low-temperature plasma technology [75]. Zhang et al. fabricated kapok/cellulose aerogels with a dual-scale porous structure modified with silane, which possess the characteristics of lightweight, excellent hydrophobicity, good mechanical property, outstanding oil absorption capacity [77]. Roslin et al. developed hydrophobic gamma-irradiated CNC/gelatin aerogel using simple chemical modification with methyltrimethoxysilane (MTMS) and gamma irradiation methods. Silane modification successfully modifies the surface of the gamma-irradiated CNC-MTMS/gelatin aerogel by changing its properties from hydrophilic to hydrophobic and shows selective absorption capability due to its highly hydrophobic and oleophilic property [78].

16.6.5 Removal of Pharmaceutical Waste

Pharmaceutical residues have become a global concern due to their widespread presence in various environmental matrices, including surface water on every continent [79]. These residues, characterized by low biodegradability, ease of bioaccumulation, high persistence, and potential adverse effects on ecosystems and human health, have garnered significant attention over the past four decades [8]. A recent assessment identified more than 600 different medications as known pollutants [80]. Notably, wastewater effluents contain a significant class of pharmaceutical residues, encompassing steroids, antibiotics, antidepressants, antacids, analgesics, and anti-inflammatories [81]. While most pharmaceutical residues are not persistent in the environment, their continuous introduction from multiple sources in large quantities renders them "pseudo-persistent" [82]. Conventional wastewater treatment plants (WWTPs), designed to handle degradable organics in the milligrams per liter ($mg\,l^{-1}$) range, fail to completely decompose pharmaceutical residues. These WWTPs prove ineffective in removing pharmaceutical residues below the gram per liter to nanogram per liter ($g\,l^{-1}$–$ng\,l^{-1}$) range [83]. Traditionally, liquid–liquid extraction (LLE) has been employed for organic analyte extraction from aqueous solutions [84]. However, due to drawbacks such as inefficiency at low analyte concentrations, the solid phase extraction (SPE) technique has gained prominence. SPE offers advantages over LLE, including lower solvent consumption, rapid extraction times, effective extraction of diverse organic analytes, enhanced extraction efficiency, higher selectivity, and improved reproducibility, even at low analyte concentrations. Unfortunately, commercially available SPE cartridges are costly and inaccessible [85]. Efforts have been directed toward synthesizing new classes of adsorbents for SPE that are efficient, affordable, and nontoxic [86]. Graphene-based nanomaterials and CNTs have demonstrated excellent potential for the removal of emerging pollutants due to their low detection limits [87]. CNTs have also been investigated for the extraction and preconcentration of developing pollutants [88]. However, their high cost limits their widespread application for preconcentration of contaminants. Other promising adsorbents include microporous framework materials like MOFs and covalent organic frameworks (COFs). However, the limited stability of MOFs and high cost of COFs hinder their application [89]. Molecularly imprinted materials (MIM), specifically molecularly imprinted polymers (MIP), have emerged as efficient sorbents for extracting pharmaceutical residues and nonsteroidal anti-inflammatory drugs (NSAIDs) from aqueous solutions. MIM utilizes a "lock and key" mechanism to address the challenge of quantifying contaminants by providing selectivity for target analytes. However, the health risks associated with the reagents used in their synthesis pose a significant drawback [90, 91]. Cellulose-based adsorbents, particularly those derived from abundant and renewable natural biopolymer cellulose, have gained interest in SPE due to their distinctive form, non-toxic nature, high functionality, and degradability. Current efforts focus on enhancing the physicochemical properties of cellulose to improve its effectiveness as an adsorbent and SPE material [92]. Several categories of cellulose-based adsorbents have been developed, including NCs materials, magnetic nanocomposites, cellulose metal-organic frameworks (cMOFs), and cellulose-based MIP [93]. Researchers have successfully prepared CNCs functionalized with graphene oxide and $FeCl_2$ via solvothermal methods for the removal of fluoroquinolones. Acid hydrolysis of CNFs with $FeCl_2$ has also shown promise in removing tetracycline, as confirmed by UV–Vis spectrophotometry [94].

16.6.6 Air Filtration

Air filters utilizing various types of NC have garnered considerable attention from researchers worldwide. The mechanical properties, synthesis process, dimensions and density of cellulose and its derivatives are crucial considerations in the fabrication of such filters. These filters are designed for the removal of PM, VOCs, and CO_2 from the air [11]. Cellulose-based filters exhibit remarkable thermal and chemical stability due to their highly crystalline nature [95]. Cellulose-based filters exhibit remarkable thermal and chemical stability due to their highly crystalline nature. The performance of these filters in highly humid environments is significantly influenced by the hydrophilic or hydrophobic properties of cellulose and its derivatives. To enhance filtration efficiency in high humidity conditions, preference is often given to hydrophobic cellulose derivatives or surface-functionalized cellulose filters [11].

16.6.7 Removal of PM and VOCs

Physiosorption and chemisorption are the two distinct mechanisms employed in cellulose-based air filters for air filtration. In the physiosorption process, harmful pollutants adhere to the surface of the filter or bind to atoms or molecules embedded within the filtration system. On the other hand, chemisorption, known as the chemically interactive sorption process, involves the chemical interaction of contaminants with the filter membrane's surface, as well as with molecules or atoms present within, resulting in the formation of simpler compounds. The effectiveness of sorption techniques relies on the availability of active binding sites within the filter system. Similar to conventional synthetic polymer-based air filters, cellulose-based air filters remove PM through mechanisms such as impaction, interception, diffusion, and electrostatic interactions. The specific filtration process is influenced by the type and size of contaminants. Smaller particles, ranging from 100–1000 nm, are primarily filtered through gravitational settling, straining, electrostatic interactions, and interactions with the fibers. Pollutants with a diameter larger than 1000 nm, such as spores and dust, are physically obstructed by the filter [11]. Nanofibers typically experience interception and diffusion rather than straining, gravitational impaction, and inertial impaction [96]. Fine PM (preferably 0.5 μm) traveling with the airflow and staying within a distance equal to or smaller than their radius from the fiber surface is intercepted. Diffusion, on the other hand, is effective for significantly smaller particles with an aerodynamic diameter of less than 100 nm, where Brownian motion plays a dominant role; causing atypical collisions with fiber surfaces and resulting in particle capture [97]. Electrostatic filtration operates by attracting and capturing charged contaminants through charged functional groups present on cellulose-based filters. This method is particularly effective for removing particles within the diameter range of 100–1000 nm [17]. Physiosorption and chemisorption can both be employed to remove VOCs and gaseous compounds. In addition to electrostatic interactions, the physiosorption of pollutants onto cellulose-based filters involves interactions such as hydrogen bonding, Van der Waals forces, hydrophobic interactions, and polar–polar interactions [17]. Filters are often designed with multiple layers to enhance their efficiency in pollutant removal. The air permeability of a filtration system or mask generally decreases with an increase in thickness or number of layers. However, stacking filters has shown significant improvements in virus suppression and moisture control. Generally, a higher dust holding capacity of a

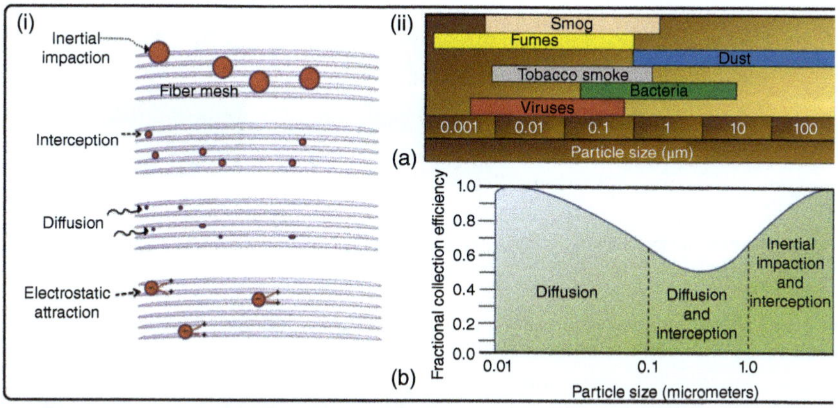

Figure 16.7 (i) Mechanism showing the adsorption of PM or VOCs using fiber mash and (ii) (a) Relative size of common air pollutants and (b) fractional removal efficacy for different mechanical filters concerning the diameter of the pollutant. Source: Rana et al. [11] Reproduced with the permission from Elsevier.

filter indicates a longer lifespan, and vice versa [11]. In the study conducted by Wu et al., the incorporation of silver nanowires (AgNWs) into bacterial cellulose (BC) resulted in improved permeability and porosity (AgNWs/BC), achieving filtration efficiencies of 99.7% and 99.8% for PM [98] (Figure 16.7).

16.6.8 CO_2 Separation or Adsorption

The gas permeability of a material depends on its porosity. So, in the case of CO_2 adsorption, the gas permeability of that material plays a significant role. This can vary depending on so many factors, such as relative humidity (RH), surface functionalization procedures, and the type of fillers used. The nanopore size between NC molecules expands when exposed to moisture, or when there is interfacial adhesion between fillers and NC. This expansion typically occurs at RH levels of 65–85%. Additionally, if fillers with suitable porosity and quantity are employed, the nanopores can enlarge to a degree where CO_2 molecules can readily diffuse through the molecular sieve made of NC. In contrast, CH_4 and H_2 molecules do not diffuse as quickly. Due to size differences between gaseous molecules (CH_4 is 0.38 nm, CO_2 is 0.34 nm, and N_2 is 0.36 nm), CH_4 and N_2 diffuse very slowly than CO_2. Because of its small size, CO_2 can easily pass through a sieve with the right pore size. Moreover, an elevated RH or increased concentration of fillers induces a greater degree of swelling in the polymeric matrix or generates additional voids within the molecular sieve. This, in turn, leads to an increased separation between the NC fibers [99, 100]. Consequently, when the pore size or gaps reach a certain threshold, allowing all gas molecules to pass through, the selectivity of the NC-based membrane is compromised. Consequently, the process of gas separation is significantly influenced by both pore size and humidity. Furthermore, apart from its role in CO_2 separation, NC-based membranes have the capability to adsorb CO_2 molecules onto their surfaces (Figure 16.8).

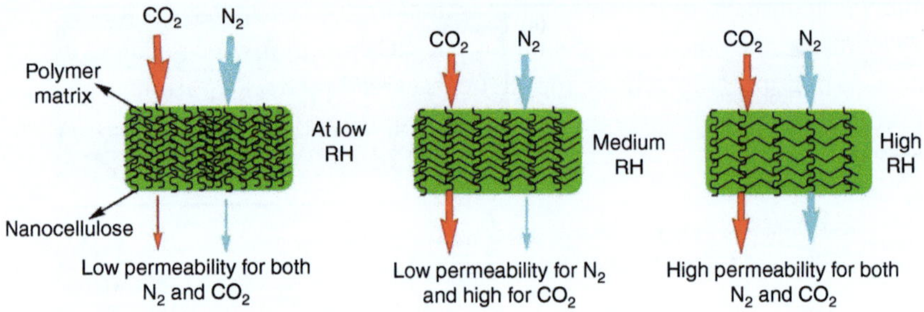

Figure 16.8 The transportation of gas molecule through membranes based on NC. Source: Rana et al. [11]/with permission of ELSEVIER.

The adsorption process occurs as a result of the interaction between CO_2 and the functional groups present on membranes composed of functionalized NC. Additional factors that influence CO_2 adsorption include the surface area, textural characteristics, and densities of the filters [11]. Sepahvand et al. successfully developed nanofilters based on CNFs and Chitosan (CNFs/CS) that exhibit an impressive capability to effectively remove PM and CO_2 from the air [101].

16.7 Conclusion

NC has a wide range of applications due to its natural abundance. A lot of functionalization can be done due to the availability of large number of hydroxyl group, high aspect ratio, low density, biodegradability, and robust. This material can be used for the preparation of NC-based aerogel for the various environmental applications. The CNF-based aerogel typically exhibits a higher Young's modulus compared to the CNC-based aerogel due to its higher aspect ratio and smaller lateral dimension compared to the CNC. These characteristics promote the development of a robust entangled network structure. The prepared aerogels can be used for various types of environmental applications such as air purification and water purification. There are so many emerging pollutant and extremely harmful chemicals such as coloring agents, heavy metals ion, oil, pharmaceutical waste are present in the water bodies. So, these pollutants can be effectively removed by the CNF-based aerogels. In the case of air purification this material can effectively filter or adsorb PM, VOCs, and CO_2. In its pristine form, cellulose adsorb/absorb oil and cationic pollutants due its high hygroscopic nature, which provides poor structural stability in water this can be sort out by the use of crosslinkers to improve its adsorption performance with the addition of functional groups. On the other hand, hydrophobic modification can be also done for the selective removal of oil/organic pollutant. By this way this material can be used for air purification in humid environments. The directional FD technology can be used for the fabrication of meso/macroporous long percolating channel, open pore structure which increases the rate of diffusion of air and water through the pores further this long channel increases the contact time in the pore surface will selectively remove the pollutant. Despite the abundance of cellulosic sources, a fundamental obstacle to its commercial production is the expensive extraction of NC. To further reduce the cost and simplify the NC extraction process, more exploration has to be done. It is encouraging all the research to extract NC from nontimber forest products for the effective extraction of NC at low cost and without ecological damage.

References

1 Kjellstrom, T., Lodh, M., Mcmichael, T., Shrestha, R., and Kingsland, S. (2009). *Air and Water Pollution: Burden and Strategies for Control*. Virginia Quarterly Review.

2 Burek, P., Satoh, Y., Fischer, G. et al. (2016). Water futures and solution. *Int. Inst. Appl. Syst. Anal.* 1–113.

3 United Nations (2021). The United Nations World Water Development Report 2021:Facts and figures. Valuing Water. *Unesco* 1–12.

4 Boretti, A. and Rosa, L. (2019). Reassessing the projections of the World Water Development Report. *npj Clean Water* 2.

5 Paul, J. and Ahankari, S.S. (2023). Nanocellulose-based aerogels for water purification: a review. *Carbohydr. Polym.* 309: 120677.

6 Chen, C., Zhu, X., and Chen, B. (2019). Durable superhydrophobic/superoleophilic graphene-based foam for high-efficiency oil spill cleanups and recovery. *Environ. Sci. Technol.* 53: 1509–1517.

7 Kasonga, T.K., Coetzee, M.A.A., Kamika, I. et al. (2021). Endocrine-disruptive chemicals as contaminants of emerging concern in wastewater and surface water: a review. *J. Environ. Manag.* 277: 111485.

8 Valdez-Carrillo, M., Abrell, L., Ramírez-Hernández, J. et al. (2020). Pharmaceuticals as emerging contaminants in the aquatic environment of Latin America: a review. *Environ. Sci. Pollut. Res.* 27: 44863–44891.

9 Rizal, S., Olaiya, F.G., Saharudin, N.I. et al. (2021). Isolation of textile waste cellulose nanofibrillated fibre reinforced in polylactic acid-chitin biodegradable composite for green packaging application. *Polymers (Basel).* 13: 325.

10 Chu, K.H., Mang, J.S., Lim, J. et al. (2021). Variation of free volume and thickness by high pressure applied on thin film composite reverse osmosis membrane. *Desalination* 520: 115365.

11 Rana, A.K., Mostafavi, E., Alsanie, W.F. et al. (2023). Cellulose-based materials for air purification: a review. *Ind. Crop. Prod.* 194: 116331.

12 Pope, C.A., Burnett, R.T., Thun, M.J. et al. (2002). Lung cancer, cardiopulmonary mortality, and long-term exposure to fine particulate air pollution. *J. Am. Med. Assoc.* 287: 1132–1141.

13 Fan, X., Wang, Y., Kong, L. et al. (2018). A nanoprotein-functionalized hierarchical composite air filter. *ACS Sustain. Chem. Eng.* 6: 11606–11613.

14 Dominici, F., Greenstone, M., and Sunstein, C.R. (2014). Particulate matter matters. *Science* 344: 257–259.

15 Ansaloni, L., Salas-Gay, J., Ligi, S., and Baschetti, M.G. (2017). Nanocellulose-based membranes for CO_2 capture. *J. Membr. Sci.* 522: 216–225.

16 Darunte, L.A., Walton, K.S., Sholl, D.S., and Jones, C.W. (2016). CO_2 capture via adsorption in amine-functionalized sorbents. *Curr. Opin. Chem. Eng.* 12: 82–90.

17 Gough, C.R., Callaway, K., Spencer, E. et al. (2021). Biopolymer-based filtration materials. *ACS Omega* 6: 11804–11812.

18 Souzandeh, H., Wang, Y., Netravali, A.N., and Zhong, W.H. (2019). Towards sustainable and multifunctional air-filters: a review on biopolymer-based filtration materials. *Polym. Rev.* 59: 651–686.

19 Iskandar, M.A., Yahya, E.B., Abdul Khalil, H.P.S. et al. (2022). Recent progress in modification strategies of nanocellulose-based aerogels for oil absorption application. *Polymers (Basel).* 14: 1–22.

20 Liu, H., Tian, X., Xiang, X., and Chen, S. (2022). Preparation of carboxymethyl cellulose/graphene composite aerogel beads and their adsorption for methylene blue. *Int. J. Biol. Macromol.* 202: 632–643.

21 Nargatti, K.I., Subhedar, A.R., Ahankari, S.S. et al. (2022). Nanocellulose-based aerogel electrodes for supercapacitors: a review. *Carbohydr. Polym.* 297: 120039.

22 Zhang, T., Xiao, S., Fan, K. et al. (2022). Preparation and adsorption properties of green cellulose-based composite aerogel with selective adsorption of methylene blue. *Polymer (Guildf)* 258: 125320.

23 Long, L.-Y., Weng, Y.-X., and Wang, Y.-Z. (2018). Cellulose aerogels: synthesis, applications, and prospects. *Polymers (Basel)* 10: 623.

24 Calvino, C., Macke, N., Kato, R., and Rowan, S.J. (2020). Development, processing and applications of bio-sourced cellulose nanocrystal composites. *Prog. Polym. Sci.* 103: 101221.

25 Chen, C. and Hu, L. (2018). Nanocellulose toward advanced energy storage devices: structure and electrochemistry. *Acc. Chem. Res.* 51: 3154–3165.

26 Xie, H., Du, H., Yang, X., and Si, C. (2018). Recent strategies in preparation of cellulose nanocrystals and cellulose nanofibrils derived from raw cellulose materials. *Int. J. Polym. Sci.* 2018: 1–25.

27 Tayeb, A., Amini, E., Ghasemi, S., and Tajvidi, M. (2018). Cellulose nanomaterials—binding properties and applications: a review. *Molecules* 23: 2684.

28 Yadav, C., Saini, A., Zhang, W. et al. (2021). Plant-based nanocellulose: a review of routine and recent preparation methods with current progress in its applications as rheology modifier and 3D bioprinting. *Int. J. Biol. Macromol.* 166: 1586–1616.

29 Ee, L.Y. and Yau Li, S.F. (2021). Recent advances in 3D printing of nanocellulose: structure, preparation, and application prospects. *Nanoscale Adv.* 3: 1167–1208.

30 Chen, Y., Zhang, L., Yang, Y. et al. (2021). Recent progress on nanocellulose aerogels: preparation, modification, composite fabrication applications. *Adv. Mater.* 33: 2005569.

31 Liu, W., Liu, K., Du, H. et al. (2022). Cellulose nanopaper: fabrication, functionalization, and applications. *Nano-Micro Lett.* 14: 1–27.

32 Lavoine, N., Desloges, I., Dufresne, A., and Bras, J. (2012). Microfibrillated cellulose – its barrier properties and applications in cellulosic materials: a review. *Carbohydr. Polym.* 90: 735–764.

33 Siró, I. and Plackett, D. (2010). Microfibrillated cellulose and new nanocomposite materials: a review. *Cellulose* 17: 459–494.

34 Naderi, A. (2017). Nanofibrillated cellulose: properties reinvestigated. *Cellulose* 24: 1933–1945.

35 Azeredo, H.M.C., Rosa, M.F., and Mattoso, L.H.C. (2017). Nanocellulose in bio-based food packaging applications. *Ind. Crop. Prod.* 97: 664–671.

36 Shankaran, D.R. (2018). Cellulose nanocrystals for health care applications. In: *Applications of Nanomaterials*, 415–459. Elsevier https://doi.org/10.1016/B978-0-08-101971-9.00015-6.

37 Blanco, A., Monte, M.C., Campano, C. et al. (2018). Nanocellulose for industrial use. In: *Handbook of Nanomaterials for Industrial Applications*, 74–126. Elsevier https://doi.org/10.1016/B978-0-12-813351-4.00005-5.

38 Nickerson, R.F. and Habrle, J.A. (1947). Cellulose intercrystalline structure. *Ind. Eng. Chem.* 39: 1507–1512.

39 Thomas, P., Duolikun, T., Rumjit, N.P. et al. (2020). Comprehensive review on nanocellulose: recent developments, challenges and future prospects. *J. Mech. Behav. Biomed. Mater.* 110: 103884.

40 Edgar, C.D. and Gray, D.G. (2003). Smooth model cellulose I surfaces from nanocrystal suspensions. *Cellulose* 10: 299–306.

41 Almeida, T., Silvestre, A.J.D., Vilela, C., and Freire, C.S.R. (2021). Bacterial nanocellulose toward green cosmetics: recent progresses and challenges. *Int. J. Mol. Sci.* 22: 2836.

42 Barhoum, A., Jeevanandam, J., Rastogi, A. et al. (2020). Plant celluloses, hemicelluloses, lignins, and volatile oils for the synthesis of nanoparticles and nanostructured materials. *Nanoscale* 12: 22845–22890.

43 Noremylia, M.B., Hassan, M.Z., and Ismail, Z. (2022). Recent advancement in isolation, processing, characterization and applications of emerging nanocellulose: a review. *Int. J. Biol. Macromol.* 206: 954–976.

44 Bäckdahl, H., Risberg, B., and Gatenholm, P. (2011). Observations on bacterial cellulose tube formation for application as vascular graft. *Mater. Sci. Eng. C* 31: 14–21.

45 Ullah, M.W., Manan, S., Kiprono, S.J. et al. (2019). Synthesis, structure, and properties of bacterial cellulose. In: *Nanocellulose*, 81–113. Wiley https://doi.org/10.1002/9783527807437.ch4.

46 Samyn, P., Meftahi, A., Geravand, S.A. et al. (2023). Opportunities for bacterial nanocellulose in biomedical applications: review on biosynthesis, modification and challenges. *Int. J. Biol. Macromol.* 231: 123316.

47 Pierre, A.C. and Pajonk, G.M. (2002). Chemistry of aerogels and their applications. *Chem. Rev.* 102: 4243–4266.

48 Kistler, S.S. (1931). Coherent expanded aerogels and jellies. *Nature* 127: 741–741.

49 Sun, Y., Chu, Y., Wu, W., and Xiao, H. (2021). Nanocellulose-based lightweight porous materials: a review. *Carbohydr. Polym.* 255: 117489.

50 Leys, S.P. and Hill, A. (2012). The physiology and molecular biology of sponge tissues. *Adv. Mar. Biol.* 62: 1–56. https://doi.org/10.1016/B978-0-12-394283-8.00001-1.

51 Robaina, M., Murillo, K., Rocha, E., and Villar, J. (2020). Circular economy in plastic waste - efficiency analysis of european countries. *Sci. Total Environ.* 730: 139038.

52 Jiang, F. and Hsieh, Y.-L. (2014). Amphiphilic superabsorbent cellulose nanofibril aerogels. *J. Mater. Chem. A* 2: 6337–6342.

53 Liu, A., Medina, L., and Berglund, L.A. (2017). High-strength nanocomposite aerogels of ternary composition: poly(vinyl alcohol), clay, and cellulose nanofibrils. *ACS Appl. Mater. Interfaces* 9: 6453–6461.

54 Nascimento, D.M., Nunes, Y.L., Figueirêdo, M.C. et al. (2018). Nanocellulose nanocomposite hydrogels: technological and environmental issues. *Green Chem.* 20: 2428–2448.

55 Benselfelt, T. and Wågberg, L. (2019). Unidirectional swelling of dynamic cellulose nanofibril networks: a platform for tunable hydrogels and aerogels with 3D shapeability. *Biomacromolecules* 20: 2406–2412.

56 Zhu, L., Zong, L., Wu, X. et al. (2018). Shapeable fibrous aerogels of metal–organic-frameworks templated with nanocellulose for rapid and large-capacity adsorption. *ACS Nano* 12: 4462–4468.

57 Nita, L.E., Ghilan, A., Rusu, A.G. et al. (2020). New trends in bio-based aerogels. *Pharmaceutics* 12: 449.

58 Guastaferro, M., Baldino, L., Reverchon, E., and Cardea, S. (2021). Production of porous agarose-based structures: freeze-drying vs. supercritical CO_2 drying. *Gels* 7: 198.

59 Li, Y., Tanna, V.A., Zhou, Y. et al. (2017). Nanocellulose aerogels inspired by frozen tofu. *ACS Sustain. Chem. Eng.* 5: 6387–6391.

60 Wu, P., Zhang, B., Yu, Z. et al. (2019). Anisotropic polyimide aerogels fabricated by directional freezing. *J. Appl. Polym. Sci.* 136: 47179.

61 Cheng, Q., Huang, C., and Tomsia, A.P. (2017). Freeze casting for assembling bioinspired structural materials. *Adv. Mater.* 29: 1703155.

62 Gurav, J.L., Jung, I.-K., Park, H.-H. et al. (2010). Silica aerogel: synthesis and applications. *J. Nanomater.* 2010: 1–11.

63 Zu, G., Shen, J., Zou, L. et al. (2016). Nanocellulose-derived highly porous carbon aerogels for supercapacitors. *Carbon N. Y.* 99: 203–211.

64 Zhang, H., Zhao, T., Chen, Y. et al. (2022). A sustainable nanocellulose-based superabsorbent from kapok fiber with advanced oil absorption and recyclability. *Carbohydr. Polym.* 278: 118948.

65 Ahmaruzzaman, M. and Gupta, V.K. (2011). Rice husk and its ash as low-cost adsorbents in water and wastewater treatment. *Ind. Eng. Chem. Res.* 50: 13589–13613.

66 Song, L., Liu, F., Zhu, C., and Li, A. (2019). Facile one-step fabrication of carboxymethyl cellulose based hydrogel for highly efficient removal of Cr(VI) under mild acidic condition. *Chem. Eng. J.* 369: 641–651.

67 Cervin, N.T., Aulin, C., Larsson, P.T., and Wågberg, L. (2012). Ultra porous nanocellulose aerogels as separation medium for mixtures of oil/water liquids. *Cellulose* 19: 401–410.

68 Liu, N., Zhang, W., Li, X. et al. (2017). Fabrication of robust mesh with anchored Ag nanoparticles for oil removal and in situ catalytic reduction of aromatic dyes. *J. Mater. Chem. A* 5: 15822–15827.

69 Beh, J.H., Lim, T.H., Lew, J.H., and Lai, J.C. (2020). Cellulose nanofibril-based aerogel derived from sago pith waste and its application on methylene blue removal. *Int. J. Biol. Macromol.* 160: 836–845.

70 Zhang, N., Zang, G.-L., Shi, C. et al. (2016). A novel adsorbent TEMPO-mediated oxidized cellulose nanofibrils modified with PEI: preparation, characterization, and application for Cu(II) removal. *J. Hazard. Mater.* 316: 11–18.

71 Azeh Engwa, G., Udoka Ferdinand, P., Nweke Nwalo, F., and Unachukwu, M.N. (2019). Mechanism and health effects of heavy metal toxicity in humans. In: *Poisoning in the Modern World - New Tricks for an Old Dog?* IntechOpen https://doi.org/10.5772/intechopen.82511.

72 Chen, G., Shah, K.J., Shi, L., and Chiang, P.-C. (2017). Removal of Cd(II) and Pb(II) ions from aqueous solutions by synthetic mineral adsorbent: performance and mechanisms. *Appl. Surf. Sci.* 409: 296–305.

73 Rong, N., Chen, C., Ouyang, K. et al. (2021). Adsorption characteristics of directional cellulose nanofiber/chitosan/montmorillonite aerogel as adsorbent for wastewater treatment. *Sep. Purif. Technol.* 274: 119120.

74 Mo, L., Pang, H., Tan, Y. et al. (2019). 3D multi-wall perforated nanocellulose-based polyethylenimine aerogels for ultrahigh efficient and reversible removal of Cu(II) ions from water. *Chem. Eng. J.* 378: 122157.

75 ben Hammouda, S., Chen, Z., An, C., and Lee, K. (2021). Recent advances in developing cellulosic sorbent materials for oil spill cleanup: a state-of-the-art review. *J. Clean. Prod.* 311: 127630.

76 Chhajed, M., Yadav, C., Agrawal, A.K., and Maji, P.K. (2019). Esterified superhydrophobic nanofibrillated cellulose based aerogel for oil spill treatment. *Carbohydr. Polym.* 226: 115286.

77 Zhang, M., Jiang, S., Han, F. et al. (2021). Anisotropic cellulose nanofiber /chitosan aerogel with thermal management and oil absorption properties. *Carbohydr. Polym.* 264: 118033.

78 Rosli, N.A., Khairudin, F.A., Kargarzadeh, H. et al. (2022). Hydrophobic-oleophilic gamma-irradiated modified cellulose nanocrystal/gelatin aerogel for oil absorption. *Int. J. Biol. Macromol.* 219: 213–223.

79 Watkinson, A.J., Murby, E.J., Kolpin, D.W., and Costanzo, S.D. (2009). The occurrence of antibiotics in an urban watershed: from wastewater to drinking water. *Sci. Total Environ.* 407: 2711–2723.

80 aus der Beek, T., Weber, F.A., Bergmann, A. et al. (2016). Pharmaceuticals in the environment-Global occurrences and perspectives. *Environ. Toxicol. Chem.* 35: 823–835.

81 Khan, A., Khuda, F., Elseman, A.M. et al. (2018). Innovations in graphene-based nanomaterials in the preconcentration of pharmaceuticals waste. *Environ. Technol. Rev.* 7: 73–94.

82 Patel, M., Kumar, R., Kishor, K. et al. (2019). Pharmaceuticals of emerging concern in aquatic systems: chemistry, occurrence, effects, and removal methods. *Chem. Rev.* 119: 3510–3673.

83 Verlicchi, P., Al Aukidy, M., and Zambello, E. (2012). Occurrence of pharmaceutical compounds in urban wastewater: removal, mass load and environmental risk after a secondary treatment—a review. *Sci. Total Environ.* 429: 123–155.

84 Cantwell, F.F. and Losier, M. (2002). Chapter 11 Liquid—liquid extraction. *Compr. Anal. Chem.* 37: 297–340. https://doi.org/10.1016/S0166-526X(02)80048-4.

85 Nouri, N., Khorram, P., Duman, O. et al. (2020). Overview of nanosorbents used in solid phase extraction techniques for the monitoring of emerging organic contaminants in water and wastewater samples. *Trends Environ. Anal. Chem.* 25: e00081.

86 Azzouz, A., Kailasa, S.K., Lee, S.S. et al. (2018). Review of nanomaterials as sorbents in solid-phase extraction for environmental samples. *TrAC Trends Anal. Chem.* 108: 347–369.

87 Ibrahim, W.A.W., Nodeh, H.R., and Sanagi, M.M. (2016). Graphene-based materials as solid phase extraction sorbent for trace metal ions, organic compounds, and biological sample preparation. *Crit. Rev. Anal. Chem.* 46: 267–283.

88 Lalović, B., Đurkić, T., Vukčević, M. et al. (2017). Solid-phase extraction of multi-class pharmaceuticals from environmental water samples onto modified multi-walled carbon nanotubes followed by LC-MS/MS. *Environ. Sci. Pollut. Res.* 24: 20784–20793.

89 Tan, S.C. and Lee, H.K. (2019). A metal-organic framework of type MIL-101(Cr) for emulsification-assisted micro-solid-phase extraction prior to UHPLC-MS/MS analysis of polar estrogens. *Microchim. Acta* 186: 165.

90 Madikizela, L.M., Tavengwa, N.T., and Chimuka, L. (2018). Applications of molecularly imprinted polymers for solid-phase extraction of non-steroidal anti-inflammatory drugs and analgesics from environmental waters and biological samples. *J. Pharm. Biomed. Anal.* 147: 624–633.

91 Martinez-Sena, T., Armenta, S., de la Guardia, M., and Esteve-Turrillas, F.A. (2016). Determination of non-steroidal anti-inflammatory drugs in water and urine using selective molecular imprinted polymer extraction and liquid chromatography. *J. Pharm. Biomed. Anal.* 131: 48–53.

92 Udoetok, I.A., Dimmick, R.M., Wilson, L.D., and Headley, J.V. (2016). Adsorption properties of cross-linked cellulose-epichlorohydrin polymers in aqueous solution. *Carbohydr. Polym.* 136: 329–340.

93 Wang, Y.-F., Wang, Y.-G., Ouyang, X.-K., and Yang, L.-Y. (2017). Surface-imprinted magnetic carboxylated cellulose nanocrystals for the highly selective extraction of six fluoroquinolones from egg samples. *ACS Appl. Mater. Interfaces* 9: 1759–1769.

94 Sun, J., Cui, L., Gao, Y. et al. (2021). Environmental application of magnetic cellulose derived from Pennisetum sinese Roxb for efficient tetracycline removal. *Carbohydr. Polym.* 251: 117004.

95 Delmer, D.P. (1999). Cellulose biosynthesis: exciting times for a difficult field of study. *Annu. Rev. Plant Physiol. Plant Mol. Biol.* 50: 245–276.

96 Kadam, V.V., Wang, L., and Padhye, R. (2018). Electrospun nanofibre materials to filter air pollutants – a review. *J. Ind. Text.* 47: 2253–2280.

97 Sinclair, D. (1976). Penetration of hepa filters by submicron aerosols. *J. Aerosol Sci.* 7: 175–179.

98 Wu, A., Hu, X., Ao, H. et al. (2022). Rational design of bacterial cellulose-based air filter with antibacterial activity for highly efficient particulate matters removal. *Nano Sel.* 3: 201–211.

99 Dai, Z., Deng, J., Yu, Q. et al. (2019). Fabrication and evaluation of bio-based nanocomposite TFC hollow fiber membranes for enhanced CO_2 capture. *ACS Appl. Mater. Interfaces* 11: 10874–10882.

100 Venturi, D., Grupkovic, D., Sisti, L., and Baschetti, M.G. (2018). Effect of humidity and nanocellulose content on polyvinylamine-nanocellulose hybrid membranes for CO_2 capture. *J. Membr. Sci.* 548: 263–274.

101 Sepahvand, S., Bahmani, M., Ashori, A. et al. (2021). Preparation and characterization of air nanofilters based on cellulose nanofibers. *Int. J. Biol. Macromol.* 182: 1392–1398.

Index

Nanomaterials for Air and Water Purification, First Edition.
Edited by Deepu A. Gopakumar, Jyotishkumar Parameswaranpillai, Jinu Jacob George, and Midhun Dominic C.D.